Paul L. Younger (Ed.)

Geothermal Energy: Delivering on the Global Potential

MDPI

This book is a reprint of the special issue that appeared in the online open access journal *Energies* (ISSN 1996-1073) in 2014 and 2015 (available at: http://www.mdpi.com/journal/energies/special_issues/geothermal-energy).

Guest Editor
Paul Younger
University of Glasgow
Scotland

Editorial Office
MDPI AG
Klybeckstrasse 64
Basel, Switzerland

Publisher
Shu-Kun Lin

Senior Assistant Editor
Guoping (Terry) Zhang

1. Edition 2015

MDPI • Basel • Beijing • Wuhan

ISBN 978-3-03842-133-7 (Hbk)
ISBN 978-3-03842-134-4 (PDF)

Table of Contents

Chapter 2: Uptake of Geothermal Energy

Chapter 3: Operational Performance of Geothermal Energy Systems

List of Contributors

Radomír Adamovský: Department of Mechanical Engineering, Faculty of Engineering, Czech University of Life Sciences Prague, Kamýcká 129, 165 21 Prague-Suchdol, Czech Republic.

Thorsten Agemar: Leibniz Institute for Applied Geophysics, Stilleweg 2, 30655 Hannover, Germany.

Reynir S. Atlason: Centre for Productivity, Performance and Processes, Department of Industrial Engineering, Mechanical Engineering and Computer Science, University of Iceland, Hjardarhagi 6, 107 Reykjavik, Iceland

Dieter Brüggemann: Zentrum für Energietechnik, Universität Bayreuth, Universitätstrasse 30, Bayreuth 95447, Germany.

Simone Carr-Cornish: Commonwealth Scientific Industrial Research Organisation (CSIRO), PO Box 883, Kenmore, QLD 4069, Australia.

Valentina Ciani: Terra Energy Srl, Spin-off of the University of Pisa, Via S. Maria 53, 56126 Pisa, Italy.

Alan J. Cresswell: The Scottish Universities Environmental Research Centre (SUERC), Rankine Avenue, Scottish Enterprise Technology Park, East Kilbride G7 0QF, UK.

Robin Curtis: GeoScience Ltd., Falmouth Business Park, Bickland Water Rd., Falmouth TR11 4SZ, UK.

Angelo De Santis: Istituto Nazionale di Geofisica e Vulcanologia-via di Vigna Murata, 605, 00143 Roma, Italy.

María B. Díaz-Aguado: Oviedo School of Mines, University of Oviedo, Independencia 13, Oviedo 33004, Spain.

Rafael Rodríguez Díez: Oviedo School of Mines, University of Oviedo, Independencia 13, Oviedo 33004, Spain.

Paolo Favali: Istituto Nazionale di Geofisica e Vulcanologia-via di Vigna Murata, 605, 00143 Roma, Italy.

Paolo Fulignati: Dipartimento di Scienze della Terra, University of Pisa, Via S. Maria 53, 56126 Pisa, Italy.

Thomas L. Harley: School of Engineering, University of Glasgow, James Watt (South) Building, Glasgow G12 8QQ, UK.

Florian Heberle: Zentrum für Energietechnik, Universität Bayreuth, Universitätstrasse 30, Bayreuth 95447, Germany.

Thomas Hermans: Applied Geophysics, University of Liege, Chemin des Chevreuils 1, 4000 Liege, Belgium; FNRS (Fonds de la Recherche Scientifique), 1000 Bruxelles, Belgium.

Seiichiro Ioka: North Japan Research Institute for Sustainable Energy, Hirosaki University, 2-1-3 Matsubara, Aomori 030-0813, Japan.

Francesco Italiano: Istituto Nazionale di Geofisica e Vulcanologia-via Ugo La Malfa 153, 90146 Palermo, Italy.

Dietmar Kuhn: Institute for Nuclear and Energy Technologies, Karlsruhe Institute of Technology (KIT), Karlsruhe 76021, Germany.

Jacek Majorowicz: Department of Physics, University of Alberta, 11322-89 Ave., Edmonton, AB T6G 2G7, Canada.

Paola Marianelli: Dipartimento di Scienze della Terra, University of Pisa, Via S. Maria 53, 56126 Pisa, Italy.

Alistair T. McCay: School of Engineering, University of Glasgow, James Watt (South) Building, Glasgow G12 8QQ, UK.

Hirofumi Muraoka: North Japan Research Institute for Sustainable Energy, Hirosaki University, 2-1-3 Matsubara, Aomori 030-0813, Japan.

Pavel Neuberger: Department of Mechanical Engineering, Faculty of Engineering, Czech University of Life Sciences Prague, Kamýcká 129, 165 21 Prague-Suchdol, Czech Republic.

Frédéric Nguyen: Applied Geophysics, University of Liege, Chemin des Chevreuils 1, 4000 Liege, Belgium.

Yodha Y. Nusiaputra: Institute for Nuclear and Energy Technologies, Karlsruhe Institute of Technology (KIT), Karlsruhe 76021, Germany; German Research Centre for Geosciences (GFZ), Potsdam 14473, Germany.

Gudmundur V. Oddsson: Centre for Productivity, Performance and Processes, Department of Industrial Engineering, Mechanical Engineering and Computer Science, University of Iceland, Hjardarhagi 6, 107 Reykjavik, Iceland.

Monia Procesi: Istituto Nazionale di Geofisica e Vulcanologia, Via di Vigna Murata 605, 00143 Roma, Italy; Unione Geotermica Italiana (UGI), Largo Lucio Lazzarino 1, 56126 Pisa, Italy.

Mario Luigi Rainone: Dipartimento di Ingegneria e Geologia-Università "G. d'Annunzio" 66100 Chieti, Italy.

Simon Rees: Institute of Energy and Sustainable Development, De Montfort University, The Gateway, Leicester LE1 9BH, UK.

Andre Revil: Department of Geophysics, Colorado School of Mines, Golden, CO 80401, USA; ISTerre (Institut des Sciences de la Terre), CNRS, UMR CNRS 5275 (Centre National de la Recherche Scientifique), Université de Savoie, 73376 Cedex, Le Bourget du Lac, France.

Tanguy Robert: Department, AQUALE SPRL, Rue Montellier 22, 5380 Noville-les-Bois, Belgium.

Lygia Romanach: Commonwealth Scientific Industrial Research Organisation (CSIRO), PO Box 883, Kenmore, QLD 4069, Australia.

Sergio Rusi: Dipartimento di Ingegneria e Geologia-Università "G. d'Annunzio", 66100 Chieti, Italy.

Graham Alexander Ryan: Institute of Earth Science and Engineering, University of Auckland, Auckland 1142, New Zealand.

Ladislaus Rybach: Institute of Geophysics, ETH Zurich, Sonneggstrasse 5, CH-8092 Zurich, Switzerland.

David C. W. Sanderson: The Scottish Universities Environmental Research Centre (SUERC), Rankine Avenue, Scottish Enterprise Technology Park, East Kilbride G7 0QF, UK.

Alessandro Sbrana: Dipartimento di Scienze della Terra, University of Pisa, Via S. Maria 53, 56126 Pisa, Italy.

Rüdiger Schulz: Leibniz Institute for Applied Geophysics, Stilleweg 2, 30655 Hannover, Germany.

Michaela Šeďová: Department of Mechanical Engineering, Faculty of Engineering, Czech University of Life Sciences Prague, Kamýcká 129, 165 21 Prague-Suchdol, Czech Republic.

Eylon Shalev: Institute of Earth Science and Engineering, University of Auckland, Auckland 1142, New Zealand.

Patrizio Signanini: Dipartimento di Ingegneria e Geologia-Università "G. d'Annunzio", 66100 Chieti, Italy.

Yota Suzuki: Graduate School of Science and Technology, Hirosaki University, 3 Bunkyo-cho, Hirosaki, Aomori 036-8561, Japan.

Chris Underwood: Faculty of Engineering & Environment, Northumbria University, Newcastle NE1 8ST, UK.

Runar Unnthorsson: Centre for Productivity, Performance and Processes, Department of Industrial Engineering, Mechanical Engineering and Computer Science, University of Iceland, Hjardarhagi 6, 107 Reykjavik, Iceland.

Josef Weber: Leibniz Institute for Applied Geophysics, Stilleweg 2, 30655 Hannover, Germany.

Simon Weides: Helmholtz Centre Potsdam GFZ—German Research Centre for Geosciences, Telegrafenberg, 14473 Potsdam, Germany.

Hans-Joachim Wiemer: Institute for Nuclear and Energy Technologies, Karlsruhe Institute of Technology (KIT), Karlsruhe 76021, Germany.

Paul L. Younger: School of Engineering, University of Glasgow, James Watt (South) Building, Glasgow G12 8QQ, UK; School of Engineering, University of Glasgow, Glasgow G23 5EB, Scotland, UK.

About the Guest Editor

Paul L. Younger holds the Rankine Chair of Energy Engineering at the University of Glasgow, and is Professor of Energy Engineering. His geothermal energy research portfolio ranges from heat-pump applications in shallow aquifers and flooded mineworkings, through novel approaches to combined heat and power developments using mid-enthalpy reservoirs, to holistic analyses of unorthodox high-enthalpy systems in Eastern Africa. Younger has a long track record of collaborative working at all levels from global boardrooms to impoverished villages in the Global South, and he was previously Pro-Vice-Chancellor for Engagement at Newcastle University, where his community-based research won the Queen's Anniversary Prize for Higher Education in 2005. He was elected a Fellow of the Royal Academy of Engineering in 2007, and holds honorary doctorates from leading universities in Spain and Peru. Younger is a founder-Director of Hotspur Geothermal, a London-based company active in the UK and Africa. He has more than 400 publications to his credit.

Preface

Geothermal Energy: Delivering on the Global Potential

Paul L. Younger

Abstract: Geothermal energy has been harnessed for recreational uses for millennia, but only for electricity generation for a little over a century. Although geothermal is unique amongst renewables for its baseload and renewable heat provision capabilities, uptake continues to lag far behind that of solar and wind. This is mainly attributable to (i) uncertainties over resource availability in poorly-explored reservoirs and (ii) the concentration of full-lifetime costs into early-stage capital expenditure (capex). Recent advances in reservoir characterization techniques are beginning to narrow the bounds of exploration uncertainty, both by improving estimates of reservoir geometry and properties, and by providing pre-drilling estimates of temperature at depth. Advances in drilling technologies and management have potential to significantly lower initial capex, while operating expenditure is being further reduced by more effective reservoir management—supported by robust models—and increasingly efficient energy conversion systems (flash, binary and combined-heat-and-power). Advances in characterization and modelling are also improving management of shallow low-enthalpy resources that can only be exploited using heat-pump technology. Taken together with increased public appreciation of the benefits of geothermal, the technology is finally ready to take its place as a mainstream renewable technology, exploited far beyond its traditional confines in the world's volcanic regions.

Reprinted from *Energies*. Cite as: Younger, P.L. Geothermal Energy: Delivering on the Global Potential. *Energies* **2015**, *8*, 11737-11754.

1. Introduction

Geothermal energy is thermal energy produced naturally in the planetary interior [1,2], principally by the decay of radioisotopes of potassium, uranium and thorium [3]. As such, it is the only renewable energy source independent of solar radiation and/or the gravitational attraction of the sun and moon [4]. Since time immemorial, geothermal energy emerging at the earth's surface as natural hot springs has been instinctively harnessed by human beings—and indeed other animals, most famously the macaque (snow monkeys) of Japan [5]—as a source of comfort and cleansing. For instance, in the ancient Roman Empire, few natural hot springs were overlooked for their potential to service the hot water demands of the public baths that were such an indispensable part of army life and wider Roman culture [2]. Natural thermal springs have also long been used for laundry purposes, and even for cooking. All of these uses—together with space heating and various industrial heating applications—are instances of direct use of geothermal resources [6].

The only other, *indirect*, use of geothermal energy is for power generation. Where high-enthalpy reservoirs exist, this is most commonly achieved using various types of flash plant, in which the pressure of hot, deep fluids is carefully manipulated to achieve quantitative conversion of hot water to high-pressure steam, which can then be used to spin conventional steam turbines [7]. The earliest plant of this type was commissioned little over a century ago at Larderello, Italy [2,7]. In cases where the temperature of the geothermal fluid is too low for flashing to steam, electricity can still be produced by means of "binary" power plants [4], in which a secondary working fluid with a far lower boiling point than water is heated via a heat-exchanger such that it is converted to a high-pressure gaseous phase, which again can spin a turbine.

As the water exiting a flash or binary geothermal power plant is typically still hot enough for myriad direct uses, geothermal energy is especially suited to combined-heat-and-power (CHP) applications [8]. If thus exploited, the overall efficiency of geothermal power conversion is far higher than for most other forms of energy. Furthermore, geothermal power plants are characterized by extremely high capacity factors (typically in excess of 90%, with many over 95%), which means that they are typically operated 24/7, producing copious amounts of baseload power and heat [6]. As geothermal power plants typically have very low carbon emissions, their ability to supply baseload puts them on a par with nuclear energy for overall performance [4,9], with none of the operational safety and hazardous waste management issues posed by nuclear.

The baseload power and heat production attributes of geothermal distinguish it from most other renewables [4]: although biomass CHP plants can perform similar service, they typically have far higher operating expenditure (opex) requirements than geothermal plants, due to the need to continually supply fuels of rather low energy-density; furthermore, their capacity factors tend to be rather lower than for geothermal, due to their greater maintenance requirements and vulnerability to interruptions of fuel supply. Solar, wind and wave notoriously suffer from intermittency, reflected in low capacity factors (<30%); and although tidal power is highly predictable, any one plant still tends to have a capacity factor less than 60%. The capacity factors for hydropower plants are seldom much greater. It is also important to note that wind, wave, tidal and hydropower cannot directly produce heat, and using the electricity they produce for conventional heating (*i.e.*, without heat-pumps) is a very wasteful use of high-grade energy. While solar thermal energy is growing in importance, it is generally restricted to producing hot water and rarely manages to provide much space heating.

However, despite all these advantages, the uptake of geothermal energy has to date been disappointing, with annual growth rates in installed capacity since 2004 averaging around 5%, which compares highly unfavourably with the equivalent rates for wind and solar PV (25%–30%) [10]. While lack of appropriate technology for deep, mid-enthalpy systems is partly to blame, and is exacerbated by a persistent lack of public understanding of invisible, subsurface phenomena [11], discussions with investors and engineers throughout the geothermal sector invariably identify two common factors inhibiting more rapid uptake of geothermal energy across all enthalpy categories:

(i) uncertainties over resource availability in poorly-explored reservoirs; and
(ii) the cost profile, in which a large proportion of the full-lifetime costs of systems are concentrated in early-stage capital expenditure (capex).

Some of the solutions to these problems surely lie, at least in part, in the domain of economic policy instruments, such as dry-hole insurance schemes [12] and long-term loan arrangements. However, there is still ample scope for technological innovation to contribute to addressing the barriers to uptake [13], particularly in non-volcanic regions where the majority of resources are low- to mid-enthalpy ("petrothermal") resources in deep strata of unexceptional natural permeability [10].

This paper critically appraises some recently-reported innovations and identifies gaps for future developments, taking a broad view across the entire spectrum of geothermal technology: from drilling and reservoir stimulation, through reservoir modelling and management, to design and operation of mechanical plant at surface that completes the energy conversion process. It also ranges across the entire range of enthalpies found in the subsurface [14], and concludes with a proposal for a whole-system research agenda to expedite realization of the full global potential of geothermal energy.

2. Historical Context and Resource Categorization

The development of modern geothermal energy technology has had at least two dimensions: from high-enthalpy to low-enthalpy resources; and from direct use, through indirect use to CHP and heat-pump applications [2,6,9,10,15]. The earliest impetus for technological development was as an alternative to imported fossil fuels in countries that lacked these in abundance. While the prime motivations related to economics and securing energy supplies, the air-quality benefits of switching from smog-producing coal and oil combustion to the near-zero particulate emissions of geothermal was soon recognized as an important auxiliary advantage [16]. By the dawn of the 21st Century, the principal motivation for developing geothermal had become its low-carbon and renewable credentials. In the case of geothermal, these credentials are not as straightforward to assure as for solar and wind. For instance, the renewability of geothermal can be compromised by poor reservoir management—especially any shortcomings in the reinjection of cooled geothermal fluids—which can lead to quite marked overdraft of the resource base, at least locally and temporarily (albeit the time-scale may be decadal). Similarly, some geothermal systems can have quite high CO_2 emissions, especially in volcanic regions where the magma conduits cut through carbonate sedimentary rocks (as in much of Italy, for instance; [17]). However, the majority of geothermal systems have very low carbon emissions, with systems used only for heating purposes having some of the lowest carbon emissions of any renewable technologies, at around 4 g of CO_2 equivalent per kWh [9,18].

As previously noted, the very earliest human use of geothermal resources was for recreational direct-use purposes [2] with electricity generation commencing only in 1912 at Larderello (Italy) [7]. These two historic uses exploit, respectively, low and high enthalpy resources. Far more recent are the various attempts to exploit very low enthalpy systems (which is solely for direct-use purposes and requires the use of heat-pumps) and mid-enthalpy systems (mainly for direct use, but also potentially for power generation—and thus CHP—by means of binary cycle power plants; [6–9,19]). Meanwhile, deep drilling in Iceland has successfully intercepted a super-critical geothermal reservoir [20], which had originally been discovered by accident. If super-critical reservoirs can be successfully engineered—without inducing pressure decreases within the reservoirs that would take them below the critical point—the rewards will be high indeed: a single super-critical geothermal well can be expected to produce an order of magnitude more energy than a well of similar dimensions accessing only sub-critical high-enthalpy resources [20].

Given these recent developments at both ends of the enthalpy spectrum, the old bipartite categorization of geothermal resources into low- and high-enthalpy systems [1,9] is no longer fit for purpose [14]. A more refined categorization of resources, which corresponds quite closely with the optimal domains for application of different energy conversion technologies, was recently proposed by Younger [14], and is further developed here in graphical form (Figure 1).

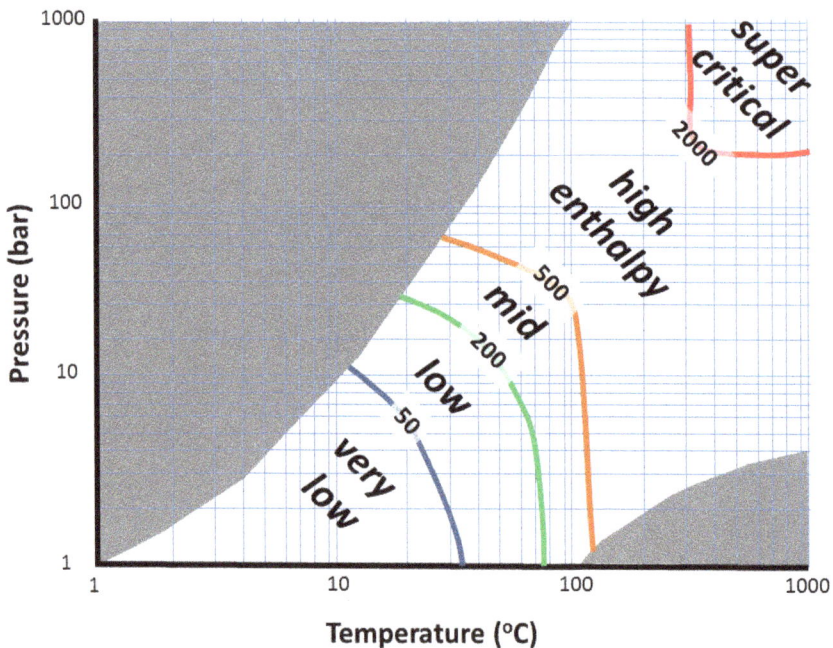

Figure 1. Categorization of geothermal resources on the basis of enthalpy. The shaded areas indicate parts of the parameter space that are rare/impossible in natural systems. The numbers on the lines dividing the different enthalpy categories are approximate values of enthalpy in kJ/kg.

3. Recent Innovations

3.1. Spheres of Endeavour

It is both the fascination and the challenge of geothermal energy that it is a multifaceted business, requiring critical inputs from a wide range of engineering, natural science and social science disciplines. As in all commercial spheres, not all significant innovations are reported in the open literature, either due to economic sensitivities, or simply due to a lack of a pressure for industrial innovators to publish. This paper is also focused on introducing and explaining the context for this geothermal Special Issue of the journal *Energies*. Hence, the account that follows will inevitably be partial. In broad terms, however, it is clear that significant innovations in geothermal energy have been made in the following areas:

- Reservoir exploration and development;
- Reservoir management and modelling;
- Design, operation and maintenance of energy conversion technologies; and
- Socio-economic constraints on geothermal energy use.

Each of these areas is explored in the following sub-sections.

3.2. Reservoir Exploration and Development

The concept of "reservoir" is seldom discussed in connection with very low-enthalpy geothermal resources exploited using closed- or open-loop heat-pump systems. There has been a tendency to tacitly assume that individual heat-pump systems are unlikely to interfere with each other, so that the overall heat (and water) balance of the "reservoir" can be neglected. Where ground-coupled heat-pump systems (GCHPS) are used for individual dwellings in rural areas, this tacit assumption may be unproblematic. However, for larger GCHPS, and wherever neighbouring systems occur in close proximity to one another, failure to characterise and manage the ground exploited by the system can lead to poor performance, manifest in coefficients of performance (COPs) well below the usual minimal target design value of 3 [15]. It can also result in mutual interference between adjoining subsurface heat-exchangers, diminishing the ability of a given volume of ground to support the desired heating/cooling load [21]. A volume of ground used for such purposes has been termed an "aestifer" [22], being a body of geological material that stores and transmits heat. As such, an aestifer is analogous to the more familiar "aquifer" that stores and transmits water. Indeed, for large open-loop systems an aestifer might be entirely identified with an aquifer. However, because heat conduction is not limited solely to permeable rocks, non-aquifer lithologies may fall within the boundaries of an aestifer, particularly where the GCHPS exploiting it is a large closed-loop system. In such cases, characterisation of an aestifer involves delineation of spatial boundaries and determination of its intrinsic thermal properties, especially thermal conductivity and specific heat capacity [22].

Clearly drilling, *in situ* testing, sample retrieval and laboratory testing all have crucial parts to play in identifying fields of thermal properties within an aestifer, and indeed of point-specific temperature as a key state variable. However, as in all other arenas of geological exploration, such direct measurement methods can never fully capture the totality of the parameter fields. It is in this connection that geophysical methods can play an important role, both in guiding the siting of the limited number of boreholes that the project can afford, and in interpolating petrophysical properties between boreholes. While routinely used in applied investigations of geology at substantial depths (e.g., for mineral prospecting, hydrocarbon exploration and high-enthalpy geothermal exploration), the overall neglect of aestifer characterization in very low-enthalpy GCHPS applications is reflected in a scarcity of geophysical investigations of shallow soils and rocks coupled to heat pumps. However, in a rigorous review of experiences to date, Hermans *et al.* [23] have found that a combination of electrical resistivity tomography (ERT), the self-potential method (SP) and distributed temperature sensing (DTS) can provide reliable sensing of variations in subsurface temperatures and, by joint inversion with other geoscientific information, powerful insights into spatial variations in thermal conductivity and specific heat capacity.

A particular category of aestifer with potentially widespread use in many old industrial conurbations in Europe and North America are flooded coalmines [24]. Large open-loop GCHPS exploiting these are operating successfully in Springhill (Nova Scotia), Heerlen (Netherlands) and Miéres (Spain) [14]. As the movement of both ground water and heat in flooded mine workings is typically complex, sometimes involving turbulent flow conditions atypical of most natural aquifers, assessment of these aestifers is particularly challenging. However, statutory compilation of mine plans in most jurisdictions means that records of former mines are generally quite good, at least for mines dating from the final quarter of the 19th Century onwards. This certainly assists in the characterization of thermal properties. Ironically, however, the amount of detail obtained from such plans can be overwhelming, and difficult to analyse over very large areas. Hence, simplified modelling approaches are often most appropriate for regional-scale evaluations of both the hydrogeology [25] and thermal behaviour [24] of flooded coal workings. For instance, *prima facie* reasoning, assuming typical values for several key thermal properties, suggests that a first approximation of the amount of thermal energy that can be extracted from abandoned coalmines can be estimated from historic coal production figures [24]. Using median parameter values, it is estimated that about 2.5 MW_{th} ought to be extractable using heat pumps for every 10 Mt of coal formerly mined from the flooded workings. While no substitute for site-specific investigations, this simplified approach can at least allow rapid screening of districts where more detailed studies seem most likely to prove fruitful. As a minimum, this suggests that 3000 MW_{th} could be sustainably produced from the former coalmines of the European Union, delivering a carbon emissions reduction equivalent to around 5 $Mt_{CO2-equiv}$ per annum [24].

Mid-enthalpy geothermal reservoirs (Figure 1) have the advantage over very low-enthalpy systems that heat pumps are not required to attain temperatures high enough for most space heating and hot water supply purposes. While a handful of studies have considered closed-loop boreholes for the exploitation of such reservoirs (e.g., [26]), most mid-enthalpy systems are predicated on open-loop pumping and reinjection of ground water that obtains its heat from the surrounding rocks. For this to be feasible, two factors are indispensable: sufficient permeability and sufficient heat flow. Frustratingly, many of the rocks with the best heat flow properties have indifferent permeability (the so-called "hot dry rock" scenario), so that reservoir stimulation techniques are necessary in order to obtain sufficient yields—this is the approach termed "engineered (or enhanced) geothermal systems" (EGS), and it has been the subject of several concerted investigations in the USA and Europe since the 1970s (e.g., [1,6,9]). In the last decade there has been increased appreciation that sufficient natural permeability can be encountered where boreholes intercept natural geological structures oriented suitably in relation to the present-day natural stress field; this was the case at Eastgate (northern England), for instance, where a geothermal exploration borehole proved the highest permeability yet recorded in deep granite anywhere in the world [27].

Whereas permeability is amenable to some degree of manipulation, the same cannot be said of petro-thermal properties. While archival heat flow estimates may well be improved by application of updated models, which more accurately allow for the effects of high topographic relief and/or the residual effects of palaeoclimatic conditions [28], the fundamental parameters of radiothermal heat production, thermal conductivity and specific heat capacity are essentially objective. That is not to say, however, that the methods for determining these parameters are beyond improvement. For instance, topography, atmospheric conditions and spatial patterns of heterogeneity can all affect measured levels of gamma-ray emissions from radiothermal source rocks. Hence enhanced data collection and inversion methods for spectral gamma surveys will facilitate more precise estimation of heat production and flow rates, helping refine selection of drilling targets, such as potassium-rich granites and thick sequences of black shales [3]. Nevertheless, quantification of heat production rates at depth is insufficient to accurately predict the spatial distribution of the warmest waters in overlying sedimentary strata—quantification of climatic influences (past and present) and convective ground water flow patterns are at least as important [29].

These factors are also important in the case of high-enthalpy systems [29], though constraints on upper-bound temperatures are also dependent on the maximum depth of hydrothermal circulation, which corresponds to the horizon of transition from brittle to plastic deformation, as revealed by an abrupt cessation of earthquake foci [30]. In the vicinity of major Quaternary volcanoes in Japan, for instance, this horizon approximates to the inferred 380 °C isotherm, beneath which seismicity, fracturing and hydrothermal convection are all observed to cease in granitic crust [30]. Within the zone of hydrothermal circulation, seismic processes may provide valuable insights into reservoir functioning. For instance, variations in mineral assemblages correlated with hydrothermal alteration are such that there is a negative correlation between reservoir temperature and seismic velocity anomalies at temperatures less than ~220 °C, whereas at higher temperatures the correlation is positive [31]. Hence interpretation of natural seismic data may provide direct estimates of reservoir temperatures, in addition to its more orthodox applications in delineating spatial boundaries and internal structures in reservoirs [31].

The overall process of evaluation of high-enthalpy resources at the exploration stage is multi-faceted, effectively triangulating the best estimate of reservoir enthalpies (and other reservoir characteristics) from a range of alternative approaches using largely independent data-sets. The case of Chile offers a compelling worked example of how such an approach can be used to estimate future energy productivity for individual fields, and thence for an entire country in which no geothermal power plant has yet been developed [32]. Already, consideration is being given to developments even further into the future, when geothermal developments might follow the historical precedent of hydrocarbons and progress to offshore exploitation of submarine hydrothermal circulation systems, such as those associated with the Marsili Seamount in the southern Tyrrhenian Sea (Italy) [33] or with sea-floor spreading ridges off the northwestern coast of the USA [34].

Nevertheless, continued success in high-enthalpy exploration will require renewal of paradigms on the part of many practitioners. Given that the majority of highly productive systems developed to date have been associated with conspicuous stratovolcanoes, it is unsurprising that the most common exploration model is predicated on the search for hydrothermal systems associated with such features. However, in many dissimilar settings heat flows are just as elevated, yet geothermal exploration has barely commenced. The non-volcanic tracts of the East African Rift system are a case in point. A more open-minded approach to exploration paradigms will be required if valuable resources are not to be overlooked [35].

One such example of a paradigm shift in exploration relates to supercritical geothermal resources, the deliberate search for which was prompted by experiences of unanticipated interception of reservoirs with supercritical properties in Italy and Iceland. The engineering challenges in accessing and harnessing such high temperature (>400 °C), high-pressure (>22 MPa) reservoirs are considerable, but have recently been substantially addressed at Krafla volcano by the Iceland Deep Drilling Project [20]. Recent theoretical analysis has clarified the conditions that give rise to supercritical conditions, as well as illuminating the likely frequency of occurrence and extent of such reservoirs [36]. The findings are encouraging, suggesting that a supercritical root zone can be expected to occur above young magmatic intrusions that underlie many well-known high-enthalpy reservoirs. Further deliberate exploration for supercritical reservoirs is currently scheduled in Iceland, Japan and New Zealand [34], with potential to develop production wells ten times more prolific than typical high-enthalpy wells. If this potential can be realized widely and at scale, the contribution of geothermal energy to the generation mix will be greatly enhanced.

3.3. Reservoir Management and Modelling

It is ironic that the management of deep, high-enthalpy geothermal reservoirs is far more advanced and exhaustively documented than that of the shallow, low-enthalpy variably-saturated soil systems ("aestifers") exploited using heat-pumps. In part this is because the uptake of GCHPS was, until the last decade or so, sufficiently modest that interference between adjoining installations could be safely overlooked. This is particularly so where GCHPS installations only serve single dwellings, with individual system capacities seldom exceeding 20 kW. However, as there is a proliferation of multi-MW installations serving large commercial premises, scope increases for mutual interference between systems, as well as for cumulative depletion of the ability of the aestifer to continue to provide the heating and/or cooling services demanded of it [21,37,38]. Detailed, site-specific investigations of such cases are increasingly being reported [37,39], often supported by numerical modelling [37,40]. Such studies are providing the basis for pro-active regulation of open-loop GCHPS developments [38], although closed-loop systems continue to evade regulatory control in most jurisdictions. This is not simply a matter of legislative loopholes: modelling of closed-loop systems is often more complicated than for open-loop because of the occurrence of multi-phase fluid flow above the water table, and because of the complex geometry of shallow, looped heat-exchangers buried in soil. Analytical solutions to the latter problem have been obtained and applied [41], though these necessarily involve rather sweeping assumptions to be made about soil properties.

The management of mid-enthalpy reservoirs has received little more attention than GCHPS aestifers, though experience with exploiting such systems for district heating and CHP applications is rapidly growing. France was the earliest entrant into the large-scale use of mid-enthalpy systems, with a cluster of systems that have now been exploiting the deep Chalk aquifer in the Paris Basin for more than forty years [9]. More recently, favourable government support programs have led to around 200 projects coming forward in Germany [42], with annual production of geothermal heat and power increasing from 60 to 530 GWh_{th} and from 0 to 36 GWh_e, respectively, over the decade 2003–2013 [42]. The German experience indicates that, even with a very supportive governance framework, most mid-enthalpy geothermal systems require around six years to proceed from initial concept to full commissioning [42].

As the most recent volcanism in Germany (In the Eifel district of the Rhine valley) only ceased 10,000 years ago, there are almost certainly high-enthalpy resources yet to be developed there, as regions with volcanism within the last million years often remain highly prospective for high-enthalpy reservoirs [35]. Nevertheless, the exploitation of high enthalpy resources is still effectively confined to countries with conspicuous active volcanism, such as Italy [43] and the countries of the circum-pacific "Ring of Fire," not least Japan [30]. The management of geothermal reservoirs requires judicious design and operation of both production and reinjection boreholes—the latter being used not only to prevent the environmental damage which discharge of hot and (usually) briny spent geothermal fluids to surface waters would cause, but also to maintain reservoir pressures at depth. In doing so, a delicate compromise must be negotiated between injecting so close to the production zones that the temperature of the produced fluids is reduced and injecting so far away that the desired pressure maintenance effects are not achieved. This balancing act is made no simpler by the tendency for thermal contraction fracturing to increase the permeability where cooler reinjectates enter the high-temperature reservoir [44]; a zone that seemed suitably remote from the production zones can become more intimately hydraulically connected with it over time, as these thermal contraction fractures propagate. A further consideration is the minimization of undesirably large seismic events. Thus the management of a geothermal reservoir is always a work in progress, with the tasks of production and reinjection being assigned to different wells over time. Additional "make-up" wells are typically required to maintain total production rates as the exploitation of a reservoir matures [45].

As further wells become available, their geological and production characteristics gradually expand the knowledge base for the reservoir, allowing refinement—or even wholesale replacement—of the prevailing conceptual model that is used to inform reservoir management decisions [43]. Furthermore, as the database of seismic events (natural and induced) in and around the reservoir grows, the identification of important geological structures and features (such as the zone of brittle/plastic transition) becomes clearer [30,43]. Indeed, as shown by Ryan and Shalev [31], correlations between seismic velocity and *in situ* temperatures could even make remote sensing of reservoir temperatures feasible. Thus an overall geothermal reservoir model comprises an assemblage of mutually consistent geological [43], geophysical [30,31] and hydrogeological models [43] should be developed in parallel, and iteratively updated to achieve harmonious coupling between them. The overall aim is a robust and constantly-evolving conceptual model of each geothermal reservoir, which is carefully adjusted to ensure that as much consistency as possible is achieved between concepts and available data. With this in place, rational management decisions are facilitated.

3.4. Design, Operation and Maintenance of Energy Conversion Technologies

The simplest geothermal direct-use systems require no more technology than conventional plumbing to deliver their benefits: this is the case, for instance, with low-enthalpy resources used for balneological purposes, or mid-enthalpy resources used for space heating. Depending on the composition of the geothermal fluid (whatever its enthalpy), heat-exchangers may be required even for these purposes, and these can be costly where fluid compositions would tend to give rise to either corrosion or clogging with mineral scales or biofilms.

It is at the extremes of enthalpy that technological requirements become most exacting for those very low enthalpy resources (Figure 1) that can only be usefully exploited using heat-pumps, robust approaches to design, installation, operation and maintenance of the pumps is essential. The introduction of legal requirements for a certain proportion of on-site renewable energy production for all new commercial buildings of a certain size led to a boom in demand for GCHPS in the UK [15]. However, the policing of this rule was weak, with the result that tokenism crept into too many designs: too many builders were content simply to obtain approval to proceed with their development, and did not care if the supposed 10% renewable technology actually worked in the long-run. This led to installation of many under-sized GCHPS, as became apparent some years later when a publicly-funded national study of system performance revealed actual COPs averaging only 2.2—significantly lower than the typical design values (>3). Detailed modelling of typical installations revealed that the government-approved standard for GCHPS design actually leads to under-sizing of the subsurface heat-exchange arrays, thereby adding an additional 20% to the electricity demand for the heat-pumps [46]. This is but another example of a situation in which political will was not sufficiently under-pinned by *a priori* engineering rigor. It also underlines the importance of fully considering future operational conditions at the design stage.

For mid-enthalpy systems, direct use of geothermal resources for space heating typically does not require use of heat-pumps; heat exchangers and circulation pumps are all that is required to deliver heat to district pipe networks, albeit the inclusion of hot water storage tanks has recently been shown to decrease reliance on peak-up or back-up boiler plant [47]. Where the temperature of mid-enthalpy systems approaches or exceeds 100 °C, it is also possible to convert at least part of the energy to electricity, by means of binary cycle power plants, in which a secondary working fluid with a far lower boiling point than water is converted into high-pressure vapour, which then spins a turbine in a closed-loop cycle. Although the secondary working fluid can be an ammonia/water mixture, as in the Kalina Cycle [48], the most widespread binary cycles use an organic compound (typically butane, pentane or a proprietary refrigerant), with the resultant systems being termed "Organic Rankine Cycle" (ORC) plants [19]. ORC technology has been increasingly used and refined since the 1980s and may now be regarded as a mature technology. As such, the design principles are now well established, and the frontiers of research currently focus on maximization of efficiency, extending the lower temperature threshold for ORC applications, and extending the applicability of the technology by reducing costs. Two examples of the latter may be cited:

(i) The development of small, modular ORC plants that can be rapidly deployed to remote areas as a pioneer power generation technology [19]. This raises the possibility of single wellhead ORC operations during geothermal field development, helping provide the power for drilling of further wells. (Hitherto, wellhead turbines have been restricted to atmospheric venting or back-pressure steam turbine units [6]).

(ii) Hybrid power plants, in which waste heat from other processes is harnessed together with mid-enthalpy geothermal energy in combined heat and power systems that are more efficient than would be the case were either heat source used in isolation [8].

Although there is no strict upper limit for ORC applications, on the grounds of capital cost the technology of choice for high enthalpy systems is, and is likely to remain, the steam turbine. Dry steam fields that can be used directly to supply turbines are globally rare [7]; far more common are flash systems, in which super-heated water is converted to high-pressure steam by pressure adjustment; the steam is separated and used in the turbine while the separated water is typically harnessed for direct use before being reunited with turbine exhaust condensate for reinjection. Single- or double-flash systems may be used, dependent on the enthalpy and desired yield of the system [6–8]. While steam turbines are a very mature technology, most applications to date have been in the fossil fuel sector, in which the purity of the steam delivered to the turbine is under the control of the plant operator. In contrast, the steam harnessed in geothermal systems is natural, and thus subject to variations in chemical composition—both between different systems and within any one systems, depending on reservoir dynamics. This means that the maintenance of geothermal flash power plant is subject to greater uncertainty and risk than tends to be the case in, say, a coal-fired power plant [49].

3.5. Socio-Economic Constraints on Geothermal Energy Use

Like any other energy resource, geothermal is subject to constraints arising from societal preferences and economic exigencies. These constraints are experienced across the full range of enthalpies and technologies. For instance, in the case of very low-enthalpy systems exploited using heat-pumps, while the levelised cost of heat may already be competitive with gas heating, the balance between capital expenditure (capex) and operating expenditure (opex) is quite different between the two: the marginal capex cost of replacing a gas boiler where the gas grid connection already exists is typically far less than the capex cost of installing a GCHPS from scratch. On the other hand, the opex costs of the GCHPS are likely to be far lower than those of the gas-fired system, because three-quarters of the heat delivered by a GCHPS is effectively sourced free of charge from the subsurface, whereas every Joule of heat delivered by a gas-fired system comes from purchased fuel. However, the scale of the capex cost can be a powerful disincentive to invest in GCHPS, despite lower total life-cycle costs. The disincentives are even more marked where the party paying for the capex (e.g., a house building firm) is different from the party who will benefit from the low opex (*i.e.*, the house occupant). Such an impasse can only be overcome where mandatory rules favour the reduction in carbon emissions that a GCHPS offers in comparison to a gas-fired system [15]. Such mandates are common in Europe, but far less socially acceptable in North America, for instance.

A similarly high capex/opex ratio affects the production of electricity from mid-enthalpy resources: without a subsidy recognizing the value of the low carbon emissions of geothermal systems, electricity produced using ORC plants in Germany cannot compete with fossil-fuelled and nuclear power production [42]. Even though the economics of direct-use are generally far more favourable, there is still a challenge in displacing incumbent energy systems, due to the initial capex penalty of deep drilling and installing a district heating network. Again, technology "lock-in" is unlikely to be overcome by market forces alone, and governmental incentives are likely to be required if the full potential of mid-enthalpy geothermal resources is to be realized.

This in turn demands that public opinion be sufficiently in favour of geothermal energy. Public approval cannot be taken for granted. There is good evidence that provision of information helps improve the approval rating of geothermal, yet perceived risks (which bear little relation to reality, and are possibly driven by the negative publicity surrounding other subsurface technologies, such as shale gas) led to a majority of participants in an Australian survey favouring siting of geothermal wellfields far from communities [11]. This is at odds with the requirements for district heating (which is, admittedly, of little concern in the warm Australian climate, where interest in geothermal has focused far more on its potential to produce electricity). It is also odds with the complete public ease with extensive, long-established urban geothermal wellfields in Paris and Reykjavik, for instance.

The higher the enthalpy, the larger-scale must be the socio-economic framework within which geothermal energy is developed. Thus issues of energy security and balance of payments can favour the development of high-enthalpy geothermal resources. A case in point is Chile, which is the world's largest producer of copper, yet is heavily-dependent on imported fossil fuels for most of its energy needs. While hydropower has significant potential in the south of the country, in the far distant north where the copper mines are located the climate is arid to hyper-arid and hydropower potential is therefore negligible. However, the same volcanic processes that gave rise to the copper orebodies continue to give rise to significant deep hydrothermal circulation systems, several of which are highly prospective for geothermal power production. As much as 30% of the total power demand in the mining region could be met by development of geothermal resources identified to date [32], significantly reducing greenhouse gas emissions at the same time.

4. Gaps Analysis of Geothermal Innovation: Towards a Whole-System Research Agenda

For all the advantages it offers, geothermal remains one of the least-developed of the renewables [10]. This is despite the fact that high-enthalpy geothermal can already out-compete wind and solar on the basis of levelised costs of energy, as well as providing baseload service that are beyond the capability of those technologies. Geothermal district heating can already have a lower levelised cost than gas in many cases. However, uptake of geothermal continues to be hindered by its particularly high capex/opex ratio, which makes it an unconventional investment opportunity. One of the principal causes of this high ratio is the scale and uncertainty of resource characterization: where resource estimation for a wind farm will typically require only 1% of the total project cost, for geothermal it averages 47% [50]. There is therefore great value in any development that streamlines geothermal exploration activities [3,24,28–30,35,51].

Drilling and completion of wells is a large component of the geothermal capex burden, which has to date resisted many attempts at cost reduction. This is partly because geothermal drilling is but one small part of global drilling market that remains dominated by the hydrocarbon sector, in which the high market value of the commodity sought has tended to lessen pressures to reduce drilling costs. Furthermore, the high temperatures and pressures encountered in geothermal wells that intercept high-enthalpy reservoirs considerable exceed those encountered in hydrocarbon wells, so that very particular challenges arise in relation to mechanical processes and the performance of engineering materials [52]. For instance, drill-bits commonly used to achieve rapid rates of penetration in deep hydrocarbon wells (such as polycrystalline diamond compact bits rotated by rotor/stator devices actuated by drill mud, and conventional air- or water-actuated down-the-hole-hammer (DTTH) devices) will not function where high losses of drilling fluid occur to the strata (which is common in buried volcanic rock sequences containing lava tubes), nor where temperatures exceed the stability range of the elastomers used in the rotating or reciprocating components. Hence recent innovations have included the development of novel materials that will allow DTTH operations at pressures in excess of 250 °C [53]. Alternative cutting technologies that are insensitive to formation temperatures have also been developed such as plasma torches that ablate the rock at temperatures as high as 4500 °C [54,55] and a "ram accelerator" technology derived from gun technology, in which high velocity projectiles are fired at the bottom of the borehole at such a velocity that both the rock and the projectile vaporize [56]. Similarly, downhole deployment of lasers could offer a rapid and controllable technique for enhancing connectivity between wells and the pore networks in the surrounding reservoir rocks [57]. Such promising technologies have the potential to more than double rates of penetration, install casing during drilling and minimize the number of time-consuming and unproductive "trips" of the downhole assembly out of the hole; together, these measures could reduce geothermal drilling and completion costs by more than 50%. However, significant technical challenges remain, including the challenge of ensuring borehole stability when the wall rocks exceed melting point [55], and the difficulties of obtaining good records of the geology encountered when there is no core or even chippings to examine. Incremental advances in more conventional reservoir stimulation approaches must also be pursued, such as "soft" thermal [44] and chemical stimulation of permeability.

At the other extreme of the enthalpy spectrum, there has been a tendency to achieve cheaper well completions for GCHPS by implementing lesser standards of casing and cementing than would be used in conventional water wells of similar depth. This tendency is particularly prevalent in the case of closed-loop GCHPS, but it should be stoutly resisted as it will often lead to undesirable (and even illegal) inter-connection of separate aquifers, to the detriment of water resources [21].

In the realm of geothermal reservoir development and management, there remains vast scope for advances in geophysical prospecting tools, both in refining existing tools for geothermal applications and in developing entirely new tools. Increasingly, it is combinations of tools with differing physical basis that is proving most useful in achieving less equivocal signatures of reservoir geometries and hydrogeothermal properties [23,30,31]. Hence advances in joint inversion of different types of geophysical data must remain a property in geothermal research. It is important to stress that the use of such multi-physics approaches to reservoir characterization should not be restricted to the exploration phase of geothermal development but can provide invaluable insights into the evolution of thermo-physical conditions within productive reservoirs [30].

In terms of geothermal energy conversion technologies, advances in plant design that allow efficient operation across a wider range of input temperatures and pressures would allow geothermal to extend its range of application from typical baseload applications to dispatchable service—something that few other renewable technologies can offer. At the simplest level, this might be achieved by smarter design and operation of storage facilities within direct-use systems; in electricity production, dynamic adjustment of working fluid properties could allow turbines to be operated closer to their optimal conditions over a wider range of input fluid conditions.

Finally, there is a continuing need for advances in the objective assessment and minimization of undesirable environmental side-effects of geothermal exploitation. To give but one example, there has as yet been very little examination of the "water–energy nexus" for the case of geothermal. Where geothermal is exploited in humid countries, with rain-fed agriculture and abundant water resources, the water–energy nexus may not yet be too pressing. However, in water-scarce countries with large and growing populations—such as most of eastern Africa and the "*Cono Sur*" of South America—a perfect storm is brewing [35]: ever-expanding demand for reliable renewable energy is prompting geothermal developments in regions where increasingly sparse ground water resources are the sole source of potable water supply. In many cases, there may be no direct hydraulic continuity between deep geothermal reservoirs and shallow groundwater recharge zones—in which case the renewability of the geothermal resource itself may be questionable. In some cases, however, hydraulic continuity will exist, albeit it may be masked by the very large spatial and temporal scales over which depletion of ground water resources might finally become apparent (reflecting the slow velocities of Darcian subsurface flow, the vast volumes of ground water stored in undeveloped aquifers, and the considerable distances between zones of aquifer recharge and natural discharge). It has long been appreciated [58] that the response of ground water systems to development depends on a case-specific interplay between (at least some) depletion of long-term storage, induced increases in recharge and/or decreases in natural discharge. Any robust appraisal of the geothermal water–energy nexus simply *must* address these hydrodynamics in a rigorous manner.

5. Conclusions

Recent advances in reservoir characterization techniques are beginning to narrow the bounds of exploration uncertainty, both by improving estimates of reservoir geometry and properties, and by providing pre-drilling estimates of temperature at depth. Advances in drilling technologies and management have potential to significantly lower initial capex, while operating expenditure is being further reduced by more effective reservoir management—supported by robust models—and increasingly efficient energy conversion systems (flash binary and heat exchange for direct-use). Advances in characterization and modelling are also improving management of shallow low-enthalpy resources that can only be exploited using heat-pump technology. Taken together with increased public appreciation of the benefits of geothermal, the technology is finally ready to take its place as a mainstream renewable technology, exploited far beyond its traditional confines in the world's volcanic regions.

Acknowledgments

This paper was prepared to provide an overview and summary of the papers published in a Special Issue of *Energies* on geothermal energy, which was guest-edited by the present author. The author gratefully acknowledges the original invitation from the editorial team of *Energies* to serve as Guest Editor, and for their excellent support throughout the process of publishing the Special Issue.

Besides placing the papers in the Special Issue in their broader context, this paper draws on the wider literature and the experiences of the author in the geothermal sector, both as a university researcher and as an adviser to industry. Most recently, this has been under the auspices of the Horizon 2020 project "LoCAL" (Low-Carbon After-Life for Coalmines: geothermal energy), funded by the European Commission under the auspices of the Research Fund for Coal and Steel (Contract No. RFCR-CT-2014-00001). However, the views expressed are solely those of the author, and cannot be construed as representing those of any of the private and public entities with which he has collaborated.

Conflicts of Interest

The author declares no conflict of interest. Although he is a founding non-executive director of Hotspur Geothermal Group (formerly Cluff Geothermal Ltd.), none of the work reported here was funded by that company; nor do the findings of this paper give that company any competitive advantage.

References

1. Armstead, H.C.H. *Geothermal Energy: Its Past, Present and Future Contributions to the Energy Needs of Man*; Spon: London, UK, 1978.
2. Stober, I.; Bucher, K. History of geothermal energy use. In *Geothermal Energy: From Theoretical Models to Exploration and Development*; Stober, I., Bucher, K., Eds.; Springer-Verlag: Berlin/Heidelberg, Germany, 2013.
3. McCay, A.T.; Harley, T.L.; Younger, P.L.; Sanderson, D.C.W.; Cresswell, A.J. Gamma-ray spectrometry in geothermal exploration: State of the art techniques. *Energies* **2014**, *7*, 4757–4780.
4. Younger, P.L. *Energy: All That Matters*; Hodder and Stoughton/John Murray: London, UK, 2014.
5. Zhang, P.; Watanabe, K.; Eishi, T. Habitual hot-spring bathing by a group of Japanese macaques (*Macaca fuscata*) in their natural habitat. *Am. J. Primatol.* **2007**, *69*, 1425–1430.
6. Dickson, M.H.; Fanelli, M. *Geothermal Energy: Utilization and Technology*; Earthscan: London, UK, 2005.
7. DiPippo, R. *Geothermal Power Plants: Principles, Applications, Case Studies and Environmental Impact*, 3rd ed.; Butterworth-Heinemann: Oxford, UK, 2012.
8. Heberle, F.; Brüggemann, D. Thermoeconomic analysis of hybrid power plant concepts for geothermal combined heat and power generation. *Energies* **2014**, *7*, 4482–4497.
9. Garnish, J.; Brown, G. Geothermal energy. In *Renewable Energy. Power for a Sustainable Future*, 3rd ed.; Boyle, G., Ed.; Open University-Oxford University Press: Oxford, UK, 2012; pp. 409–459.
10. Rybach, L. Geothermal power growth 1995–2013—A comparison with other renewables. *Energies* **2014**, *7*, 4802–4812.
11. Carr-Cornish, S.; Romanach, L. Differences in public perceptions of geothermal energy technology in Australia. *Energies* **2014**, *7*, 1555–1575.

12. Geothermal Risk Mitigation Facility for Eastern Africa. Available online: http://www.grmf-eastafrica.org (accessed on 20 March 2015).

13. Glassley, W.E. *Geothermal Energy: Renewable Energy and the Environment*, 2nd ed.; CRC Press: Boca Raton, FL, USA; Taylor & Francis: London, UK, 2015.

14. Younger, P.L. Hydrogeological challenges in a low-carbon economy. *Q. J. Eng. Geol. Hydrogeol.* **2014**, *47*, 7–27.

15. Rees, S.; Curtis, R. National deployment of domestic geothermal heat pump technology: Observations on the UK experience 1995–2013. *Energies* **2014**, *7*, 5460–5499.

16. Björnsson, S. (Ed.) *Geothermal Development and Research in Iceland*; Orkustofnun: Reykjavik, Iceland, 2010; p. 40. Available online: http://www.nea.is/media/utgafa/GD_loka.pdf (accessed on 29 September 2015).

17. Gambardella, B.; Cardellini, C.; Chiodini, G.; Frondini, F.; Marini, L.; Ottonello, G.; Vetuschi Zuccolini, M. Fluxes of deep CO_2 in the volcanic areas of central-southern Italy. *J. Volcanol. Geotherm. Res.* **2004**, *136*, 31–52.

18. Holm, A.; Jennejohn, D.; Blodgett, L. *Geothermal Energy and Greenhouse gas Emissions*; Geothermal Energy Association: Washington, DC, USA, 2012. Available online: http://geo-energy.org/reports/GeothermalGreenhouseEmissionsNov2012GEA_web.pdf (accessed on 29 September 2015).

19. Nusiaputra, Y.; Wiemer, H.; Kuhn, D. Thermal-economic modularization of small, Organic Rankine Cycle power plants for mid-enthalpy geothermal fields. *Energies* **2014**, *7*, 4221–4240.

20. Friðleifsson, G.Ó.; Elders, W.A.; Albertsson, A. The concept of the Iceland deep drilling project. *Geothermics* **2014**, *49*, 2–8.

21. Younger, P.L. Ground-coupled heating-cooling systems in urban areas: How sustainable are they? *Bull. Sci. Technol. Soc.* **2008**, *28*, 174–182.

22. Banks, D. *Thermogeology: Ground-Source Heating and Cooling*, 2nd ed.; Wiley Blackwell: Chichester, UK, 2012; p. 526.

23. Hermans, T.; Nguyen, F.; Robert, T.; Revil, A. Geophysical methods for monitoring temperature changes in shallow low enthalpy geothermal systems. *Energies* **2014**, *7*, 5083–5118.

24. Rodríguez Díez, R.; Díaz-Aguado, M.B. Estimating limits for the geothermal energy potential of abandoned underground coalmines: A simple methodology. *Energies* **2014**, *7*, 4241–4260.

25. Adams, R.; Younger, P.L. A strategy for modeling ground water rebound in abandoned deep mine systems. *Ground Water* **2001**, *39*, 249–261.

26. Law, R. *Deep Geothermal Heat Production—Single Well Trial*; Ground-Source Heat Pump Association: Uxbridge, UK, 2014. Available online: http://www.gshp.org.uk/DeMontfort/GeothermalEngineeringLtd.pdf (accessed on 20 August 2015).

27. Younger, P.L.; Manning, D.A.C. Hyper-permeable granite: Lessons from test-pumping in the Eastgate Geothermal Borehole, Weardale, UK. *Q. J. Eng. Geol. Hydrogeol.* **2010**, *43*, 5–10.

28. Westaway, R.; Younger, P.L. Accounting for palaeoclimate and topography: A rigorous approach to correction of the British geothermal dataset. *Geothermics* **2013**, *48*, 31–51.

29. Weides, S.; Majorowicz, J. Implications of spatial variability in heat flow for geothermal resource evaluation in large foreland basins: The case of the Western Canada Sedimentary Basin. *Energies* **2014**, *7*, 2573–2594.

30. Suzuki, Y.; Ioka, S.; Muraoka, H. Determining the maximum depth of hydrothermal circulation using geothermal mapping and seismicity to delineate the depth to brittle-plastic transition in northern Honshu, Japan. *Energies* **2014**, *7*, 3503–3511.

31. Ryan, G.; Shalev, E. Seismic velocity/temperature correlations and a possible new geothermometer: Insights from exploration of a high-temperature geothermal system on Montserrat, West Indies. *Energies* **2014**, *7*, 6689–6720.

32. Procesi, M. Geothermal potential evaluation for northern Chile and suggestions for new energy plans. *Energies* **2014**, *7*, 5444–5459.

33. Italiano, F.; de Santis, A.; Favali, P.; Rainone, M.; Rusi, S.; Signanini, P. The Marsili Volcanic Seamount (southern Tyrrhenian Sea): A potential offshore geothermal resource. *Energies* **2014**, *7*, 4068–4086.

34. Elders, W.A. The potential for on- and off-shore high-enthalpy geothermal systems in the USA. In Proceedings of the 40th Workshop on Geothermal Reservoir Engineering, Stanford University, Stanford, CA, USA, 26–28 January 2015.

35. Younger, P.L. Missing a trick in geothermal exploration. *Nat. Geosci.* **2014**, *7*, 479–480.

36. Scott, S.; Driesner, T.; Weis, P. Geologic controls on supercritical geothermal resources above magmatic intrusions. *Nat. Commun.* **2015**, *6*, doi:10.1038/ncomms8837.

37. Ferguson, G.; Woodbury, D. Observed thermal pollution and post development simulations of low-temperature geothermal systems in Winnipeg, Canada. *Hydrogeol. J.* **2006**, *17*, 1206–1215.

38. Fry, V.A. Lessons from London: regulation of open-loop ground source heat pumps in central London. *Q. J. Eng. Geol. Hydrogeol.* **2009**, *42*, 325–334.

39. Birks, D.; Younger, P.L.; Tavendale, L.; Coutts, C.; Parkin, G.; Button, P.; Whittall, S. Groundwater reinjection and heat dissipation: Lessons from the operation of a large groundwater cooling system in Central London. *Q. J. Eng. Geol. Hydrogeol.* **2015**, *48*, 94–103.

40. Gandy, C.J.; Clarke, L.; Banks, D.; Younger, P.L. Predictive modelling of groundwater abstraction and artificial recharge of cooling water. *Q. J. Eng. Geol. Hydrogeol.* **2010**, *43*, 279–288.

41. Neuberger, P.; Adamovský, R.; Šeďová, M. Temperatures and heat flows in a soil enclosing a slinky horizontal heat exchanger. *Energies* **2014**, *7*, 972–987.

42. Agemar, T.; Weber, J.; Schulz, R. Deep geothermal energy production in Germany. *Energies* **2014**, *7*, 4397–4416.

43. Fulignati, P.; Marianelli, P.; Sbrana, A.; Ciani, V. 3D geothermal modelling of the Mount Amiata hydrothermal system in Italy. *Energies* **2014**, *7*, 7434–7453.

44. Gholizadeh Doonechaly, N.; Abdel Azim, R.R.; Rahman, S.S. A study of permeability changes due to cold fluid circulation in fractured geothermal reservoirs. *Groundwater* **2015**, doi:10.1111/gwat.12365.

45. Sanyal, S.S.; Morrow, J.W. An investigation of drilling success in geothermal exploration, development and operation. *Geotherm. Resour. Counc. Trans.* **2011**, *35*, 233–237.

46. Underwood, C. On the design and response of domestic ground-source heat pumps in the UK. *Energies* **2014**, *7*, 4532–4553.

47. Kyriakis, S.A. Effect of heat storage in geothermal district heating. In Proceedings of the ASME-ATI-UIT 2015 Conference on Thermal Energy Systems: Production, Storage, Utilization and the Environment, Naples, Italy, 17–20 May 2015.

48. Kalina, A.I.; Leibowitz, H.M. Application of the Kalina Cycle technology to geothermal power generation. *Geotherm. Resour. Counc. Trans.* **1989**, *13*, 605–611.

49. Atlason, R.; Oddsson, G.; Unnthorsson, R. Geothermal power plant maintenance: Evaluating maintenance system needs using quantitative Kano Analysis. *Energies* **2014**, *7*, 4169–4184.

50. Harber, A. *Study into the Potential for Deep Geothermal Energy in Scotland*; Scottish Government Project Number: AEC/001/11; AECOM/Scottish Government: Edinburgh, UK, 2013; Volume 1, p. 216. Available online: http://www.gov.scot/Publications/2013/11/2800/9 (accessed on 29 September 2015).

51. Younger, P.L.; Feliks, M.E.J.; Westaway, R.; McCay, A.; Harley, T.L.; Elliott, T.; Stove, G.D.C.; Ellis, J.; Watson, S.; Waring, A. Renewing the exploration approach for mid-enthalpy systems: Examples from northern England and Scotland. In Proceedings of the World Geothermal Congress 2015, Melbourne, Australia, 19–25 April 2015; p. 7.

52. Finger, J.; Blankenship, D. *Handbook of Best Practices for Geothermal Drilling*; Report SAND2010–6048; Sandia National Laboratory: Albuquerque, NM, USA, 2010; p. 84. Available online: http://www1.eere.energy.gov/geothermal/pdfs/drillinghandbook.pdf (accessed on 1 October 2015).

53. United States Department of Energy. Percussive Hammer Enables Geothermal Drilling. Available online: http://energy.gov/eere/success-stories/articles/percussive-hammer-enables-geothermal-drilling (accessed on 1 October 2015).

54. Kristofič, T. PLASMABIT™ technology presentation. In Proceedings of the 2012 Geothermal District Heating (GeoDH) Project Conference: Brussels, Belgium, 9 October 2012; p. 11. Available online: http://geodh.eu/wp-content/uploads/2012/10/PlasmaBit-Kristofic.pdf (accessed on 1 October 2015).

55. Bazargan, M.; Gudmundsson, A.; Meredith, P.; Forbes, N.; Soliman, M.; Habibpour, M.; Rezaee, A. Wellbore instability during plasma torch drilling in geothermal reservoirs. In Proceedings of the 49th US Rock Mechanics/Geomechanics Symposium, American Rock Mechanics Association, San Francisco, CA, USA, 28 June–1 July 2015.

56. Russell, M.C. Ram Accelerator System; Patent No. WO2014149173 A1/US20140260930, 25 September 2014. Available online: http://www.google.com/patents/WO2014149173A1 (accessed on 1 October 2015).

57. United States Department of Energy, High power laser innovation sparks geothermal power potential. Available online: http://energy.gov/eere/geothermal/articles/high-power-laser-innovation-sparks-geothermal-power-potential (accessed on 1 October 2015).

58. Theis, C.V. The source of water derived from wells: Essential factors controlling the response of an aquifer to development. *Civil Eng.* **1940**, *10*, 277–280.

Chapter 1:
Enhancing Geothermal Reservoir Characterisation

Gamma-ray Spectrometry in Geothermal Exploration: State of the Art Techniques

Alistair T. McCay, Thomas L. Harley, Paul L. Younger, David C. W. Sanderson and Alan J. Cresswell

Abstract: Gamma-ray spectrometry is a surveying technique that allows the calculation of the heat produced during radioactive decay of potassium, uranium, and thorium within rock. Radiogenic heat producing rocks are often targets for geothermal exploration and production. Hence, refinements in gamma-ray spectrometry surveying will allow better constraint of resources estimation and help to target drilling. Gamma-rays have long half-lengths compared to other radiation produced during radiogenic decay. This property allows the gamma-rays to penetrate far enough through media to be detected by airborne or ground based surveying. A recent example of ground-based surveying in Scotland shows the ability of gamma-ray spectrometry to quickly and efficiently categorize granite plutons as low or high heat producing. Some sedimentary rocks (e.g., black shales) also have high radiogenic heat production properties and could be future geothermal targets. Topographical, atmospheric and spatial distribution factors (among others) can complicate the collection of accurate gamma-ray data in the field. Quantifying and dealing with such inaccuracies represents an area for further improvement of these techniques for geothermal applications.

Reprinted from *Energies*. Cite as: McCay, A.T.; Harley, T.L.; Younger, P.L.; Sanderson, D.C.W.; Cresswell, A.J. Gamma-ray Spectrometry in Geothermal Exploration: State of the Art Techniques. *Energies* **2014**, *7*, 4757-4780.

1. Introduction

In this paper, we review gamma-ray spectroscopy as a survey tool for geothermal resource exploration. We hope the paper will also be useful as a practical guide for those unfamiliar with gamma-ray surveying, who might benefit from using it in geothermal exploration.

Gamma-ray spectroscopy allows determination of concentrations of selected radioelements from which the heat being produced from radioactive decay can be calculated. This may be by counting gamma-rays produced either in a rock sample during a laboratory test or an area of land during an *in-situ* survey. However, the relationship between recorded gamma fluence and radioelemental concentration in the geosphere is complex. Factors such as decay series disequilibria, topographical errors, and atmospheric influence during surveying can lead to results that are not representative of the underlying rock. The radioelements of interest for geothermal resources are potassium (K), uranium (U), and thorium (Th). Rocks of high concentrations of these radioelements can be characterised by high heat flow, and the geothermal gradient can thus be favourably enhanced. Such enhancement creates useable heat at shallower depths than would otherwise be the case, thus reducing the drilling costs of a geothermal project.

Many granites are enriched in the radioelements potassium, thorium and uranium, and thus typically have higher radioactivity than many other rocks. Granite is therefore a favoured target in geothermal exploration worldwide, e.g., USA [1], Japan [2], UK [3], France [4], Switzerland [5], Australia [6]. This heat producing property of granite is particularly effective when the pluton is buried beneath layers of low heat conductivity "duvet rocks" such as coal or shale [7]. There can be crossover between classifications of duvet rocks and caprocks (*i.e.*, reservoir topseals), where the rocks both have low thermal conductivity and permeability. However, some potential duvet rocks such as the Clyde Plateau Lavas in the Midland Valley of Scotland would likely not be effective caprocks, in this case due to extensive fracturing. Where such duvet rocks cap highly radiogenic granite, vastly enhanced heat can be obtained [8–10]. Radiogenic heat production is not just a phenomenon peculiar to granite as all rocks contain some concentration of radio-elements. Depending on the depositional environment, mudstones can have elevated concentrations of radio-elements compared to other sedimentary rocks. Due to their low thermal conductivities (because of their low quartz content) this heat can remain in place within mudstones over geological time, which may result in viable geothermal resources. Metamorphic rocks, on the other hand, tend to be depleted in radio-elements [11,12]; such depletion is actually part of the process that feeds the upper crust with relatively higher concentrations of radioelements [13].

Within geothermal exploration, gamma-ray surveying can be put to a number of uses beyond heat production investigations. In geothermal investigations, gamma-ray surveying is also useful for fracture identification. Fractures in the subsurface have previously been associated with elevated uranium concentrations [14,15] due to the mobility of uranium in subsurface fluid circulation. Such mobility can cause a significant issue for gamma-ray spectrometry survey interpretation known as disequilibrium (discussed in Section 4.2). Fractures can be a source of significantly enhanced permeability [15–17] providing key conduits for fluid extraction in a geothermal system; thus, it is advantageous to accurately characterise the fluid flow properties of a fracture network during resource evaluation. The duvet layers of low heat conduction, e.g., mudstone, can also be detected by their higher gamma-ray output compared to surrounding formations. These gamma-ray counts show up during wire-line logging of boreholes.

Gamma-ray surveying also has a wide range of applications beyond geothermal exploration including: uranium exploration [18,19], sedimentary facies identification for oil and gas exploration [20–22], detection of radioactive contamination [23,24], and mineral exploration [25]. It can also be used for pure earth science discoveries, e.g., constraining deep crustal processes from potassium, uranium, and thorium concentrations in modern day outcrops [26,27].

2. Revision of Physical Concepts

2.1. Gamma-ray Formation and Detection

Gamma-ray is the name generally given to high-energy photons emitted during decay of atomic nuclei. Gamma-rays have frequencies greater than 10^{19} Hz, wavelengths less than 10^{-12} m, and have energies above 10^4 eV; gamma-rays are generally the highest energy photons in the electro-magnetic spectrum. Radioelements spontaneously decay leading to emission of alpha, beta, and/or gamma radiation depending on the decaying element. These radio-elements are naturally present in most rocks, but tend to be concentrated at higher levels in certain types (e.g., granite, mudstone). Potassium, uranium, and thorium are of particular interest for geothermal production because they contribute significantly to the heat produced during radioactive decay in the rock. The concentrations of these elements show an approximate trend to increase with silica content [28]; the same relationship has been found for gamma ray intensity in volcanic rocks [29].

Gamma-rays penetrate through materials (e.g., rock and air) much further than the other forms of radiation (alpha or beta). This penetrating ability is what makes gamma-rays useful for detecting radioelement concentrations within rock. They can penetrate up to 0.5 m through rock, allowing a sample to be collected by a portable gamma-surveyor which is large enough to not be grossly biased by local concentration heterogeneity. However, the half-length of a gamma-ray in rock such as granite is much less than 0.5 m; half-length is the distance through a material where half the gamma-rays will be attenuated. During an *in-situ* survey, most of the gamma-rays detected will effectively come from the top 0.15 m of material. For example, a portable detector placed on a rock surface will sample gamma-rays from approximately a 0.15 m deep by 1.0 m diameter disc (Figure 1); with a small contribution from deeper sources. Penetration through air can be up to several hundreds of metres; therefore, aerial surveys are typically conducted 30–300 m above the surface [30,31]. Although the penetrating property of gamma rays allow surveys to be conducted, having most gamma radiation coming from the top 0.15 m at the surface does present problems. Aerial surveys (explained in Section 3) sample a wide area during each reading; such a sample may be a mixture of bare rock, peat cover, and water courses. The portion of the measured gamma-rays that originated from bare rock can be a significant uncertainty when interpreting the results. A further issue is weathering can alter the concentration of potassium, uranium, and thorium at the rock surface [28]. These issues can be compensated for by calibrating aerial results with direct surveying of a freshly created rock surface or testing borehole samples in the laboratory [32,33].

Figure 1. Schematic showing approximate areas of rock sampled by portable gamma surveyor placed on rock surface.

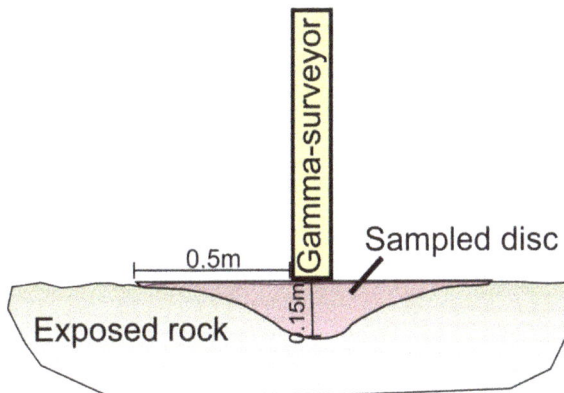

The concentrations of specific radioelements can be determined as they each impart a specific energy signature onto the photon produced during decay. The isotope ^{40}K produces photons with energy of 1.46 MeV. However, uranium and thorium are detected by their daughter products; therefore, uranium and thorium concentrations are detected as equivalent uranium (eU) and equivalent thorium (eTh). The early spectrometers used the daughter products [214]Bi (1.76 MeV) for uranium and [208]Tl (2.62 MeV) for thorium. These daughter products were originally used because their produced gamma-ray energy signatures are relatively large (e.g., 0.8 MeV for Bi^{214} compared with 0.2 MeV for U^{235} or no gamma-ray produced for U^{238}) and can be more easily distinguished. Modern spectrometers are not limited to solely these daughter products to estimate eU and eTh as improvements in spectrometers means that many of the lines in the uranium or thorium decay series can be distinguished. This allows adequate confidence in estimates of potassium, uranium, and thorium derived from gamma-ray sources. Additionally, Compton Scattering affects gamma-rays as they pass through rock due to gamma-rays "bouncing off" electrons which absorb some of the energy from the gamma-ray. This Compton Scattering means that when a photon is detected by the surveyor it may have much less energy than when it was created during decay. This diminishing energy results in photons created by thorium decay arriving at the detector with energy expected from uranium or potassium decay, in addition to U photons arriving with the expected potassium decay energy. However, these scattering affects can be compensated for in spectral analysis.

The photoelectric effect [34,35] is utilised to detect gamma-rays [36] with many detectors made from material which undergo scintillation; i.e., visible light is produced when struck by gamma-rays (the use of this effect is described further in Section 3). Detectors made from sodium-iodine are typically used in in-situ surveys [20,26,37,38]. Many other materials are used for scintillators such as bismuth germinate; ceasium-iodide detectors may also be used but these have poor resolution. Alternatively, lanthanum bromide detectors provide good resolution but have self-dose issues; cerium bromide has less self-dose problems but remains expensive. Also in use are semiconductor detectors such as intrinsic germanium. This is used for lab studies as it requires cryogenic cooling.

2.2. Heat Production from Radioelements

The heat from radioactive decay is produced in accordance with the well-known Einsteinian expression $E = mc^2$ [39]. This summarises the fact that energy cannot be created or destroyed, but mass can be converted into energy and *vice-versa*. During the decay processes, some mass must be converted into the energy that produces the heat. Initially, this energy exists in the form of the emitted particle's kinetic energy. During subsequent collisions, this energy is absorbed and converted to heat.

It is important to note that, although gamma-rays are used to determine the quantities of potassium, uranium, and thorium in rock, the gamma-rays themselves are not actually responsible for significant quantities of the heat produced. Alpha and Beta components of decay produce much more heat than gamma-rays; in particular, the alpha decay of uranium [40]. Such heat producing decays come from different parts of the decay series to the detected gamma-rays, thus disequilibrium of the decay series (further discussed in Section 4.2) can lead to radioelemental concentrations that do not represent the radiogenic heat being produced by the rock. Neutrinos are also produced during decay but pass through the planet, thus some energy of the Earth is lost to outer space [41].

3. Instrumentation

Several different types of gamma spectrometers can be used *in-situ* in the field or laboratory, but all share a common basic architecture (Figure 2). Detectors are either based on scintillation or semiconductors. Scintillation detectors consist of both a scintillator and photomultiplier. The scintillator is made out of material which reacts with gamma-ray photons producing photons of visible light. The visible light forces electrons to be ejected from the photomultiplier which are then multiplied [42]. The electrons strike an anode, which produces a negative voltage pulse which is proportional to the energy of the photon which struck the scintillator. This proportionality is how the energy can be determined by the spectral analyzer and thus the origin of the gamma-ray can be determined. Semiconductor detectors are diodes in which incident radiation generates electron-hole pairs which migrate to electrodes due to a high voltage across the diode. This produces a current pulse proportional to the energy deposited by the incident radiation.

Figure 2. Block diagram showing the main sections which are common to most gamma-ray spectrometers. Figure adapted from International Atomic Energy Agency (IAEA) 2003 report [42].

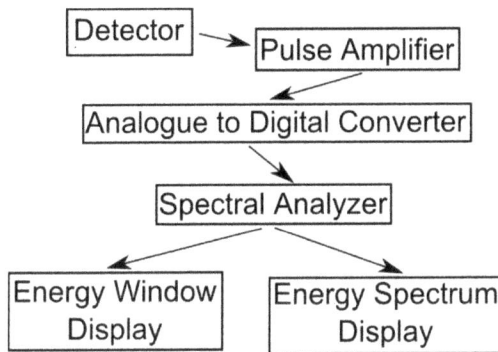

Many geothermal exploration studies utilise lab-based gamma-spectrometers. These gamma-spectrometers can be very precise, such as intrinsic germanium semiconductor counters. They deliver reliable results because they can be regularly checked against standards and the surrounding environment remains largely stable. To conduct the gamma readings, a sample of approximately 100 g is crushed (less material may be used for more highly radioactive samples). The crushed sample is then put in the chamber with the counter; the chamber being housed in minimizes influences of externally created gamma rays. Lab techniques are often used during geothermal drilling since small drill cutting samples can be used for the analysis. Using the drill cuttings has the advantage that drilling does not have to be interrupted for wire-line logging to gain data about the heat production of the geothermal exploration target rocks. Laboratory measurements have typical errors of 0.03%–0.1% K, 0.5–2 ppm eU, and 0.1–0.3 ppm eTh [42].

Large-scale *in-situ* gamma-ray surveys can be conducted using airborne spectrometers mounted on aeroplanes [43] or helicopters [33]. The aircraft are fitted with special mounts to suspend the scintillation counters; sampling time for airborne surveys can vary from 1 s to min [32]. Such aerial surveys exploit the favourable penetration of gamma-rays through air [42]. However, this does mean that the survey flights must be conducted within a few hundred metres of the ground surface [31]. Their advantage is the huge expanse of terrain that can be covered in a few days of surveying. The height of the airborne surveys means that each reading represents an average of a wide area; e.g., at 100 m altitude the sample area may have a diameter of approximately 190 m [44]. The survey sample area will therefore be spatially variable [45]; e.g., in Scotland, patches of exposed high heat production-granite will be recorded alongside areas of peat cover masking the heat production-granite below. However, case studies show that areas of high gamma-ray intensity and therefore high heat production can still be identified despite this averaging of properties [32]. The results from airborne surveys are not trivial to analyse and require a range of careful corrections for influences such as topography and altitude [31,42]; these influences can even be further compounded by material heterogeneity. Such information is, however, readily collectable during airborne gamma-ray readings [23]. Count rates during airborne surveys typically have standard deviations of 6.3% for potassium, 12.3% for uranium, and 13.7% for thorium if the surveyed material had concentrations of 2% K, 2.5 ppm eU, and 9 ppm eTh [42].

In-situ surveys can also be conducted using hand-held portable spectrometers. These surveys have the advantage of being extremely flexible; to cover a wider area in minimal time, for instance, readings can be taken at 100 s of metres spacing, while for collection of detailed information, readings can be taken every 0.5 m [20–22,38,46]. In addition to lateral spacing, the counter can be placed on rock surfaces for small volume sampling (see Figure 1) or else held above rock surfaces to sample significantly wider areas at once [18,47]. Sampling time also varies in surveys from seconds [48] to several minutes [46,49] depending on the surveyor and survey design. Typically in areas of lower radioactivity rock, longer sample times are needed for suitably accurate results [22]. An alternative approach used by some surveys is to monitor continuously then integrate the count times in intervals, e.g., every 5–10 s [50]. Rock and soil samples can also be collected during these *in-situ* surveys, to compare with the *in-situ* gamma results or to conduct more general rock mineral analysis [32,33,44,51] and be able to quantify near surface geometrical effects [52]. Modern hand-held surveyors are very portable, weighing in at a few kilograms and being small enough to fit inside a small backpack. Indeed some designs have been specifically mounted on backpacks and readings taken at automatic intervals [32,50]. Such portable spectrometers have precisions of approximately 0.1%–0.14% K, 0.6–0.8 ppm eU, and 0.6–1.5 ppm eTh [42,53].

Car-borne surveys can offer a useful compromise between wide area, low resolution airborne surveys and high resolution, narrow area hand-held portable surveys. However, the car-borne surveys are limited to locations that permit vehicular access. Even so, given the right settings, car-borne surveys can effectively survey a much larger area in a shorter space of time compared with walkover surveys and could also provide a valuable mix of surveying scales [32,33]. Car-borne surveys have similar survey times as airborne surveys of several seconds [32] but these could be increased if the needs of a survey warranted longer survey times.

A comprehensive survey may include several of these techniques. Walkover and car-borne surveys can be run at complementary scales with airborne surveys [23], to calibrate the airborne surveys [32]. Each of these techniques is best suited to different desired outcomes of a survey, so thought must be given as to which would be most suited to the survey needs.

4. Calculation

4.1. Data Corrections

The collected data require correcting prior to any analysis of the results. The corrections depend upon which of the survey modes were utilized. In commercially available instruments, some of these corrections are done automatically; otherwise the corrections and calculations must be completed by the surveyor. Geothermal explorationists need to understand the transformations of the raw data to K, eU, and eTh concentrations, if they are to use the results confidently, and be able to engage sufficiently with survey physicists to help design field campaigns. An outline of where corrections may be needed in gamma-ray spectrometry surveying is provided in this section.

Due to operating at heights hundreds of metres above ground, airborne surveys are particularly susceptible to influence from gamma-rays produced by cosmic-rays. These cosmic-rays interact with the Earth's atmosphere and produce gamma-rays as secondary radiation [42]. Cosmic ray intensity gets higher with altitude, doubling almost every 2000 m from an intensity of about 32 nGy/h at sea level [54]. Additionally, increases in altitude results in decreasing fluence of gamma radiation originating from the ground surface, as these are progressively scattered and absorbed by the atmosphere. For these reasons, airborne surveys are usually conducted within a 30–300 m altitude [30,31]. Such a height gives each airborne survey measurement a ground sample area of approximately 300 by 300 m with the sample area increasing proportionally with altitude. Surveys are also susceptible to influence from gamma-rays originating in the atmosphere due to radon decay, the intensity of such are variable both spatially and temporally. The gamma-ray count associated with cosmic rays and radon can be found by flying at several heights over a large body of water; as the water shields the aircraft from the gamma-radiation that is produced by the ground surface below. This atmospheric gamma-ray count can then be subtracted from the results as appropriate. The body of water should preferably be several metres deep, and should also be fresh-water because sea-water has a modest uranium content. Consideration may also be given for the count produced by the vessel on which measurement is taken over the body of water.

Airborne surveys are strongly affected by undulating terrain, as this affects how the area of ground surface is exposed to the gamma spectrometer on the aircraft. Such influences can increase the count rates by 100% in valleys and decrease by 10%–30% over mountain ridges [31]. Corrections for topography can be conducted [31] but can assume a homogeneous medium for airborne surveys. Additionally information about the underlying topography at the moment each measurement is acquired must also be collected. Small scale topographical features have been found to show variations of radioactivity by up to six times due to source redistribution by natural processes [55]; which shows the issue of a homogeneous assumption during topographical corrections. Such varying topography can also be an issue for maintaining a constant survey height above the ground surface. Another potential issue is the variation of half-lengths gamma-rays in different materials; surveys may be weighted towards material of lower density in which gamma-rays can penetrate more easily. Further corrections may be needed for biomass, as vegetation affect gamma-ray data [56,57] due to covering exposed rock and emitting their own gamma-rays.

Portable surveys are typically calibrated assuming that the surveyor is taking readings from an area that is 2π. A 2π area is where there is a solid angle with the surface, *i.e.*, the rock is flat, $>2\pi$ would be where the surveyor is placed in a depression leading to overestimation, $<2\pi$ where surveyor is placed on a mound or at edge of rock leading to underestimation (Figure 3). Due to gamma-rays travelling for hundreds of metres through air, note must be taken that even distant topographical features can influence the results. Figure 4 shows how, as readings are taken approaching a small granodiorite cliff (10 m high), there is a steady increase in the total gamma count due to the influence of the cliff. This phenomenon is particularly important where some readings may be taken in valleys or cirques surrounded by slopes of 100 s of metres; on the other hand, readings on ridge crests may not be as subjected to such sources of error. Careful noting of any field conditions that may affect the results should be taken, and then compared with the data during analysis to avoid any spurious conclusions over anomalously high results.

Figure 3. Schematic cross-section of rock outcrop showing different possibilities for the locations of gamma-ray readings. Location (a) would collect readings from an area of $>2\pi$ so would overestimate results, location (b) is next to a ledge so would collect readings from an area $<2\pi$ and underestimate gamma-ray counts. Location (c) is a relatively flat section of outcrop more than a meter away from ledges; this would likely be a 2π area where the results are not affected by topography.

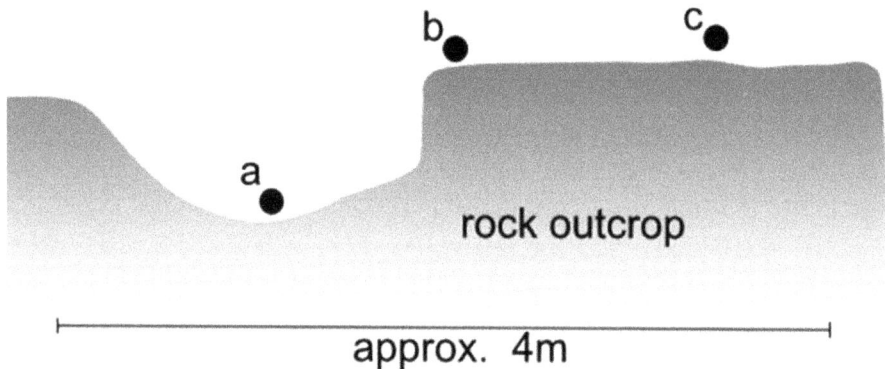

The data also need to be corrected for the interference of photons derived from the decay of thorium and uranium in the "count windows" of the other heat producing radioelements. Correcting for this is done in spectral analysis, one method is "stripping" [30,58–60] but principle component analysis and least square fitting analytical techniques are also used regularly. Count windows are the energy levels at which photons from a particular element in the decay series create distinctive peaks. An example of such peaks that may be used are shown in Figure 5, and correspond to [208]Tl (2.62 MeV) in the thorium decay series, [214]Bi (1.76 MeV) in the uranium decay series, and [40]K (1.46 MeV) for potassium. Figure 5 also highlights how photons from the decay series of thorium interfere in the uranium and potassium windows, and how photons form the decay series of uranium interfere in the potassium window. This interference is partly due to photons scattering as they travel through a medium; the scattering reduces the energy of the moving photons, and/or creates new photons of lower energy. Interference also occurs due to other gamma-ray emissions from the decay series. For uranium, stripping can be done by assessing the ratio of the count of scattered thorium photons in the uranium window (1.76 MeV) with the count in the thorium window (2.62 MeV). The same process strips the scattered uranium and thorium photons in the potassium window (1.46 MeV). These scattered photon counts are subtracted from the total window counts to get the true count produced by ^{214}Bi in the uranium window and ^{40}K in the potassium window.

Figure 4. Total gamma counts taken on a beach at the Solway Firth, SW Scotland (54°51'15''N, 3°40'59''W). Red dots indicate locations that gamma-ray readings were taken; red line indicates gamma-ray dose rates at each location showing gradual increase in counts as granodiorite cliff approached.

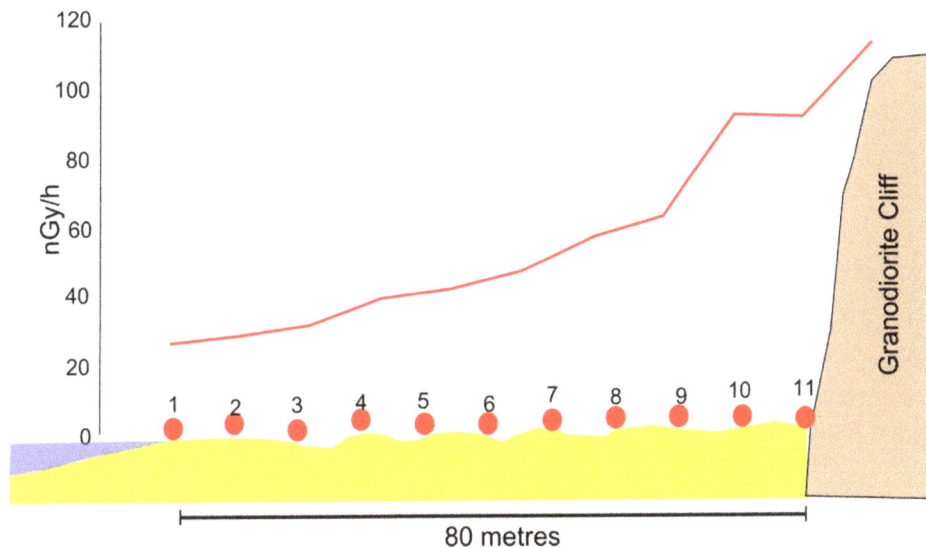

The counts corrected by stripping in the respective windows can then be used to estimate the concentrations in parts per million of uranium (Uppm) and thorium (Thppm) and the percentage by weight (K%) of potassium. To do this, gamma-ray surveyors are calibrated at concrete pads which are doped with a known concentration of potassium, uranium, or thorium [30]. These pads are used both to determine the stripping characteristics of a scintillation crystal and to estimate its sensitivity. Such calibration is required because each scintillation crystal will react differently to bombarding photons; producing different counts for the same radioelement concentration.

Figure 5. Figure showing typical counts of different energies produced by scattering of photons produced by the decay of thorium and uranium with daughters, and potassium. The distributions of energy photons demonstrate how thorium daughters produce photons in the uranium and potassium window and uranium daughters produce photons also in the potassium window. Figure adapted from [49].

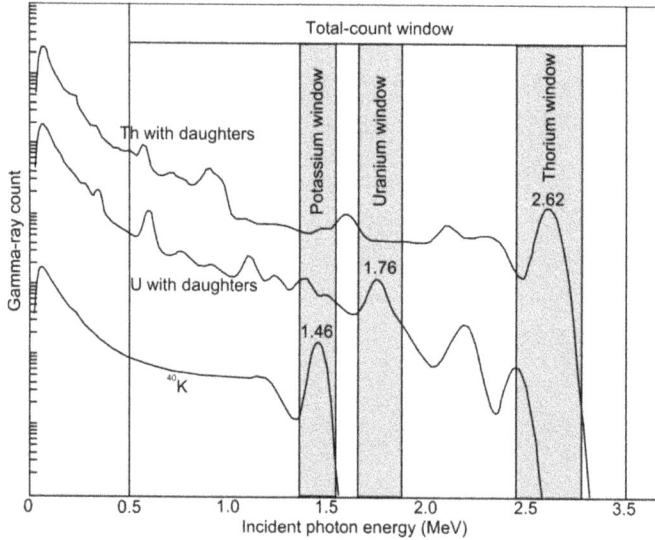

4.2. Heat Production

Once reliable values for K%, Uppm, and Thppm have been obtained, these values can be used to calculate the heat that is being produced by the radioactive decay in the rock (*i.e.*, the radiogenic heat production). Heat production (HP) can be found using Equation (1) which was developed by calculating the energy released during alpha, beta, and gamma decay of the radioelements [40,41]:

$$HP(\mu W m^{-3}) = \rho \, (0.035C_K + 0.097C_U + 0.026C_{Th}) \tag{1}$$

where: ρ is rock density (kg m^{-3}), C_K is concentration of potassium by % weight, C_U and C_{Th} are concentration of uranium and thorium in ppm.

In Equation (1) each of the radioelement concentrations are multiplied by a numerical constant. These constants reflect the differing contributions to the radiogenic heat production of each radioelement; in nW per kg of rock per unit of potassium, uranium or thorium. The constant for uranium (0.097) is more than double the constants for potassium (0.035) or thorium (0.026); reflecting the dominant role that uranium has in producing heat compared with thorium or potassium. In fact, it is the alpha decay of uranium which provides most of the radiogenic heat production [41]. This means that often granites with high U/Th ratios tend to have favourable radiogenic heat production properties [10,61]. However, when U/Th ratios are 0.25, then cumulatively U and Th produces similar amounts of heat. It is important to note that Equation (1) relies on the assumption that there is a fixed ratio between the daughter products used to estimate eU and eTh. However, the various daughter products of uranium and thorium have differing mobility properties under reducing or oxidizing conditions; *i.e.*, some daughter products may be transported away from the rock over time. This would result in disequilibrium meaning that there could truly be a higher or lower concentration of uranium or thorium than is indicated by the gamma surveyor. Disequilibrium occurs when discrepancies exist in the ratios between parent isotopes and daughter products. Due to differing leaching rates from the subsurface, certain daughter products can be preferentially removed or remain relative to the parent atom (U^{235}, U^{238}, Th^{232}). Such mobilization and leaching of daughter products can mean the detected radioactive decay not be proportional to the amount of uranium or thorium in the rock. This effect is most prominent in the U decay series which is mobile under oxidizing conditions but is precipitated under reducing conditions [62] (resulting in some ocean originated black shales having very large U concentrations [41]). Radium [28] and radon in particular due to it being a gas can also be causes of disequilibrium due to both being mobile and part of the uranium and thorium decay series. It is important to stress that the gaseous highly mobile state of radon means if a post radon decay of the uranium series is used to determine eU, then the likelihood of disequilibrium is high enough that it makes it questionable whether it is accurate to use the full decay series for the calculation of heat production. Supplementary work may be required to examine the state of radon loss in the decay series to produce a reliable heat production value. Uranium is of particular interest for disequilibrium because it is the dominant producer of heat compared to potassium or thorium.

The dominant role that uranium plays in heat production is highlighted in the graphs in Figure 6. Figure 6 shows three graphs showing K%, U (ppm), or Th (ppm) against calculated heat production. The data was taken from the survey described in Section 6. The graphs in Figure 6 demonstrate the strong correlation between uranium concentration and heat production, compared with the weaker correlations with thorium and potassium concentration.

Figure 6. Heat production *versus* potassium, uranium, and thorium concentrations, for data collected during the Scottish case study described in Section 6.

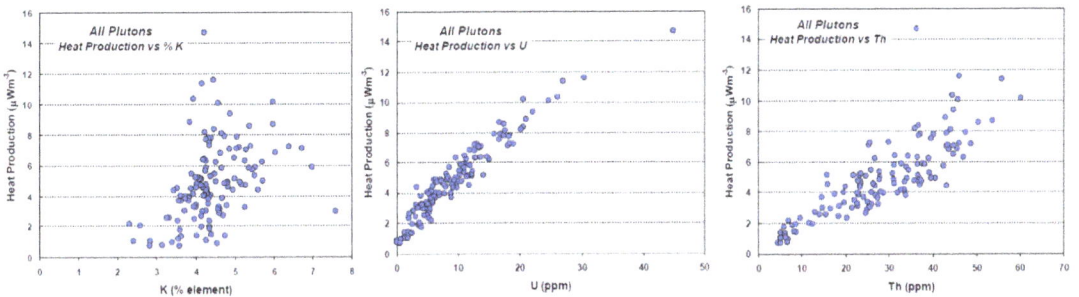

5. Case Study from Scotland

During July 2013, we conducted an *in-situ* survey over several Scottish granite plutons using a portable gamma-ray surveyor. The aims of the survey were: (i) to re-evaluate the radiogenic heat production of the granites; and (ii) to allow comparison between results from the portable gamma-ray surveyor and previous lab-based investigations.

A GAMMA SURVEYOR II (GSII) instrument (made by GF Instruments in Brno, Czech Republic) was used for the *in-situ* survey. The detector in the GSII is a Bismuth Germanate Oxide with a volume of 20 cm^3. The analyser measures 1024 different channels between 0.03 and 3 MeV. The surveyor weighs 1.8 kg and is compact and lightweight enough for it to be readily carried in a small rucksack [50]which is important in the highly mountainous terrains which granite often gives rise to.

Seven plutons were visited in total, six of the plutons in the Grampian region of Scotland; Monadhliath, Cairngorm, Lochnagar, Ballater, Grantown, and Strathspey, and one pluton on the Isle of Mull (the Ross of Mull granite). These plutons were selected for their previously identified high heat production [3], close to areas of high heat demand and with clear areas of exposure visible from aerial photographs.

To minimise topography-related errors (e.g., Figure 3), sample locations were chosen for having several square metres of exposed granite which were relatively flat. The GSII was placed on the surface of the granite as far as possible away from large open fissures and other voids that could influence results. Figure 7 shows a typical fracture outcrop where the GSII is placed in the centre of an intact block of granite away from fissures. Notes were made during measurements of any identifiable features which might influence measurements. No average point density was aimed for during the surveys because survey points were dictated by suitably exposed intact granite and accessibility. Therefore, in some areas, a higher density of points (spaced at tens of metres) could be achieved and in other areas exposures were separated by several hundred metres of peat cover.

Figure 7. An example of where the largest section of intact granite was chosen to place the gamma surveyor II (GSII), away from the perpendicular fractures. For scale, the GSII is 28 cm long and 9 cm wide.

Three measurements were taken at each location to ensure the results were not affected by anomalies in the internal algorithms in the GSII. The vast majority of times this was not necessary, but the repeated measurements did provide extra confidence in the results particularly when readings were unusually low or high. Each measurement lasted 3 min, a period previously established as adequate for a reliable sample of rocks with similar counts per second as granite [46,63], albeit measurement times can be shorter in high activity areas or longer in lower activity areas such as metamorphic basement [26]. To ensure 3 min measurements were long enough to be reliable we checked against half hour measurements (Figure 8), obtaining reassuringly similar results. All 3 min measurements were within 10% of the 30 min measurement; except for one 3 min reading which was 11% lower than its equivalent 30 min reading. Such close correlation confirms that there would be no useful improvement in accuracy to taking significantly longer for such moderately radioactive material.

Figure 8. Graph showing excellent correlation between the 30 min and 3 min readings taken at the same location.

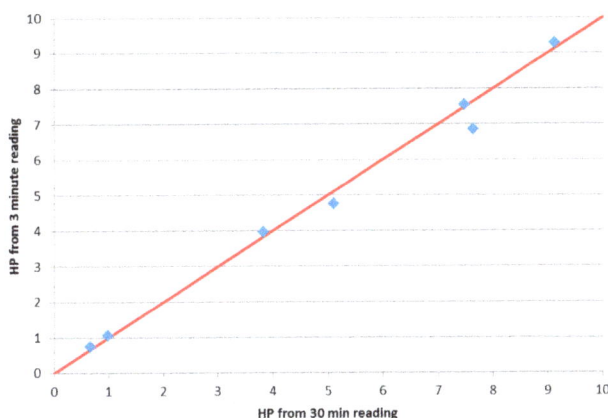

The survey identified the Cairngorm and the Ballater granites as particularly high in heat production (Figure 9) with values of 5.7 ± 2.6 $\mu W/m^3$ and 8.2 ± 1.5 $\mu W/m^3$, respectively. A general convention in geothermal exploration is that anything above 4 $\mu W/m^3$ is considered as high heat production and thus a potentially economic heat resource. We believe this threshold is derived from old imperial units of radiogenic heat production (10^{-13} cal/cm^3·s), as in those units 10×10^{-13} cal/cm^3 is equivalent to 4.18 $\mu W/m^3$ (*i.e.*, ≈ 4 $\mu W/m^3$). However, such a convention may not be useful in many circumstances as local geology can mean a viable heat resource exists even with lower values of radiogenic heat production, due to covering "duvet layers". The Lochnagar and Monadhlaith granites both have median heat values of HP above 4 $\mu W/m^3$ so could also host viable heat resources. The Grantown granite has a low heat production value because it is an "S-type" granite, *i.e.*, one which formed primarily by the melting of sedimentary rocks. The Strathspey and Isle of Mull plutons both show low heat production, and thus are unlikely to be good targets for further geothermal resource investigation. Ultimately, it is the heat flow and geothermal gradient, in addition to permeability, which would determine the suitability of a rock for geothermal production. However, rocks of high heat production have been correlated with areas of high heat flow [3,64], for example in North West Scotland and South East England zones of high heat flow exist over high heat production granites, and so can be considered an important aspect of the exploration and appraisal process.

The spread of heat production values in Figure 10 demonstrates the importance of gathering numerous data from a pluton during surveys. More data means that any outliers (whether low or high) can be identified, so they do not unduly skew results, preventing a pluton from being wrongly categorized as having either high or low heat production. Such anomalous values of heat production could sometimes be attributed to observable features such as hydrothermal alteration which may have leached radioelements or dykes of other material intruded into the granite. However, granite plutons are not homogeneous but have varying composition due to magma mixing, assimilation of country rocks, fractional melting, fractional crystallization, water activity, and the pressure and temperature pathways of magma evolution [65–69]. Such differing composition results in variation of radiogenic properties across the pluton. This survey did not have sufficient sample density coverage to be able to determine zones in the plutons of higher or lower heat production related to past geological processes; such as crystallisation. An aim of future investigations targeting the granites of higher heat production could be to explore the heat production variation within the granite; if such information was considered favourable to characterising the geothermal resource.

Figure 9. Locations of the studied granite plutons: Monadhlaith (A), Cairngorm (B), Lochnagar (C), Ballater (D), Grantown (E), Strathspey (F), Ross of Mull (G).

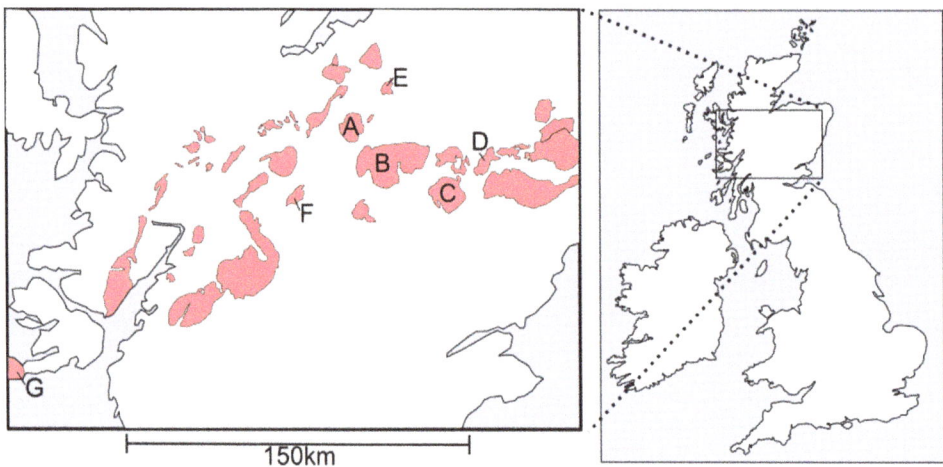

The results of the July 2013 portable gamma-ray spectrometry survey show good correlation with collated results from previous lab-based surveys [3] (Figure 11). The data from these laboratory test were collected by taking rock samples from shallow (<300 m depth) boreholes and outcrops, which were then analysed using lab-based gamma-ray spectrometry techniques [36,70]. Although the two surveys show variation of the heat production for many of the plutons, both agree on which plutons have high heat production of >4 $\mu W/m^3$ (Ballater, Cairngorm, Monadhliath, Lochnagar) and those with low heat production of <4 $\mu W/m^3$ (Strathspey, Grantown, Ross of Mull). This establishes that although *in-situ* studies may lack the precision of lab based work, they can quickly and simply provide an accurate portrayal of the heat production in a geothermal exploration area.

Figure 10. Box plot showing measured heat production from Scottish granite plutons. The horizontal center line in each box shows the median heat production from each pluton, the edges of the boxes are the first and third quartiles, *i.e.*, 50% of the data lies within the box. The whiskers extending beyond the boxes contain data which are within 1.5 times the interquartile range, and data out with this range are plotted as hoops. There are 37 readings from the Cairngorm Pluton, 34 from the Lochnagar Pluton, 22 from the Monadhliath Pluton, 19 from the Ballatar Pluton, seven from the Grantown Pluton, five from the Strathspey Pluton, and seven from the Ross of Mull Pluton.

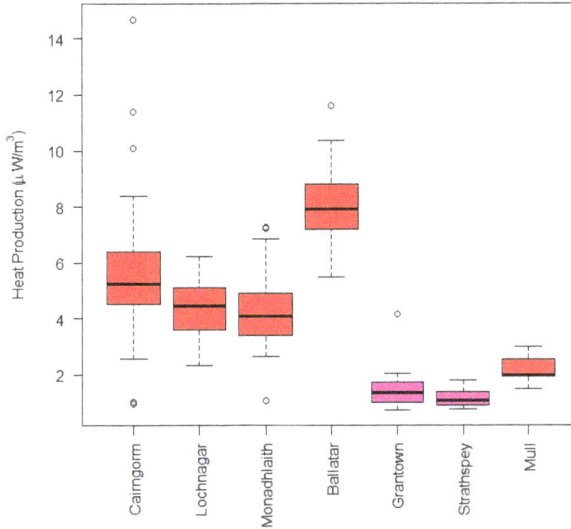

Figure 11. Graph showing correlation between original study of Scottish granites (Downing and Gray: *x*-axis) and new data from July 2013 portable gamma spectrometer survey (y-axis). Blue triangles are plutons where Downing and Gray (1986) [3] cautioned that not enough data were collected for confidence in the calculated heat production value. Red line is $x = y$, for ease of comparison of results.

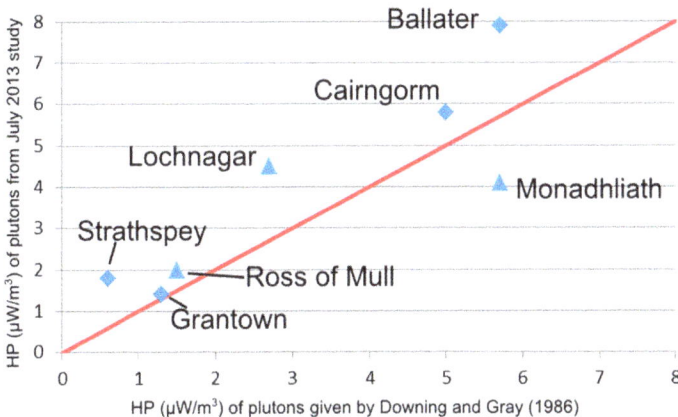

The July 2013 survey demonstrates how portable gamma-ray spectrometry can be used to gain quick results that give an initial indication of which plutons may have high radiogenic heat producing properties. The survey was conducted over one month by two people and would have been able to provide reliable first estimates of radiogenic heat productions of the granites in a previously unexplored area. This information could be used to target more comprehensive studies later on.

6. Discussion

6.1. The Geothermal Targets of Gamma-ray Surveying?

Section 6 showed an example of how portable gamma-ray spectrometry can be used to screen granite plutons for further evaluation of their geothermal potential. However, gamma-ray spectrometry can have more of a role to play; even just within analysis of the radiogenic heat production of a single granite pluton. For example, plutons commonly comprise concentric rings of different types of granite, e.g., the Criffel Pluton in Southeast Scotland [65,67]. These zones may have significantly different geochemistry due to fractionation processes during emplacement; such zones could therefore have higher or lower radiogenic heat production properties. This could result in a single radiogenic heat production value for a pluton being fairly meaningless. Even within a relatively homogeneous granite pluton there are likely to be small discrete zones of unusually high or low radiogenic heat production. This is shown in Figure 10, where the high outlier in the Cairngorm granite faded to a median value several meters away but there was no visual clue as to why this should be such a hot spot. By increasing the density of readings over a larger expanse of granite, such outliers can be better identified ensuring they do not inaccurately skew the calculated radiogenic heat production of the granite pluton upwards. Such hot spots could be of particular concern for determining the radiogenic heat production of concealed granite; where samples are confined to the borehole track through the granite. There remains opportunity for further research to be able to improve understanding of the link between heat production properties with chemistry and pluton genesis; with one aim being improved targeting of high heat production zones in concealed granite.

Sedimentary basins can be areas of elevated heat flow [3] which coupled with the favorable permeability of sandstone layers can make potential mid/low enthalpy geothermal targets. Such sedimentary basins will also typically have a significant argillaceous component; that is mudstone or shale layers. Mudstone and shale can contribute to the geothermal prospects of sedimentary basins in two ways: Firstly, they may act as a "duvet rock" allowing heat to build up in the sandstone below due to the low thermal conductivity of mudstone or shale [7]. Secondly, mudstone and shale can have higher radiogenic heat production than most other sedimentary rocks [71], possibly due to unusually high uranium concentrations [41]. Gamma-ray spectrometry would be able to identify heat producing mudstones from wire-line logging in boreholes or from surface surveys where outcrops are available. We found no reports in the literature of research into or development of the geothermal potential of such high heat producing sedimentary systems. There is further opportunity for basic research into high heat producing sedimentary systems to determine whether they may have potential as a viable geothermal resource.

These examples of survey targets show the adaptable and variable way gamma-ray spectrometry surveys can be used. It is also clear that there are further improvements and research to be made in geothermal resource evaluation using gamma-ray spectrometry. When heat production is likely to be important to the geothermal resource of an area, then a gamma-ray survey is likely to be able to provide useful data on the heat production properties.

6.2. General Guidelines for Gamma-ray Surveying in Geothermal Exploration

Gamma-ray spectrometry surveys can seem a daunting task with the myriad of options available for surveying and all the potential sources of bias. However, gamma-ray spectrometry has an established history, during which many changes and improvements have been made. Sensitivity improvements in the 1940s were made when scintillation detectors were developed [42]. Soon after this, the first airborne surveys were conducted for uranium exploration in the late 1940s and 1950s [72]. Lab and *in-situ* surveys were conducted for mineral exploration and environmental monitoring [59,73–75]. Further improvements in multi-channel analyzers, digitization, and data processing increased the ease of use of spectrometers as well as improved portability allowing detailed surveys to be made of complicated rocks [20,46] with real time data analysis [76]. For airborne surveys, improvements allowed rapid calibration of aerial data with calibration sites and improved spectral analysis [45]. Such an established history means that prior to conducting a survey using gamma-ray spectrometry techniques then previous experience can be called upon to ensure new surveys gain the most accurate data possible.

For ground-based portable gamma-ray spectrometry surveys, Table 1 shows specific tasks that should be taken into account when planning a survey. These are partly based on experience gained during the Scottish case study example in this paper.

For an airborne gamma-ray spectrometry study, general outlines have been previously described [42] with a wealth of literature [77]. Many considerations in Table 1 also relate to airborne surveys. In addition to these, Table 2 shows a sample of tasks more specific to airborne surveys.

As with many forms of surveying, the precise nature and scope of a gamma-ray survey depends upon the aims, objectives, and available outcrops in addition to budgetary constraints. Which is why "Specify Aims" is first in the list of checkpoints; the rest of the study design is dependent upon what these aims are. In this paper, we showed an example of a portable survey which aimed to generally categorize Scottish granite plutons of lower or higher radiogenic heat production. The results corroborated a previous lab based study; showing the reliability of a rapid surface study to categorize the radiogenic heat production of granite plutons. However, if the aim of the survey was to categorize in detail only, say, the Cairngorm Pluton, then choices for the reading density, lithology targets *etc.* would have been quite different. If the target lithology of a survey is a concealed granite (buried under several hundred metres of sediment [8,10]), then a borehole survey or collecting drill-core for lab analysis are the only available options, since any gamma-ray radiation given off the concealed granite will be shielded by only a few metres of sediment cover. Spectral gamma-ray logging is routinely performed by service companies. Airborne surveying gains data from a large area in a relatively short amount of time. The costs of chartering aircraft are not trivial and some surveyors stress the need for calibration of airborne *in-situ* tests with ground-based or lab tests [23,45,78,79]. Due to sediment cover, then airborne studies may estimate radiogenic heat production to be around half that of lab or ground-based surveys [78]. Nevertheless, airborne studies which had designated calibration sites showed self-consistency between airborne surveys and accompanying traditional ground surveys [32,33].

Table 1. Tasks worth considering during a ground based portable gamma-ray spectrometry study with examples of where the decisions may have an impact.

Task	Example
Specify Aims	Is this survey as a first estimate of radiogenic heat production or to gain more details of its distribution within a single pluton?
Extent of survey area	Aerial surveys may be favourable if the survey area is particularly extensive.
Sizes of individual sample areas	For portable surveys the surveyor can be placed on the ground gaining an effective circular sample area with a diameter of one metre. Holding the surveyor one metre above ground gains a sample area with a diameter of 10 metres [59].
Key lithologies to be targeted	Are all the rock types that may have radiogenic heat production included in the survey plan?
Availability of rock exposure	In the Scottish case study, higher altitude plutons generally had much more exposed area than lower plutons, which tended to be mantled with peat bog.
Easy access routes to exposure	Tracks due to other land use can be used to get to exposure, use of these can be incorporated into the survey design e.g., sample spoke lines coming from a driveable track.
Land access	Gamma-ray spectrometry surveys may cover an area which has different land uses or owners; in Scotland it is not advisable to conduct a portable survey near deer hunting areas in the shooting season.
Repeated readings and length of readings	Should all readings be repeated or only a small sub-sample to check reliability of results? Depending on dose rate longer or shorter count times may be appropriate.
Features to survey near (e.g., faults)	Some features may have an influence on the radiogenic heat production, e.g., hydrothermal alteration around faults. Depending on the aims of the survey these could specifically be targeted or avoided so these results do not interfere with gaining an overall representative value of a pluton's radiogenic heat production.
Target areas for background readings	Identify bodies of freshwater, if available, to get background readings.
Density of readings/resolution of survey	If there is a limited time, to gain an overall value for radiogenic heat production of a pluton, readings should be sparser. If there is need to understand the varied distribution of radiogenic heat production across a pluton then a tighter survey grid may be more appropriate.

Table 2. Tasks worth considering during an airborne gamma-ray spectrometry study.

Task	Example
Determine distance between flight lines	Higher concentrations of flight lines may cover the survey area more comprehensively but will decrease the area that can be covered in a limited time.
Ground Speed	As for line spacing, survey speed is a compromise between data quality and available time.
Altitude of survey	Reduced ground clearance results in more spectral information—you get less atmospheric scatter and higher count rates. Generally, higher surveys can be flown faster (less worries for the pilot re: ground obstacles such as power lines), there is usually less radon at height (though not always) and the data are less susceptible to topographic effects and small variations in altitude.
Refuel points	If refuel points near to the survey area can be arranged with local landowners, then more time can be spent conducting the survey rather than journeying back and forth to base.
Ground calibration sites	When conducting an airborne survey then local calibration areas allow checking of the instrument sensitivity to ensure it is not drifting during the survey [32,44,45].
Detector background	This comprises internal activity in the detector and aircraft, cosmic radiation and radon. Flying over clean bodies of water allows this background to be recorded but there is still scope for radon background to vary with location. "Upward" facing detectors help with this by measuring radiation from the air above the aircraft due to radon.
Topography	Helicopters may be better choice in rough terrain than aircraft as they can more effectively follow the topographical changes.

In this paper, we have discussed some of the issues surrounding accounting for inaccuracies created by topographical [31], distributional [44] and series disequilibrium [28] effects during gamma-ray surveying. Topographical corrections [31] rely on a homogenous medium assumption which suffers when the spatial distribution of gamma-ray production is investigated [44]. Further work could bring together these different influences as a useful improvement in the accuracy of *in-situ* gamma-ray spectrometry, particularly if it is possible to account for the varying gamma-ray half-lengths introduced by heterogeneous material. Additionally, disequilibrium appears to often be acknowledged during gamma-ray spectrometry but less often can be quantitatively accounted for during the scope of a study. There is additional scope for research to constrain which geological processes may make different series disequilibrium more likely and from this provide simplified estimation for accounting for disequilibrium during gamma-ray spectrometry surveys.

7. Conclusions

Gamma-rays are particularly useful, when surveying for radioelements contained within rock, due to their penetrating properties. This allows collection of a sample of the concentrations of potassium, uranium, and thorium from which the heat production ($\mu W/m^2$) can be calculated.

Many different types of gamma-spectrometers may be used; use may depend on whether the survey is *in-situ*—either ground based or airborne—or samples collected and analysed in a laboratory.

Portable gamma-ray surveying has been deployed as a quick but effective technique for determining granite plutons of high heat production in Scotland. The survey allowed high heat production granite to be identified which may warrant further investigation.

Gamma-ray spectrometry will be vital for further research into the zonation of heat production in granite. In addition, the technique will be deployed when investigating sedimentary rocks which may have high heat production (e.g., some mudstones) enhancing the heat flow within basin settings.

Gamma-ray spectrometry has been shown to have played a useful role in past geothermal exploration. The technique is likely to stay relevant in the future as it remains a quick and cost effective way to assess the radiogenic heat production properties of any rock. When compared with the costs of a poorly placed drill-site, the surveys more than show their worth.

Author Contributions

Primary drafting of the paper was by Alistair T. McCay (60%), with contributions of material and editing by all other authors (10% each).

Conflicts of Interest

The authors declare no conflict of interest.

References

1. Shevenell, L.A.; Garside, L.J.; Hess, R.H.; Chaney, R.L.; Tingley, S.L.; Snow, J.H.; Meeuwig, R.O. *Nevada Geothermal Resources*; Nevada Bureau of Mines and Geology: Reno, NV, USA, 2003.
2. Mogi, T.; Okada, S. Gamma-ray spectra survey in geothermal area. *J. Geotherm. Res. Soc. Jpn.* **1990**, *12*, 295–309.
3. Downing, R.A.; Gray, D. *Geothermal Energy: The Potential in the United Kingdom*; Her Majesty's Stationery Office: London, UK, 1986.
4. Kappelmeyer, O.; Gérard, A.; Schloemer, W.; Ferrandes, R.; Rummel, F.; Benderitter, Y. European HDR project at soultz-sous-forêts: General presentation. *Geotherm. Sci. Technol.* **1991**, *2*, 263–289.
5. Häring, M.O.; Schanz, U.; Ladner, F.; Dyer, B.C. Characterisation of the basel 1 enhanced geothermal system. *Geothermics* **2008**, *37*, 469–495.
6. Wyborn, D. Update of development of the geothermal field in the granite at Innamincka, South Australia. In Proceedings of the World Geothermal Congress, Bali, Indonesia, 25–30 April 2010; pp. 25–29.
7. Midttomme, K.; Roaldset, E.; Aagaard, P. Thermal conductivity of selected claystones and mudstones from england. *Clay Miner.* **1998**, *33*, 131–145.

8. Chopra, P.; Wyborn, D. Australia's first hot dry rock geothermal energy extraction project is up and running in granite beneath the cooper basin, ne south australia. In Proceedings of the Ishihara Symposium: Granites and Associated Metallogenesis, Sydney, Australia, 22–24 July 2003; pp. 43–45.

9. Baisch, S.; Weidler, R.; Vörös, R.; Wyborn, D.; de Graaf, L. Induced seismicity during the stimulation of a geothermal HFR reservoir in the Cooper Basin, Australia. *Bull. Seismol. Soc. Am.* **2006**, *96*, 2242–2256.

10. Manning, D.; Younger, P.; Smith, F.; Jones, J.; Dufton, D.; Diskin, S. A deep geothermal exploration well at Eastgate, Weardale, UK: A novel exploration concept for low-enthalpy resources. *J. Geol. Soc.* **2007**, *164*, 371–382.

11. Cohen, A.; O'Nions, R.; O'Hara, M. Chronology and mechanism of depletion in lewisian granulites. *Contrib. Mineral. Petrol.* **1991**, *106*, 142–153.

12. Kumar, P.S.; Reddy, G. Radioelements and heat production of an exposed Archaean crustal cross-section, Dharwar craton, south India. *Earth Planet. Sci. Lett.* **2004**, *224*, 309–324.

13. Heier, K.; Adams, J. Concentration of radioactive elements in deep crustal material. *Geochim. Cosmochim. Acta* **1965**, *29*, 53–61.

14. Sanyal, S.; Che, M.; Dunlap, R.E.; Twichell, M.K. Qualitative response patterns on geophysical well logs from the Geysers, California. *Trans. Geotherm. Resour. Counc.* **1982**, *6*, 313–316.

15. Quinn, T.; Suzukilll, N.-I.M.; Takagim, S. Mineralogy evaluation in a geothermal well using statistical probabilistic log evaluation techniques. *Geotherm. Resour. Counc. Trans.* **1989**, *13*, 277–287.

16. Younger, P.; Manning, D. Hyper-permeable granite: Lessons from test-pumping in the Eastgate Geothermal Borehole, Weardale, UK. *Q. J. Eng. Geol. Hydrogeol.* **2010**, *43*, 5–10.

17. Nelson, S.T.; Mayo, A.L.; Gilfillan, S.; Dutson, S.J.; Harris, R.A.; Shipton, Z.K.; Tingey, D.G. Enhanced fracture permeability and accompanying fluid flow in the footwall of a normal fault: The Hurricane fault at Pah Tempe hot springs, Washington County, Utah. *Geol. Soc. Am. Bull.* **2009**, *121*, 236–246.

18. Killeen, P. Gamma ray spectrometric methods in uranium exploration—Application and interpretation. *Geophys. Geochem. Search Met. Ores* **1979**, *31*, 163–230.

19. Grasty, R. Gamma ray spectrometric methods in uranium exploration—Theory and operational procedures. *Geophys. Geochem. Search Met. Ores* **1979**, *31*, 147–155.

20. Myers, K.; Bristow, C. Detailed sedimentology and gamma-ray log characteristics of a namurian deltaic succession II: Gamma-ray logging. *Geol. Soc. Lond. Spec. Publ.* **1989**, *41*, 81–88.

21. Bristow, C.; Williamson, B. Spectral gamma ray logs: Core to log calibration, facies analysis and correlation problems in the Southern North Sea. *Geol. Soc. Lond. Spec. Publ.* **1998**, *136*, 1–7.

22. Davies, S.; McLean, D. Spectral gamma-ray and palynological characterization of Kinderscoutian marine bands in the Namurian of the Pennine Basin. *Proc. Yorks. Geol. Polytech. Soc.* **1996**, *51*, 103–114.

23. Rybach, L.; Schwarz, G.F. Ground gamma radiation maps: Processing of airborne, laboratory, and *in situ* spectrometry data. *First Break* **1995**, *13*, 97–104.

24. Sanderson, D.; East, B.; Scott, E. *Aerial radiometric survey of parts of North Wales in July 1989*; Scottish Universities Research and Reactor Centre: Glasgow, UK, 1989.

25. Mero, J.L. Uses of the gamma-ray spectrometer in mineral exploration. *Geophysics* **1960**, *25*, 1054–1076.

26. Ray, L.; Roy, S.; Srinivasan, R. High Radiogenic Heat Production in the Kerala Khondalite Block, Southern Granulite Province, India. *Int. J. Earth Sci.* **2008**, *97*, 257–267.

27. Weaver, B.L.; Tarney, J. Lewisian gneiss geochemistry and archaean crustal development models. *Earth Planet. Sci. Lett.* **1981**, *55*, 171–180.

28. Dickson, B.; Scott, K. Interpretation of aerial gamma-ray surveys-adding the geochemical factors. *AGSO J. Aust. Geol. Geophys.* **1997**, *17*, 187–200.

29. Stefansson, V.; Gudlaugsson, S.T.; Gudmundsson, A. Silica Content and Gamma Ray Logs in Volcanic Rocks. In Proceedings of the World Geothermal Congress, Kyushu–Tohoku, Japan, 28 May–10 June 2000; pp. 2893–2897.

30. Lovborg, L. The calibration of portable and airborne gamma-ray spectrometers—Theory, problems, and facilities. *Rise Natl. Lab.* **1984**, *2456*, 3–207.

31. Schwarz, G.; Klingelé, E.; Rybach, L. How to handle rugged topography in airborne gamma-ray spectrometry surveys. *First Break* **1992**, *10*, 11–17.

32. Sanderson, D.; Cresswell, A.; Lang, J.; Scott, E.; Lauritzen, B.; Karlsson, S.; Strobl, C.; Karlberg, O.; Winkelmann, I.; Thomas, M. *An International Comparison of Airborne and Ground Based Gamma Ray Spectrometry. Results of the Eccomags 2002 Exercise Held 24th May to 4th June 2002, Dumfries and Galloway, Scotland*; University of Glasgow: Glasgow, UK, 2004.

33. Sanderson, D.; Cresswell, A.; Scott, E.; Lang, J. Demonstrating the European capability for airborne gamma spectrometry: Results from the ECCOMAGS exercise. *Radiat. Prot. Dosim.* **2004**, *109*, 119–125.

34. Hertz, H. Ueber einen einfluss des ultravioletten lichtes auf die electrische entladung. *Ann. Phys.* **1887**, *267*, 983–1000. (In German)

35. Einstein, A. Über einen die erzeugung und verwandlung des lichtes betreffenden heuristischen gesichtspunkt. *Ann. Phys.* **1905**, *322*, 132–148. (In German)

36. Ewan, G.; Tavendale, A. High-resolution studies of gamma-ray spectra using lithium-drift germanium gamma-ray spectrometers. *Can. J. Phys.* **1964**, *42*, 2286–2331.

37. Myers, K. The origin of the lower jurassic cleveland ironstone formation of North-East England: Evidence from Portable gamma-ray spectrometry. *Geol. Soc. Lond. Spec. Publ.* **1989**, *46*, 221–228.

38. Parkinson, D. Gamma-ray spectrometry as a tool for stratigraphical interpretation: Examples from the western European Lower Jurassic. *Geol. Soc. Lond. Spec. Publ.* **1996**, *103*, 231–255.

39. Einstein, A. Ist die trägheit eines körpers von seinem energieinhalt abhängig? *Ann. Phys.* **1905**, *323*, 639–641. (In German)

40. Rybach, L. Radioactive heat production: A physical property determined by the chemistry of rocks. In *The Physics and Chemistry of Minerals and Rocks*; Stems, R.G.J., Ed.; Wiley-Interscience: New York, USA, 1976; pp. 309–318.

41. Birch, F. Heat from radioactivity. In *Nuclear Geology*; Wiley: New York, NY, USA, 1954; pp. 148–174.

42. Erdi-Krausz, G.; Matolin, M.; Minty, B.; Nicolet, J.; Reford, W.; Schetselaar, E. *Guidelines for Redioelement Mapping Using Gamma Ray Spectrometry Data*; International Atomic Energy Agency: Vienna, Austria, 2003.

43. Rybach, L.; Bucher, B.; Schwarz, G. Airborne surveys of swiss nuclear facility sites. *J. Environ. Radioact.* **2001**, *53*, 291–300.

44. Tyler, A.; Sanderson, D.C.; Scott, E.M.; Allyson, J. Accounting for spatial variability and fields of view in environmental gamma ray spectrometry. *J. Environ. Radioact.* **1996**, *33*, 213–235.

45. Sanderson, D.; Allyson, J.; Tyler, A. Rapid quantification and mapping of radiometric data for anthropogenic and technologically enhanced natural nuclides, application of uranium exploration data and techniques in environmental studies. In Proceedings of a Technical Committee Meeting, Vienna, Austria, 9–12 November 1993; pp. 197–216.

46. Davies, S.; Elliott, T. Spectral gamma ray characterization of high resolution sequence stratigraphy: Examples from Upper Carboniferous fluvio-deltaic systems, County Clare, Ireland. *Geol. Soc. Lond. Spec. Publ.* **1996**, *104*, 25–35.

47. Kogan, R.; Nazarov, I.; Fridman, S.D. *Gamma Spectrometry of Natural Environments and Formations*; Israel Program for Scientific Translations: Jerusalem, Palestine, 1971.

48. Evans, R.; Mory, A.; Tait, A. An outcrop gamma ray study of the Tumblagooda Sandstone, Western Australia. *J. Pet. Sci. Eng.* **2007**, *57*, 37–59.

49. Løvborg, L.; Bøtter-Jensen, L.; Kirkegaard, P.; Christiansen, E. Monitoring of natural soil radioactivity with portable gamma-ray spectrometers. *Nucl. Instrum. Methods* **1979**, *167*, 341–348.

50. Cresswell, A.; Sanderson, D.; Harrold, M.; Kirley, B.; Mitchell, C.; Weir, A. Demonstration of lightweight gamma spectrometry systems in urban environments. *J. Environm. Radioact.* **2013**, *124*, 22–28.

51. International Commission on Radiation Uits and Measurements. *Gamma-Ray Spectrometry in the Environment*; International Commission on Radiation Uits and Measurements: Oxford, UK, 1994.

52. Sanderson, D.; Placido, F.; Tate, J. Scottish vitrified forts: Tl results from six study sites. *Int. J. Radiat. Appl. Instrum. Part D. Nucl. Tracks Radiat. Meas.* **1988**, *14*, 307–316.

53. GF-Instruments. *Gamma Surveyor II User Manual*; GF-Instruments: Brno, Czech Republic, 2013.

54. Grasty, R.; Carson, J.; Charbonneau, B.; Holman, P. *Natural Background Radiation in Canada*; Geological Survey of Canada: Ottawa, ON, Canada, 1984; Volume 360.

55. Tyler, A. Broadening the scope and environmental applications of *in situ* gamm ray spectrometry. Recent Applications and Development Sin Mobile and Airborne Gamma Spectrometry. In Proceedings of the Radmags Symposium, University of Stirling, 15–18 June 2000; Sanderson, D.C., McLeod, J., Eds.; University of Glasgow: Glasgow, UK, 2000.

56. Ahl, A.; Bieber, G. Correction of the attenuation effect of vegetation on airborne gamma-ray spectrometry data using laser altimeter data. *Near Surf. Geophys.* **2010**, *8*, 271–278.

57. Sanderson, D.; Cresswell, A.; Hardeman, F.; Debauche, A. An airborne gamma-ray spectrometry survey of nuclear sites in Belgium. *J. Environ. Radioact.* **2004**, *72*, 213–224.

58. Killeen, P.; Carmichael, C. Gamma-ray spectrometer calibration for field analysis of thorium, uranium and potassium. *Can. J. Earth Sci.* **1970**, *7*, 1093–1098.

59. Beck, H.L.; Decampo, J.; Gogolak, C. *In Situ Ge (Li) and Nai (Tl) Gamma-Ray Spectrometry*; Health and Safety Lab.: New York, NY, USA, 1972.

60. Allyson, J. *Environmental Gamma-Ray Spectrometry: Simulation of Absolute Calibration of in-situ and Airborne Spectrometers for Natural and Anthropogenic Sources*; University of Glasgow: Glasgow, UK, 1994.

61. Lee, M.; Wheildon, J.; Webb, P.; Brown, G.; Rollin, K.; Crook, C.; Smith, I.; King, G.; Thomas-Betts, A. Hot dry rocks prospects in caledonian granites: Evaluation of results from the bgs-ic-ou research programme (1981–1984). In *Investigation of the Goethermal Potential of the UK*; British Geological Survey: Keyworth, UK, 1984.

62. Sherman, H.M.; Gierke, J.S.; Anderson, C.P. Controls on spatial variability of uranium in sandstone aquifers. *Ground Water Monit. Remediat.* **2007**, *27*, 106–118.

63. Hampson, G.J.; Davies, W.; Davies, S.J.; Howell, J.A.; Adamson, K.R. Use of spectral gamma-ray data to refine subsurface fluvial stratigraphy: Late Cretaceous strata in the Book Cliffs, Utah, USA. *J. Geol. Soc.* **2005**, *162*, 603–621.

64. Roy, R.F.; Blackwell, D.D.; Birch, F. Heat generation of plutonic rocks and continental heat flow provinces. *Earth Planet. Sci. Lett.* **1968**, *5*, 1–12.

65. Miles, A.J.; Graham, C.M.; Hawkesworth, C.J.; Gillespie, M.R.; Hinton, R.W. Evidence for distinct stages of magma history recorded by the compositions of accessory apatite and zircon. *Contrib. Mineral. Petrol.* **2013**, *166*, 1–19.

66. DePaolo, D.J. Trace element and isotopic effects of combined wallrock assimilation and fractional crystallization. *Earth Planet. Sci. Lett.* **1981**, *53*, 189–202.

67. Stephens, W.E.; Whitley, J.E.; Thirlwall, M.F.; Halliday, A.N. The criffell zoned pluton: Correlated behaviour of rare earth element abundances with isotopic systems. *Contrib. Mineral. Petrol.* **1985**, *89*, 226–238.

68. Gardner, J.E.; Carey, S.; Sigurdsson, H.; Rutherford, M.J. Influence of magma composition on the eruptive activity of Mount St. Helens, Washington. *Geology* **1995**, *23*, 523–526.

69. Kemp, A.; Hawkesworth, C.; Foster, G.; Paterson, B.; Woodhead, J.; Hergt, J.; Gray, C.; Whitehouse, M. Magmatic and crustal differentiation history of granitic rocks from Hf-O isotopes in zircon. *Science* **2007**, *315*, 980–983.

70. Freck, D.; Wakefield, J. Gamma-ray spectrum obtained with a lithium-drifted p-i-n junction in germanium. *Nature* **1962**, *1983*, 669.

71. Ehinola, O.; Joshua, E.; Opeloye, S.; Ademola, J. Radiogenic heat production in the cretaceous sediments of yola arm of Nigeria benue trough: Implications for thermal history and hydrocarbon generation. *J. Appl. Sci.* **2005**, *5*, 696–701.

72. Berbezier, J.; Blangy, B.; Guiton, J.; Lallemant, C. Methods of car-borne and air-borne prospecting: The technique of radiation prospecting by energy discrimination. In Proceedings of the 2nd UN International Conference Peaceful Use of Atomic Energy, Genova, Switzerland, 1–13 September 1958.

73. Beck, H.L.; Condon, W.J.; Lowder, W.M. *Spectrometric Techniques for Measuring Environmental Gamma Radiation*; Health and Safety Lab., New York Operations Office (AEC): New York, NY, USA, 1964.

74. Beck, H.; Lowder, W.; McLaughlin, J. *In situ External Environmental Gamma-Ray Measurements Utilizing Ge (Li) and Nai (Tl) Spectrometry and Pressurized Ionization Chambers*; Atomic Energy Commission: New York, NY, USA, 1971.

75. Lowder, W.; Condon, W.; Beck, H.L. *Field Spectrometric Investigations of Environmental Radiation in the USA*; Atomic Energy Commission: New York, NY, USA, 1968.

76. Sanderson, D.; Allyson, J.; Tyler, A.; Scott, E. Environmental applications of airborne gamma spectrometry. In Proceedings of the IAEA Technical Committee Meeting on the Use of Uranium Exploration Data and Techniques in Environmental Studies, Vienna, Austria, 9–12 November, 1993; pp. 9–12.

77. Sanderson, D.; McLeod, J.; Ferguson, J. A european bibliography on airborne gamma-ray spectrometry. *J. Environ. Radioact.* **2001**, *53*, 411–422.

78. Richardson, K.; Killeen, P. *Regional Radiogenic Heat Production Mapping by Airborne Gamma-ray Spectrometry*; Geological Survey of Canada: Ottawa, ON, Canada, 1980; pp. 227–232.

79. Thompson, P.; Judge, A.; Charbonneau, B.; Carson, J.; Thomas, M. *Thermal Regimes and Diamond Stability in the Archean Slave Province, Northwestern Canadian Shield, District of Mackenzie, Northwest Territories*; Geological Survey of Canada: Ottawa, ON, Canada, 1996; pp. 135–146.

Seismic Velocity/Temperature Correlations and a Possible New Geothermometer: Insights from Exploration of a High-Temperature Geothermal System on Montserrat, West Indies

Graham Alexander Ryan and Eylon Shalev

Abstract: In 2013, two production wells were drilled into a geothermal reservoir on Montserrat, W.I. (West Indies) Drilling results confirmed the main features of a previously developed conceptual model. The results confirm that below ~220 °C there is a negative correlation between reservoir temperature and seismic velocity anomaly. However, above ~220 °C there is a positive correlation. We hypothesise that anomalous variations in seismic velocity within the reservoir are controlled to first order by the hydrothermal mineral assemblage. This study suggests a new geophysical thermometer which can be used to estimate temperatures in three dimensions with unprecedented resolution and to indicate the subsurface fluid pathways which are the target of geothermal exploitation.

Reprinted from *Energies*. Cite as: Ryan, G.A.; Shalev, E. Seismic Velocity/Temperature Correlations and a Possible New Geothermometer: Insights from Exploration of a High-Temperature Geothermal System on Montserrat, West Indies. *Energies* **2014**, *7*, 6689-6720.

1. Introduction

Montserrat is a small volcanic island approximately 16 km long and 11 km wide located in the northern part of the Lesser Antilles archipelago of volcanic islands in the Eastern Caribbean. The island is home to the Soufrière Hills volcano which has been in eruption since 18 July 1995 [1].

The Lesser Antilles has some of the highest electricity costs in the world ranging up to 50 cents/kwh depending on the oil price [2]. The majority of the Lesser Antilles is dependent on fossil fuels for electricity generation and Montserrat is no exception as it relies solely on diesel generation to supply the 1.7 MW (peak load) [2] required by the population of ~5000. The costs of electrical power for the country (~2.5 M$ per annum) represent a significant portion of the island's budget and uses up valuable foreign currency reserves.

The idea of using geothermal power for electricity generation on Montserrat was first proposed by Meidav in 1972 [3]. Since that time there have been several studies which have explored the question of whether a suitable resource exists on the island [4–7]. After four decades, two production wells 500 m apart were drilled between March and September of 2013 [8].

In 2009, a geothermal exploration project consisting of geological, geochemical and geophysical surveys was conducted on the island. Using these data, and data from previous studies, high priority areas for exploratory drilling were proposed by EGS Inc. [6] (Figure 1). In 2013, the geophysical data were reanalyzed to better pinpoint drilling targets within the priority zones [9]. In this study, geophysical data; magnetotelluric, earthquake hypocenters and seismic tomography data were used in conjunction with other geological information to construct a 3D conceptual model of the reservoir. The model was then used to develop a numerical index that identified areas of greater and lower prospectivity. This prospectivity index was based on the product of model proxies for subsurface temperature and permeability inferred from the conceptual model. A key component of the study was the hypothesis that the low P-wave seismic velocity anomaly located in the southwestern area of the island was related to subsurface temperature variations in the (then unproven) geothermal system. This hypothesis was based in part on laboratory experiments on reservoir rocks carried out by Kristinsdottir *et al.* [10]. The results of the current study suggest that rather than being directly related to temperature as suggested by the work of Kristinsdottir *et al.* [10], slow seismic velocity anomalies in the Montserrat geothermal reservoir are controlled by the temperature dependent hydrothermal mineral assemblage and particularly by the relative abundance of phyllosilicate clays which tend to cause a slowing of seismic velocity [11].

The seismic anomaly data came from the SEA-CALIPSO (Seismic Experiment with Airgun-source-Caribbean Andesite Lava Island Precision Seismo-geodetic Observatory) study designed to image the magmatic plumbing system beneath the Soufrière Hills Volcano [12]. This study yielded a well-constrained high resolution P-wave velocity model of the subsurface down to about three kilometers.

Shalev *et al.* [13] first identified three anomalously low velocity zones around the island of Montserrat, one of which was coincident with the possible location of a high temperature geothermal system as identified by EGS Inc. [6] and Ryan *et al.* [14]. Shalev *et al.* [13] conjectured that the low velocity could be related to hydrothermal alteration in that area due to coincidence with low resistivity zones interpreted to be due to clay alteration in the same area [6,14]. The detailed structure of these low velocity zones was not investigated. The detailed spatial inter-relation between the low velocity and low resistivity zones was explored by Ryan *et al.* [9] who determined that there was an offset between the two anomalies with the low velocity zone sitting just beneath the low resistivity zone. Ryan *et al.* [9] proposed the hypothesis that the seismic velocity anomaly was related to temperature variations in the geothermal reservoir as suggested by laboratory experiments on reservoir rocks presented by Kristinsdottir *et al.* [10]. This current study suggests that this initial hypothesis should be superseded by the hypothesis that velocity in the reservoir is largely controlled by temperature dependent variations in the hydrothermal mineral assemblage.

Hydrothermal mineral assemblages have long been used as geothermometers and variations in the conductivity of clay minerals associated with hydrothermal alteration is used to interpret resistivity models and determine temperature profiles in high temperature geothermal systems [15]. Although it is known that seismic properties vary as a function of hydrothermal alteration [11], limitations of resolution have prevented seismic tomography data from being used as geothermometers in the past.

The SEA-CALIPSO data set for Montserrat is a uniquely high resolution seismic tomography data set for a high temperature geothermal system. In this preliminary study, we take data from geothermal exploration and drilling to hypothesise a model which relates modeled seismic velocity anomalies to temperature dependent variations in the mineral assemblages. Should this technique prove to be robust, it opens up a new geophysical technique for estimating reservoir temperatures using seismic tomography data similar to the interpretation schemes used for resistivity data [15]. It also opens the possibility of determining subsurface temperature patterns and fluid flow pathways. This study allows us the unique possibility of testing our fairly detailed conceptual model against the results obtained from drilling of the first wells into the geothermal reservoir on Montserrat.

Figure 1. High priority drilling areas (areas outlined in red and green) identified by EGS Inc. [6]. Wells MON-1 and MON-2 are represented as green and purple dots, respectively. Figure modified from EGS Inc. [6].

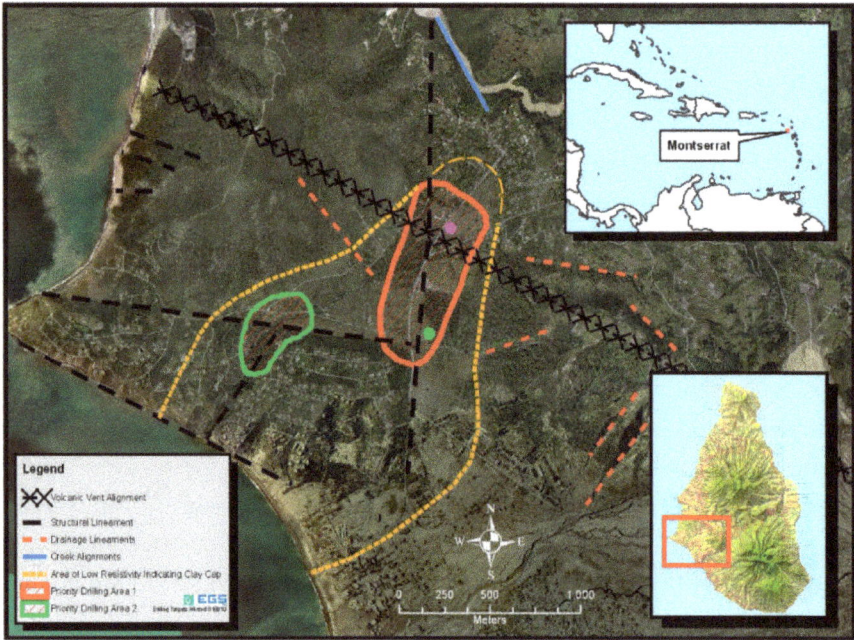

2. Data

2.1. Well Log Data

The detailed well completion results can be found in Brophy *et al.* [8] and in EGS Inc. [16]. Preliminary flow test data from well MON-1 show recorded flow rates of up to 22 kg·s⁻¹ and recorded temperatures of up to 230 °C. Flow rates of up to 12 kg·s⁻¹ for well MON-2 along with a maximum bottom hole temperature of 265 °C were recorded. Data from the cutting logs in MON-1 [16] indicate a zone of smectite clay between 600 and 1210 m depth. The percentage of smectite in the clay samples was determined using the methylene blue test [17].

Temperature data were collected via wireline log at a range of depths as the probe was lowered and raised in the wells. To obtain punctual temperature estimates we have averaged the temperature data every 5 m during both the downward and upward going surveys. Due to thermal inertia of the thermometer there is a discrepancy between the temperature measurements made in the downward and upward parts of the survey. This difference is usually below 1 °C. The difference is below 2 °C for the vast majority of the depth intervals. However, in depth ranges where the thermal gradients are the largest the difference can be as high as 12 °C. In well MON-2, there is a significant perturbation in the well temperatures between 1325 and 2195 m. This reduction in temperature is thought to be due to incomplete warming up of the well after cooling during the circulation of drilling fluids. To compensate for this perturbation, we calculated a linear temperature fit across the perturbed zone which is thought to approximate the unperturbed temperatures. Figure 2 shows the temperature data.

Figure 2. (**a**) Average temperature data for MON-1; (**b**) Average temperature data for MON-2. Average temperature for up and down runs from wireline temperature logs. The temperatures have been averaged over each 5 m interval and the averages for the upward and downward going sections of the survey have been averaged. The difference between the estimates obtained between the upward and downward sections of the survey are also shown. A linear fit across the perturbed region in well MON-2 is also shown.

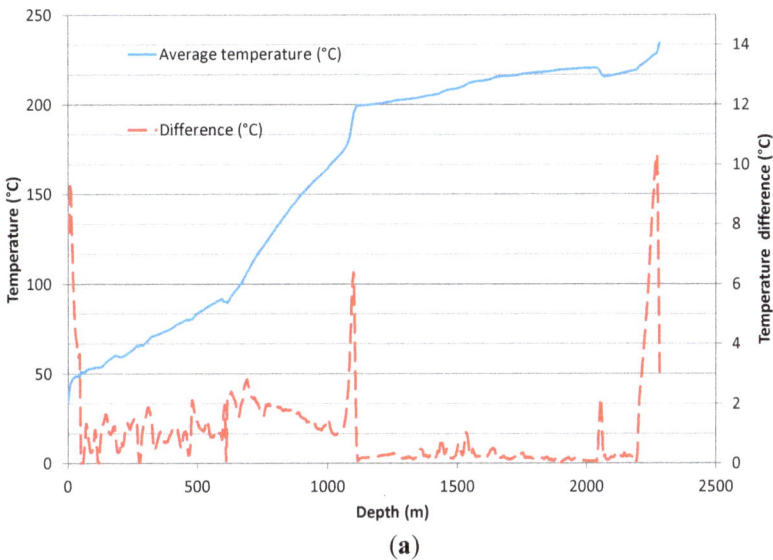

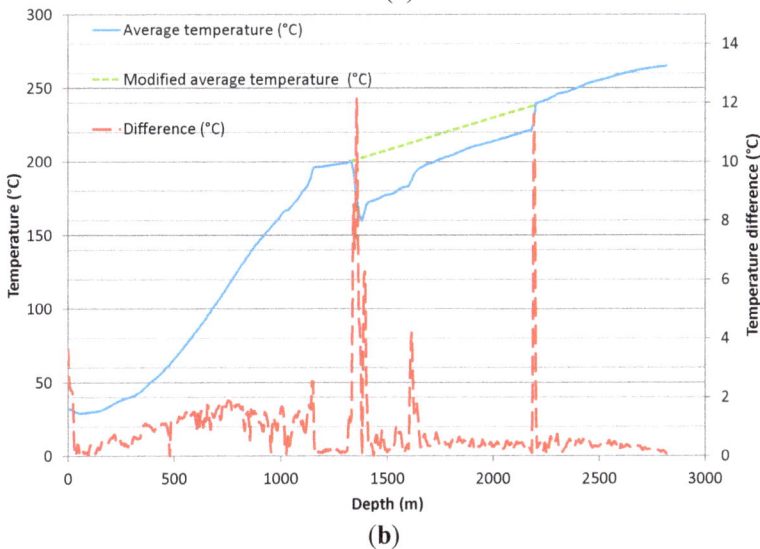

(**a**)

(**b**)

2.2. Seismic Tomography Results

As referred to in the introduction, a detailed active seismic tomography experiment was conducted on the island of Montserrat. The results of this experiment are given in Shalev *et al.* [13]. The appendix (Figure A1) gives some images of the seismic velocity perturbation model produced by Shalev *et al.* [13]. Of particular note is the slow seismic velocity anomaly observed in the SW of the island which has a maximum lateral dimension of about 5 km. The experiment utilized an active airgun source towed behind the NERC (Natural Environment Research Council) research ship the RRS (Royal Research Ship) James Cook which circumnavigated the island several times. Data from 4413 airgun shots recorded by 58 receivers, including seven Ocean-Bottom seismometers, were utilized for the tomographic inversion. Figure A2 shows the number of observations at each seismometer in the network. The geometry of the source and receiver network is shown in Figure A3.

The inversion domain which consisted of an inner high resolution 12 km × 17 km × 6 km cuboid volume was made up of individual nodes with 500 m horizontal and vertical spacing. The full domain was 50 km × 45 km × 8 km with node spacings of up to 5 km at the edges of the domain. First arrival data were inverted for a P-wave velocity model using the method of Shalev and Lees [18] which uses a Cubic B-Spline description of the 3D volume and the LSQR (Least Squares QR factorization) algorithm [19,20] to invert the data to create an inverse model which depends solely on the seismic data and the inversion algorithm. No *a priori* information was used to generate the model. The amplitude of the inversion result was controlled by two separate damping parameters for the unknowns of velocity model and station correction. The smoothness of the inversion result was controlled by a Laplacian smoothing parameter. The starting model for the inversion was constructed using two 1-D models for land and ocean which were derived from the data using the Levenberg-Marquardt nonlinear minimization algorithm [21].

Seismic velocity is dominated by large vertical variations in velocity with on-island P-wave velocities at Montserrat varying from around 3.5 km·s^{-1} at the surface to about 6.25 km·s^{-1} at 10 km depth [13]. Seismic velocity generally increases with increasing depth, in the case of Montserrat, both experimental and modelling results [22] indicate that P-wave velocity in the top 10 km is, to first order, a function of porosity and pore geometry mediated by increasing pressure. To visualise variations which are not primarily associated with the depth-related increase seismic velocity, models are often displayed in terms of velocity perturbation models. The perturbation model used here is created by calculating the percentage variation from the mean velocity at each depth in the model. This procedure brings out anomalous variations which would be difficult to detect when looking at velocity alone. Figure A1 shows the velocity perturbation model for the SEA-CALIPSO tomographic data at 2 km depth.

Two 1-D vertical columns through the inversion domain at the locations of MON-1 and MON-2 are shown in Figure 3. In Figure 3b, the seismic velocity anomaly is normalized by subtracting the anomaly value (V_p) from the maximum value (V_{Pmax}) and dividing by the difference between the maximum value and the minimum value (V_{Pmin}) in the shallow region of the model (Equation (1)).

$$V_{pn} = \frac{\left(V_p - V_{pmin}\right)}{\left(V_{pmax} - V_{pmin}\right)} \tag{1}$$

The results in Figure 3 show the percentage variation of the seismic velocity model from the average value at each depth. The sign of the anomaly has been reversed to make it positive. We will define this positive quantity as the seismic velocity anomaly henceforth. Figure 3 shows a distinctive anomaly associated with the geothermal system. Seismic velocity anomaly increases with depth to around 1750 m and then decreases down to a depth of 3000 m. The exact shape of the anomaly is determined by the damping and smoothing parameters which trade-off between data-fit and smoothness of the inversion result. This trade-off is necessary to stabilize an inversion result which is the solution to an ill-conditioned problem. The exact shape and intensity of anomalies will be affected by the choice of inversion parameters. However, the general shape and particularly the location of anomaly maxima will be fairly robust regardless of chosen inversion parameters. This robustness is due to the fact that the tomographic model is well-constrained by the data in the depth range of interest; as can be observed from ray hit plots and checkerboard tests shown in the appendix.

The southwest region of the island is particularly well constrained down to a depth of 3 km. Figure A4 shows the ray hit plot for a series of depths in the model domain. This plot illustrates that the SW portion of the island is well-constrained by observations in the 1–3 km depth range with each domain block constrained by several hundred seismic ray observations. Checkerboard test results shown in Figure A5 illustrate the resolving power of the tomography experiment. The high number of receivers both on and off the island coupled with the 360 degree azimuthal coverage of the towed airgun source at a range of different off-set distances yield an exceptional data set that constrains the velocity model to a depth of 3 km very well. The checkerboard test results show that the geometry of seismic anomalies with lateral dimensions as small as 1.5 km are easily recovered given the source and receiver geometry (the southwestern low velocity anomaly has a maximum lateral dimension of the order of 5 km). To minimize the effect of the choice of damping parameter (which affects the amplitude of the anomaly in the inverse model) the normalized seismic velocity anomaly (V_{pn}) is also shown in Figure 3. The normalised anomalies for MON-1 and MON-2 are very similar.

To illustrate the effect of inversion on the sampled anomaly we can see the effect on the checkerboard test. Figure 4 shows a 1-D section through the checkerboard test anomaly. Both the starting model and its inversion result are shown. From Figure 4 we can see the kind of "distortion" created by the inversion process. The anomaly maximum is reduced significantly from 28.29% to 13.28% and the width of the anomaly is increased slightly. Another significant "distortion" effect is the undershoot of the anomaly from 0% in the original model to a maximum of −3.93% in the inversion result below 2.75 km depth.

Figure 3. (**a**) Tomographic inversion result at position of MON-1 and MON-2 (% perturbation from mean velocity at each depth; sign reversed); (**b**) Seismic velocity anomalies for MON-1 and MON-2 have been normalised. The average of the two normalised anomalies is also shown.

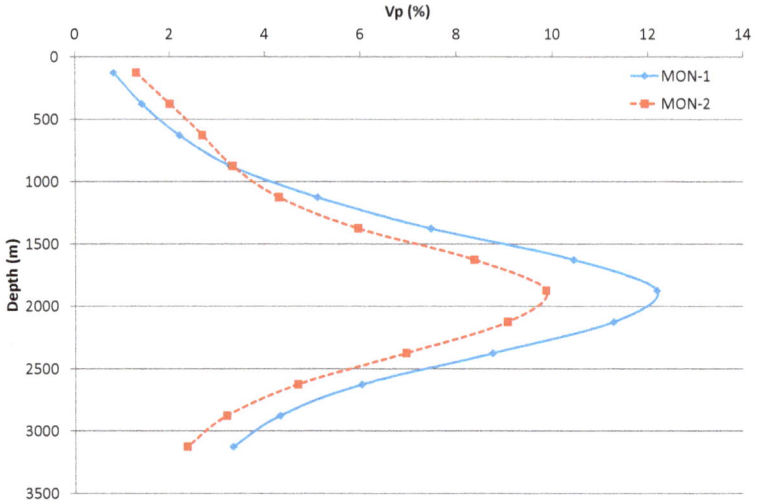

(a)

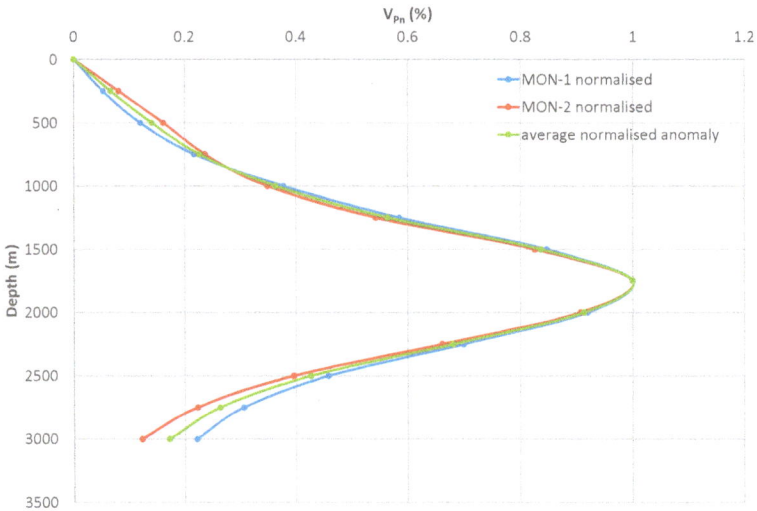

(b)

Figure 4. Comparison between checkerboard test model and inversion result.

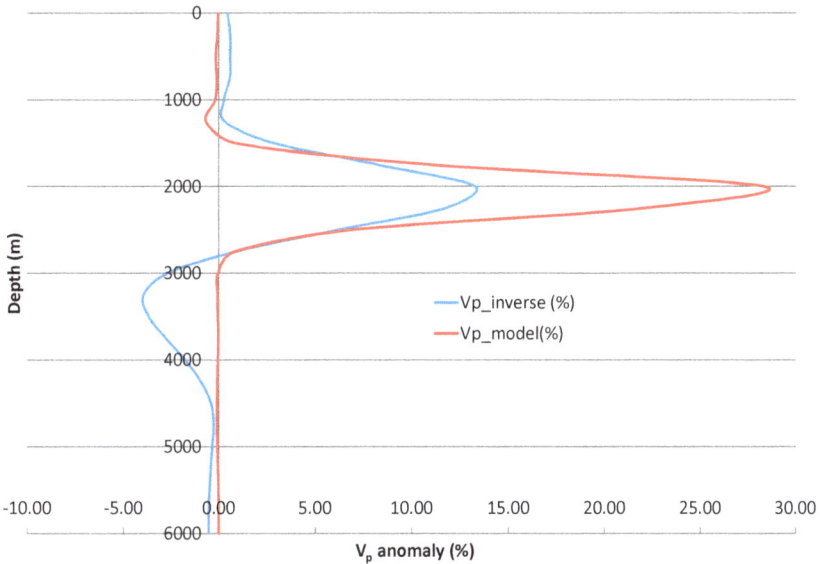

2.3. Lithology and Petrology

Cuttings from wells MON-1 and MON-2 were logged during drilling at approximately 10 m intervals. During drilling the cuttings were described in near real time with the aid of a binocular microscope and the smectite content of clays was determined by using the methylene blue test [17]. More detailed analysis of the cuttings was conducted after drilling using a petrographic microscope and prepared thin sections. The identity and concentration of the crystalline components of select samples was determined using X-ray diffraction [16]. The use of drill cuttings which are small and subject to mixing as the cuttings are recovered during drilling means that gross formation characteristics cannot be recovered to determine source sequence and how the units were emplaced [16]. There are also limitations due to incomplete sampling due to substantial loss of material in the mud circulation system.

The results of cutting analysis indicate that the lithology down to 2870 m consists predominantly of volcaniclastics probably related to block and ash and debris flows. In well MON-1 the section between 530 and 1210 m depth is comprised of sandstones, mudstones and smectite-containing clays. This region occurs at a similar depth to the low resistivity zone interpreted as the reservoir clay cap by Ryan *et al.* [14]. The section from 1210 to 2298 m is thought to be predominantly reworked volcaniclastics with small amounts of intercalated andesite or basalt flows this is thought to be the main reservoir zone [16]. Preliminary results from well MON-2, which terminates at 2870 m, indicates that it has a similar lithology to well MON-1. However, detailed analysis of petrographic samples has not been completed to date. Figure 5 shows the preliminary lithology logs. The spatially discontinuous nature of volcaniclastic deposits on a small volcanic island with multiple eruption centers and variations in depositional environment are the probable cause of the differences between the logs.

Figure 5. (**a**) Generalised lithology for well MON-1; (**b**) Generalised lithology for well MON-2 adapted from EGS Inc. [16].

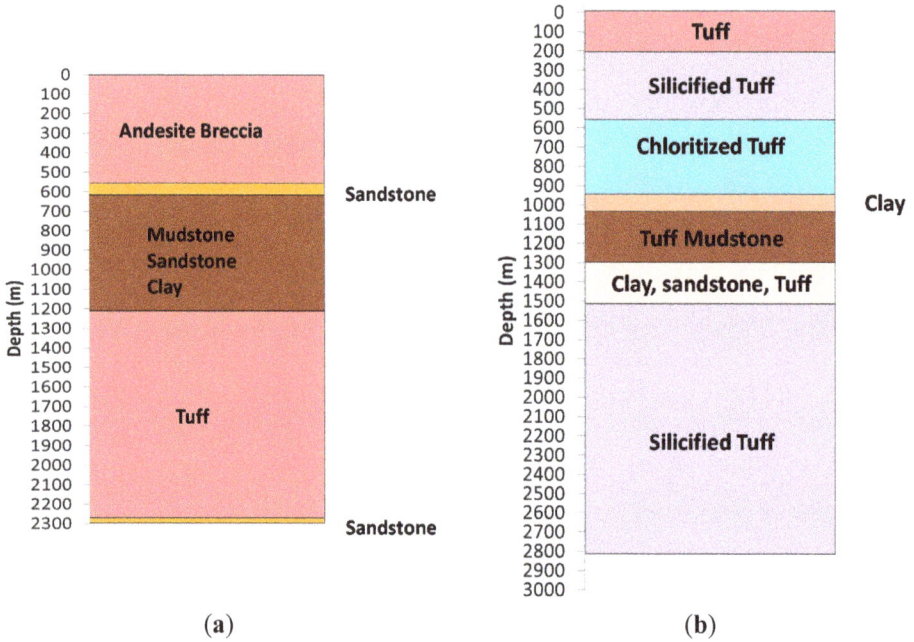

(a)   (b)

3. Discussion

3.1. Variations in Seismic Velocity with Depth and Temperature at Well-Locations

Extrapolating from the work of Kristinsdottir [10], Ryan *et al.* [9] hypothesised that the magnitude of the seismic anomaly would be proportional to the reservoir temperature. The current study indicates that rather than a direct dependence on temperature, seismic velocity is likely mediated by temperature dependent variations in the hydrothermal mineral assemblage. The data show that while seismic velocity anomaly increases with temperature below 220 °C, above this temperature the relationship reverses in a way that was not anticipated by Ryan *et al.* [9]. Figure 6 shows the variation in measured temperature with normalized seismic velocity anomaly (V_{pn}) as derived from the tomographic model of Shalev *et al.* [13] (appendix). Tomographic data relating to the vertical series of model blocks which encompass the vertical well tracks for MON-1 and MON-2 have been used.

V_{pn} increases with temperature up to around 220 °C and then decreases with temperature.

Looking closely at the well temperature logging results shown in Figure 6 we can see features in the data which support the general model of Ryan *et al.* [9].

There is a low temperature gradient between 1150 and 2800 m of about 50 °C/km. This low gradient is consistent with a high temperature convective upflow with a heat source below 2800 m depth and is unlikely to be associated with an outflow which would be associated with a temperature inversion at depth.

Between 1105 and 1150 m, there is a rapid increase from 50 to 190 °C/km in thermal gradient which indicates a switch from convective to conductive heat transfer. The rapid change in gradient indicates an impermeable barrier which impedes fluid flow across the depth at which the gradient change occurs. This impermeable barrier is consistent with the impermeable smectite clay cap observed in this depth range.

Figure 6. Plot showing variation in well temperature and variation in normalized seismic velocity anomaly (V_{pn}) with depth as derived from the tomographic model of Shalev *et al.* [13]. The V_{pn} values used are the average value for wells MON-1 and MON-2. (**a**) Values for well MON-1; (**b**) Values for MON-2.

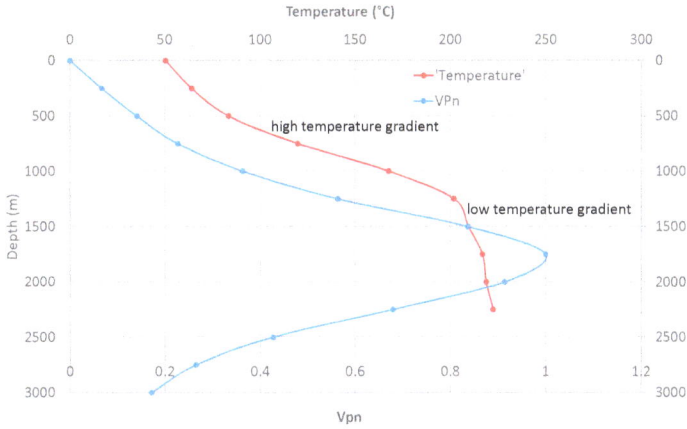

(**a**)

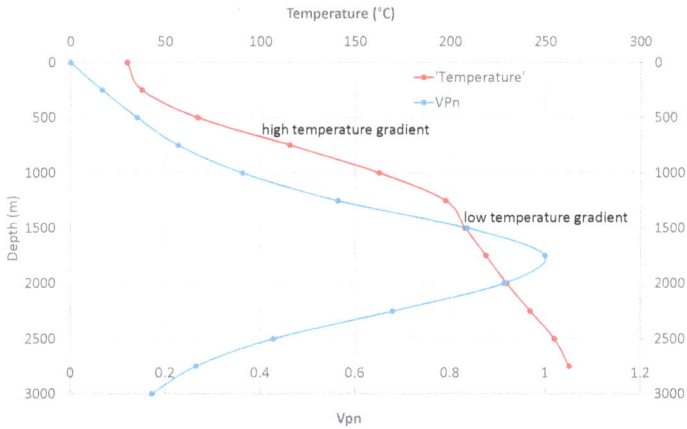

(**b**)

Seismic velocity variation across high temperature geothermal reservoirs has been observed at several locations [23–25]. In most of these studies, seismic velocity inversion is performed on data sets consisting of natural seismic events. These data sets were generated by relatively small numbers of sources and receivers. It is also impossible to control source location and density. Models generated using passive seismic data therefore generally have limited resolution usually on the scale of kilometers; larger than the length scale of individual wells in the fields. This limited resolution makes detailed comparison between measured reservoir characteristics and seismic models impossible. In these studies, variations in seismic velocity are often ascribed to subsurface steam zones or highly saturated porous regions in the reservoir which would lower seismic velocity.

In some laboratory studies on reservoir rocks, the effects of the temperature of the saturating fluid and variations in porosity and clay concentration on measured seismic velocities were investigated [11,26]. These studies were limited in the spatial scale of the samples they could study and also in the timescale of the experiments which was (typically seconds to minutes). The temperature ranges were often limited as well. There were few measurements above 200 °C. Jaya *et al.* [26] explored variation in seismic velocity of highly porous (20%) liquid saturated samples. The results showed a significant decline in velocity as temperature increased. The results are explained in terms of a modified Gassman equation [27]. Boitnott [11] measured the seismic velocities of liquid saturated samples and used linear regression to determine the relative effects of porosity and clay concentrations. He found porosity to have the strongest effect on seismic velocity, with the concentration of illite having the second most significant effect. Seismic velocity was found to decrease with both increasing porosity and illite concentration.

Kiddle [22] performed a series of experiments on rock samples from Montserrat collected from the surface and upper 200 m. Both dry andesite and andesite breccia samples show little velocity variation with temperature, samples were heated up to 600 °C. However, velocity was found to increase significantly with pressure. Pressure was increased up to 70 MPa. A large variation in seismic velocity was measured between the andesite and andesite breccia samples. P-wave velocities of $2.5–5.5$ km·s^{-1} were measured for solid andesite samples at room temperature and pressure whilst significantly lower values of $1.6–3.6$ km·s^{-1} where measured for andesite breccia samples under the same conditions. This variation is reflected in the seismic velocity anomaly model shown in Figure A1. The dense andesite cores beneath the volcanic centers, particularly the Soufrière Hills and Centre hills, have high seismic velocities compared to the flanks which are composed largely of andesite breccia. Kiddle, however, does not address the significant variations in the velocities of the flank deposits.

The direct effect of temperature on seismic velocity seems to be an unlikely explanation for velocity variations observed in Montserrat as temperature increases monotonically with depth but the seismic velocity anomaly decreases at a depth of around 1750 m and a temperature of around 220 °C. Similarly, for porosity to be responsible for the observed pattern would require a steady increase in porosity to a depth of 1750 m after which porosity rapidly decreases. This again seems unlikely although Brophy *et al.* [8] do suggest there may be some stratigraphically controlled permeability/porosity at around 2100 m. The temperature pressure profile indicates that the reservoir fluid is in a liquid state as it does not cross the boiling point for depth curve.

The hypothesis that hydrothermal clay alteration may be responsible for the observed velocity anomaly is feasible since cutting logs show smectite clay alteration in the 600–1200 m depth range. In assessing the hydrogeology of Montserrat, Geotermica Italiana [4] outlined the likely effects of hydrothermal alteration on reservoir rocks on the island and its effect on the mechanical properties of the rock. The effects of hydrothermal alteration at the water–rock interfaces give a possible explanation for the observed variations in seismic velocity.

At intermediate temperatures between 90 and 130 °C superficial alteration processes (weathering) are greatly enhanced and smectite clay is formed. Rather than being removed via erosional processes as it would be at the surface this swelling clay accumulates along preferential flow pathways reducing porosity to form an impermeable barrier. At the higher temperatures of 130–200 °C, phyllic alteration occurs with the formation of first interlayered illite-smectite and finally illite at temperatures above 200 °C. This process causes the rock to become more plastic [4].

Between 200 and 300 °C, we have propylitic alteration in which minerals such as quartz, epidote and adularia are formed; as the proportion of these framework silicates increases the reservoir rocks become more rigid and crystalline. Whilst permeability in these rocks can be reduced by deposition of silica, this can be easily reversed by tectonic activity.

This sequence of alteration gives a possible explanation for the observed velocity anomaly. Formation of phylosillicate clays, in particular illite, acts to plasticise the rock and reduce seismic velocity (increase seismic velocity anomaly) up to a temperature of around 200 °C; above this temperature propylitic alteration progressively increases the rigidity of the rocks causing an increase in seismic velocity (decrease in seismic velocity anomaly). This pattern of hydrothermal alteration represents the commonly observed pattern of alteration observed in high-temperature geothermal systems hosted in andesite rocks [28]. Similar patterns of hydrothermal alteration are reported in the geologically similar neighbouring island of Guadeloupe [29,30]. This pattern of alteration is also used to interpret the resistivity anomalies associated with high temperature geothermal systems [15,31]. Further detailed studies of the mineralogy of existing samples along with samples, including possible core, from future wells in Montserrat is required to verify the petrology.

One of the difficulties in trying to infer reservoir seismic properties from the laboratory results is that separate phenomena are likely convolved. Variations in fluid temperature change its contribution to the P-wave velocity. At the same time, the properties of the rock matrix vary with temperature as a function of changes in the equilibrium assemblage of hydrothermal minerals. Experiments of Jaya *et al.* [26] take place on much too short a timescale for equilibrium alteration to occur and the experiments of Boitnott [11] which investigate the effect of variations in porosity and variations in the concentration of hydrothermal clays all take place at a single temperature.

Experiments conducted by Kiddle [22] which investigate the effect of temperature and pressure on rock from Montserrat were conducted on dry rock samples from the surface and shallow subsurface and are not representative of saturated altered reservoir rocks. Experiments on saturated samples were also conducted but these were performed at room temperature and pressure. These experiments showed little effect of temperature on P-wave velocity but a large velocity variation was measured between andesite and andesite breccia with P-wave velocities for breccia being significantly lower. The high velocity zones seen in the velocity perturbation maps in the appendix is clearly related to the dense andesite cores associated with the volcanic centers [13,22]. However, there is a significant variation in P-wave velocity in the flank deposits. We suggest that the flank deposits with the anomalously low velocities, particularly the anomaly in the southwest of the island which is co-located with a verified geothermal system, are associated with hydrothermal alteration.

The lateral variation in the P-wave velocity of the flank deposits coupled with the fact the seismic velocity anomalies peak at a depth of 1750 m or so (within the zone of potential phyllic alteration within the reservoir) rather than being anomalous from the surface to depth supports the hypothesis that the velocity variations are dominated by changes in the hydrothermal mineral assemblage.

3.2. Correlation between Seismic Anomaly and Temperature

Figure 7 suggests, as was postulated in Ryan *et al.* [9], that there is a robust correlation between reservoir temperature and seismic velocity. To test this hypothesis, we recast the data to explore the correlation between the two. For this analysis, we combine the data from wells MON-1 and MON-2 to obtain the maximum number of data points under the assumption that the wells, which are approximately 500 m apart, sample similar regions of the reservoir. This assumption is supported by the similarity between the temperature profiles and the normalized seismic velocity anomalies for both wells. The value of the percentage P-wave velocity perturbation is multiplied by −1 to create positive values for the seismic velocity anomaly.

Figure 7. Correlation plot for normalised seismic velocity anomaly (V_{pn}) and reservoir temperature. Data for wells MON-1 and MON-2 are combined. Three different correlation regimes (R1a, R1b and R2) are suggested by the data. One point has been rejected as an outlier. This point which relates to the bottom of well MON-1 is thought to be slightly distorted.

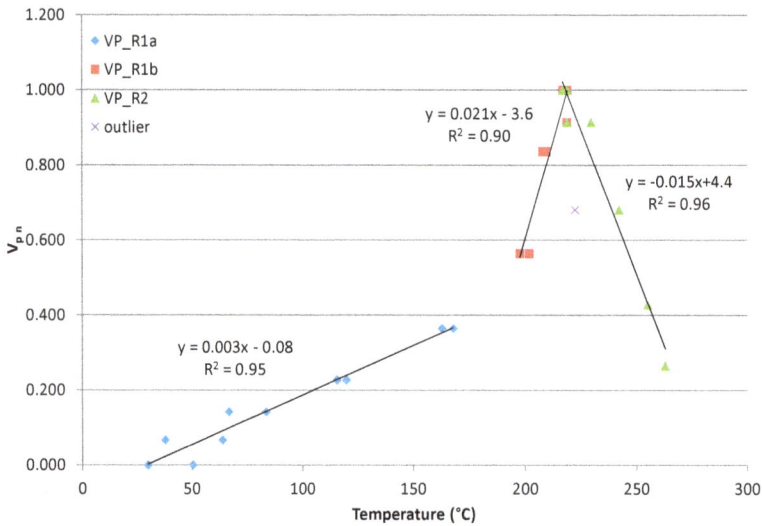

To produce this correlation plot, the resolution of the temperature data was reduced to match the resolution of the resampled seismic velocity model. The temperatures were averaged over 250 m intervals. The data suggest three dominant regimes (V_{pn} refers to normalised seismic velocity anomaly and T refers to reservoir temperature in °C):

(1) A slow increase in seismic velocity anomaly with temperature between 29 and 192 °C characterized by the following Equation (2) which fits the data with an R^2 value of 0.95.

$$V_{pn} = 0.003T - 0.08 \qquad (2)$$

(2) A rapid increase in seismic velocity anomaly with temperature between 192 and 219 °C characterized by Equation (3) which fits the data with an R^2 value of 0.90.

$$V_{pn} = 0.021T - 3.6 \tag{3}$$

(3) A rapid decrease in seismic velocity anomaly with temperature between 219 and 263 °C characterized by Equation (4) which fits the data with an R^2 value of 0.96.

$$V_{pn} = -0.015T + 4.4 \tag{4}$$

The decision to parameterize the correlation data in the way described above is purely empirical and is strictly only valid for the local area and for the specific set of tomographic inversion parameters used. The linear fits were chosen for their simplicity and because they provide a good fit to the data. The correlation could have been parameterized using different functions.

The observed correlation is between temperature data obtained from well logs and the magnitude of V_{pn} obtained from inversion of active seismic data [13]. As with all geophysical inversions, the inverse model here is non-unique [32] as discussed in Section 2.2. The exact form of the inverse model is dependent on the damping parameters used to stabilize the inversion. The values were chosen to maximize the structural information contained within the seismic data whilst minimizing the effects of errors inherent in the data which make it difficult to obtain a solution to the ill-conditioned inverse problem [32].

The correlations between V_{pn} and temperature we obtain are not solely a function of subsurface geology but also depend on the inversion parameters chosen. A related consequence of the inversion process is the tendency for geological anomalies to be "smeared out" which limits the ability of the inversion to resolve small anomalies. Checkerboard test results shown in the appendix show that the resolving power of the data remains good down to a depth of 3 km. Structures at the sub-kilometer scale are resolvable at this depth in the region of the geothermal reservoir. The checkerboard test data (appendix) also illustrates how the initial model is "distorted" by the inversion process. The intensity of the anomaly decreases and the anomaly spreads out in space. Whilst the intensity of the anomaly varies, its general shape, although blurred tends to be stable. For this reason, we use the normalised seismic velocity anomaly in our analysis to reduce the effect of the particular choice of damping parameters on the magnitude of velocity anomaly.

Our favoured interpretation of the variation in seismic velocity with temperature is that it is related to the assemblage of hydrothermal minerals at different temperatures with clays being formed between 70 and 220 °C which progressively plasticise the rock matrix and reduce bulk and shear moduli of the rock and thus reduce P-wave velocity. Above 220 °C propyllitic alteration occurs which increases the proportion of framework silicate minerals such as quartz, epidote and adularia. This change increases the rigidity of the rock and increases the elastic moduli, again increasing P-wave velocity.

Alternate explanations for the observed low velocity anomalies such as variations in lithology, porosity, fluid saturation and cementation are also possible. None of these alternatives, however, directly explain the variation of the anomaly with depth, in particular why it peaks at a depth of around 1750 m in the region of the two wells. Alternate explanations cannot be totally rejected without further investigation, particularly of the cuttings from the existing wells. Core from future wells would be particularly valuable in future studies.

3.3. Creating a 3-D Temperature Model

Having determined that a correlation between seismic velocity anomaly and temperature exists, the next step is to use this correlation to estimate the subsurface temperature field from the seismic tomographic data. Knowing how subsurface temperatures vary in 3D will shed light on subsurface fluid flow pathways that would be of interest in geothermal energy exploitation.

Assuming our hypothesis explaining the cause of seismic velocity variation is sufficiently correct, it would be straightforward to convert the seismic anomaly data to temperature estimates if they were uniquely known. As discussed in Section 2.2, inversion of the seismic arrival data leads to "distortion" of the measured anomaly leading to variations in the shape and intensity of the modeled anomaly. If we consider the seismic anomaly data as being made up of a series of 1-D vertical columns through the model space the "distortion" will vary as we move from the edge of the anomaly to the center. In particular, the intensity of the anomaly will appear to decrease as it is "smeared out" at the edges. However, using the assumption that the maxima of the anomaly should be relatively undisturbed even if its absolute value varies and that the maximum is related to a unique process *i.e.*, a change in hydrothermal mineral assemblage which occurs at a fixed temperature in this geothermal reservoir we propose using the normalised seismic velocity anomaly (V_{pn}) to "undistort" the seismic data in the region close to the wells.

The seismic velocity anomalies as measured at the locations of wells MON-1 and MON-2 which are 500 m apart have very similar, somewhat Gaussian, shapes and similar amplitudes (Figure 3). If we look at 1-D anomalies as we move away from the well locations to the edges of the anomaly we see that the shape of the anomaly remains similar but the amplitude decreases (Figure 8). We suggest compensating for this amplitude reduction in the region close to the wells, at which reservoir conditions are similar, by using V_{pn}. We deem this a logical approach because conditions are similar in the region close to the measurement point, and the seismic velocity anomaly has a similar cause and shape. We postulate that the anomaly maximum occurs at a fixed temperature and is caused by a change in the regime of hydrothermal alteration. We therefore suggest that the maximum occurs at the same temperature in locations close to the well. Although this is not a fully rigorous approach, it has the benefit of simplicity and transparency. Our assumption, that nearby regions are similar, should be most robust closest to the zone where temperatures were measured and becomes less certain with distance from that location.

Figure 8. Variation in seismic velocity anomaly as one moves from south to north near the location of well MON-1.The graphs relate to vertical columns through the seismic tomography perturbation model going from 5000 to 9500 m·N at 6000 m·E. Each block in the resampled model domain is a cube with side 250 m. The horizontal axis indicates the magnitude of the anomaly in %. Note also how the magnitude of the anomaly increases from the southern edge to a maximum and then decreases again.

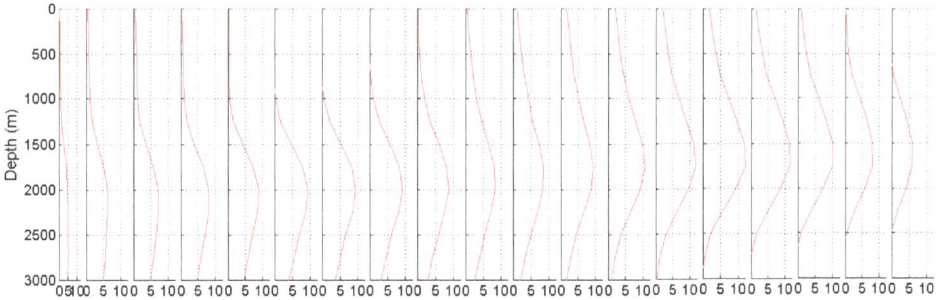

Figure 8 shows the variation in the seismic anomaly from the tomographic model going from south to north near well MON-1. This line correlates most closely to the N-S cross-section at 6250 m·E through the estimated temperature model in Figure 9. Figure 3 shows the bell-shaped anomaly seen in the tomographic model in the vicinity of the wells. The peak of this curve is associated in our interpretative framework with a change in hydrothermal mineral assemblage above 220 °C which causes an increase in the rigidity of the rock. An interesting feature of Figure 8 is that we can see the peak of the curve indicating the change in hydrothermal mineral assemblage shallowing from south to north from about 2000 m depth to just under 1500 m. This shallowing is reflected in the estimated temperature model of Figure 9 at 6250 m·E. Figure 9 shows a shallowing of the high temperature region of the model at around 7500 m·N.

To complete this analysis efficiently we created a MATLAB™ (Mathworks, Natick, MA, USA) script to automatically analyse the seismic anomaly model. The steps carried out by the script are outlined below:

(1) Create a master anomaly by averaging the anomalies at MON-1 and MON-2. Normalise this master anomaly.
(2) Use the master anomaly data and the averaged well temperature data to create three linear correlations between V_{pn} and temperature as described in Section 3.1.
(3) Loop over each vertical 1-D column in the seismic anomaly model volume.
(4) For each column, the anomaly is analysed to determine its similarity to the master anomaly's roughly Gaussian shape. This is accomplished using a weighted least squares algorithm to fit an offset Gaussian function to the anomaly (Appendix).

(5) RMS (Root Mean Square) value of the residuals between the best-fit Gaussian function and the seismic anomaly data is determined for each column of the seismic tomography model. If the RMS error is below 1/8th of the maximum value of the anomaly this is taken as a necessary but insufficient condition for similarity to the master anomaly. Figure 10 shows the fit of seismic anomaly data to an offset Gaussian. A perfect fit is not necessary. The purpose of the fitting is simply to gauge general similarity of shape and to estimate the amplitude of the anomaly.

(6) If the amplitude of the seismic velocity anomaly is greater than 4% as well as having an acceptable RMS error the anomaly is deemed similar.

(7) All "similar" anomalies are "undistorted" by normalising them to vary between 0 and 1.

(8) The three correlations functions are used to derive temperature estimates from the "undistorted" anomaly. Care is taken to break each vertical anomaly into regions so that the region in which V_{pn} increases with temperature is not confused with regions in which V_{pn} decreases with temperature (see Figure 7).

Figures 9, 11 and 12 show the results of the temperature estimation algorithm. Although the results of the estimation are, according to our theoretical framework, most justifiable at the areas closest to the wells the algorithm was allowed to operate over the entire seismic anomaly model domain. Results in regions far from the wells must be looked at skeptically.

Looking at the seismic velocity perturbation model (Figure A1) we can see that temperature estimates have been made in the areas corresponding to the three regions in which there are slow seismic anomalies in the northwest, northeast and in the region of interest which is known to host a geothermal system, the southwest. The regions to the northwest and southwest cannot be interpreted with any degree of confidence since they are far from the wells that provide the correlating data. However, the analysis shows that the seismic anomalies in these three regions are similar to each other in the details of their structure.

The method described in this study presents the possibility of a method for estimating subsurface temperatures in a geothermal system with unprecedented resolution. Such a method would be of great use in geothermal exploration and well targeting. The interpretation framework described here requires further work to verify its validity but if it proves to be valid it would deliver another interpretation framework similar to the interpretation scheme used in interpreting resistivity anomalies in high temperature reservoirs [33]. The areas of the model domain where no temperature estimates were made do not necessarily indicate low temperature. It simply means that the form of the seismic anomaly in these regions was dissimilar to that of the master anomaly and no interpretation was made.

Figure 9. (**a–f**) N-S Cross-sections through the southwestern region of the temperature model estimated from the seismic velocity anomaly data. This region corresponds to seismic anomaly associated with the known geothermal system. Black line indicates the track of well MON-1 and the red line indicates the track of well MON-2. SHV (Soufrière Hills Volcano); SGH (St. George's Hill). The colour scale indicates temperature in degrees Celsius.

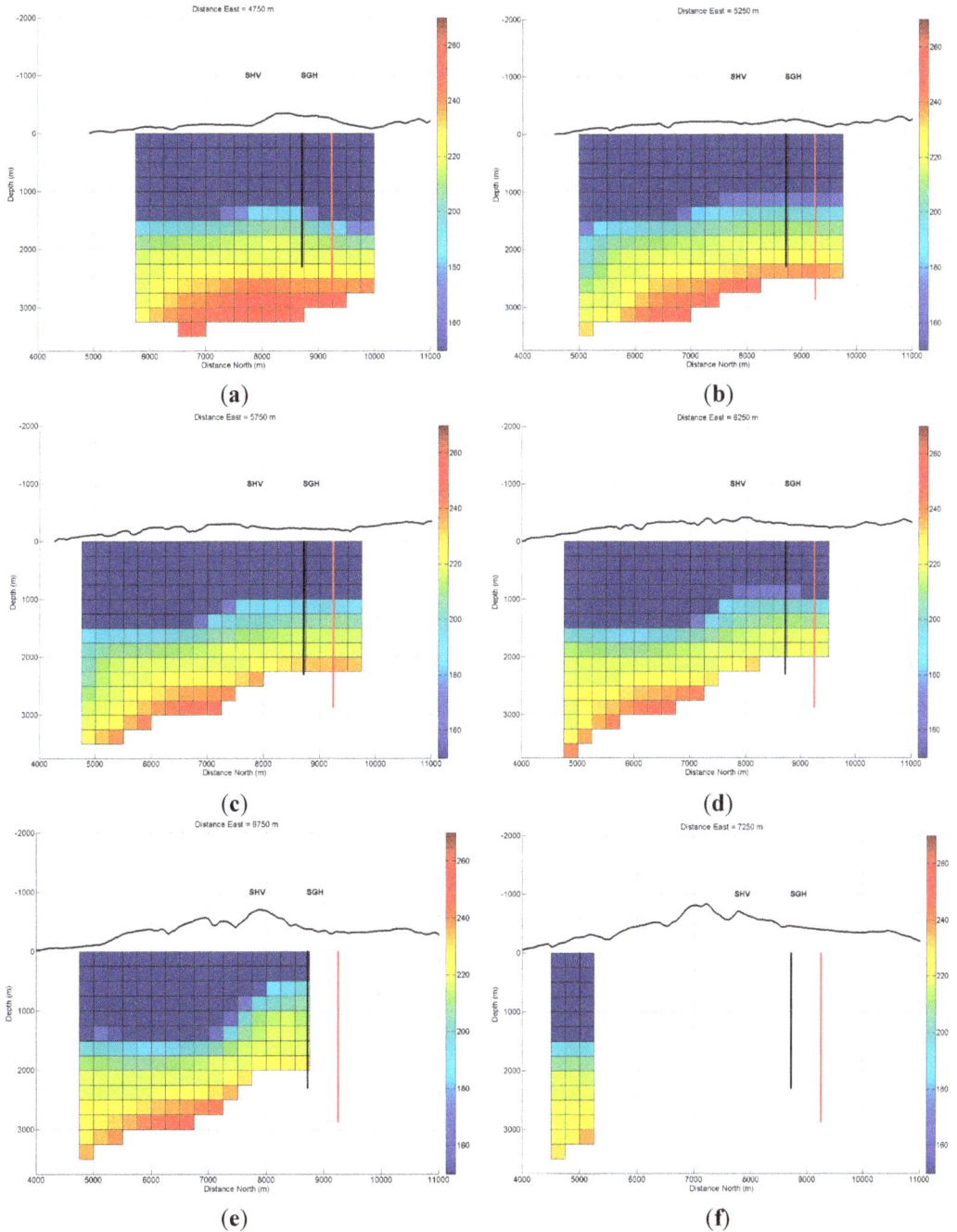

(**a**)

(**b**)

(**c**)

(**d**)

(**e**)

(**f**)

Figure 10. Fit of seismic anomaly data to an offset Gaussian function. The equation constants are $\Delta = -0.57$, $k = 8.76$, $\mu = 10.26$, $\sigma = 3.95$, the rms error is 0.80 (see appendix for details). A perfect fit is not necessary. The purpose of the fitting is simply to gauge general similarity of shape and to estimate the amplitude of the anomaly. This anomaly was at location 4500 m·E and 8750 m·N in the model.

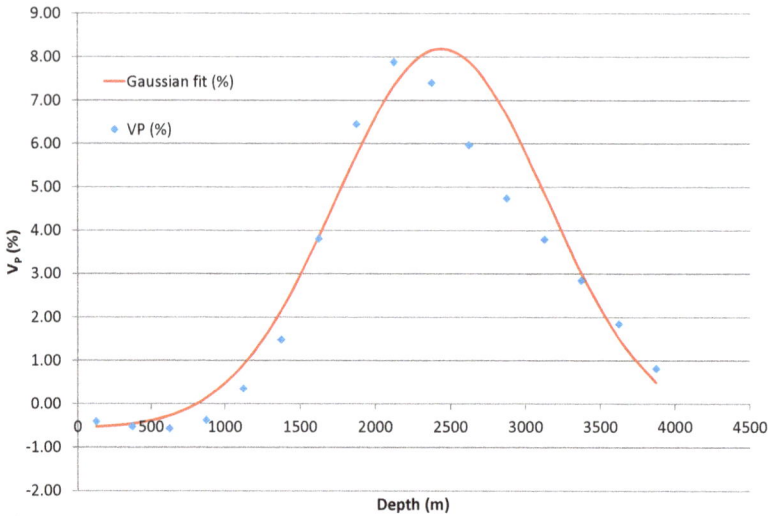

Focusing on the southwest anomaly we see that the estimation algorithm makes testable predictions about the geothermal system. The model indicates that the regions of shallowest high temperature occur beneath St. George's Hill. This is consistent with an upflow in this area. The zone of high shallow temperatures can be seen particularly well in Figure 11 at the 1500 m depth. The shallowing of the high temperatures beneath St. George's Hill is seen clearly in Figure 9 in the 6250 m east cross-section.

This finding which indicates the existence of a geothermal system with an upflow beneath St. George's Hill away from the Soufrière Hills volcano is supported by earthquake data and resistivity data described in Ryan *et al.* [9] and is consistent with early results from the well logging.

The temperature anomalies indicated to the NW and NE and seen in Figure 11h are not corroborated by any evidence of active geothermal systems in these areas. However, no extensive studies have been carried out in these areas to date. A possible explanation for these zones is that they are related to zones of relict hydrothermal alteration which has modified the seismic velocities but is no longer related to a real temperature anomaly. This illustrates the constant necessity for multiple sources of information when interpreting geophysical data.

Figure 11. (a–g) Planar views through the SW region of the temperature model estimated from the seismic velocity anomaly data. This region corresponds to the seismic anomaly associated with the known geothermal system; (**h**) Planar view through the entire temperature model at a depth of 2000 m. Apparent temperature anomalies to the NW and NE may only be indicative of relict hydrothermal alteration in these locations or some other process altogether. The NW trending black line indicates a zone of structural weakness identified by Wadge and Isaacs [34]. The NE trending black line indicates a fault plane identified from earthquake hypocenters beneath St. George's Hill [9]. The magenta dots relate to the locations of wells MON-1 and MON-2. The cyan dot relates to the location of the alkali-chloride hot spring at Hot Water pond. GBH (Garibaldi Hill); SGH (St. George's Hill); GM (Gage's Mountain); CP (Chances Peak); SHV (Soufrière Hills Volcano); RM (Roche's Mountain); SSH (South Soufrière Hills); RE (Roche's Estate); CH (center Hills); SH (Silver Hills). The colour scale indicates temperature in degrees Celsius.

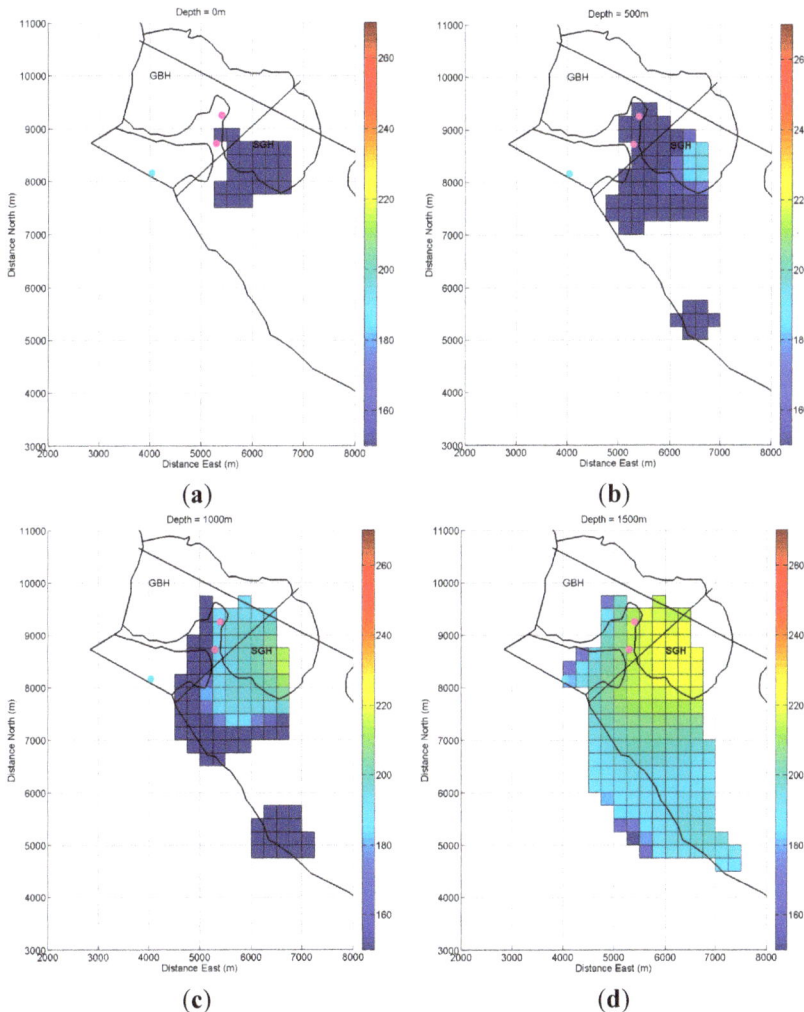

(a)

(b)

(c)

(d)

48

Figure 11. *Cont.*

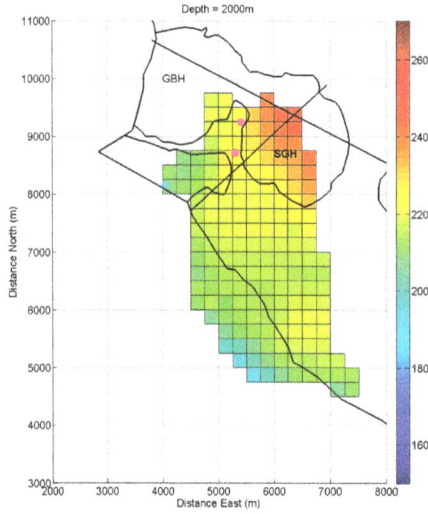

(e)

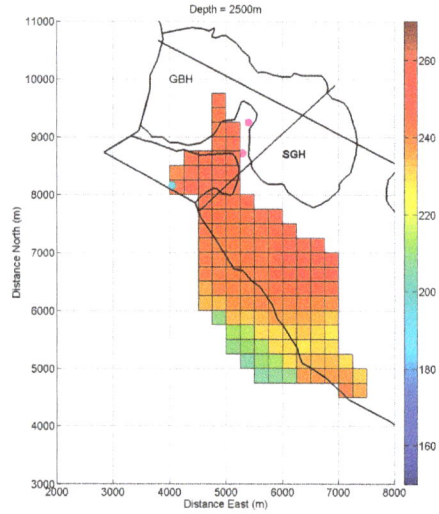

(f)

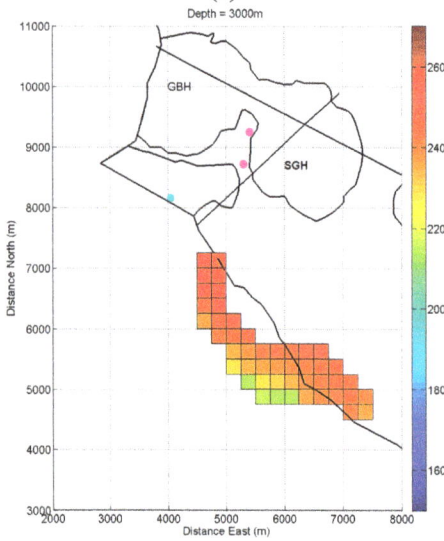

(g)

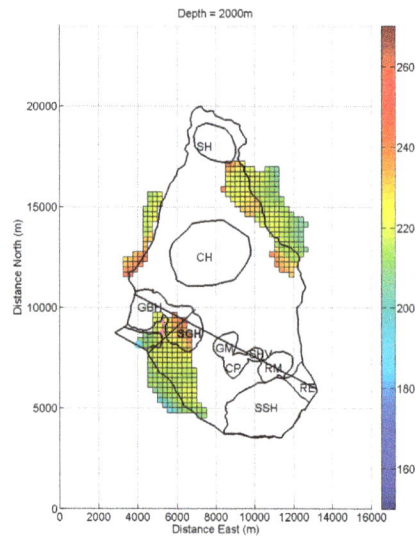

(h)

Figure 12. (a–f) E-W Cross-sections through the southwestern portion of the temperature model estimated from the seismic velocity anomaly data. This region corresponds to seismic anomaly associated with the known geothermal system. Black line indicates the track of well MON-1 and the red line indicates the track of well MON-2. SGH (St. George's Hill). The colour scale indicates temperature in degrees Celsius.

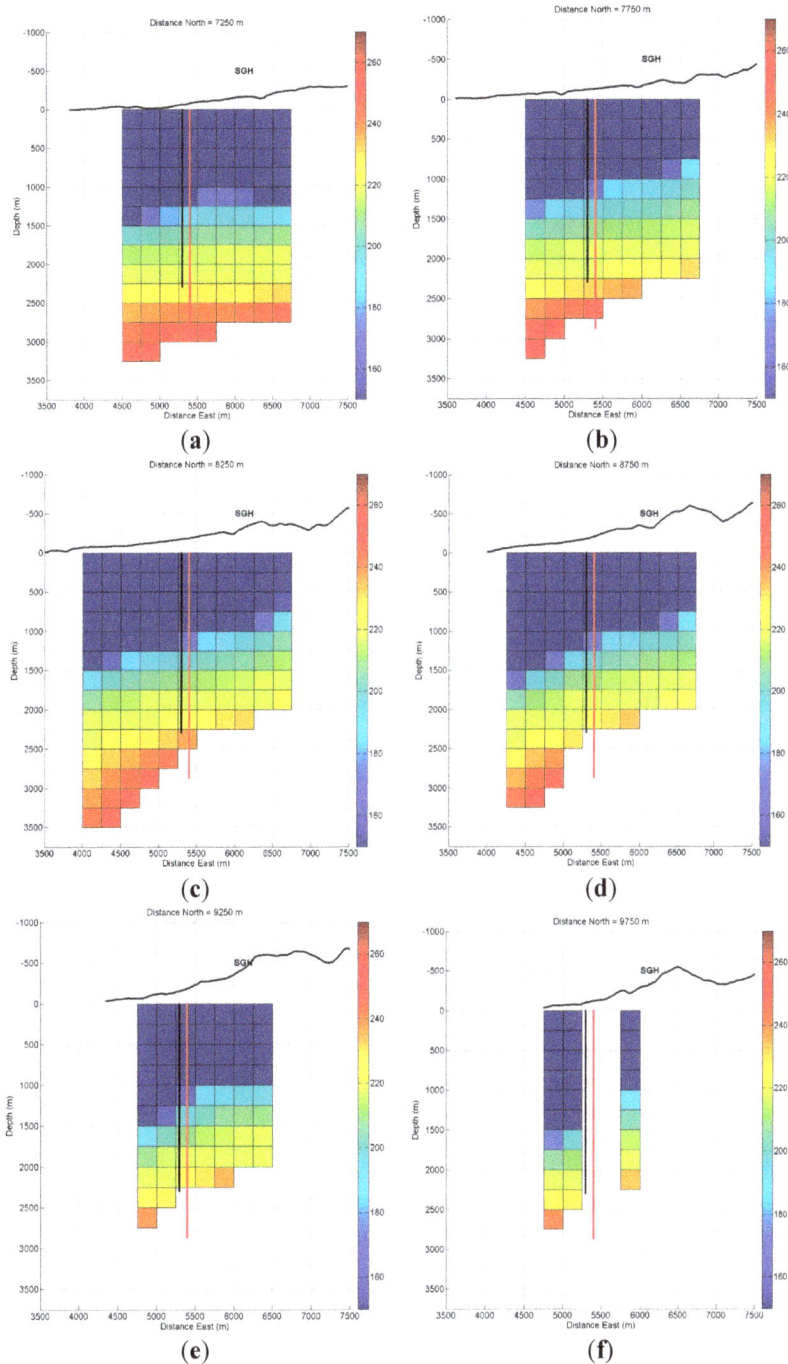

4. Conclusions and Further Work

The main conclusion of this study is that there is persuasive evidence to suggest that seismic velocities in a high temperature geothermal reservoir are strongly influenced by temperature dependent hydrothermal mineral assemblages. We suggest that this influence may allow seismic velocity anomalies to be used to estimate reservoir temperatures. Further work, however, is required to fully reject alternate possibilities.

Sub-surface temperature estimation suggests the existence of an upflow zone beneath St. George's Hill in the geothermal system in the southwest of Montserrat. Estimated "temperature anomalies" in the northwest and northeast of the island are possibly due to relict hydrothermal alteration related to extinct geothermal systems.

The current interpretation scheme for the seismic velocity anomaly data is based on a pattern of hydrothermal alteration commonly seen in high temperature geothermal systems [15]. However, further work on the petrology of samples from Montserrat's geothermal wells is required to determine whether the details of hydrothermal alteration in this case are consistent with the seismic interpretation.

The methods used to correlate temperature log data to the seismic anomaly data, and particularly to "undistort" the seismic anomaly data, are empirical and crude. A more rigorous approach should be sought in future work.

In a similar vein, it may be possible to obtain a more theoretically robust method for correlating seismic anomalies to hydrothermal alteration by using Gassman's equation [27] in tandem with experimental results on the physical properties of hydrothermally altered rocks.

Acknowledgments

We thank the government of Montserrat for permission to publish the results of this study. We are also very grateful to those who supported the SEA-CALIPSO study including NSF, NERC, Discovery Channel, British Geological Survey, Foreign and Commonwealth office, PASSCAL Instrument centre and the Montserrat Volcano Observatory. Although originally for another purpose this data set has proved to be a unique and valuable addition to research in geothermal exploration; the kind of happy accident which often occurs in science. We are also grateful to Pat Browne and Bridget Lynne for many informative discussions on hydrothermal alteration in geothermal systems. The Institute of Earth Science and Engineering is also acknowledged for providing us with time to work on this study. We gratefully acknowledge the three anonymous reviewers whose thoughtful work helped to greatly improve the clarity of this manuscript.

Author Contributions

Graham Ryan formulated the project's original concept and wrote the manuscript. He is also responsible for developing the algorithms used to analyse the well temperature log and tomographic model data, and wrote the scripts used to generate the 3-D temperature model. Eylon Shalev was responsible for creating the V_p tomographic model along with data for the V_p perturbation model, ray hit and checkerboard test plots. He also reviewed the manuscript.

Appendix

A.1. Seismic Tomography

Figure A1. P-wave tomography results from Shalev *et al.* [13] displayed as perturbation from the average velocity at each depth. Blue represents faster velocities and red represents slower velocities. (**a**) map view slice through the target volume at 2000 m depth; (**b**) E-W cross-sections through target volume at 8750 m north; and (**c**) 6750 m north; (**d**) N-S cross-section through target volume at 5250 m east; and (**e**) 6250 m east. The NW trending black line indicates a zone of structural weakness identified by Wadge and Isaacs [34]. The NE trending black line indicates a fault plane identified from earthquake hypocenters beneath St. George's Hill [9]. The magenta dots relate to the locations of wells MON-1 and MON-2. The cyan dot relates to the location of the alkali-chloride hot spring at Hot Water pond. GBH (Garibaldi Hill); SGH (St. George's Hill); GM (Gage's Mountain); CP (Chances Peak); SHV (Soufrière Hills Volcano); RM (Roche's Mountain); SSH (South Soufrière Hills); RE (Roche's Estate); CH (center Hills); SH (Silver Hills). The colour scale indicates % velocity perturbation.

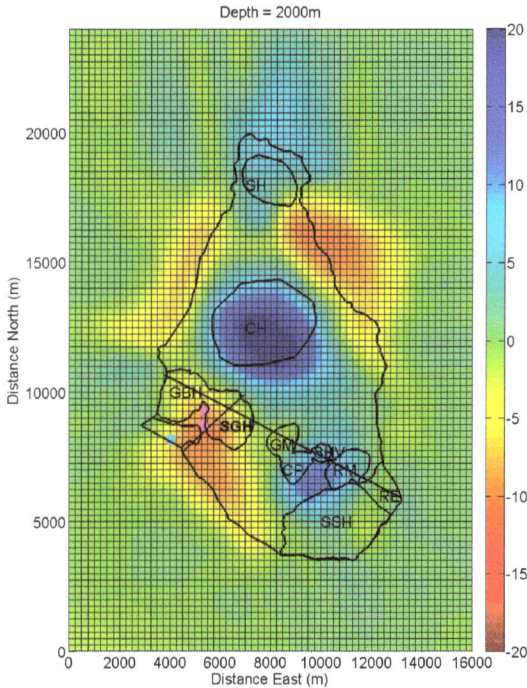

(**a**)

Figure A1. *Cont.*

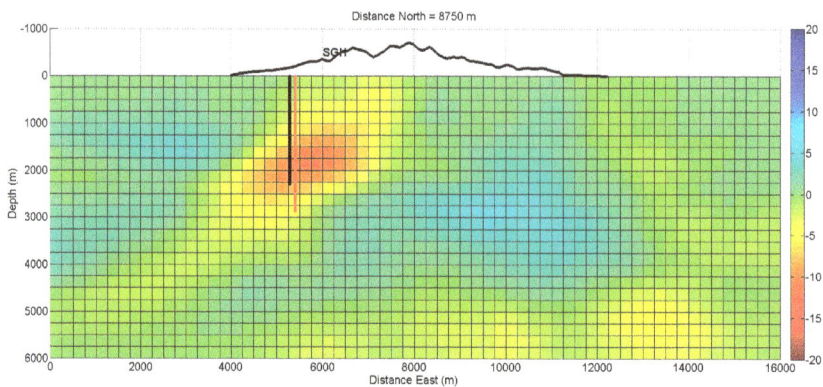

(b)

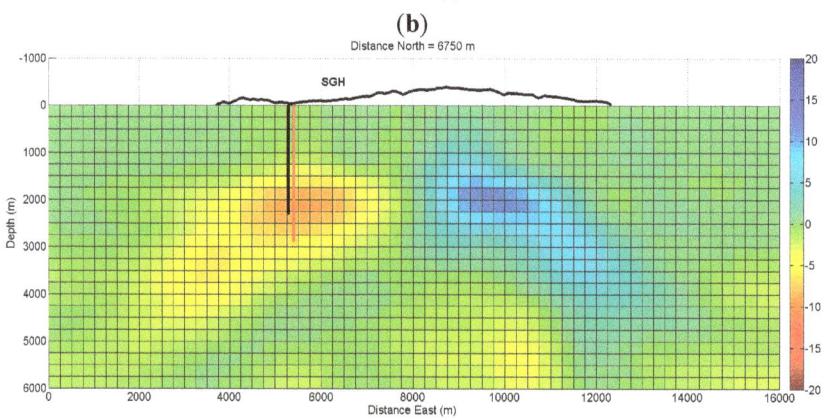

(c)

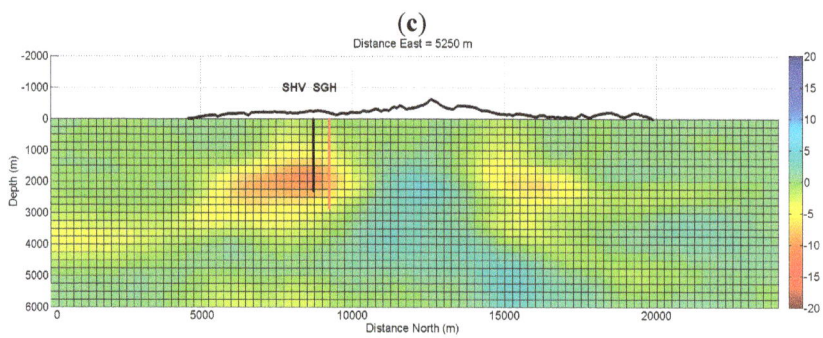

(d)

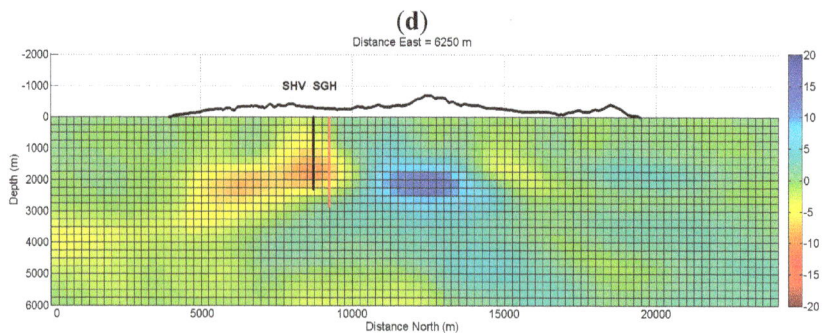

(e)

Figure A2. Number of observations used in the seismic tomography inversion shown for each on-shore seismometer. Red dots indicate Reftek seismometers. Green dots indicate Texan miniature seismic recorders and blue dots indicate seismometers which are part of the permanent monitoring network of the Montserrat Volcano Observatory.

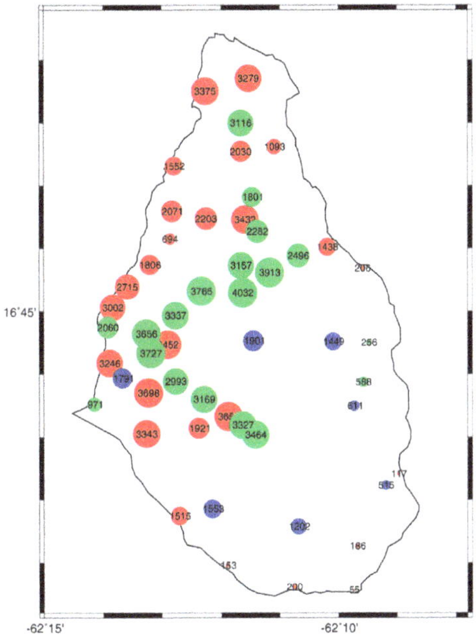

Figure A3. Location of seismometers (red triangles) used in this study located on-island and off-shore. The green line indicates the track of the vessel towing the air-gun seismic source. The green and purple dots mark the positions of wells MON-1 and MON-2, respectively. The blue outline surrounding the island indicates the extent of a shallow submarine platform.

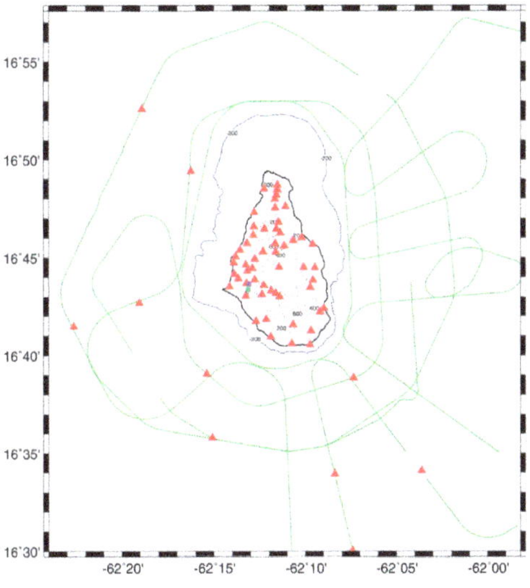

Figure A4. Ray hit counts for the tomographic inversion domain including. The panels show the number of rays that pass through 250 m × 250 m × 250 m boxes at (**a**) 0 m; (**b**) 500 m; (**c**) 1000 m; (**d**) 2000 m and (**e**) 3000 m depth. Only the center portion of the tomography box is plotted, with the outline of Montserrat in black. Colors indicate the $\log_{10}$ of the number of rays that track through each box.

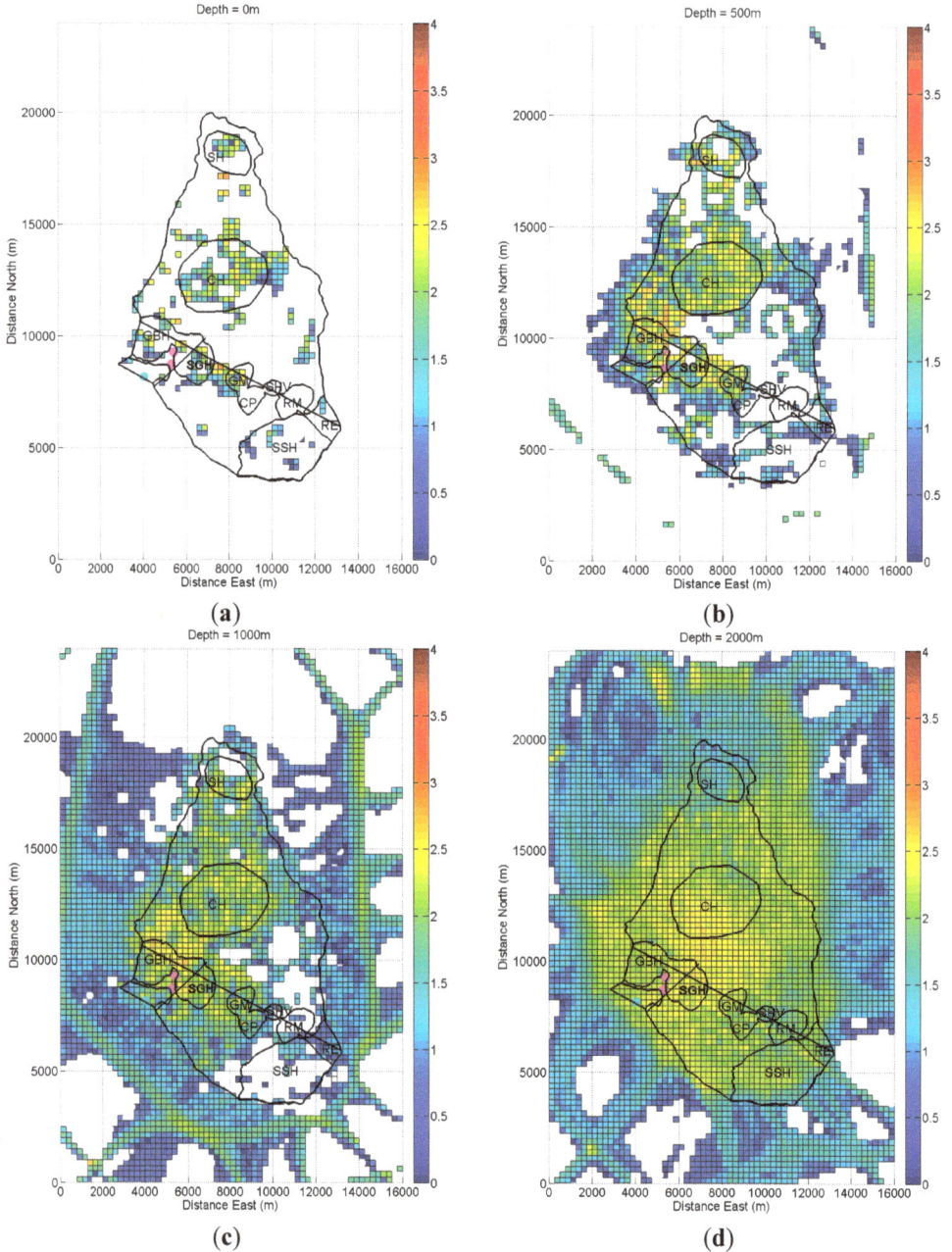

Figure A4. *Cont.*

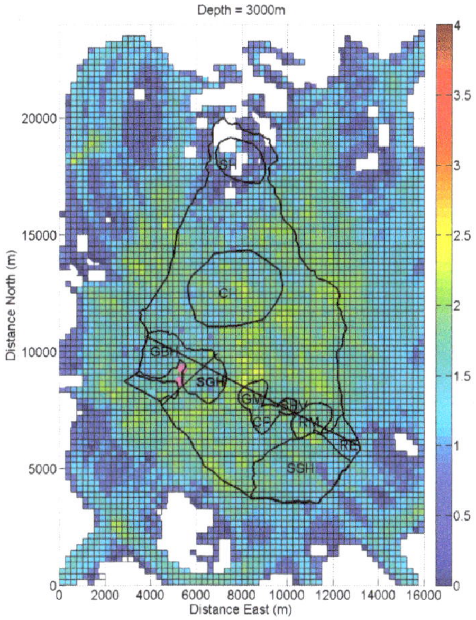

(e)

Figure A5. Checkerboard test results for anomalies at (**b**) 0 km; (**c**) 0.5 km; (**d**) 1 km; (**e**) 2 km; (**f**) 3 km and (**g**) 4 km depths; (**a**) is the source pattern for the checkerboard test (shown at 3 km depth). The *x* and *y*-axes indicate distance in km east and north respectively. The colour scale indicates percentage velocity perturbation. Synthetic noise with a zero-mean Gaussian distribution and 0.02 s standard deviation was added to the travel-time data. Note the variation in the amplitudes of anomalies.

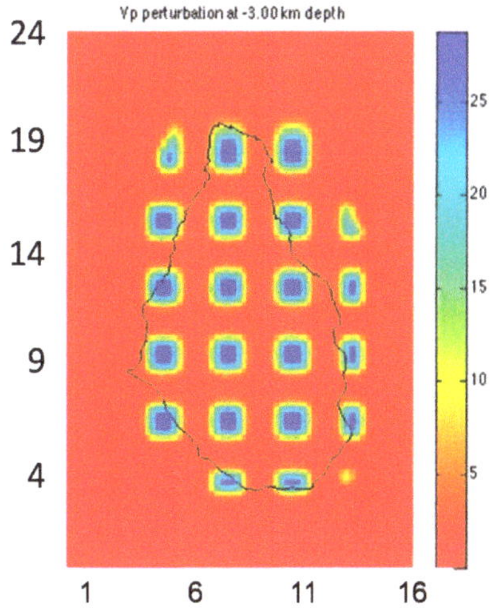

(a)

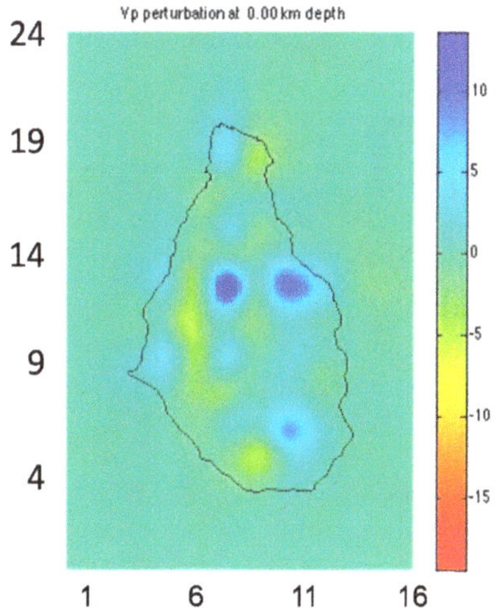

(b)

Figure A5. *Cont.*

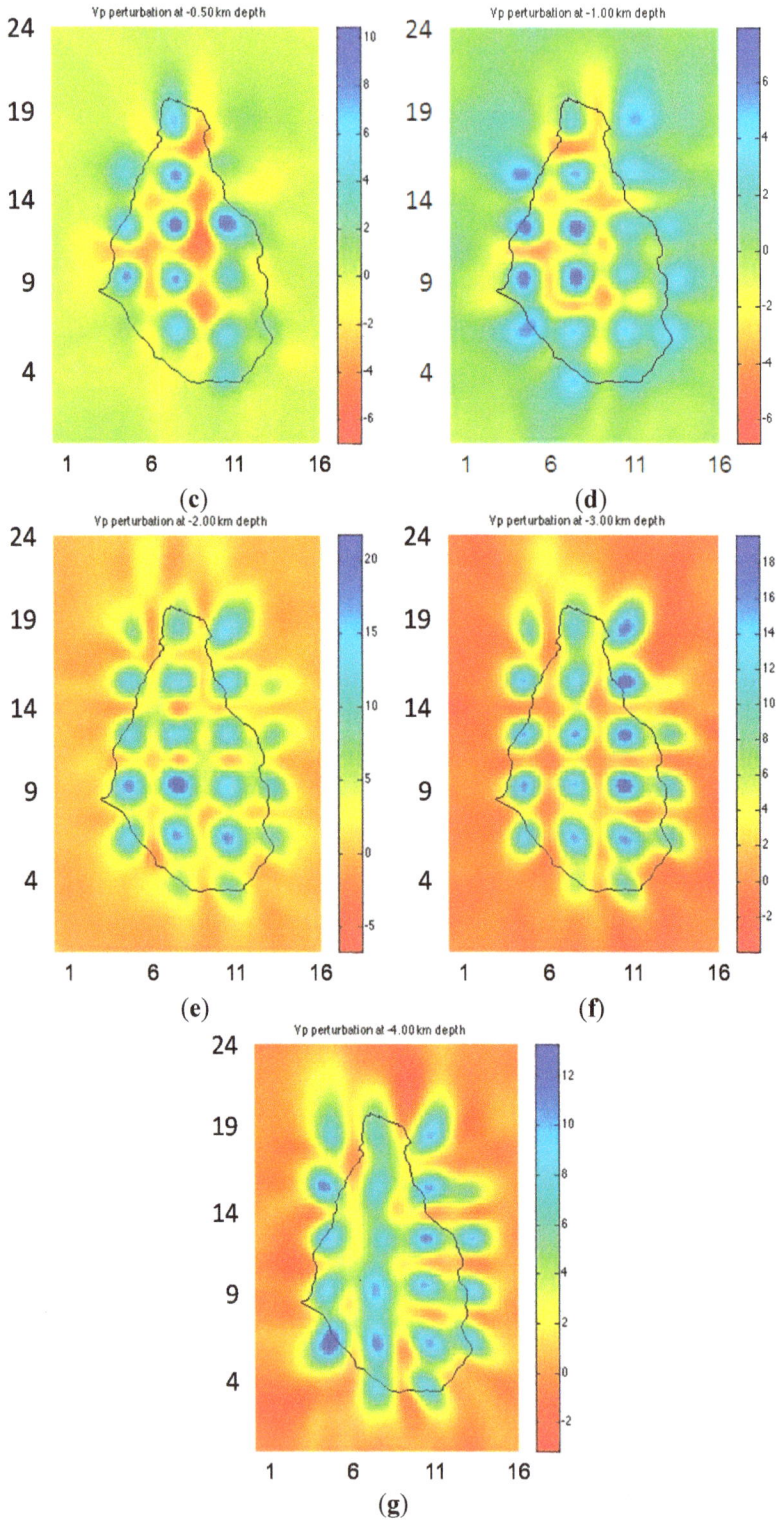

(c)

(d)

(e)

(f)

(g)

A.2. Gaussian Fitting Algorithm

In order to determine the similarity of 1-D columns through the seismic velocity anomaly model to the seismic anomalies observed at the locations of wells MON-1 and MON-2, we exploited the similarities of the seismic anomalies as a function of depth to offset Gaussian functions following the equation:

$$V_p = k \exp{-\left(\frac{(d-\mu)^2}{\sigma^2}\right)} + \Delta \qquad (A1)$$

where V_p is the magnitude of the seismic anomaly, k is a scaling factor, d is depth below sea level in meters, μ is the depth at which the peak V_p magnitude occurs, σ is the characteristic width of the Gaussian function and Δ is a scalar offset.

In order to obtain a simple linear function we assume that Δ is a known value and simply minimize the remaining Gaussian function once it has been subtracted:

$$V_p = k \exp{-\left(\frac{(d-\mu)^2}{\sigma^2}\right)} \qquad (A2)$$

To obtain a linear equation, we take the natural logarithm of both sides of the equation to obtain a polynomial in d:

$$\ln V_p = \left(-\frac{1}{\sigma^2}\right) d^2 + \left(\frac{2\mu}{\sigma^2}\right) d + \left(\ln k - \left[\frac{\mu^2}{\sigma^2}\right]\right) \qquad (A3)$$

Treating the variables d and d^2 as linear variables, we can do a simple weighted least squares parameter estimation using Equation (A4).

In order to bias the fitting algorithm to fit the peaks of the anomalies at the expense of the tails, we employed a weighted least squares algorithm. In which the errors varied monotonically from 1/10th to 1/100th of the average value of $\ln V_p$ from the initial low value to the peak value of the anomaly and then back to 1/10th. This error information is captured in the diagonal matrix C where the leading diagonal contains the squared reciprocals of the corresponding error values:

$$\langle x_w \rangle = [A^T CA] A^T Cb \qquad (A4)$$

where $\langle x_w \rangle$ is the vector of linear best-fit coefficients of Equation (A3), A is the matrix of linear variables in Equation (A3) and b is the vector of $\ln V_p$ values. The parameters μ, σ and k can be recovered from simple rearrangement of the elements of $\langle x_w \rangle$. To generate the best-fit parameters from data from a 1-D seismic anomaly, we simply use Equation (A4) using the values of $\ln V_p$, d and d^2. A description of weighted least squares estimation can be found in any book on elementary linear algebra, e.g., [35].

The RMS error of the fit of the data to the Gaussian model is obtained using the Equation (A5):

$$RMSerror = \left[\frac{\sum_{1}^{n} \left(V_{pmeas} - V_{pfit}\right)^2}{n} \right]^{1/2} \tag{A5}$$

where V_{pmeas} is the value from the seismic velocity anomaly model and V_{pfit} is the estimate derived from the best fit model. n is the number of data points in the 1-D column.

Conflicts of Interest

The authors declare no conflict of interest.

References

1. Wadge, G.; Voight, B.; Sparks, R.S.J.; Cole, P.D.; Loughlin, S.C.; Robertson, R.E.A. An overview of the eruption of Soufrière Hills Volcano, Montserrat from 2000 to 2010. *Geol. Soc. Lond. Mem.* **2014**, *39*, 1–40.
2. PKF. *Feasibility Study into Montserrat Geothermal Energy Final Report*; PKF Accountants & Business Advisers: London, UK, 2012; p. 56.
3. Wright, E.P.; Murray, K.H.; Bath, A.H. *Draft Report on Geothermal Investigations in Montserrat*; WD/OS/76/012; Institute of Geological Sciences, Hydrogeological Department: London, UK, 1976; p. 62.
4. Geotermica Italiana. *Exploration for Geothermal Resources in the Eastern Caribbean*; TCDCON15/90-RLA/87/037; United Nations Department of Technical Cooperation for Development: Pisa, Italy, 1991.
5. Principe, C. *Geothermal Potential in Montserat: Scoping Survey Report*; Instituto di Geoscienze e Georisorse, CNR: Pisa, Italy, 2008; p. 24.
6. EGS Inc. *Final Report Geothermal Exploration in Montserrat, Caribbean*; EGS Inc.: Santa Rosa, CA, USA, 2010; p. 189.
7. Younger, P.L. Reconnaissance assessment of the prospects for development of high-enthalpy geothermal energy resources, Montserrat. *Q. J. Eng. Geol. Hydrogeol.* **2010**, *43*, 11–22.
8. Brophy, P.; Poux, B.; Suemnicht, G.; Hirtz, P.; Ryan, G. Preliminary results of deep geothermal drilling and testing on the island of Montserrat. In Proceedings of the Thirty-Ninth Workshop on Geothermal Reservoir Engineering, Stanford, CA, USA, 24–26 February 2014; p. 11.
9. Ryan, G.A.; Peacock, J.R.; Shalev, E.; Rugis, J. Montserrat geothermal system: A 3D conceptual model. *Geophys. Res. Lett.* **2013**, *40*, 2038–2043.
10. Kristinsdóttir, L.H.; Flóvenz, O.G.; Árnason, K.; Bruhn, D.; Milsch, H.; Spangenberg, E.; Kulenkampff, J. Electrical conductivity and P-wave velocity in rock samples from high-temperature Icelandic geothermal fields. *Geothermics* **2010**, *39*, 94–105.
11. Boitnott, G.N. *Core Analysis for the Development and Constraint of Physical Models of Geothermal Reservoirs*; DE-FG07–99ID13761; New England Research, Inc.: White River Junction, VT, USA, 2003; p. 52.

12. Voight, B.; Sparks, R.S.J. Introduction to special section on the Eruption of Soufrière Hills Volcano, Montserrat, the CALIPSO Project, and the SEA-CALIPSO Arc-Crust Imaging Experiment. *Geophys. Res. Lett.* **2010**, *37*, 4, doi:10.1029/2010GL044254.

13. Shalev, E.; Kenedi, C.L.; Malin, P.; Voight, V.; Miller, V.; Hidayat, D.; Sparks, R.S.J.; Minshull, T.; Paulatto, M.; Brown, L.; *et al.* Three-dimensional seismic velocity tomography of Montserrat from the SEA-CALIPSO offshore/onshore experiment. *Geophys. Res. Lett.* **2010**, *37*, 6, doi:10.1029/2010GL042498.

14. Ryan, G.A.; Onacha, S.A.; Shalev, E.; Malin, P.E. *Imaging the Montserrat Geothermal Prospect Using Magnetotelluric (MT) and Time Domain Electromagnetic Induction (TDEM) Measurements*; 16366.00-2009.01; Institute of Earth Science and Engineering: Auckland, New Zealand, 2009; p. 77.

15. Anderson, E.; Crosby, D.; Ussher, G. Bulls-eye—simple resistivity imaging to reliably locate the geothermal reservoir. In Proceedings of the World Geothermal Congress, Kyushu-Tohoku, Japan, 28 May–10 June 2000; pp. 909–914.

16. EGS Inc. *Well Completion Report MON-1 and MON-2*; Government of Montserrat, Public Works Department: Woodlands, Montserrat, 2014; p. 310.

17. Gunderson, R.; Cumming, W.; Astra, D.; Harvey, C. Analysis of smectite clays in geothermal drill cuttings by the methylene blue method: For well site geothermometry and resistivity sounding correlation. In Proceedings of the World Geothermal Congress, Kyushu-Tohoku, Japan, 28 May–10 June 2000; pp. 1175–1181.

18. Shalev, E.; Lees, J.M. Cubic B-splines tomography at Loma Prieta. *Bull. Seismol. Soc. Am.* **1998**, *88*, 256–269.

19. Paige, C.C.; Saunders, M.A. LSQR: An algorithm for sparse linear equations and sparse least squares. *Trans. Math. Softw.* **1982**, *8*, 43–71.

20. Reidel, D.; Holland, D. Seismic wave propagation and seismic tomography. In *Seismic Tomography*; Nolet, G., Ed.; Reidel Publishing Company: Dordrecht, The Netherlands, 1987; pp. 1–47.

21. Press, W.H.; Teukolsky, S.A.; Vetterling, W.T.; Flannery, B.P. *Numerical Recipes in C: The Art of Scientific Computing*, 2nd ed.; Cambridge University Press: Cambridge, UK, 1992.

22. Kiddle, E.J. The Structure of the Crust and Magmatic System at Montserrat, Lesser Antilles. Ph.D. Thesis, University of Bristol, Bristol, UK, 2011.

23. Hauksson, E.; Unruh, J.R. Interaction of shallow and deep geothermal reservoirs at Coso, Eastern California, as inferred from 3-D seismic velocity models and seismicity. *Geotherm. Resour. Counc. Trans.* **2004**, *28*, 659–662.

24. Nakajima, J.; Hasegawa, A. Tomographic imaging of seismic velocity structure in and around the Onikobe volcanic area, northeastern Japan: Implications for fluid distribution. *J. Volcanol. Geotherm. Res.* **2003**, *127*, 1–18.

25. Gunasekera, R.C.; Foulger, G.R.; Julian, B.R. Reservoir depletion at the Geysers geothermal area, California, shown by four-dimensional seismic tomography. *J. Geophys. Res. B Solid Earth* **2003**, *108*, 11, doi:10.1029/2001JB000638.

26. Jaya, M.S.; Shapiro, S.A.; Kristinsdóttir, L.H.; Bruhn, D.; Milsch, H.; Spangenberg, E. Temperature dependence of seismic properties in geothermal rocks at reservoir conditions. *Geothermics* **2010**, *39*, 115–123.

27. Avseth, P.; Mukerji, T.; Mavko, G. *Quantitative Seismic Interpretation: Applying Rock Physics Tools to Reduce Interpretation Risk*; Cambridge University Press: Cambridge, UK, 2010; p. 371.

28. Inoue, A. Formation of clay minerals in hydrothermal environments. In *Origin and Mineralogy of Clays*; Velde, B., Ed.; Springer: Berlin, Germany, 1995; pp. 268–329.

29. Bouchot, V.; Sanjuan, B.; Traineau, H.; Guillou-Frottier, L.; Thinon, I.; Baltassat, J.M.; Fabriol, H.; Bourgeois, B.; Lasne, E. Assessment of the bouillante geothermal field (Guadeloupe, French West Indies): Toward a conceptual model of the high temperature geothermal system. In Proceedings of the World Geothermal Congress, Bali, Indonesia, 25–29 April 2010; p. 8.

30. Mas, A.; Guisseau, D.; Patrier Mas, P.; Beaufort, D.; Genter, A.; Sanjuan, B.; Girard, J.P. Clay minerals related to the hydrothermal activity of the Bouillante geothermal field (Guadeloupe). *J. Volcanol. Geotherm. Res.* **2006**, *158*, 380–400.

31. Simmons, S.; Browne, P.R.L. A Three dimensional model of the distribution of hydrothermal alteration minerals within the ohaaki-broadlands geothermal field. In Proceedings of the 12th New Zealand Geothermal Workshop, Auckland, New Zealand, 7–9 November 1990; pp. 25–30.

32. Parker, R.L. *Geophysical Inverse Theory*; Princeton University Press: Princeton, NJ, USA, 1994; p. 400.

33. Ussher, G.; Harvey, C.; Johnstone, R.; Anderson, E. Understanding the resistivities observed in geothermal systems. In Proceedings of the World Geothermal Congress, Kyushu-Tohuku, Japan, 28 May–10 June 2000; pp. 1915–1920.

34. Wadge, G.; Isaacs, M.C. Mapping the volcanic hazards from Soufriere Hills Volcano, Montserrat, West Indies using an image processor. *J. Geol. Soc.* **1988**, *145*, 541–552.

35. Strang, G. *Introduction to Linear Algebra*, 4th ed.; Wellesley Cambridge Press: Wellesley, MA, USA, 2009.

Determining the Maximum Depth of Hydrothermal Circulation Using Geothermal Mapping and Seismicity to Delineate the Depth to Brittle-Plastic Transition in Northern Honshu, Japan

Yota Suzuki, Seiichiro Ioka and Hirofumi Muraoka

Abstract: This paper defines the maximum possible vertical extent of hydrothermal circulation in granitic crust, and thus the maximum depth within which geothermal reservoirs can be encountered. To evaluate prospective geothermal fields we constructed a geothermal database in northern Honshu, Japan that includes 571 points of thermal data of existing wells and hot springs. Depth-temperature curves were normalized by the Activity Index for three-dimensional extrapolation and a depth contour map of the 380 °C isotherm was plotted as an assumed brittle-plastic transition for granitic crust. Shallower-depth anomalies of the brittle-plastic transition on this map are closely coincident with the Quaternary volcanoes and their prospective geothermal fields. It should be noted that the bottom of the spatial distribution of seismicity in the volcanic fields shows strong correlation to the 380 °C isotherm. This result indicates reliability of the subsurface three-dimensional thermal map and suggests that the 380 °C isotherm strongly constrains the bottom surface of seismicity, fracturing and hydrothermal convection in granitic crust.

Reprinted from *Energies*. Cite as: Suzuki, Y.; Ioka, S.; Muraoka, H. Determining the Maximum Depth of Hydrothermal Circulation Using Geothermal Mapping and Seismicity to Delineate the Depth to Brittle-Plastic Transition in Northern Honshu, Japan. *Energies* **2014**, *7*, 3503-3511.

1. Introduction

Since the unprecedented disastrous Great East Japan Earthquake at 11 March 2011, geothermal energy is being re-evaluated as a more crucial energy resource, not only as green energy, but also as indigenous energy in northern Japan. When compared to other renewable energy sources, the main strength of geothermal energy among the variety of renewable energy sources is that it can be produced stably in all weather conditions, whereas its main weakness is that it uses invisible underground resources prolonging development lead time as well as increasing the initial investment risks and costs. In order to explore subsurface geothermal resources, geological, geochemical and geophysical exploration techniques have been developed, but one of the efforts that should be done is the estimation of subsurface thermal structure from thermal data such as existing wells and hot springs.

Fortunately 27,219 hot spring sources are distributed all over Japan [1], and numerous hot springs are concentrated in Aomori Prefecture, a northern end of Honshu Island, Japan (Figure 1). Most of these Japanese hot springs are artificially developed by shallow wells. We constructed a geothermal database in Aomori Prefecture that includes available information on geothermal exploratory wells, hot spring wells, and natural hot springs.

Figure 1. Map showing the location of the Aomori Prefecture, a northern end of Honshu Island, Japan. Red triangles show the main Quaternary volcanoes and calderas.

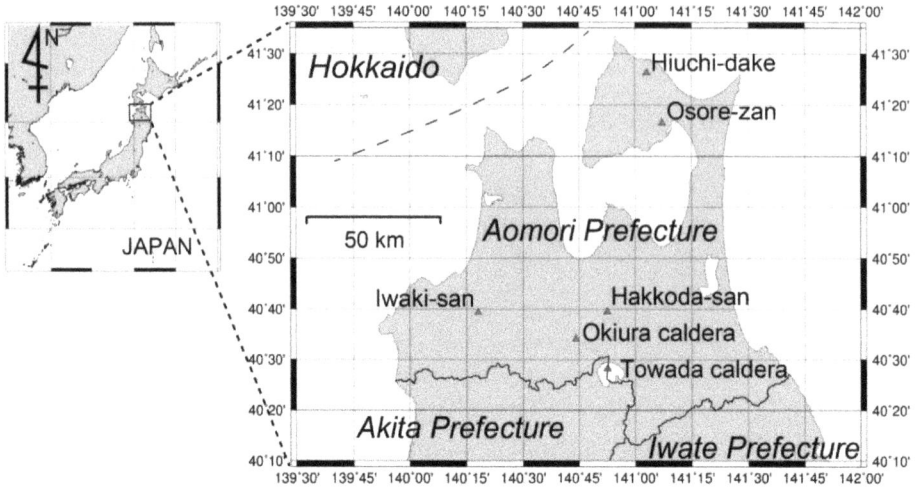

The neighboring Iwate Prefecture and Akita Prefecture have three installed geothermal power plant units, respectively, including the oldest Matsukawa plant in Japan that has been working for an almost half century since 1966, but unfortunately no geothermal power plants have been developed in Aomori Prefecture so far. The geothermal database we have set up, will assist geothermal development in Aomori Prefecture before detailed geothermal exploration surveys are carried out in individual geothermal fields.

The geothermal database can be applied to draw a variety of subsurface thermal structures. As the spatial density of the point data is still relatively limited, it will be adequate to apply the data to a broad but fundamental structure. One of the objectives would be to map the depth of the brittle-plastic transition because it is synonymous with the bottom of the permeability and comprises the broadest as well as most fundamental structure delineating a bottom of the entire hydrothermal convection sphere.

This paper describes a geothermal database in Aomori Prefecture and draws a map of the depth of the brittle-plastic transition based on some depth-temperature normalization techniques that allow three-dimensional extrapolation in up-flow zones. Then, this map of the depth of the brittle-plastic transition is verified by the bottom of seismicity which is far independent destruction phenomena from the thermal structure concerned. We used the Generic Mapping Tools 4.5.9 for drawing most of figures in this paper [2].

2. Geothermal Database

Our geothermal database consists of 35 points of geothermal exploratory well data, 510 points of hot spring well data and 26 points of natural hot spring data. Our geothermal database was based on two main data sources, both of which are open to the public. One is *"Record of Hot Spring Geology in Aomori Prefecture"* published by the Department of Health and Welfare, Aomori

Prefectural Office in 1997 [3]. The other is *"Atlas of Hydrothermal Systems in Japan"* published from the Geological Survey of Japan, the National Institute of Advanced Industrial Science and Technology (AIST) in 2007 [4].

The *"Record of Hot Spring Geology in Aomori Prefecture"* describes 434 points of hot spring data in Aomori Prefecture by their locations, well depths, discharge temperatures, discharge rates, major chemical constituents and geological columns along the wells. Among them 26 hot springs are natural hot springs with no artificial drilling. We used the 26 points of natural hot spring data and 395 points of hot spring data developed by wells of which both the well depth and discharge temperature are known. We digitized all the paper-based data from the report.

The *"Atlas of Hydrothermal Systems in Japan"* describes 7203 points of hot spring data and 3066 well data points from all over Japan for drawing a variety of geothermal maps. We used 115 points of hot spring data and 35 points of geothermal exploratory well data from Aomori Prefecture. They were also originally paper-based data, but a digital-version on CD-ROM was published later [5].

3. Estimate of Subsurface Geothermal Structure

Determining accurate subsurface three-dimensional thermal structure is one of the ultimate goals on geothermal exploration. However, depth-temperature curves obtained by temperature logging in geothermal wells show usually complicated patterns due to the permeability variation of given geological formations with depth and three-dimensional geothermal structure that cannot easily be acquired. Then, normalization techniques for depth-temperature curves are necessary. One of the normalization techniques is the Activity Index originally proposed for evaluating temperature ranks of geothermal fields by Hayashi [6].

Geothermal exploration mainly concerns up-flow zones of the hydrothermal convection system rather than down-flow zones. The highest temperature curves in up-flow zones are normally limited by the boiling point temperature curve [7]. The lowest temperature curves are normally limited by the linear thermal conduction curve with an average continental geothermal gradient 30 °C/km. The Activity Index (*AI*) is defined by the following equation:

$$AI = \frac{a}{b} \times 100 = \left(1 - \frac{Tb - Tm}{Tb - Tg}\right) \times 100 \tag{1}$$

where Tm is the maximum temperature at the observed depth, Tb is the boiling point temperature curve of pure water at the observed depth, and Tg is an average geothermal gradient (30 °C/km) at the observed depth. In addition, a and b show the $Tm - Tg$ and $Tb - Tg$ respectively. The concept of the Activity Index is graphically represented (Figure 2).

Figure 2. A concept of the Activity Index. Six solid curves show $AI = 0$, $AI = 20$, $AI = 40$, $AI = 60$, $AI = 80$ and $AI = 100$, respectively. Red circles show 571 points of thermal data in Aomori Prefecture.

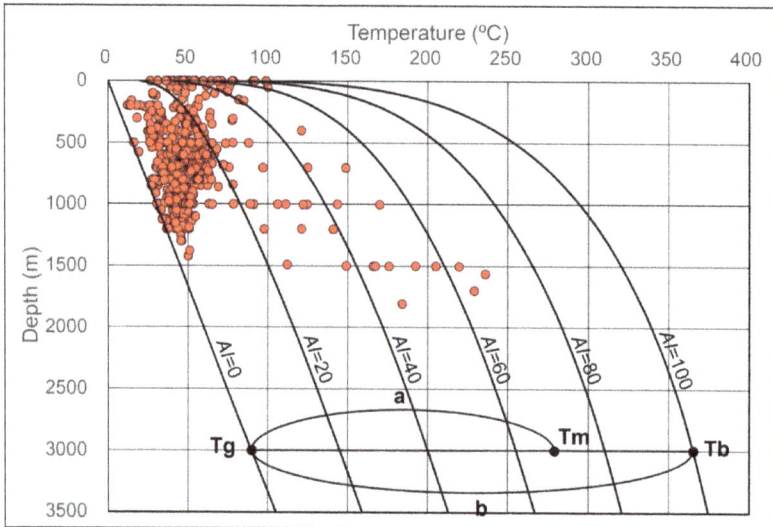

The AI 100 curve exactly coincides with the boiling point temperature curve of pure water and is here calculated in the approximation equation by Haas [7]. When the Activity Index can be determined at a given depth, we can extrapolate the normalized temperature curve to the arbitrary depth using the Equation (1) [6]. Therefore, we can easily plot the subsurface three-dimensional thermal structure by this normalization technique. When the temperature logging data are available, the borehole temperature is used as Tm at the given depth. However, only discharge temperature is available on most of hot spring wells. Tm was approximated by the discharge temperature on these hot springs. It is convenient that the discharge temperature in Celsius degree of natural hot springs at the ground surface can be adopted as the same numbers of AI as discussed by Hayashi [6]. As natural hot springs represent the surface manifestation of a hydrothermal up-flow, their temperatures tend to form a cluster around 100 °C. However, natural hot springs are only 26 pieces among the entire 536 hot springs in this database and no biases are found in the normalization by the Activity Index. Chemical geothermometry is useful to estimate reservoir temperatures, but we did not use the method because it does not provide the depth information. As a result, we can use 571 point data for AI mapping in Aomori Prefecture (Figure 2).

Pure water reaches a critical point temperature near the depth of 3500 m. However, the critical point of water dramatically shifts to the higher temperature with increasing salinity in the natural brine system so that we simply extrapolated the AI curves to the depth of 15 km. Then, we can plot the subsurface three-dimensional thermal structure of upper crust in Aomori Prefecture.

We here try to plot the brittle-plastic transition which is critically important to delineate the bottom surface of hydrothermal convection, particularly to extract the shallower apices of the isothermal surface associated with high-level magmatic intrusions. Deformed rocks within fault zones indicate that the transition from purely brittle to purely plastic deformation processes may occur over a relatively broad range of temperature and pressure [8]. This range is often denoted as a semi-brittle region between the brittle-ductile transition (BDT) and the brittle-plastic transition (BPT) [9]. For quartz and feldspar rich rocks in continental crust this range occurs at an approximate temperature in the range from 250 to 400 °C.

The temperature of the *in-situ* brittle-plastic transition in an actual borehole was too high to be measured by the conventional temperature logging tools but it was successfully measured by a variety of techniques in the well WD-1a which was drilled into the depth of 3729 m in the Kakkonda geothermal field, Iwate Prefecture, northern Japan in 1995 [10]. The bottom-hole temperature exceeded 500 °C (Figure 3) [10,11]. The temperature profile includes a temperature inflection point at 380 °C and at the depth of 3100 m where a boiling point curve above the depth shows a hydrothermal convection zone and a steep gradient thermal conduction curve below the depth shows a non-fracture and non-hydrothermal convection zones [10,11]. The temperature inflection point at 380 °C is thus considered to be the brittle-plastic transition (BPT) on granitic crust [12] (Figure 3). On the other hand, the maximum strength zone was estimated at the temperature 340 °C and at the depth of 2000 m by the high density of fractures, measurements of the differential strain curve analysis (DSCA) [13] on oriented cores and the theory of the strength envelope of lithosphere [12]. This maximum strength zone can be denoted as the brittle-ductile transition (BDT; Figure 3). The brittle-ductile transition (BDT) is important for a nest of earthquakes and fractures because of the preferred stress concentration on the maximum strength zone [12]. From a geothermal point of view, the brittle-plastic transition (BPT) is crucial because it is a bottom surface of seismicity, fracturing and hydrothermal convection.

Figure 3. Synoptic models of the brittle-plastic transition (BPT) and the brittle-ductile transition (BDT). (**a**) Depth-temperature profile along the well WD-1a at the Kakkonda geothermal field, Iwate, Japan [10,11]. (**b**) Depth-strength envelope along the well WD-1a at the Kakkonda geothermal field, Iwate, Japan [12]. λ shows pore pressure.

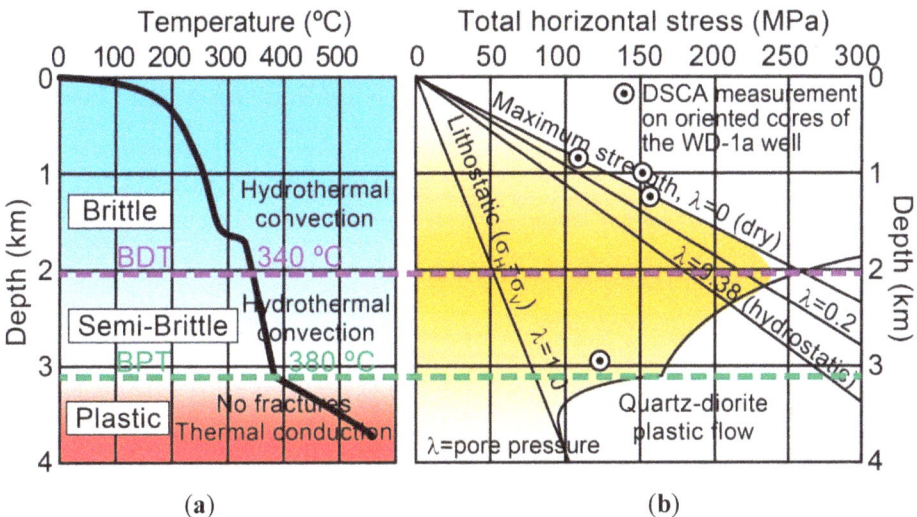

(a)   (b)

Thus, the most reliable temperature of the brittle-plastic transition is 380 °C on granitic crust [10–12], and we tried to draw the depth of the 380 °C isothermal surface from the Activity Index. Figure 4 shows the calculated results of the assumed brittle-plastic transition. Most of the depth of the brittle-plastic transition showed deeper than 12 km below sea level. However, the prominent shallow depth anomalies coincide with the Quaternary clustered volcanoes such as Hiuchi-dake and Osore-zan in the Shimokita Peninsula to the north and Hakkoda-san to the south. The map is consistent with known prospective geothermal fields.

Figure 4. Map of the depth of the 380 °C isotherm below sea level as a brittle-plastic transition in Aomori Prefecture, Japan. Solid dots show 571 points of thermal data in Aomori Prefecture.

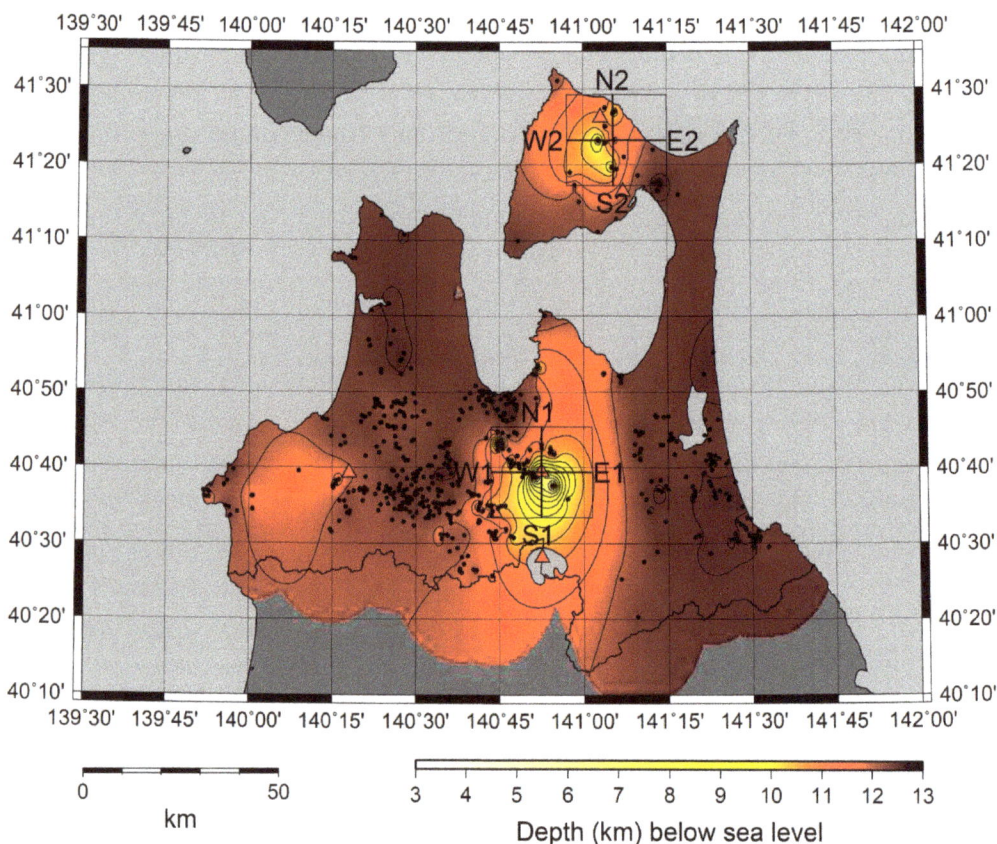

4. Discussion: Verification by Seismicity

Shallow-depth anomalies of the 380 °C isotherm are exactly consistent with the areas of the Quaternary volcanoes and their prospective geothermal fields (Figures 1 and 4). The shallowest apex is observed at a depth about 5 km in the Hakkoda-san clustered volcano area that seems not necessarily shallow as a magmatic heat source region. It is due to the regional-scale smooth contouring from the scarce random point data and 0.2 min (about 370 m) gridding. The map would provide the subsurface three-dimensional thermal structure and seems useful for subsurface geothermal assessments. However, further verification would be expected from a scientific point of view. To verify the reliability of the subsurface three-dimensional thermal structure on this map seismicity is useful because the seismic phenomenon is primarily independent from thermal phenomena.

We used the earthquake catalog provided by the Japan Meteorological Agency (JMA) [14]. The JMA earthquake data were collected in Aomori Prefecture during the period from January 1990 to July 2012. As these data include deep subduction zone earthquakes, we selected earthquake data with their epicenters to be shallower than the depth of 20 km. To verify the reliability of the map (Figure 4), the spatial distribution of the earthquakes is compared with the thermal structure (Figure 5). The thermal point data such as wells and hot springs are restricted on the land areas. Therefore, we selected earthquake data on the land area too.

Figure 5. Comparison between the 380 °C isotherms and seismicity on cross sections. (**a**) Hakkoda volcano area; the upper shows the east-west cross section and the lower shows the north-south cross section (Figure 4); (**b**) Shimokita Peninsula area; the upper shows the east-west cross section and the lower shows the north-south cross section (Figure 4). Red triangles show the main Quaternary volcanoes and calderas. Solid lines show the 380 °C isotherm and open circles show epicenters of earthquakes enlarging the diameter with magnitude during January 1990 to July 2012.

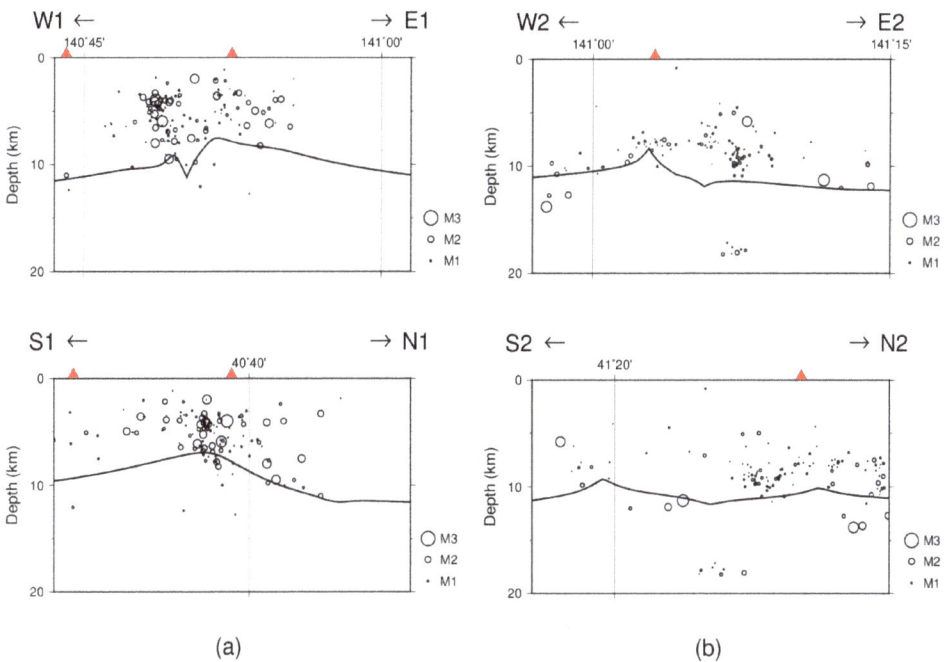

(a)   (b)

Figure 5 shows the estimated 380 °C isotherm produced from the Activity Index. Hakkoda volcano area and Shimokita Peninsula area show the prominent anomalies that are characterized by the dramatically shallower depth of the brittle-plastic transition than other areas. The bottom of the spatial distribution of seismicity shows the strong correlation to the 380 °C isotherm. This result indicates that the subsurface three-dimensional thermal map in Figure 4 is basically reliable. This result also suggests that the 380 °C isotherm strongly correlates to the bottom surface of seismicity, fracturing and hydrothermal convection. Geothermal database and the Activity Index are useful to draw the subsurface three-dimensional thermal structure.

5. Conclusions

Based on the geothermal database for available thermal data such as wells and hot springs in Aomori Prefecture, northern Honshu, Japan, depth-temperature curves were normalized by the Activity Index for three-dimensional extrapolation and a depth contour map of the 380 °C isotherm was plotted as an assumed brittle-plastic transition for granitic crust. Shallower-depth anomalies of the brittle-plastic transition on this map are closely coincident with the Quaternary volcanoes and their prospective geothermal fields. It should be noted that the bottom of the spatial distribution of seismicity in the volcanic fields shows strong correlation to the 380 °C isotherm. This result indicates reliability of the subsurface three-dimensional thermal map and suggests that the 380 °C isotherm strongly constrains the bottom surface of seismicity, fracturing and hydrothermal convection in granitic crust.

Acknowledgments

This work was supported by the Japan Society for the Promotion of Science (JSPS) Kakenhi (Grant-in-Aid for Scientific Research B) Grant Number 26281053. We thank the Japan Meteorological Agency (JMA) and the Ministry of Education, Culture, Sports, Science and Technology in Japan (MEXT) for providing the earthquake epicenter data. This paper benefited from the comments and suggestions of three anonymous reviewers.

Author Contributions

Yota Suzuki is a principal investigator of this work so that he has made the geothermal database in Aomori Prefecture, the first manuscript and all the figures. Seiichiro Ioka advised on hydrology and the data quality of the geothermal database. Hirofumi Muraoka co-operatively provided many papers on the geothermal database and the Kakkonda project.

Conflicts of Interest

The authors declare no conflict of interest.

References

1. Ministry of the Environment, Japan. The State of Use of Hot Springs in Japan in the Fiscal Year 2012. Available online: http://www.env.go.jp/nature/onsen/data/riyo_h24_1.pdf (accessed on 1 May 2014). (In Japanese)

2. Wessel, P.; Smith, W.H.F. New, improved version of the generic mapping tools released. *Eos Trans. Am. Geophys. Union* **1998**, *79*, 579.

3. *Record of Hot Spring Geology in Aomori Prefecture*; Department of Health and Welfare, Aomori Prefectural Office: Aomori, Japan, 1997. (In Japanese)

4. Muraoka, H.; Sakaguchi, K.; Tamanyu, S.; Sasaki, M.; Shigeno, H.; Mizugaki, K. *Atlas of Hydrothermal Systems in Japan*; Geological Survey of Japan, the National Institute of Advanced Industrial Science and Technology (AIST): Tsukuba, Japan, 2007. (In Japanese)

5. *Geothermal Potential Map in Japan*; Digital Geoscience Map GT-4 (CD-ROM); Geological Survey of Japan, the National Institute of Advanced Industrial Science and Technology (AIST): Tsukuba, Japan, 2009. (In Japanese)

6. Hayashi, M.; Taguchi, S.; Yamasaki, T. Activity index and thermal history of geothermal systems. *Geotherm. Resour. Counc. Trans.* **1981**, *5*, 177–180.

7. Haas, J.L., Jr. The effect of salinity on the maximum thermal gradient of a hydrothermal system at hydrostatic pressure. *Econ. Geol.* **1971**, *66*, 940–946.

8. Sibson, R.H. Fault rocks and fault mechanisms. *J. Geol. Soc. Lond.* **1977**, *133*, 191–213.

9. Kohlstedt, D.L.; Evans, B.; Mackwell, S.J. Strength of the lithosphere: Constraints imposed by laboratory experiments. *J. Geophys. Res.* **1995**, *100*, 17587–17602.

10. Ikeuchi, K.; Doi, N.; Sakagawa, Y.; Kamenosono, H.; Uchida, T. High-temperature measurements in well WD-1a and the thermal structure of the Kakkonda geothermal system, Japan. *Geothermics* **1998**, *27*, 591–607.

11. Muraoka, H.; Uchida, T.; Sasada, M.; Yagi, M.; Akaku, K.; Sasaki, M.; Yasukawa, K.; Miyazaki, S.; Doi, N.; Saito, S.; *et al.* Deep geothermal resources survey program: Igneous, metamorphic and hydrothermal processes in a well encountering 500 °C at 3729 m depth, Kakkonda, Japan. *Geothermics* **1998**, *27*, 507–534.

12. Muraoka, H. Geothermal Energy. In *Handbook of Climate Change Mitigation*; Chen, W., Seiner, J., Suzuki, T., Lackner, M., Eds.; Springer: Berlin, Germany, 2012; pp. 1325–1353.

13. Strickland, F.G.; Ren, N.K. Use of differential strain curve analysis in predicting *in situ* stress state for deep wells. In Proceedings of the 21st U.S. Symposium of Rock Mechanics, Rolla, MI, USA, 27–30 May 1980; pp. 523–532.

14. Okada, Y.; Kasahara, K.; Hori, S.; Obara, K.; Sekiguchi, S.; Fujiwara, H.; Yamamoto, A. Recent Progress of Seismic Observation Networks in Japan—Hi-Net, F-Net, K-NET and KiK-net. *Earth Planets Space* **2004**, *56*, 15–28.

3D Geothermal Modelling of the Mount Amiata Hydrothermal System in Italy

Paolo Fulignati, Paola Marianelli, Alessandro Sbrana and Valentina Ciani

Abstract: In this paper we build a subsurface model that helps in visualizing and understanding the structural framework, geology and their interactions with the Mt. Amiata geothermal system. Modelling in 3D provides the possibility to interpolate the geometry of structures and is an effective way of understanding geological features. The 3D modelling approach appears to be crucial for further progress in the reconstruction of the assessment of the geothermal model of Mt. Amiata. Furthermore, this model is used as the basis of a 3D numerical thermo-fluid-dynamic model of the existing reservoir(s). The integration between borehole data and numerical modelling results allows reconstructing the temperature distribution in the subsoil of the Mt. Amiata area.

Reprinted from *Energies*. Cite as: Fulignati, P.; Marianelli, P.; Sbrana, A.; Ciani, V. 3D Geothermal Modelling of the Mount Amiata Hydrothermal System in Italy. *Energies* **2014**, *7*, 7434-7453.

1. Introduction

The Mt. Amiata volcano-geothermal area, located in Southern Tuscany (Figure 1), is characterized by very high heat flux (up to 600 mW/m^2) and hosts two high-temperature geothermal fields (Bagnore and Piancastagnaio) that, together with Larderello-Travale, are the only economically exploited fields in Italy for power production.

Mt. Amiata is the largest Tuscan volcano, consisting of dacitic, rhyodacitic and minor olivine-latitic lavas that erupted in a period ranging from 300 to 190 ka [1,2]. It is located on a structural high delimitated to the East by the normal (border) faults of the western side of Siena-Radicofani Neogene basin. Mt. Amiata volcanic products are mainly effusive and of silicic composition. Volcanics cover an area of about 90 km^2 (Figure 1) lying on "Tuscan" units (metamorphic rocks of Palaeozoic age forming the basement, Triassic-Jurassic carbonate evaporitic successions and Cretaceous-Oligocene terrigenous formations), Oligocene-Mesozoic Ligurian units (shale, sandstones and marles), Palaeogene Subligurian units and Neogene sediments [3–8]. The volcanic activity was triggered by the activation of regional transcurrent faults following the Mio-Pliocene extensional phase that induce an important crustal thinning. This latter and the associated tectonic favoured the emplacement of a magma body at 6–7 km depth [9–11]. The relatively shallow intrusion of low-density magma was at the beginning of the uplift (up to 1000 m) of this area during the Quaternary [6,11]. The main fault system in the Mt. Amiata volcanic complex has an ENE-WSW direction. This direction represents the main fissure system along which magma raised from the shallow magma chamber. Nearly all the lava domes are aligned along this axial fissure system [1] known as the Mt. Amiata fault (see the partly coinciding location of faults and eruptive centres in Figure 1). Other transcurrent fault systems occur North and South of the volcano. The active geothermal fields as well as Hg-(Sb) mineralization of Mt. Amiata area are linked to these fractured systems and to the thermal anomaly induced by the shallow intrusion (Figure 1).

Figure 1. Geological sketch map of the area of Mt. Amiata and its surroundings [8]. The whole area coincides with the *X-Y* extents of the 3D geological model; the rectangle in red coincides with the *X-Y* extents of the numerical model.

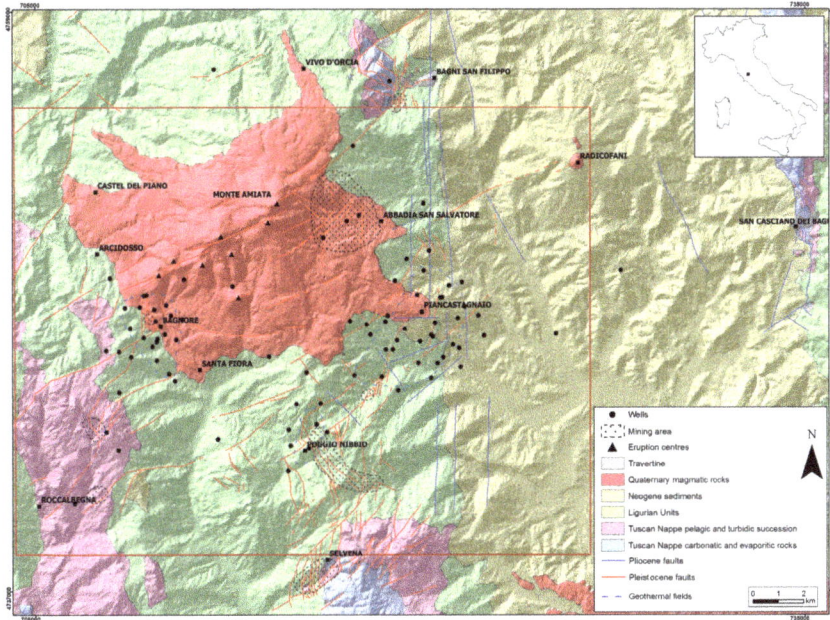

Geothermal exploration of the Mt. Amiata area started at the beginning of the 50s, and the first overview of the geothermal system was carried out in 1970 [3]. Previous studies were developed for mining activities linked to cinnabar production [12]. All available data (more than 100 drillings from 500 to about 4000 m depth, as well as many geophysical, geological, structural, geochemical and hydrogeological studies), derived from the exploration and exploitation activities mainly carried out for geothermal purposes, allow developing a well-constrained geothermal model.

The goal of this paper is to build a fit-for-purpose 3D geological model of Mt. Amiata geothermal system. Modelling in 3D provides the possibility to interpolate the geometry of geological structures and is an effective way of understanding the subsoil geology. As a consequence, a 3D modelling approach appears to be crucial to make further progress in the understanding of the geothermal-geological system of Mt. Amiata. Furthermore, this geological model is used as the basis of a 3D numerical thermo-fluid-dynamic simulation of the volcanic geothermal system that led to an integrated geothermal model of the area.

2. Mt. Amiata Geothermal System

Mt. Amiata is a classic volcanic-intrusive geothermal system. This is characterized by clay-rich sedimentary units (Ligurian alloctonous nappes and Neogene units) that represent a very efficient impermeable cover. Two confined water-dominated geothermal reservoirs occur. The shallower one (500–1500 m of depth) is hosted in carbonate-evaporitic rocks belonging to the Tuscan Nappe, in particular in the "Calcare Cavernoso" formation and locally in fractured carbonate units. Calcare

Cavernoso is a vacuolar carbonate breccia characterized by dolostone porphyroclasts (often dissolved) cemented by carbonate cement resulting from alteration of gypsum evaporite layers. This formation is often very permeable due to fracturing and porosity. The mean temperatures found in this reservoir range from 150 to 220 °C. The reservoir hosts a two-phase water dominated aquifer. In correspondence of structural highs (traps) this reservoir presents pressurized gaseous caps in which gas and steam accumulate [13]. The gaseous phase is mainly constituted by CO_2, in accordance with the very high CO_2 degassing that characterizes the whole Mt. Amiata area [14,15]. Shallow reservoir outcrops in two different areas (North East and South of Mt. Amiata). These probably represent the main recharge areas of the geothermal fields.

The deeper reservoir (below 2500–3000 m of depth) is hosted in thermometamorphic fractured phyllites and metasandstones of the Tuscan Metamorphic Complex, modified by thermometamorphic processes [16]. These belong to the termometamorphic aureole of the recent silicic intrusion. In the deeper reservoir temperatures up to 300–350 °C are actually measured. The fluid in this reservoir is a two-phase too [3,10]. This represents the main reservoir industrially exploited at present [13] with a total power production of 88 MW_e.

Below the deeper reservoir, reflection seismic profiles reveal the occurrence of a geopressurized system evidenced by discontinuous "bright spot"-type reflections, named K-horizon [17–21], at depth between 4 and 7 km. This is assumed to be the root of the hydrothermal systems in the Mt. Amiata area. The temperatures estimated for the K-horizon are considered to be above 400 °C [17,22]. This horizon is present also outside the geothermal region but at much greater depth [18]. The interpretation of K horizon is still debated: some authors [18,21] suggested, basing on the earthquake hypocentres distribution, that the K-horizon corresponds to a ductile-brittle boundary; others [22] suggested that this ductile-brittle transition occurs in coincidence with high fluid pressure; others [23] explain the K-horizon as the roof of over-pressurized reservoirs on the basis of the interpretation of deep crustal seismic lines.

The Mt. Amiata intrusion, present at about 6 km of depth, represents the heat source of the hydrothermal system described above.

3. 3D Geological and Geothermal Model

3.1. Available Data

Geological models in three dimensions integrate different kinds of data (*i.e.*, geological maps, cross-sections, seismic profiles, gravity profiles, logs of deep wells *etc.*) much more easily than traditional techniques and allow the three-dimensional (3D) structure to be represented more coherently and exhaustively [24–26]. The available data implemented in this work are reported as follow:

- Geological maps and cross-sections
- Data from boreholes
- Data from geophysical investigation

The geological maps used in this work derive from the recent 1:10,000 maps of the CARG (CARtografia Geologica) project [27]. The cross sections are taken from CARG project and from literature [3,4,6,8,27–29].

More than 100 deep geothermal boreholes (from 153 to 4836 m of depth) occur (UNMIG, Ufficio Nazionale Minerario per gli Idrocarburi e le Georisorse geothermal database) in the Mt. Amiata area. Geothermal wells (Figures 1 and 2) are mostly located in correspondence with the active geothermal fields of Bagnore and Piancastagnaio. Well data concern stratigraphy, temperatures, pressure of the fluid phase, flow rates, technical drilling profiles, depth of the water table. Unfortunately no geophysical borehole logs are available, with exception of temperature logs, because in some cases they were not carried out and in others the data were not published by the owner companies. Many other shallower wells (from 100 to 750 m of depth) were drilled in the past for mining activities. All data concerning these wells were also used in this work.

Data from geophysical investigation, derived from past geothermal and hydrocarbon (this latter in the Radicofani graben) exploration, are available (Figure 3). The geophysical data used in this work are: the map of the resistive substratum [3] and seismic profiles [4,20,29,30]. Regional gravity data were also incorporated in the study [31–33].

3.2. Data Integration Toward a 3D Geological Model

All data were implemented in the 3D modelling Petrel 2011 platform (Schlumberger Limited, Houston, TX, USA). Data include both hard copies (digitalized and georeferenced) and GIS (geographical information system) format data. Geological data were elaborated in the geodatabase. Geological units were grouped on the basis of their "geothermal" characteristics, essentially in terms of cover type and reservoir type formations, taking into account their permeability (Figure 4).

Six units were derived as follow: (1) the volcanic complex: fractured and permeable lavas; (2) the Neogene and Quaternary mainly impermeable sedimentary deposits: clay, sands and minor conglomerate; (3) the Ligurian units mainly impermeable: shale, sandstones and marls; (4–5) the Tuscan Nappe, subdivided into two sub-units: (4) an upper sub-unit denoted as TN 2, represented by the mainly impermeable terrigenous successions and (5) a lower sub-unit that is denoted as TN 1 and represented by the fractured and permeable carbonate-evaporitic successions (shallow geothermal reservoir); (6) the Palaeozoic basement, formed by the Tuscan Metamorphic Complex constituted by low permeability phyllites and permeable thermometamorphosed fractured phyllites (this latter is the deep geothermal reservoir). The geological map of the area (Figures 1 and 2) and the geological cross sections (Figure 5) [3,4] were elaborated on the basis of the above reported criteria, digitalized and uploaded in the model.

Figure 2. 3D view of geological map spread on DEM (digital elevation model) of the Mt. Amiata area (vertical dimension is exaggerated by factor of 2). Z-axis in meters, X-axis and Y-axis: the coordinate system used is WGS 84/UTM zone 32N (EPSG: 32632). Arrow points to the North. Legend as reported in Figure 1.

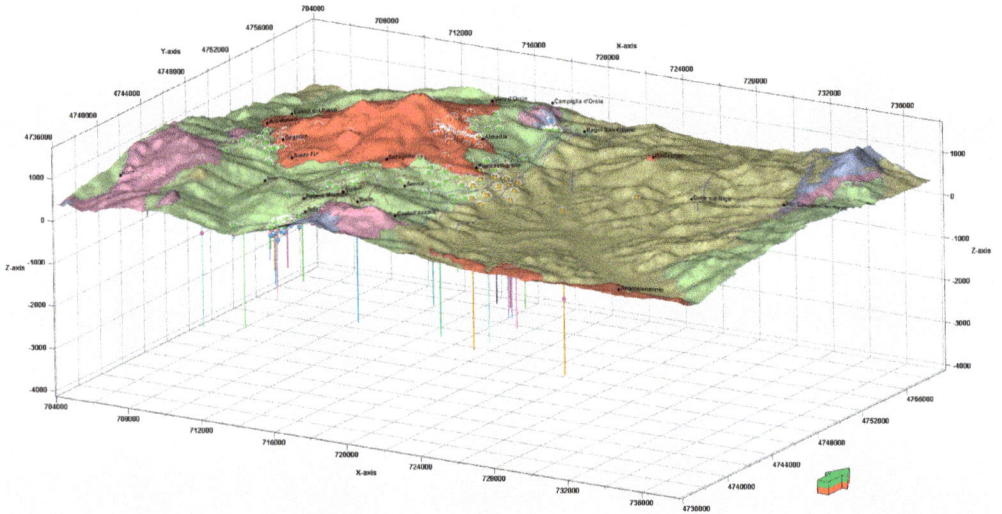

Figure 3. Map of the Mt. Amiata area with the location of the deep boreholes and the traces of geological, seismic and gravimetric sections.

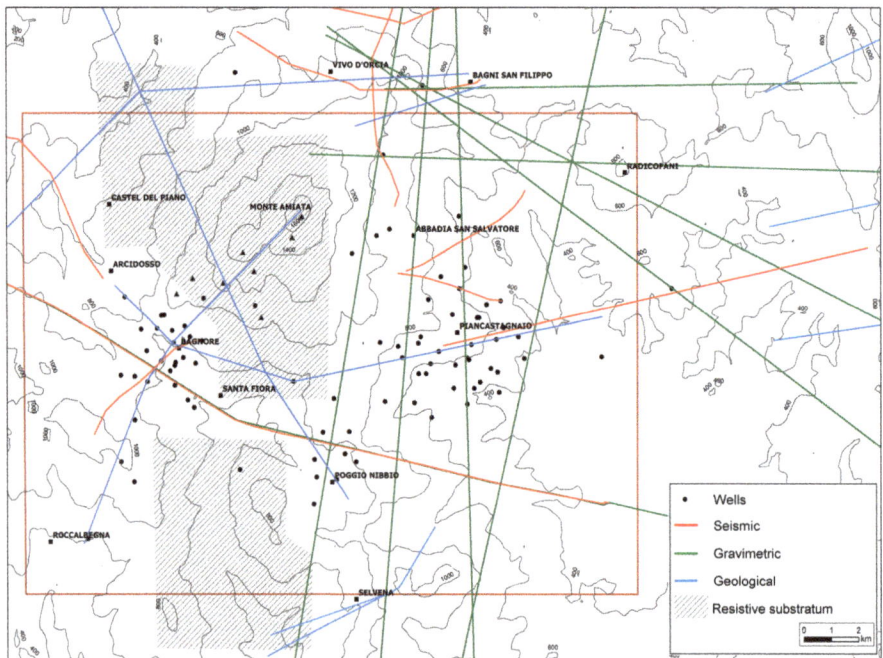

Figure 4. Sketch of the geological units used in this work (right) and the complete geological succession taken from literature. Stratigraphic succession in the Monte Amiata area (modified after [34,35]: see the references for further information). Ligurian and Subligurian Units; Ophiolitic Unit; APA: "Argille a Palombini" Fm.; CCL: "Calcari a calpionelle" Fm.; DSA: "Diaspri" Fm.; PRN, GBB, BRG: Ophiolites (serpentinites, gabbros, basalts); "S. Fiora" Unit; MLL: "Monte Morello" Fm.; AVR: "Santa Fiora" Fm.; PTF: "Pietraforte" Fm.; Subligurian Unit; ACC: "Canetolo" Fm.; Tuscan Nappe: MAC: "Macigno" Fm.; STO: "Scaglia Toscana" Fm.; MAI: "Maiolica" Fm.; DSD: "Diaspri" Fm.; POD: "Marne a Posidonia" Fm.; LIM: "Calcare Selcifero" Fm.; RSA: "Calcare Rosso Ammonitico" Fm.; MAS: "Calcare Massiccio" Fm.; RET: "Calcare a Rhaetavicula contorta" Fm.; CCA: "Burano and Calcare cavernoso" Fm.; Tuscan metamorphic units; "Monticiano-Roccastrada" Unit (MRU); MRU3: "Verrucano" Group; MRU2: Palaeozoic Phyllite-Quartzite Group.

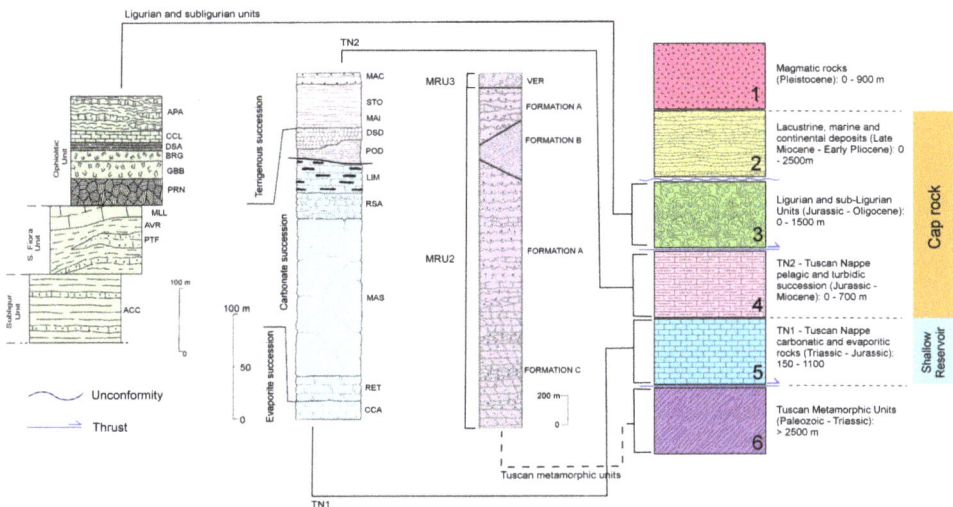

Also the well stratigraphy was generalized on the basis of the simplified stratigraphy described above (see Figure 4 right). As regard the data from geophysical investigations, they were implemented in the geodatabase as follows:

Interpreted 2D reflection seismic data were added to geodatabase and used for constraining geometries and characteristics of the subsoil units. In Figure 3 traces of interpreted seismic lines, derived from [4–6,10,13,19–22,29,30], are reported. In particular, seismic lines define the structural setting of the main geological units, evidencing reverse faults (main thrusts) involved in Miocene compressive phase of North Apennine chain formation, normal faults formed during the successive Pliocene-Pleistocene tensional phase.

Geoelectrical data, covering wide areas of the volcano and carried out during the early exploration phase of the Mt Amiata geothermal fields [3], are used to define geometries and thickness of resistive substratum, coinciding with carbonate and evaporite formations (shallow geothermal reservoir).

Figure 5. Fence diagram built from geological cross sections (vertical dimension is exaggerated by factor of 2). Z-axis in meters; X-axis and Y-axis: the coordinate system used is WGS 84/UTM zone 32N (EPSG: 32632). Arrow points to the North.

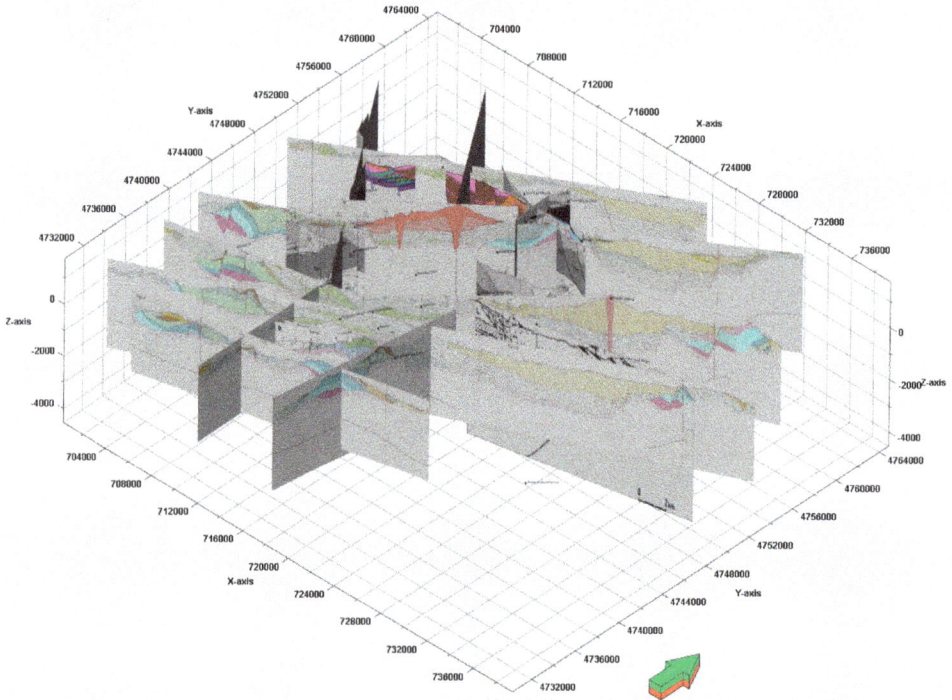

Gravity data reported in the Bouger anomaly map were elaborated through 2.5 D numerical modelling and implemented in geodatabase to constrain geometries and thickness of subsoil formations, including deep intrusive low-density bodies and thickness of Neogene sedimentary basin. All geophysical interpretations and modelling were carefully calibrated using deep and shallow wells, previously implemented in our geodatabase.

The 3D geological model is 28 km × 20 km wide, with its longer wide oriented EW, and has a vertical extension of approximately 4 km. A 3D grid, which forms the skeleton of the geological model, was then generated and used for the property modelling stage, in which 3D surface (such as lithology, resistivity and temperature) were created.

The main horizons, corresponding to the top of the simplified geological units discussed above, were obtained by using the "convergent interpolation algorithm" built in Petrel platform (*i.e.*, see Figure 6).

Figure 6. 3D view of geological map spread on DEM and of the horizon corresponding to the top of the first geothermal reservoir in the Mt Amiata area (vertical dimension is exaggerated by factor of 2). Z-axis in meters; X-axis and Y-axis: the coordinate system used is WGS 84/UTM zone 32N (EPSG: 32632). Arrow points to the North. Legend as reported in Figures 1 and 4.

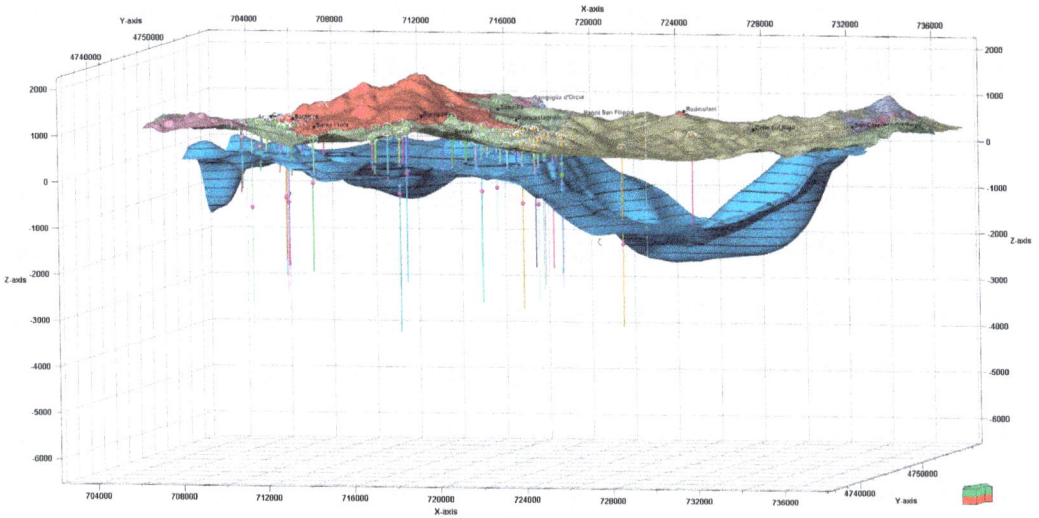

Shallower horizons are more accurately located and modelled because more wells encounter them. The reconstruction of deeper horizons is more problematic because there are fewer data; obviously, the horizons are more accurately modelled in the areas of Bagnore and Piancastagnaio active geothermal fields where density of deep wells and geophysical data is considerably higher compared to other areas (Figures 1, 2 and 6). The outputs of the 3D model contain information about depths of stratigraphic units and other geological information about faults, erosive surfaces and geothermal information such as reservoir volumes, used for geothermal potential assessment, and to constrain the conceptual model of the geothermal system. In correspondence of the Mt. Amiata volcano, geoelectric survey data were used to reconstruct surfaces between cover and reservoir units. Seismic reflection lines and gravimetric modelling are used to constrain the surface shapes where deep wells are not present. The top surface of Ligurian cover unit is well constrained in correspondence of geothermal fields (Figure 7) where stratigraphy of deep boreholes occurs. Particular attention was also dedicated to the reconstruction of surfaces delimitating the shallow reservoir and the metamorphic basement (Figure 7). The abrupt deepening of all surfaces toward East reflects the occurrence of the normal master fault, delimitating the Western side of Radicofani Pliocene sedimentary basin (Figures 1, 6 and 7). In this particular situation low angle normal fault offset reaches 200–400 m. Radicofani basin (eastward) have about 2400 m of sediments in its depocentre, as indicates by deep wells.

Figure 7. 3D subsurface horizons of the study area that are incorporated in the model, as reconstructed from available surface and subsurface data (vertical dimension is exaggerated by factor of 2). In figure three of these modelled horizons are shown: (**a**) green = top of cover formations (Ligurian units); (**b**) Blue = top of shallow geothermal reservoir; (**c**) Purple = top of metamorphic basement (see Figure 4 for details).

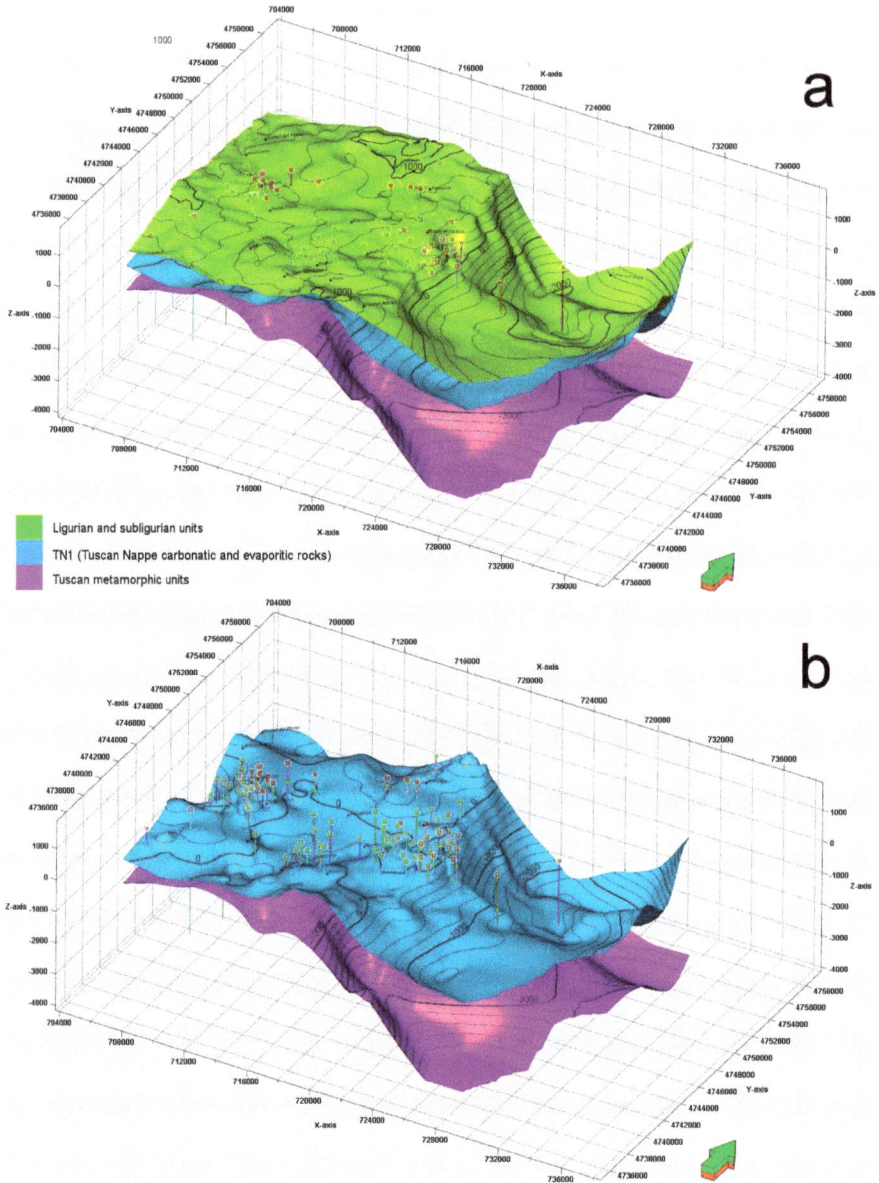

Figure 7. *Cont.*

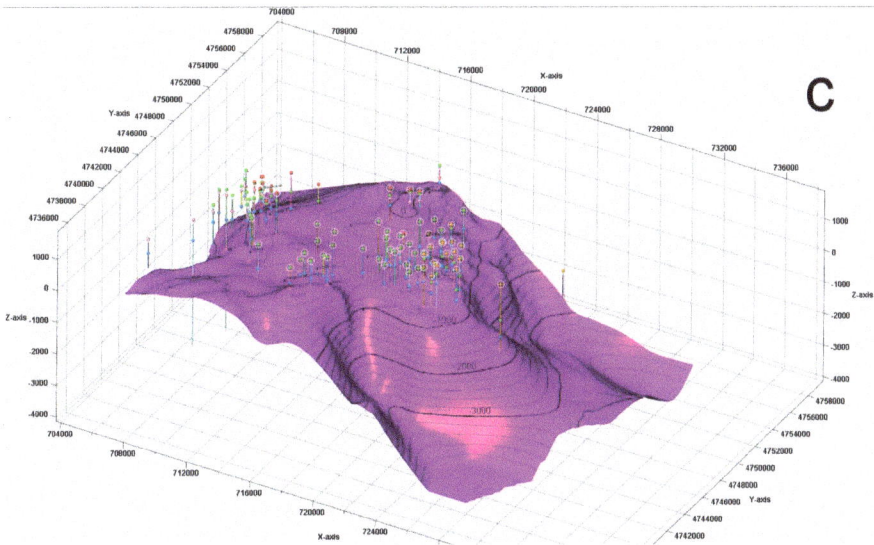

4. Numerical Modelling

The output of the 3D geological modelling was used as input data for the numerical simulation of Mt. Amiata hydrothermal system. The geometry of the top of deeper geothermal reservoir, which is hosted within the Palaeozoic phylladic basement (unit 6 of Figure 4), was defined on the basis of the boreholes that encountered it. No data constrain the base of this reservoir; therefore in the numerical model we assumed a thickness of 500 m.

The Mt. Amiata hydrothermal system was simulated using the numerical simulator TOUGH2 (Lawrence Livermore Laboratories, Berkeley, CA, USA), a general-purpose code for modelling multi-dimensional, multiphase/multi-component flow and heat transport in porous and fractured media [36]. The modelling was performed using the "equation of state" module (EOS1) for pure water. Corey curves were adopted for the relative permeability. Capillarity, adsorption, and double porosity were not considered. The simulation domain has an area of 23 km × 17 km (391 km², Figure 1) and a depth of 5500 m below sea level (*i.e.*, the model height ranges from 7160 to 5750 m, considering local topographic relief). The model is centred on the active geothermal fields where available data allow constraining a reliable physical model. The numerical grid is subdivided into 22 layers and consists of 8228 cells (Figure 8); each horizontal layer consists of 374 (22 × 17) cells with variable sizes. A cell size of 1 km × 1 km is used. Along the domain boundary, where great detail is not necessary, cell size of 2 km × 2 km has been adopted. The layer thickness varies from a maximum of a few thousand meters to only 60 m as a function of the horizons derived from 3D geological modelling.

The physical parameters used in the model (permeability, porosity, density, thermal conductivity under saturated condition, specific heat) are listed in Table 1 and based on literature values [37–40].

Figure 8. 3D simulation grid. Cap rock corresponds to units 2, 3, 4 of Figure 4; Shallow reservoir corresponds to unit 5 of Figure 4; Phylladic basement, Deep reservoir and Basement rock correspond to unit 6 of Figure 4.

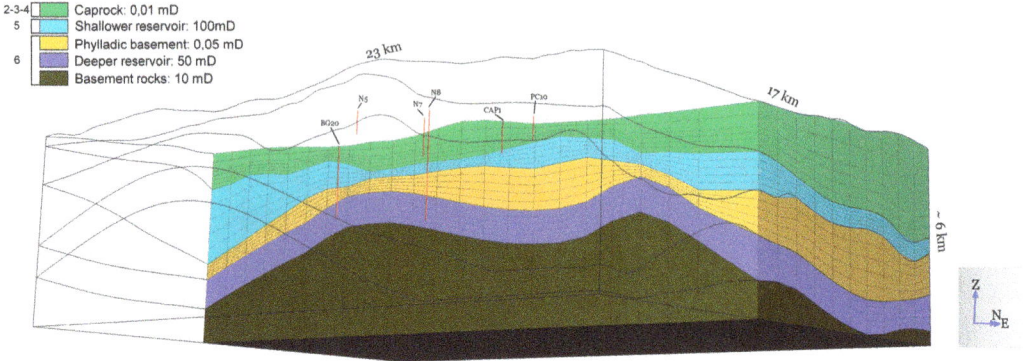

Table 1. Petrophysical parameters of main rock types.

Rock types	Density (Kg/m³)	Porosity	Permeability (Darcy)	Thermal Cond. (W/m * K)	Specific Heat (J/kg * K)
Cap rock	2400	0.0055	0.00001	2.4	883
Shallower reservoir (carbonates)	2660	0.02	0.1	2.4	836
Phylladic basement	2570	0.01	0.00005	4.0	1000
Deeper reservoir (basement)	2570	0.013	0.05	4.0	1000
Basement	2570	0.013	0.005	4.0	1000

Temperature and pressure are considered time invariant in the cells at the top and bottom of the simulation grid. A fixed temperature of 10 °C and atmospheric pressure were set for cells along the upper boundary. A variable temperature between 400 and 500 °C was assumed for the cells in the bottom layer, in accordance with the temperatures assumed for K-horizon. The lateral boundaries are considered no-flow insulating boundaries. The simulation of the natural state of the geothermal system (steady state) covers a period of 200,000 years but in some simulations the steady state was reached before. During the simulation some parameters (mainly permeability and temperatures at the bottom layer) were tuned to match the temperature distribution at depth. To verify the reliability of the simulated values of temperatures, simulated well temperature profiles were compared with temperature profiles of wells drilled in the area. A good match between simulated and measured temperature in deep wells was obtained (Figure 9), suggesting that a satisfactory simulation was achieved. The shape of the profiles from temperature modelling is linear or stepped. A vertical shape of the profiles indicates the occurrence of a convective heat transfer (reservoir zones), whereas an oblique shape is suggestive of a mainly conductive heat transfer (cover and impermeable units). The simulated temperature profiles of PC 30, BAGN 13 bis and BAGN 3 bis deep boreholes are indicative on the occurrence of both geothermal reservoirs of Mt. Amiata area (deep and shallow) and this is in agreement with the well logs. The simulated temperature profile of Nibbio 8 borehole is in good agreement with the measured temperatures at the level of the deeper reservoir whereas it does not agree at the level of the shallower one. The simulated profile suggests a convective heat transfer whereas the measured temperatures indicate a mainly conductive heat transfer. This disagreement is probably due to scarce permeability of the carbonate units of the Tuscan Nappe in the area of Nibbio 8 well; this would locally prevent the efficient hydrothermal fluid circulation and the consequent convective heat transfer. Capannacce 1 and PC 24 boreholes reach only the shallower geothermal reservoir.

Figure 9. Simulated and observed temperature vertical profiles along hole for selected deep geothermal wells in Bagnore and Piancastagnaio active geothermal fields of the Mt. Amiata area. On the left of each temperature profile the simplify stratigraphy (see Figure 4) of the well is reported.

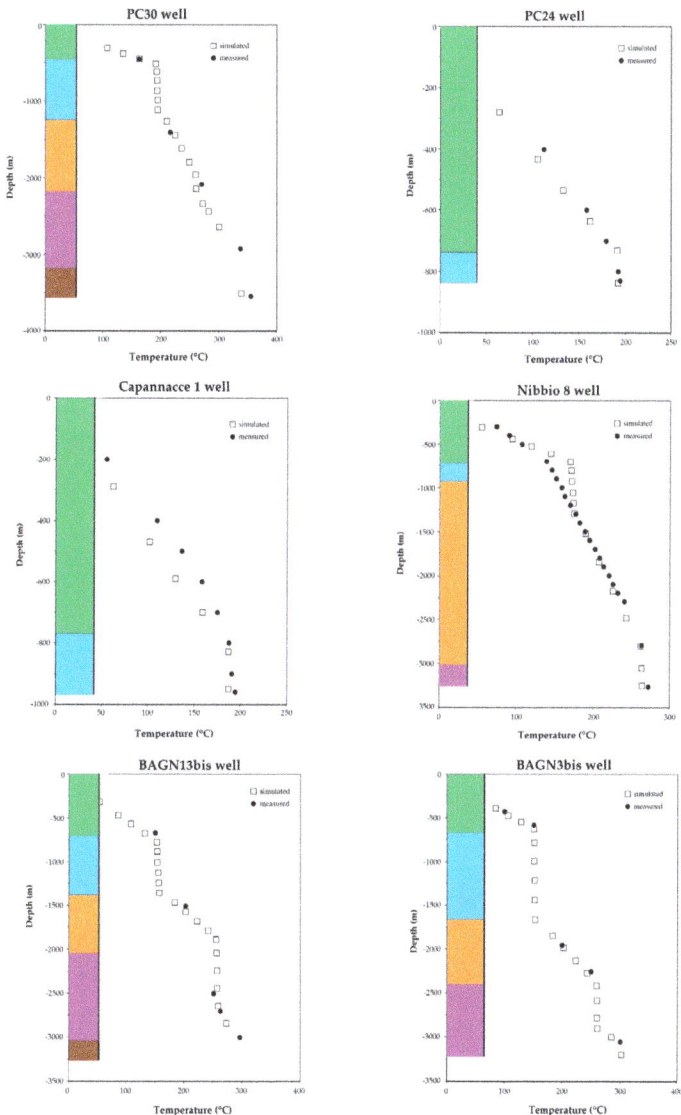

The main results of numerical modelling are shown in Figure 10. At the level of first geothermal reservoir (0 m sea level), numerical modelling results evidence the occurrence of three high temperature zones centred in the Piancastagnaio (East), Bagnore (West) and Nibbio (South-East) geothermal areas. The higher temperatures observed in the Piancastagnaio area (East) are due to the normal fault system that characterizes the Eastern sector of Mt. Amiata. On the other hand, the highest temperature and highest productivity of the deep boreholes recorded in the area are found in the Piancastagnaio geothermal field. At level of the second and deeper geothermal reservoir (−2000, −3000 m sea level) the only two significant high temperature zones occur in correspondence of the Piancastagnaio and Bagnore geothermal fields (Figure 10). This agrees with the temperature measured in deep geothermal boreholes.

Figure 10. Thermal state of Mt. Amiata area at 0 m sea level (sl) (**a**), −1000 m sl (**b**), −2000 m sl (**c**), −3000 m sl (**d**) respectively, as deduced from numerical modelling. Dotted circles indicates the two active industrially exploited geothermal fields of Bagnore (left) and Piancastagnaio (right).

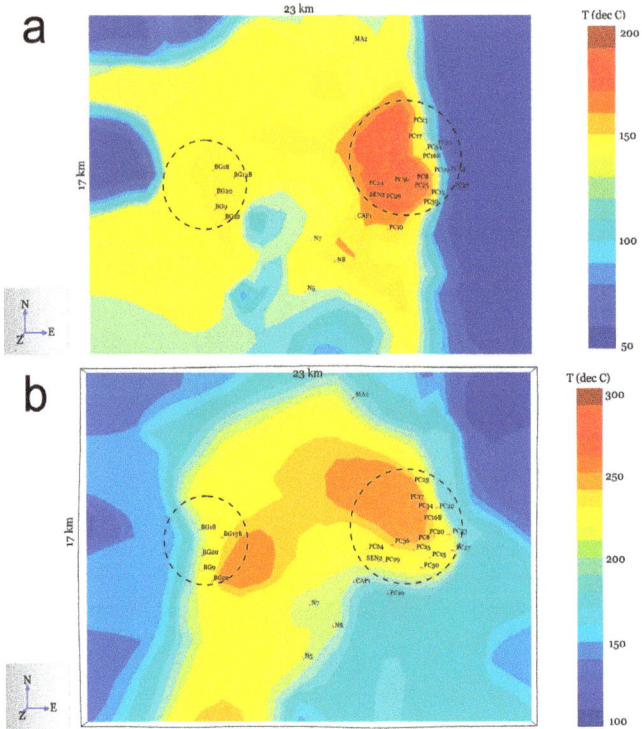

Figure 10. *Cont.*

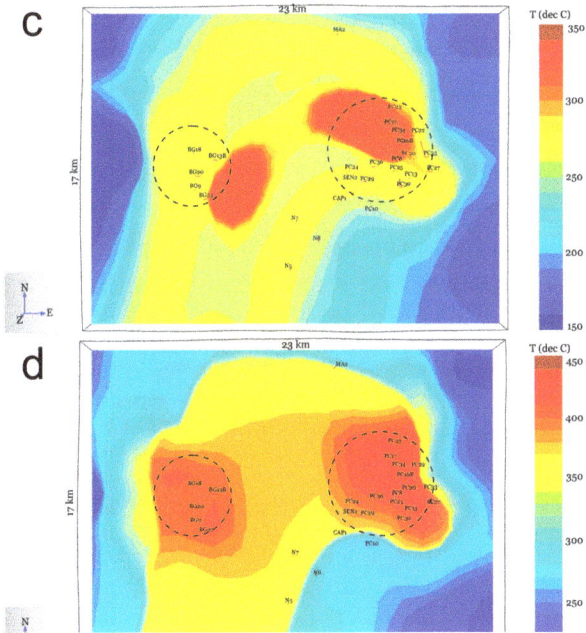

5. Discussion and Conclusions

The use of a 3D subsurface modelling represents an effective aid to better comprehend the geologic framework of the Mt. Amiata geothermal system. After combining the information, we received a 3D geological/stratigraphic model and outlined a conceptual model of the geothermal system of Mt. Amiata area.

The Mt. Amiata geothermal system is formed by two distinct water dominated reservoirs [3,13,41]. The shallower one is hosted in the Mesozoic, carbonate-evaporitic formations of the Tuscan Nappe at depths ranging between 500–1500 m, the second, and deeper one in the Palaeozoic, metamorphic basement at depths below 2500–3000 m. These two reservoirs are characterized by different temperatures (150–220 °C for the shallower one, 300–350 °C for the deeper one) and, although they are in hydrostatic equilibrium, they are separated by a low permeable layer as evidenced by the temperature profiles of deep wells (Figure 9).

3D integrated modelling allows the reconstruction of the shallow geothermal reservoir geometries (Figure 7). Figure 7a evidences the irregular shape of the top surface of Ligurian cover units and highlights the sharp deepening of this surface that occurs in correspondence of the western master fault of the Radicofani basin. In Figure 7b the top surface of carbonate fractured shallow reservoir is shown. This has a complex geometry due to the tectonic framework of the area, particularly characterized by an important uplift caused by the emplacement (between 130 and 300 ka) of the low-density intrusion that constituted the volcano feeding system. In the northern side of Mt Amiata, the reservoir outcrops in Bagni S. Filippo area (Figure 1), this represents one of the recharge zones of the geothermal fields. The carbonate units deepen abruptly (more than 30° dip westward) reaching about 1000 m of deepening westward (Figure 11). Shallow reservoir outcrops also in Selvena-Castell'Azzara hills and deep northward toward the volcano edifice (Figures 7b and 11).

This represents another probable recharge area of the geothermal systems. The top surface of the reservoir shows structural highs in correspondence with the Piancastagnaio, Bagnore and Poggio Nibbio geothermal fields and Bagni S. Filippo low-temperature hydrothermal system, where thermal and gas pools are widespread. These highs correspond to highs either in the Palaeozoic basement and intrusive body and K-horizon. The 3D geological model takes in evidence that the thickness of the shallow geothermal reservoir is strongly variable (150–1000 m) with the thinner portions that often correspond to the structural highs (Figure 7b). In Figure 7c the top surface of the metamorphic basement is represented. The dome shape of the metamorphic basement influences the thickness of the carbonate-evaporitic shallow reservoir that is laminated in correspondence of the metamorphic basement highs. This is due to the displacement of the ductile carbonate-evaporitic units of the Tuscan Nappe as a consequence of the emplacement and uplift of the low-density intrusion below Mt. Amiata volcano.

Figure 11. Isobaths (m sl) of the top of the shallow geothermal reservoir in the Mt. Amiata area.

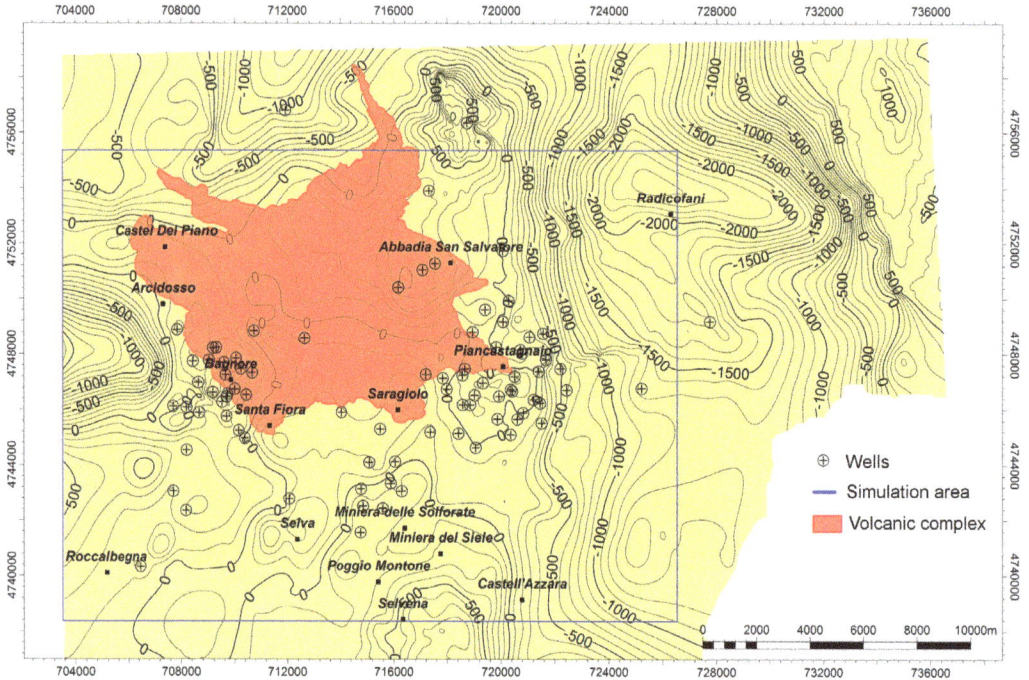

It is impossible to reconstruct with the same accuracy the top of the deeper geothermal reservoir. This is due not only to the paucity of borehole data that encountered this horizon but also because the top of the deeper reservoir does not correspond with a specific geologic unit but it depends on the occurrence of thermometamorphic and metasomatic processes that affected the phylladic basement in the past, modifying its rheology. This is analogous to what observed in the similar Larderello-Travale geothermal system [16].

The occurrence of the active geothermal systems in the south-central portion of the Mt. Amiata may suggest that the intrusion (the heat source of the geothermal systems) is probably asymmetric with respect to the Mt. Amiata volcano. Further evidence supporting this is represented by the occurrence of considerable mercury mineralization [12,42] that is linked to the hydrothermal activity of the area and mainly distributed in the central-southern part of Mt. Amiata (Figure 1).

The subsequent combination of borehole data and numerical modelling results (in the areas where borehole data of temperature are not present), implemented in the Petrel geological model platform, allows reconstructing the temperature behaviour at the top of the shallower geothermal reservoir that is an industrial target for geothermoelectric power production (Figure 12). This is an important issue accomplished thank to this approach. In previous studies on Mt. Amiata geothermal system [13], numerical modelling experiments reconstructed the isotherms along horizontal slices in the subsoil of the geothermal system, as we also carried out in Figure 10. This of course does not take into account the 3D behaviour of geological units that are cut by horizontal slices (Figure 10). The isotherms reported in Figure 12 show the highest values in correspondence of Piancastagnaio area, where they reach temperatures in excess of 230 °C, and Bagnore area where temperatures up to 180 °C are shown. These very high temperatures could be related to highs of the intrusive body and, particularly in the Piancastagnaio area, to the fractured gauge zone of the normal master fault delimiting the western side of the Radicofani basin, which favours the rise of hot fluids from depth toward the shallow geothermal system.

Figure 12 shows that, in the northern and southern portion of the area, the temperatures sharply decrease. This occurs in correspondence with the outcropping of the evaporite-carbonatic units of the Tuscan Nappe (Figures 1 and 11) that form the shallow geothermal reservoir. This is in agreement with the fact that these outcrops are assumed as the main hydrogeological recharge areas of the Mt. Amiata geothermal system.

The obtained 3D geothermal model of Mt. Amiata area can be used for a first theoretical evaluation of the geothermal potential of the shallow reservoir (where we have more data for constraining its volume) [43]. As an output of the 3D model, a volume of 293 km^3 has been estimated for the shallow geothermal reservoir, considering the area of the numerical modelling simulation (23 km × 17 km), and an average temperature of the reservoir of 148 °C. The estimate of geothermal energy stored underground above the mean annual temperature (15 °C), obtained by the volume method [43], is 2840 GW$_y$ (thermal). Assuming a mean value of the recovery factor of 10%, the exploitable energy of the first reservoir can be quantified in 284 GW$_y$.

The quality of the elaborated model and the possibility of using it for prediction purposes will be improved when data from new deep wells will be available. An integrated approach with 3D tools makes geologic information more accessible to a wider audience of non-geologic specialists and decision-makers. The model is also suitable for the planning of new boreholes and further exploration and exploitation steps to proceeds with the usage of geothermal energy in Mt. Amiata area.

86

Figure 12. Map of the isotherms at the top of the shallower geothermal reservoir as elaborated by the combining of 3D geological and numerical modelling (up). Map of the isobaths of the top of the shallower geothermal reservoir from 3D geological model (down).

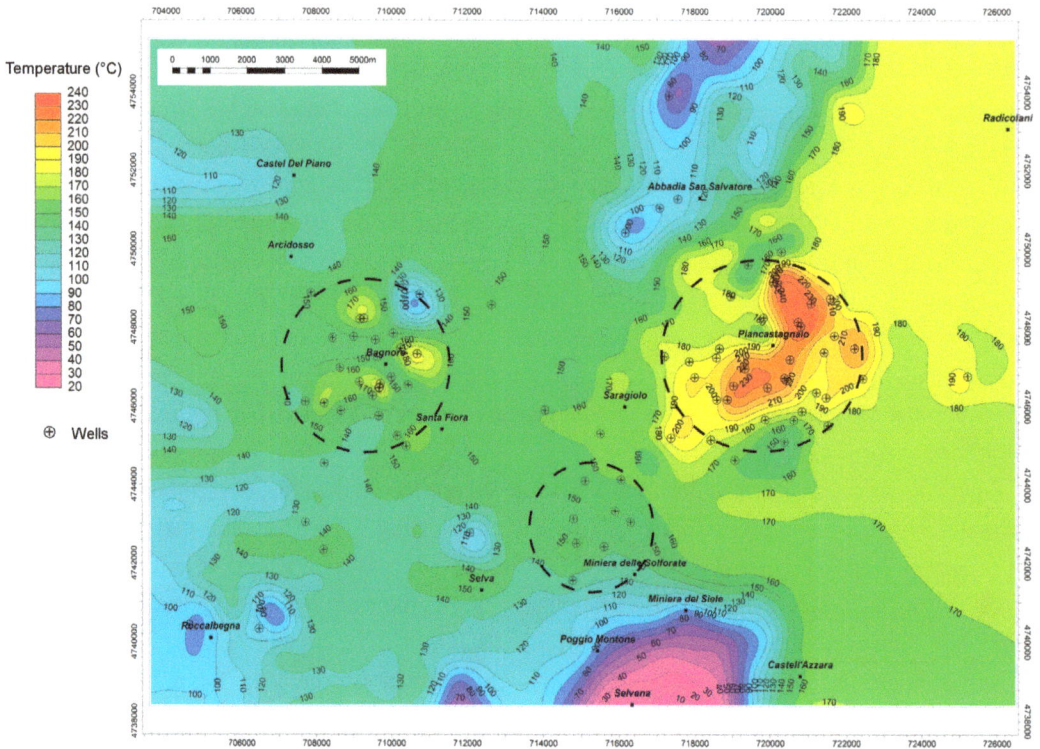

Acknowledgments

We thank SORGENIA GEOTHERMAL Srl (Milan, Italy) for access to the data and for permission to publish them and Matteo Ceroti and Sefano Scazzola for encouraging this work. We are in debt with two anonymous referees whose comments and suggestions significantly contribute to improve the quality of this work.

Author Contributions

Alessandro Sbrana had the original idea for the study and, with all co-authors contributed to the conceptualization and writing of the paper.

Conflicts of Interest

The authors declare no conflict of interest.

References

1. Ferrari, L.; Conticelli, S.; Burlamacchi, L.; Manetti, P. Volcanological evolution of the Monte Amiata, Southern Tuscany: New geological and petrochemical data. *Acta Vulcanol.* **1996**, *8*, 41–56.

2. Conticelli, S.; D'Antonio, M.; Pinarelli, L.; Civetta, L. Source contamination and mantle heterogeneity in the genesis of Italian potassic and ultrapotassic volcanic rocks: Sr-Nd-Pb isotope data from Roman Province and Southern Tuscany. *Mineral. Petrol.* **2002**, *74*, 189–222.

3. Calamai, A.; Cataldi, R.; Squarci, P.; Taffi, L. Geology, geophysics and hydrogeology of the Monte Amiata geothermal field. *Geothermics* **1970**, *1*, 1–9.

4. Brogi, A. Seismic reflection and boreholes logs as tools for tectonic and stratigraphical investigations: New geological data for the Tuscan Nappe exposed in the northeastern Mt. Amiata area (Northern Apennines, Italy). *Boll. Soc. Geol. Ital.* **2004**, *123*, 189–199.

5. Brogi, A. Neogene extension in the Northern Apennines (Italy): Insights from the southern part of the Mt. Amiata geothermal area. *Geodin. Acta* **2006**, *19*, 33–50.

6. Brogi, A. The structure of the Monte Amiata volcano-geothermal area (Northern Apennines, Italy): Neogene-Quaternary compression *versus* extension. *Int. J. Earth Sci.* **2008**, *97*, 677–703.

7. Cadoux, A.; Pinti, D.L. Hybrid character and pre-eruptive events of Mt Amiata volcano (Italy) inferred from geochronological, petro-geochemical and isotopic data. *J. Volcanol. Geotherm. Res.* **2009**, *179*, 169–190.

8. Brogi, A.; Fabbrini, L.; Liotta, D. Sb-Hg ore deposit distribution controlled by brittle structures: The case of the Selvena mining district (Monte Amiata, Tuscany, Italy). *Ore Geol. Rev.* **2011**, *41*, 35–48.

9. Marinelli, G. Genèse des magmas du volcanisme Plio-Quaternaire des Apennins. *Geol. Rundsch.* **1967**, *57*, 127–141. (In French)

10. Gianelli, G.; Puxeddu, M.; Batini, F.; Bertini, G.; Dini, I.; Pandeli, E.; Nicolich, R. Geological model of a young volcano-plutonic system: The geothermal region of Monte Amiata (Tuscany, Italy). *Geothermics* **1988**, *17*, 719–734.

11. Marinelli, G.; Barberi, F.; Cioni, R. Sollevamenti neogenici e intrusioni acide della Toscana e del Lazio settentrionale. *Mem. Soc. Geol. Ital.* **1993**, *49*, 279–288. (In Italian)

12. Arisi Rota, F.; Brondi, A.; Dessau, G.; Franzini, M. Monte Amiata SPA Stab. Minerario del Siele, Stea, B., Vighi, L. I Giacimenti minerari della Toscana meridionale. *Rend. Soc. It. Miner. Petr.* **1971**, *27*, 357–544. (In Italian)

13. Barelli, A.; Ceccarelli, A.; Dini, I.; Fiordelisi, A.; Giorgi, N.; Lovari, F.; Romagnoli, P. A review of the Mt. Amiata geothermal system (Italy). In Proceedings of the World Geothermal Congress, Bali, Indonesia, 25–29 April 2010; pp. 1–6.

14. Frondini, F.; Caliro, S.; Cardellini, C.; Chiodini, G.; Morgantini, N. Carbon dioxide degassing and thermal energy release in the Monte Amiata volcanic-geothermal area (Italy). *Appl. Geochem.* **2009**, *24*, 860–875.

15. Ceroti, M.; Fulignati, P.; Marianelli, P.; Sbrana, A.; Scazzola, S. Integrated exploration of geothermal systems in Southern Tuscany for medium enthalpy resources detecting and possible exploiting through very low impact power plants. In Proceedings of the European Geothermal Congress, Pisa, Italy, 6 June 2013; pp. 1–4.

16. Carella, M.; Fulignati, P.; Musumeci, G.; Sbrana, A. Metamorphic consequences of Neogene thermal anomaly in northern Apennines (Radicondoli-Travale area, Larderello geothermal field—Italy). *Geodin. Acta* **2000**, *13*, 345–366.

17. Gianelli, G.; Manzella, A.; Puxeddu, M. Crustal models of the geothermal areas of southern Tuscany (Italy). *Tectonophysics* **1997**, *281*, 221–239.

18. Decandia, F.A.; Lazzarotto, A.; Liotta, D.; Cernobori, L.; Nicolich, R. The Crop 03 traverse: Insights on post-collisional evolution of Northern Apennines. *Mem. Soc. Geol. Ital.* **1998**, *52*, 427–439.

19. Barchi, M.; Minelli, G.; Pialli, G. The CROP 03 profile: A synthesis of results on deep structures of the Nothern Appennines. *Mem. Soc. Geol. Ital.* **1998**, *52*, 383–400.

20. Brogi, A.; Lazzarotto, A.; Liotta, D.; Ranalli, G.; CROP 18 Working Group. Crustal structures in the geothermal areas of southern Tuscany (Italy): Insights from the CROP 18 deep seismic reflection lines. *J. Volcanol. Geotherm. Res.* **2005**, *148*, 60–80.

21. Cameli, G.M.; Dini, I.; Liotta, D. Brittle/ductile boundary from seismic reflection lines of southern Tuscany (Northern Apennines, Italy). *Mem. Soc. Geol. Ital.* **1998**, *52*, 153–163.

22. Liotta, D.; Ranalli, G. Correlation between seismic reflectivity and rheology in extended lithosphere: Southern Tuscany, inner Northern Apennines, Italy. *Tectonophysics* **1999**, *315*, 109–122.

23. Accaino, F.; Tinivella, U.; Rossi, G.; Nicolich, R. Geofluid evidence from analysis of deep crustal seismic data (Southern Tuscany, Italy). *J. Volcanol. Geotherm. Res.* **2005**, *148*, 46–59.

24. Caumon, G.; Collon-Drouaillet, P.; le Carlier de Veslund, C.; Viseur, S.; Sausse, J. Surface-based 3D modelling of geological structures. *Math. Geosci.* **2009**, *41*, 927–945.

25. Holden, L.; Mostad, P.F.; Nielsen, B.F.; Gjerde, J.; Townsend, C.; Ottesen, S. Stochastic structural modelling. *Math. Geosci.* **2003**, *35*, 899–914.

26. Kaufman, O.; Martin, T. 3D geological modelling from boreholes, cross-sections and geological maps, applicationsover former natural gas storages in coal mines. *Comput. Geosci.* **2008**, *34*, 278–290.

27. Regione Toscana, Cartografia Geologica Regionale 1:10.000. Available online: http://www.regione.toscana.it/-/carta-geologica (accessed on 6 November 2014).

28. Pandeli, E.; Bertini, G.; Castellucci, P.; Morelli, M.; Monechi, S. The sub-Ligurian units of Mt. Amiata geothermal Region (south-eastern Tuscany): New stratigraphic and tectonic data and insight into their relationships with the Tucsan Nappe. *Boll. Soc. Geol. Ital.* **2005**, *3*, 55–71.

29. Liotta, D. Structural features of the Radicofani basin along the Piancastagnaio (Mt. Amiata)–S. Casciano dei Bagni (Mt. Cetona) cross section. *Mem. Soc. Geol. Ital.* **1994**, *48*, 401–408.

30. Brogi, A.; Lazzarotto, A.; Liotta, D.; CROP 18 Working Group. Structural features of southern Tuscany and geological interpretation of the CROP 18 seismic reflection survey (Italy). *Boll. Soc. Geol. Ital.* **2005**, *3*, 213–236.

31. ISPRA Gravimetric Map of Italy. Available online: http://www.isprambiente.gov.it/it/progetti/cartografia-gravimetrica-digitale/index (accessed on 6 November 2014).

32. Bernabini, M.; Bertini, G.; Cameli, G.M.; Dini, I.; Orlando, L. Gravity interpretation of Mt. Amiata geothermal area (Central Italy). In Proceedings of the World Geothermal Congress, Florence, Italy, 18–31 May 1995; pp. 859–862.

33. Orlando, L.; Bernabini, M.; Cameli, G.M.; Dini, I.; Bertini, G. Interpretazione preliminare del minimo gravimetrico del Monte Amiata. *Stud. Geol. Camerti Spec.* **1994**, *1*, 175–181. (In Italian)

34. Elter, F.M.; Pandeli, E. Structural features of the metamorphic Paleozoic-Triassic sequences in deep geothermal drillings of the Mt. Amiata area (SE Tuscany, Italy). *Boll. Soc. Geol. Ital.* **1991**, *110*, 511–522.

35. Batini, F.; Brogi, A.; Lazzarotto, A.; Liotta, D.; Pandeli, E. Geological features of Larderello-Travale and Mt. Amiata geothermal areas (southern Tuscany, Italy). *Episodes* **2003**, *26*, 239–244.

36. Pruess, K.; Oldenburg, C.; Moridis, G. *TOUGH2-User's Guide, Version 2.0*; Report LBNL-43134; Lawrence Livermore Laboratory: Berkeley, CA, USA, 1999; p. 198.

37. Della Vedova, B.; Bellani, S.; Pellis, G.; Squarci, P. Deep temperature and surface heat flow distribution. In *Anatomy of an Orogen: The Apennines and Adjacent Mediterranean Basins*; Vai, G.B., Martini, L.P., Eds.; Springer: Berlin, Germany, 2001.

38. Viganò, A.; Della Vedova, B.; Ranalli, G.; Martin, S.; Scafidi, D. Geothermal and rheological regime in the Po plain sector of Adria (Northern Italy). *Int. J. Geosci.* **2011**, *131*, 228–240.

39. Gualteri, L.; Zappone, A. Hypothesis of ensialic subduction in the Northern Apennines: A petrophysical contribution. *Mem. Soc. Geol. Ital.* **1998**, *52*, 205–214.

40. Trippetta, F.; Collettini, C.; Vinciguerra, S.; Meredith, P.G. Laboratory measurements of the physical properties of Triassic Evaporites from Central Italy and correlation with geophysical data. *Tectonophysics* **2010**, *492*, 121–132.

41. Bertini, G.; Cappetti, G.; Dini, I.; Lovari, F. Deep drilling results and updating of geothermal knowledge on the Monte Amiata area. In Proceedings of the World Geothermal Congress, Florence, Italy, 18–31 May 1995; pp. 1283–1286.

42. Marinelli, G. Il magmatismo recente in Toscana e le sue implicazioni minerogenetiche. *Mem. Soc. Geol. Ital.* **1983**, *25*, 111–124. (In Italian)

43. Muffler, P.; Cataldi, R. Methods for regional assessment of geothermal resources. *Geothermics* **1978**, *7*, 53–89.

The Marsili Volcanic Seamount (Southern Tyrrhenian Sea): A Potential Offshore Geothermal Resource

Francesco Italiano, Angelo De Santis, Paolo Favali, Mario Luigi Rainone, Sergio Rusi and Patrizio Signanini

Abstract: Italy has a strong geothermal potential for power generation, although, at present, the only two geothermal fields being exploited are Larderello-Travale/Radicondoli and Mt. Amiata in the Tyrrhenian pre-Apennine volcanic district of Southern Tuscany. A new target for geothermal exploration and exploitation in Italy is represented by the Southern Tyrrhenian submarine volcanic district, a geologically young basin (Upper Pliocene-Pleistocene) characterised by tectonic extension where many seamounts have developed. Heat-flow data from that area show significant anomalies comparable to those of onshore geothermal fields. Fractured basaltic rocks facilitate seawater infiltration and circulation of hot water chemically altered by rock/water interactions, as shown by the widespread presence of hydrothermal deposits. The persistence of active hydrothermal activity is consistently shown by many different sources of evidence, including: heat-flow data, gravity and magnetic anomalies, widespread presence of hydrothermal-derived gases (CO_2, CO, CH_4), $^3He/^4He$ isotopic ratios, as well as broadband OBS/H seismological information, which demonstrates persistence of volcano-tectonic events and High Frequency Tremor (HFT). The Marsili and Tyrrhenian seamounts are thus an important—and likely long-lasting-renewable energy resource. This raises the possibility of future development of the world's first offshore geothermal power plant.

Reprinted from *Energies*. Cite as: Italiano, F.; De Santis, A.; Favali, P.; Rainone, M.L.; Rusi, S.; Signanini, P. The Marsili Volcanic Seamount (Southern Tyrrhenian Sea): A Potential Offshore Geothermal Resource. *Energies* **2014**, *7*, 4068-4086.

1. Introduction

A large amount of data on the geological, geophysical and geochemical features of submarine volcanic activity and seamounts have been acquired in the last three decades. The close proximity of magma chambers to the seafloor, in conjunction with tectonic activity due to plate motion, deformation and cooling of erupted lavas results in convective circulation of dense, cold seawater through the cracked and fissured upper portions of the lithosphere; this circulation promotes the formation of venting sites that release hot hydrothermal fluids and dissolved elements [1–4].

Submarine hydrothermal activity has been studied so far as an energy source for free-living and symbiotic chemosynthetic bacteria, which form the base of the food chain in these unique habitats [5]. Because of their huge, long-lasting recharge and high-temperature/high-pressure characteristics, the submarine hydrothermal fluids are now investigated also as a potentially exploitable geothermal energy source. Candidate areas for offshore geothermal exploitation have been identified in the Gulf of California, the Juan de Fuca Ridge, the Japan Sea, the Okhotsk Sea, the Andaman Sea and the Tyrrhenian Sea since late 1970s [6]. However, the technology at that time was not still achievable at competitive costs beyond environmental, legal and institutional problems which had to be overcome. Nowadays, the continuous and growing developments in oil and gas exploration and exploitation techniques allow for an easier and economically competitive approach to the investigation and the energy potential assessment of submarine hydrothermal systems. However, before any reliable quantification for energetic exploitation of any offshore geothermal reservoir, a multidisciplinary submarine exploration has to be done, including geological, geophysical geochemical, oceanographic and biological investigations.

At the beginning of the last century, Italy became the first country worldwide to exploit high-temperature geothermal resources for power generation, with the result that, nowadays, Larderello-Travale/Radicondoli and Mt. Amiata (Bagnore and Piancastagnaio) are exploited geothermal fields [7]. They are both set in a continental extensional tectonic environment (lithospheric thickness between 20 and 30 km), characterised by deep and shallow volcanic systems and high heat-flow values (regional value of 120 mW/m^2, with maxima up to 1 W/m^2 [8]). In both the geothermal areas, magmatic bodies provide the necessary heat source to deep and shallow reservoirs hosted in local metamorphic, carbonate and anhydritic formations [9,10]. In order to improve the Italian geothermal energy production, in 1970s and 1980s an intensive exploration program (geological, geophysical and geochemical surveys, as well as drilling activities) took place in Latium (about 100-deep wells were drilled in the Latera caldera, Vico Lake, Cesano, Bracciano Lake and Alban Hills), Campania (Phlaegrean Fields and Ischia Island) and Sicily (Vulcano and Pantelleria Islands) [7]. Several unfavourable physical, chemical and logistic features strongly limited the use of these potential geothermal fields. In particular, the geothermal exploitation was hindered by the low permeability of the reservoirs, the high salinity and acidity of the hot fluids, the sluggish and low amount of groundwater recharge, the high potential explosivity of these magmas (rhyolites, trachytes and phonolites) implying a high volcanic hazard as well as the strong urbanisation and/or tourist use of those areas.

The Southern Tyrrhenian Sea might represent the future target for geothermal energy exploration and exploitation due to the generally high heat-flow [11,12] and the widespread presence of seamounts for which the presence of hydrothermal deposits has been well documented. Among them, the Marsili seamount exhibits some features that make it a potential site hosting exploitable submarine geothermal systems.

This paper accounts for the recently collected information of geological, geochemical and geophysical features of the Marsili seamount (Figure 1), the largest European and Mediterranean volcanic edifice, making the seamount a likely large geothermal energy resource that could significantly improve the Italian geothermal power generation, nowadays providing 5.5 TWh per year that cover only 1.6% of national mean electricity production [13].

Figure 1. Location of the study area. MV: Marsili Volcano; MB Marsili Basin; ST: volcanoes of Stromboli; VU: volcanoes of Vulcano; FL: Filicudi Island; SIS: Sisifo submarine Volcanoes; LM: Lametini Seamounts; PA: Palinuro Seamount; VB: Vavilov Basin. The map on the right is modified after [14].

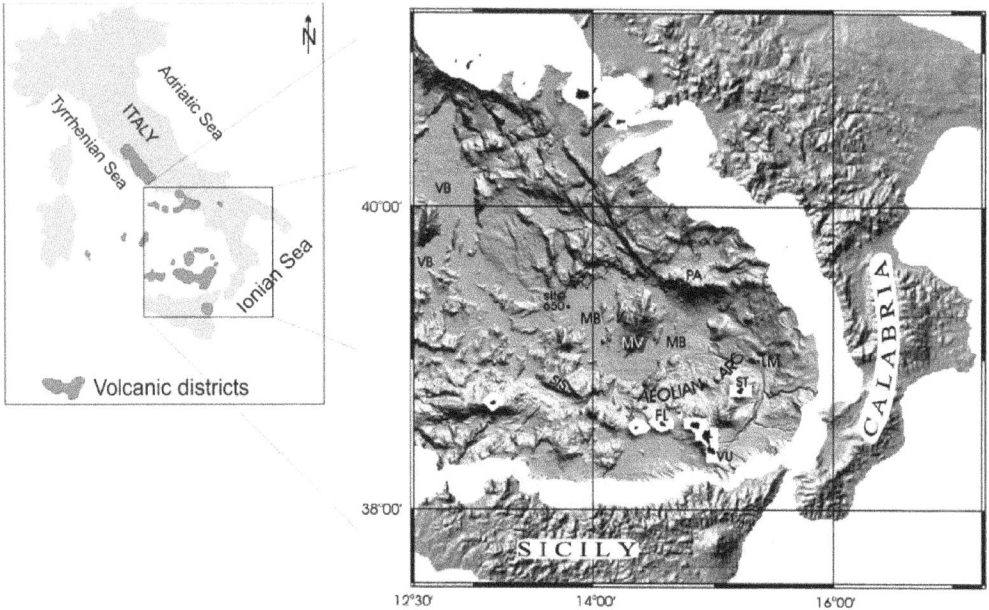

2. Results

This section summarizes the results of a large number of multidisciplinary data collected by several scientific cruises carried out by Italian Research Institutions during the last 20 years and describes the new results collected during two campaigns carried out in 2007 and 2010.

2.1. Geothermal Potential of Southern Tyrrhenian Basin and Marsili Seamount

The Southern Tyrrhenian Sea is a back-arc basin developed from Miocene to Present in the frame of the coeval formation of Apennine-Maghrebides chain, structured above the subducting north-western Ionian oceanic slab [15,16]. Its evolution has been characterised by volcanic activity induced by a wide tectonic extension from North-West (Sardinia) to South-East (Aeolian Arc) [17–19]; deep and shallow seismicity occurring in this area has been also well documented (e.g., [20–22]). The best morphologic evidences of these geodynamic processes are the two oceanic crust floored Vavilov and Marsili sub-basins and the homonymous seamounts, placed at north-west and south-east portions of the Southern Tyrrhenian basin, respectively [14,23]. Other geodynamic-related morphologic features are the numerous volcanic structures forming an arc all along the south-eastern margin of the basin: the Aeolian Arc, representing the emerged part; Palinuro, Glabro, Alcione and Lametini, representing the northern submerged arm; Eolo, Enarete and Sisifo, representing the north-western submerged arm (Figure 2; [23]).

Figure 2. Bathymetry of South-Eastern Tyrrhenian Basin. M: Marsili; MB: Marsili Basin; P: Palinuro; A: Alcione; L: Lametini; Sc: Stromboli Canyon; Eo: Eolo; En: Enarete; S: Sisifo (modified after [23]).

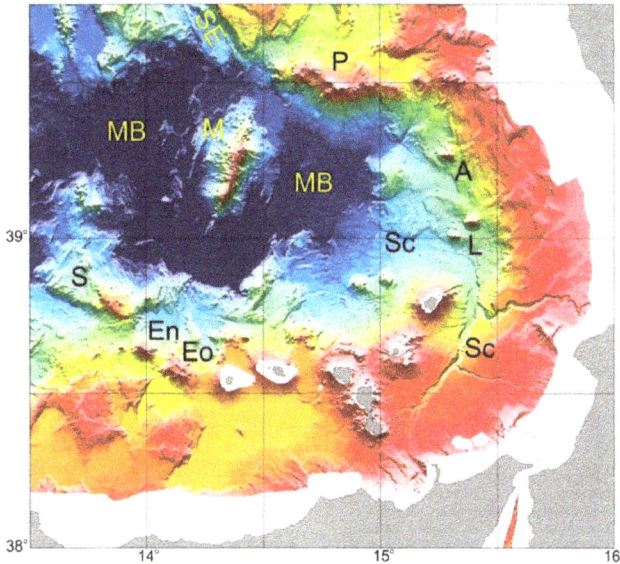

The heat-flow anomalies of Southern Tyrrhenian Sea [11,12] are always high in correspondence with the volcanic structures; the regional mean value is higher than 120 mW/m^2 and comparable to the Italian onshore volcanic district. More precisely, the highest values have been recorded close to the Vavilov (140 mW/m^2) and Marsili (250 mW/m^2) seamounts. Heat-flow rates as high as 300 mW/m^2 and 500 mW/m^2 were detected on the uppermost and central portion of Vavilov and Marsili seamounts respectively [24]. Those values fit with the geophysical data, collected in the last decades such as:

- Moho depth located 15–20 km below the Tyrrhenian abyssal plains and about 10 km beneath Vavilov and Marsili sub-basins [25,26];
- Gravity anomalies with positive values greater than 250 mGals, interpreted as due to lithosphere thinning [27], as well evidenced by high resolution reflection seismic profiles [28];
- A lithosphere mean conductivity beneath Marsili basin about one order of magnitude greater than beneath Ustica island (South-western Tyrrhenian basin) [29].

All the above-mentioned data strongly suggest the following inferences:

- Southern Tyrrhenian basin is affected by numerous and distributed heat sources, generally represented by hot magmatic bodies at shallow depths (<10 km) in a strong extensional geodynamic setting;
- A high primary permeability field is inferred, due to emplacement and cooling of magmas, successively increased by the intense and recent tectonic activity [30,31];
- A virtually infinite fluid recharge is available, supplied by pressurised seawater;
- A relatively low amount of dissolved salts with mild acidity is expected;

- A low explosivity due to the presence of lesser involved magmas with a lower amount of dissolved water is expected.

In this frame, Marsili seamount, presenting the highest aforementioned anomalies, is here proposed as the best prone site of sustained submarine geothermal potential. Marsili seamount is a recent volcanic structure (1–0.1 Ma), mainly composed of basalts and, to a lesser extent, andesites and trachy-andesites with calc-alkaline affinity [32–34]. This structure has been interpreted as an inflated, small-scale spreading centre, since bio- and magneto-stratigraphic data show features comparable to those typical of mid-ocean ridges [35–38].

Dekov and Savelli [39], summarising all the observations since the late 1960s on the rocks sampled from Marsili and surrounding volcanic centres, affirmed that these areas are affected by hydrothermal fluid circulation; in their model, cold seawater enters into fractured rocks and then is superheated by magmatic bodies at crustal depths. Clues of active venting of hydrothermal fluids over the Marsili seamount have been provided by the injection of magmatic-type volatiles into the sea water as demonstrated by the anomalous ^{3}He content and the whole chemical composition of the gases dissolved in water column [40]. However, direct thermal measurements on Marsili seamount are still lacking. This latter discovery strongly points to Marsili still being hydrothermally active, in agreement with the previous mineralogical data, and supported by a significant contribution by juvenile fluids.

2.2. Morphology

The physiographic features of Marsili seamount are shown in Figure 3. The Marsili seamount rises 3500 m from the abyssal plain to 489 m minimum depth. The volcanic edifice is about 60 km long and about 20 km wide; two other smaller structures run parallel to the volcano and are located on its left and right part, with heights on the order of several hundred metres. Marsili is elongated mainly along a NNE-SSW axis; however, this extensional axis is not perfectly linear but it shows a sigmoid trend, with the southernmost and northernmost axial directions both north-eastern trending, while the central axis is closer to the N-S direction. This deformational feature is commonly associated with strike-slip faulting under volcanoes [41], but it can also be related to long kilometric normal faults orthogonal to the volcanic system spreading direction, with transverse faults at the tips that accommodate the regional strain [42]. In both cases, this observation implies a strong tectonic control of the whole volcanic edifice. The volcano summit is characterised by a narrow crest, 20 km long and 1-km wide, over the 1000 m isobaths, cut by linear structures, mainly disposed parallel to the extensional axis. They appear as small ridges, up to 100 m high and 750 m long, not fault bounded, formed by alignments of small monogenic volcanic mounds. These segments may result from the feeding of magma to the seafloor along dikes that produce, or follow, discontinuous, en-echelon crack systems. Steep bathymetric gradients separate the crest portions from the deeper volcano flanks, forming very long and lesser sloped scarps extending to the basin. In several places, these lower steep scarps terminate with gently dipping terraces, elevated for several hundreds of meters from the abyssal Marsili basin. Spherical cones are present on the flanks of the volcano with diameters ranging between a few kilometres to several hundred metres.

These flanks are also cut by several, well evidenced kilometric valleys; the largest of them is located near to the central and top portion of Marsili. It has an amphitheatre shape in the uppermost part edged by roughly vertical walls; such geomorphologic feature suggests a flank collapse of this portion of the Marsili flank, with landslide debris accumulated distally on the Marsili basin, a kind of process already found elsewhere [43,44].

Figure 3. Detailed bathymetry of Marsili seamount. Red dashed lines: linear structures; violet dotted lines: main circular cones and terraces; yellow dashed line: major landslide (modified after [45]).

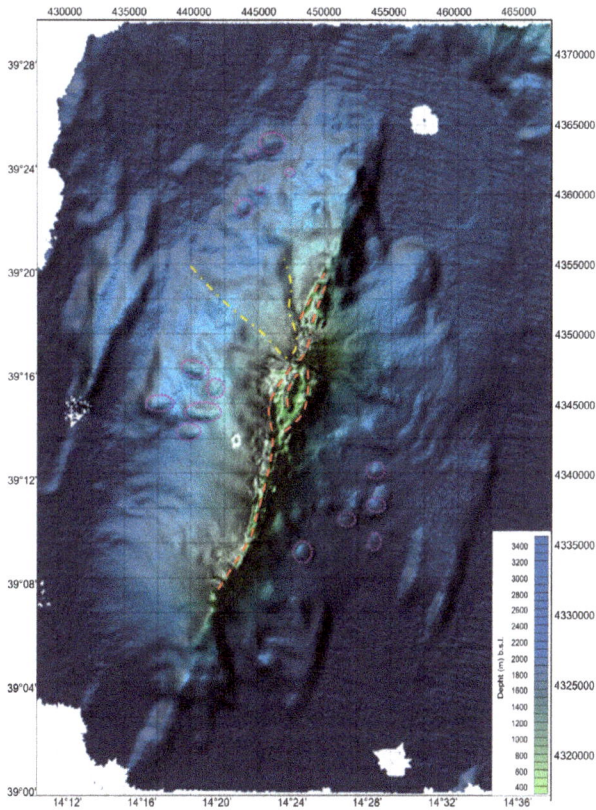

2.3. Magnetic Data

Figure 4 shows the total field magnetic anomaly map, reduced to the magnetic pole, recorded above Marsili seamount. In agreement with literature data [35,37,38], positive magnetic anomaly maxima are located along the central sectors of the volcanic structure. At northern and southern tips the highest positive anomalies of Marsili are present, with maximum values around 1500 nT; in the central and highest portion of volcano and in the middle of these two former anomalies, the values are around 0 nT, reaching a minimum of −100 nT. Negative anomalies are located at the base of the western flank, with a mean value around −200 nT; the base of the eastern flank again shows negative values, achieving a local minimum of −500 nT in the central and southern portions, whereas the north-eastern part of this sector is characterised by small positive values.

Figure 4. Magnetic anomaly field (reduced to magnetic pole) of Marsili; values of the magnetic anomaly field in nT (modified after [45]).

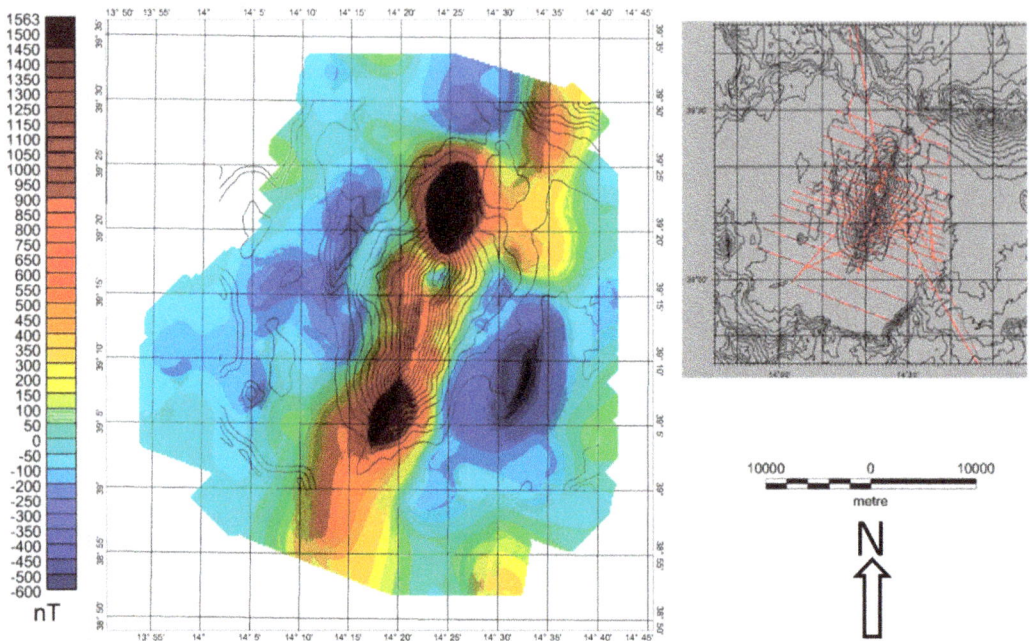

2.4. Gravity Data

According to the map of Faye gravity anomalies of Marsili seamount (Figure 5), positive maxima, reaching more than 80 mGals, are present on the crest portion of volcano, above the 1000 m isobaths, and progressively decrease to about 20 mGals on the volcano base. On the whole, the sea-surface free-air anomaly field largely reflects the topography of the volcano; moreover, the relative low amplitude of the positive gravity anomalies suggests the presence of a low density zone below the seamount.

2.5. Seismological Recordings

The broadband OBS/H (Ocean Bottom Seismometer with Hydrophone), deployed on the Marsili flat top (39°16,383', 14°23,588') at a depth of 790 m, recorded more than 1000 seismo-volcanic and hydrothermal signals [3]. By comparing the signals recorded with typical volcanic seismic activity, the recorded signals were grouped into: Volcano-Tectonic type A (1 event), Volcano-Tectonic type B (817 event), High Frequency Tremor (159 events) and quasi-monochromatic Short Duration Events (32 event).

Figure 5. Faye anomaly field of Marsili seamount; values of the gravity field in mGal (modified after [45]).

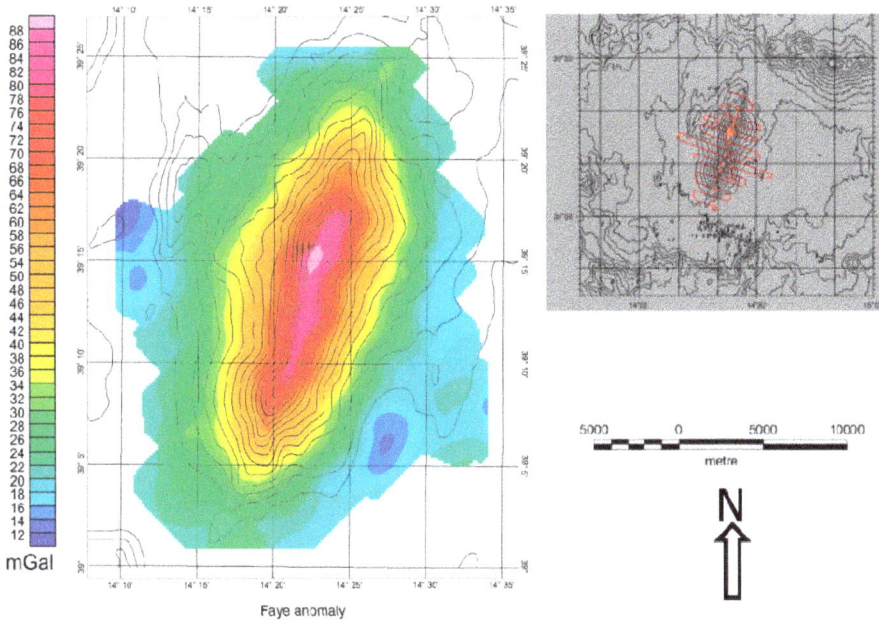

Faye anomaly

An intriguing feature of the seismic noise is its spectral content presenting progressively growing energy levels in a broad-band frequency range from 4 to 60 Hz; diffuse spectral peaks are also present between 4 and 30 Hz. The seismic events are characterized by an indiscernible S-phase. On the basis of preliminary frequency content observations they can be distinguished in two main groups: about 720 events with frequency content between 4 and 10 Hz and about 80 events with very high frequency content, between 40 and 80 Hz [3].

2.6. Hydrothermal Fluids

In order to recover information on the type of vented fluids and to constrain their origin, we carried out water column surveys performing casts and tows across the seamount using a rosette equipped with a CDT (SeaBird 911+ Conductivity Temperature Depth device) and Niskin bottles. The surveys have been carried out during two scientific cruises in November 2007 (by R/V Urania) and July 2010 (by R/V Astrea).

Sea-water samples were collected to extract the dissolved gas phase by further laboratory procedures. The extracted gases were analyzed for the chemical composition calculated taking into account the solubility coefficients (Bunsen coefficients "β") of each gas specie, the volume of the gas extracted (in cm^3), the volume of the extracted water sample (in cm^3) and the equilibration temperature as described in [46,47].

Table 1 lists the results (expressed in cm^3 STP/L_{H2O}, namely millilitres of gas per litre of water at standard temperature and pressure conditions of 1bar and 25 °C) showing that CO_2, with concentrations ranging from 0.53 to 1.9 cm^3 STP/L_{H2O}, is by far the most abundant component of the dissolved gas phase besides oxygen and nitrogen, in the range 2.7–4 and 8.3–10 cm^3 STP/L_{H2O}, respectively. Among the other components CH_4 is always detected in concentrations ranging from 3.9×10^{-5} to 2.1×10^{-4} cm^3 STP/L_{H2O}, two orders of magnitude above the equilibrium with the atmosphere (ASSW value equal to 1×10^{-6} cm^3 STP/L_{H2O}).

Table 1. Analytical results of the dissolved gas phase in samples collected during two different cruises (2007 and 2011). Hydrothermal-derived gases are released besides components from the atmosphere. Analytical results in cm^3 STP/L_{H2O} (see text for details). Depth in meters. Bdl = below detection limits. Data of ASSW (Air Saturated Sea Water) and from the bottom sample of a vertical cast carried out over the Tyrrhenian Abyssal plain are reported for comparison. Coordinates are reported as they were collected during the cruises.

ID Sample	Depth	Date	Latitude	Longitude	H_2	O_2	N_2	CH_4	CO_2
				Marsili					
M1	676	13 July 2011	39°16.840'	14°23.220'	3.7×10^{-4}	3.50	9.25	9.5×10^{-5}	0.77
M1	500	13 July 2011	39°16.840'	14°23.220'	3.1×10^{-3}	3.60	10.04	5.2×10^{-5}	1.07
M1	400	13 July 2011	39°16.840'	14°23.220'	1.2×10^{-3}	4.03	12.92	1.3×10^{-4}	0.60
M1	300	13 July 2011	39°16.840'	14°23.220'	-	2.99	9.61	1.5×10^{-4}	0.63
M2	668	13 July 2011	39°17.159'	14°23.410'	-	3.23	9.10	3.9×10^{-5}	0.53
M2	500	13 July 2011	39°17.159'	14°23.410'	2.0×10^{-3}	3.42	10.21	1.3×10^{-4}	1.26
M2	400	13 July 2011	39°17.159'	14°23.410'	-	2.78	9.01	2.1×10^{-4}	1.89
M2	300	13 July 2011	39°17.159'	14°23.410'	-	3.28	9.35	1.1×10^{-4}	1.72
M3	610	13 July 2011	39°16.799'	14°23.999'	2.5×10^{-4}	3.12	9.16	5.8×10^{-5}	2.71
M3	500	13 July 2011	39°16.799'	14°23.999'	1.3×10^{-4}	2.74	8.29	1.0×10^{-4}	0.57
M3	400	13 July 2011	39°16.799'	14°23.999'	-	3.34	9.61	1.1×10^{-4}	1.72
M4	673	13 July 2011	39°16.342'	14°23.196'	4.8×10^{-4}	2.97	10.15	9.7×10^{-5}	1.29
M4	500	13 July 2011	39°16.342'	14°23.196'	2.9×10^{-4}	3.31	9.89	9.4×10^{-5}	0.80
M4	400	13 July 2011	39°16.342'	14°23.196'	1.1×10^{-3}	3.46	10.39	8.4×10^{-5}	0.98
				Tow-yow Marsili					
TY1	500	02 November 2007	39.28163°	14.38428°	bdl	2.42	7.77	2.1×10^{-4}	0.47
TY2	702	02 November 2007	39.28165°	14.38447°	bdl	2.98	8.80	3.3×10^{-4}	0.45
TY3	457	02 November 2007	39.28763°	14.42715°	bdl	3.26	9.16	2.2×10^{-4}	0.44
				Data for comparison					
Vertical cast	3164	02 November 2007	39.40733°	14.51067°	bdl	2.47	8.42	4.4×10^{-5}	0.55
	400	02 November 2007	39.11475°	14.32298°	bdl	2.45	7.58	2.1×10^{-4}	0.43
ASSW	-	-	-	-	4.1×10^{-5}	4.80	9.60	1.0×10^{-6}	0.24

3. Discussion

The collected data provide the first useful information on the characterisation of this huge offshore structure that might be a potential site for offshore geothermal exploitation. The large variety of morphological forms, *i.e.*, linear structures, lava sheets, spherical cones, fault scarps and landslides, suggests that Marsili volcanic seamount has encompassed a long evolution, composed of different stages. The whole volcanic edifice shows evidence of mature volcanism, since almost all the volcanic forms look diffusely marked by further tectonics and erosion. The most mature forms, represented by circular cones and terraces, lava sheets, *etc.*, are predominantly located on the lower flanks of Marsili volcano; the youngest ones, typically represented by linear structures, are along the crest portion. Fault scarps are easily observable, pointing out the influence of tectonics on the whole edifice. As a consequence, the volcano likely has a well-developed network of fractures and cavities that play an important role in defining the permeability field of the volcano; these provide a pathway for the inflow of seawater and the consequent hydrothermal circulation and are primarily responsible for determining the magnitude of submarine geothermal resources.

The basaltic to basaltic-andesite rock compositions reported in [34] are in agreement with the strength of magnetic anomalies measured on the northern and southern portions of the Marsili volcanic edifice. Generally, freshly erupted basalts (also rapidly cooled) are strongly magnetic [48,49] because they have not been exposed to demagnetisation processes for a long time. However, the most intriguing feature of the magnetic anomaly field are the very low values in the central sector of the crest, testifying the presence of rocks with very low and/or a-magnetic properties, for which hydrothermal activity is considered as the main cause. The hydrothermal circulation of fluids may interact with the source rocks and reduce the magnetisation by breaking down the magnetic minerals [50,51]; this evidence corroborates with the occurrence of hydrothermal processes, possibly still active or recorded by geothermal deposits on the crest of Marsili volcano [39,52]. The pattern of the observed anomalies fits with a Curie isotherm mean depth located at around 4–5 km below Marsili crest, indicating a temperature of more than 600 °C at the volcano base and highlighting the possible presence of magmatic bodies. The high internal temperatures are also supported by the extremely high heat-flow measurements carried out on Marsili [11,24]. All those complementary data coherently suggest that Marsili seamount contains an intense and shallow heat source.

Although hydrothermal minerals are generally less dense than primary minerals, their effect on gravity is negligible. The free-air anomaly high observed on the top of Marsili seamount occurs over the topographic high over which the hydrothermal system is supposed to be located. So, the gravity anomalies are controlled by local geologic structure. Generally, the observed gravity anomaly field can be fitted only assuming a mean density of the Marsili volcanic structure of about $2 \, g/cm^3$. Taking into account the petrographic features of Marsili rocks [34] as well as the magnetic data, such values can be attributed to rock porosity/permeability, possibly filled with aqueous and volatile phases. On this basis, it can be tentatively inferred that the Marsili volcano should have a significant porosity, possibly more than 10% by volume.

Caratori Tontini *et al.* [53] proposed a model for the volcano summit consisting of a large altered region, with a vanishing magnetisation of 0 A/m and a density of $2.0 \, g/cm^3$, and an active, a-magnetic hot magma chamber 3 km below the summit, with density $2.3 \, g/cm^3$ based on volatile concentrations [54].

Seismic observations by means of the OBS/H deployed on the crest of Marsili seamount [3] fit with the presence of active hydrothermal discharge highlighted by the geochemical features of the dissolved gases (Table 1). The continuous high frequency seismic and acoustic noise recorded on the Marsili crest is in agreement with historic seismic noise recordings in geothermal areas. In particular, very shallow hydrothermal manifestations (active venting sites) usually produce signals with maximum energy in the frequency band over 50 Hz [55–57], as observed on Marsili seamount.

Moreover, the seismicity characterised by a frequency content between 4 and 10 Hz could be connected to hydrothermal fluids circulation, in analogy with long-period (LP) events (e.g., [58–60]), while those with 40–80 Hz frequency content could be generated by surface degassing phenomena ([3] and references therein).The results of the chemical and isotopic analyses of gas extracted from the sampled sea-waters allow us to confirm the existence of a dissolved gas phase different from the atmospheric. The results clearly show that besides the expected atmospheric gases, a significant content of CO_2 marks all the gas samples. Figure 6 shows the contemporary presence of atmospheric components (represented by O_2 and N_2) as well as CO_2 typically originated by an endogenic source.

Figure 6. O_2-N_2-CO_2 ternary diagram. The samples plot on a straight mixing line between atmospheric components O_2 and N_2 and CO_2 typically of magmatic/hydrothermal origin The Air Saturated Sea Water mark (grey star, ASSW) indicates where dissolved gases in ASSW (air-saturated sea water) sea water plot. Blue triangles = samples from above the area marked by physical anomalies (see text); grey crosses = samples from a tow-yow across the seamount; black filled circles = samples from a vertical cast over the Tyrrhenian batial plain. The numbers indicate the sampling depths.

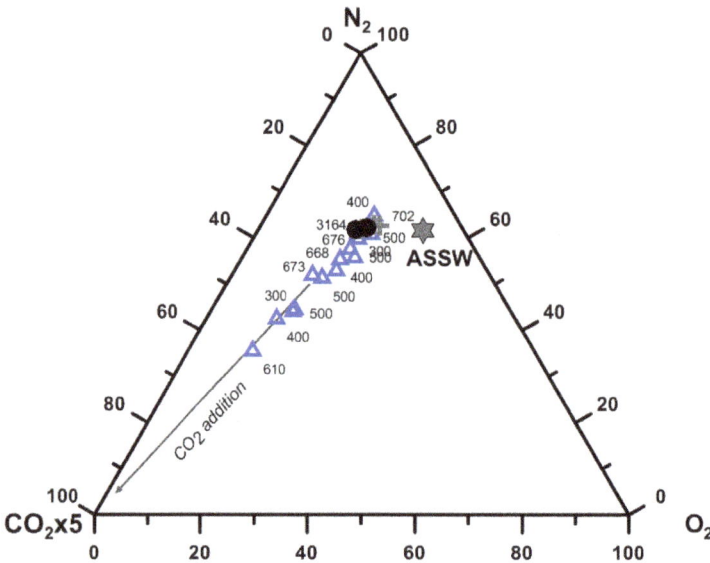

The plot along the Air-Saturated Sea Water (ASSW)-CO_2 mixing line highlights the injection of CO_2-dominated volatiles. Samples marked by the blue triangles come from the area marked by the largest anomalies of physical parameters (magnetic, gravimetric, seismic). Those samples plot along a line with a constant CO_2/N_2 ratio showing and increasing trend of hydrothermal fluids. Data from waters sampled by a tow-yow across the whole seamount, show a lower, although significant CO_2 content, due to their provenance from other areas clearly marked by the presence of hydrothermal activity although of less intensity. Figure 7 plots atmospheric gas (typically represented by oxygen) together with CO_2 and CH_4 representative for hydrothermal-type components. The plot shows the occurrence of dissolution processes responsible for CO_2 loss and enhancement of the less soluble species such as CH_4. Both figures plot data collected during two cruises in 2007 (grey crosses) and 2011 (blue triangles). The latter samples have been collected over the area supposed to be the main geothermal source of the whole Marsili seamount where the low energy volcanic blasts might have occurred in recent times [61].

Figure 7. CH_4-O_2-CO_2 triangular diagram. The plot shows the relative concentrations of hydrothermal-derived components (CO_2-CH_4) besides the atmospheric component (here represented by oxygen) normally dissolved in sea water. It is easy to recognize the large enrichment in deep-originated CO_2 and CH_4 to the respect to ASSW. The arrows show the trends of modification of the dissolved gas assemblage due to the injection of hydrothermal fluids and the gas/water interaction phenomena leading to the dissolution of the very soluble CO_2 and enrichment of less soluble gas species like CH_4. Symbols as in Figure 6.

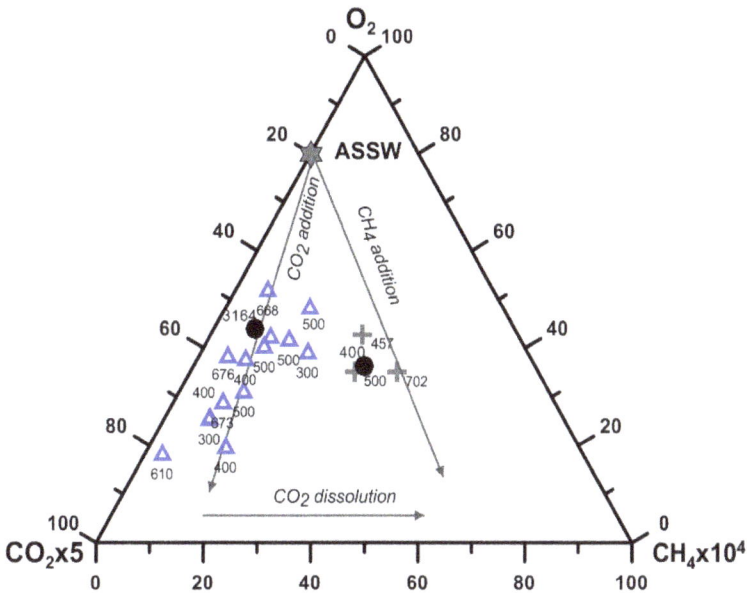

The available scientific data set has been considered a good starting point for an Italian company (Eurobuilding S.p.A., Servigliano, Italy) to propose a cutting-edge project to drill the first offshore geothermal well on the Marsili seamount, [62,63]. The estimates they performed bring to an effective electrical power generation of 200 MW. The offshore well should be connected to steam turbines that together with condenser systems, power generators and tension elevators could be hosted on the power plant platform. A high capacity cable will drive the electricity from the platform to the power grid located on the Italian coast.

4. Conclusions

The comprehensive and multidisciplinary (geological, geophysical and geochemical) dataset provided results coherently pointing to the Marsili seamount as a possible target for offshore geothermal exploration and exploitation developments.

The main scientific key points supporting future evaluations of the Marsili seamount as a possible geothermal energy resource can be summarized as:

(1) Marsili has a shallow and strong heat source;
(2) An active geothermal fluid circulation is expected as suggested by the first permeability field evaluations;
(3) The present state of Marsili volcanic activity is still controversial: recent radiometric data indicate the occurrence of recent magmatic blasts [61] although no indications come from other observations (e.g., seismic activity) and only sporadic monitoring activity has been carried out [64];
(4) The presence of solid deposits of hydrothermal origin indicates that geothermal fluids permeate the edifice and are vented into the seawater. The evidence that hydrothermal fluids as well as magmatic-type helium are injected in the deep sea waters indicates that the hydrothermal activity is still ongoing.

The possibility that a large geothermal energy reservoir exists about 100 km off the Italian coasts and that such a reservoir is able to provide a significant, long-lasting amount of exploitable energy, makes the Marsili seamount a potentially attractive target for private and public investors.

To summarize the case for developing the world's first offshore geothermal power plant in this region we consider a model of the edifice with constant average density of basalts equal to 2.67 g/cm^3, namely the difference between the gravimetric anomaly calculated from the model and that coming from the measurement leads to a density value between 1.7 and 2.3 g/cm^3. Considering also the petrologic and magnetic features of the rocks, the estimated density values can be a consequence of a high porosity of the rocks filled by volatiles. Assuming an average porosity of 10% and the total estimated volume of the geothermal reservoir >100 km^3, the volume of the geothermal fluids circulating in the central and upper part of the Marsili seamount, it results that about 10 km^3 of exploitable hot fluids can be recovered [45].

A first exploration drilling is needed to verify the proposed results and to recover (for the first time) direct information on the physical, chemical and isotopic features of the deep hydrothermal vents. In fact a hydrothermal fluid with a temperature of 250 °C and pressure of 30 bar with a flow rate of 1 m^3/s allows the energy production of 8 MW, while fluids marked by a temperature of 375 °C and pressure of 250 bar allow an energy production ten times bigger for the same flow rate.

It is absolutely clear that new data and further multidisciplinary investigations are needed, including the assessment of the natural risks (volcanic and seismic) related the huge and almost unknown submarine volcano. The development of micro-biological studies as well as research and development activities aimed at planning and deploying seafloor observatories for long-term monitoring activities [65] are also considered as mandatory. The results will improve the scientific knowledge useful to better constrain the available and exploitable geothermal resources inside the volcano. The future geothermal explorations have at least to include:

- The location of active venting sites and hydrothermal fluid release by direct (by ROVs) explorations and detailed CTD and nephelometry surveys;
- The distribution of the parameters that determine the permeability field of the volcano by geophysical tomographic methods;
- The characterization of volcanic-type ongoing processes, using seafloor observatories planned to carry out long-term monitoring of temporal variation of both physical and chemical parameters;
- The assessment of all the environmental aspects related to geothermal exploitation activities including studies to forecast and to face the impact of the human activities on the natural deep sea environment;
- An estimation of the thermal energy budget based on the hydrothermal fluids enthalpy and flow rate evaluations.

It is noteworthy to consider that, although the offshore geothermal energy exploration and exploitation have not yet been considered a feasible option, nowadays the technologies required for on-site geothermal exploitation and production of electric energy are already available, and can be adopted in view of the depth of seawater, about 600–800 m, and the minimum distance from the Italian coasts, less than 100 km. Presently, some projects are evaluated in European countries, for instance in Iceland, to develop offshore geothermal power plants to use high-temperature fluids lying along the mid-ocean ridges. Iceland is also developing the world's longest subsea power cable (about 1000 km), to export geothermal energy to Scotland. An eventual Marsili submarine power cable would "only" be around 150 km long.

Once the potentially exploitable geothermal energy of the Marsili seamount is well-constrained, that submarine volcano may become home to the world's first offshore geothermal power plant, potentially capable of doubling the current geothermal power output of Italy [45].

The increasing request for electrical power, the energy prices and the increasing know-how of the utilization of this resource makes the Marsili seamount an attractive scientific and industrial target. Several other seamounts of the Tyrrhenian sea, belonging to the same geological framework, share Marsili characteristics, among the most important: high heat-flow anomalies, relative shallow depth of the crest and rock porosity.

Acknowledgments

The Authors wish to thank the Eurobuilding S.p.A. for the fruitful cooperation during the data collection. The authors are indebted with Diego Paltrinieri for his comments and discussions during the development of the scientific activities.

Author Contributions

Patrizio Signanini and Paolo Favali were scientific coordinators of the geophysical projects; Mario L. Rainone and Sergio Rusi collected geophysical data during the 2006 scientific cruise and took care of the editing of the paper; Angelo De Santis e Francesco Italiano coordinated the geophysical and geochemical (respectively) data elaboration. All the authors cooperated to the writing of the paper.

Conflicts of Interest

The authors declare no conflict of interest.

References

1. Baker, E.T.; German, C.R.; Elderfield, H. Hydrothermal plumes over spreading-centre axes global distribution and geological inferences. In *Seafloor Hydrothermal Systems, Geophysical Monograph*; Humphris, S.E., Zierenberg, R.A., Mullineaux, L.S., Thomson, R.E., Eds.; American Geophysical Union: Washington, DC, USA, 1995; Geophysical Monograph Series, Volume 91, pp. 47–71.
2. Elderfield, H.; Schultz, A. Mid-ocean ridge hydrothermal fluxes and the chemical composition of the ocean. *Ann. Rev. Earth Planet. Sci.* **1996**, *24*, 191–224.
3. D'Alessandro, A.; D'Anna, G.; Luzio, D.; Mangano, G. The INGV new OBS/H: Analysis of the signals recorded at the Marsili submarine volcano. *J. Volc. Geoth. Res.* **2009**, *183*, 17–29.
4. Lupton, J.; de Ronde, C.; Sprovieri, M.; Baker, E.T.; Bruno, P.P.; Italiano, F.; Walker, S.; Faure, K.; Leybourne, M.; Britten, K.; *et al.* Active Hydrothermal Discharge on the Submarine Aeolian Arc: New Evidence from Water Column Observations. *J. Geophys. Res.* **2011**, *116*, B02102, doi:10.1029/2010JB007738.
5. Jannasch, H.W.; Mottl, M.J. Geo-microbiology of deep-sea hydrothermal vents. *Science* **1985**, *229*, 717–725.
6. Williams, D.L. Submarine geothermal resources. *J. Volc. Geoth. Res.* **1976**, *1*, 85–100.
7. Buonasorte, G.; Cameli, G.M.; Fiordelisi, A.; Parotto, M.; Perticone, I. Results of geothermal exploration in Central Italy (Latium-Campania). In Proceedings of the World Geothermal Congress, Florence, Italy, 18–31 May 1995; Volume 2, pp. 1293–1298.
8. Baldi, P.; Bertini, G.; Cameli, G.M.; Decandia, F.A.; Dini, I.; Lazzarotto, A.; Liotta, D. La tettonica distensiva post-collisionale nell'Area Geotermica di Larderello (Toscana meridionale). *Stud. Geol. Camerti* **1994**, *1*, 183–193.
9. Batini, F.; Brogi, A.; Lazzarotto, A.; Liotta, D.; Pandeli, E. Geological features of Larderello-Travale and Mt. Amiata geothermal areas (southern Tuscany, Italy). *Episodes* **2003**, *26*, 239–244.
10. Bertini, G.; Cappetti, G.; Fiordelisi, A. Characteristics of geothermal fields in Italy. *Giornale di Geologia Applicata* **2005**, *1*, 247–254.
11. Della Vedova, B.; Bellani, S.; Pellis, G.; Squarci, P. Deep temperatures and surface heat flow distribution. In *Anatomy of an Orogen: The Apennines and Adjacent Mediterranean Basin*; Vai, G.B., Martini, I.P., Eds.; Kluwer Academic Publishers: Dordrecht, The Netherlands, 2001; pp. 65–76.
12. Mongelli, F.; Zito, G.; de Lorenzo, S.; Doglioni, C. Geodynamic interpretation of the heat-flow in the Tyrrhenian Sea. *Mem. Descr. Carta. Geol. Ital.* **2004**, *LXIV*, 71–82.

13. Bertani, R. World geothermal generation in 2007. *GHC Bull.* **2007**, 8–19.

14. Marani, M.P.; Gamberi, F. Distribution and nature of submarine volcanic landforms in the Tyrrhenian Sea: The arc vs the back-arc. *Mem. Descr. Carta. Geol. Ital.* **2004**, *LXIV*, 109–126.

15. Doglioni, C.; Innocenti, F.; Morellato, C.; Procaccianti, D.; Scrocca, D. On the Tyrrhenian sea opening. *Mem. Descr. Carta. Geol. Ital.* **2004**, *LXIV*, 147–164.

16. Rosenbaum, G.; Lister, G.S. Neogene and Quaternary rollback evolution of the Tyrrhenian Sea, the Apennines, and the Sicilian Maghrebides. *Tectonics* **2004**, *23*, TC1013, doi:10.1029/2003TC001518.

17. Barberi, F.; Bizouard, H.; Capaldi, G.; Ferrara, G.; Gasparini, P.; Innocenti, F.; Joron, J.L.; Lambert, B.; Treuil, M.; Allegrè, C. Age and nature of basalts from the Tyrrhenian Abyssal Plain. In: *Initial Reports of the Deep-Sea Drilling Project*; Kenneth, H.J., Montadert, L., Bernoulli, D., Bizon, G., Cita, M., Erickson, A., Fabricius, F., Garrison, R.E., Kidd, R.B., Mélières, F., Müller, C., Wright, R.C., Eds.; IODP, 1978; doi:10.2973/dsdp.proc.42-1.1978, Publication date: May 2007, pp. 509–514. Available online: http://www.deepseadrilling.org/42_1/volume/dsdp42pt1_18.pdf (accessed on 25 June 2014).

18. Savelli, C. Late Oligocene to Recent episodes of magmatism in and around Tyrrhenian Sea: Implications for the processes of opening in a young inter-arc basin of intra-orogenic (Mediterranean) type. *Tectonophysics* **1988**, *146*, 163–181.

19. Beccaluva, L.; Coltorti, M.; Galassi, B.; Macciotta, G.; Siena, F. The Cainozoic calc-alkaline magmatism of the western Mediterranean and its geodynamic significance. *Boll. Geof. Teor. App.* **1994**, *36*, 293–308.

20. Selvaggi, G.; Chiarabba, C. Seismicity and P-wave velocity image of the Southern Tyrrhenian subduction zone. *Geophys. J. Int.* **1995**, *121*, 818–826.

21. Neri, G.; Caccamo, D.; Cocina, O.; Montalto, A. Geodynamic implications of earthquake data in the southern Tyrrhenian Sea. *Tectonophysics* **1996**, *258*, 233–249.

22. Favali, P.; Beranzoli, L.; Maramai, A. Review of the Tyrrhenian seismicity: How much is still to be known? *Mem. Descr. Carta. Geol. Ital.* **2004**, *LXIV*, 57–70.

23. Marani, M.P.; Gamberi, F. Structural framework of the Tyrrhenian Sea unveiled by seafloor morphology. *Mem. Descr. Carta. Geol. Ital.* **2004**, *LXIV*, 97–107.

24. Verzhbitskii, E.V. Heat flow and matter composition of the lithosphere of the world ocean. *Oceanology* **2007**, *47/4*, 564–570.

25. Steinmetz, L.; Ferrucci, F.; Hirn, A.; Morelli, C.; Nicolich, R. A 550-km-long Moho traverse in the Tyrrhenian Sea from OBS recorded P_n waves. *Geophys. Res. Lett.* **1983**, *10*, 428–431.

26. Locardi, E.; Nicolich, R. Geodinamica del Tirreno e dell'Appennino Centro Meridionale: La nuova carta della Moho. *Mem. Soc. Geol. Ital.* **1988**, *6*, 121–140.

27. Cella, F.; Fedi, M.; Florio, G.; Rapolla, A. Gravity modelling of the litho-asthenosphere system in the Central Mediterranean. *Tectonophysics* **1998**, *287*, 117–138.

28. Finetti, I.R. Innovative seismic highlights on the Mediterranean region. In *Geology of Italy, Special Volume of the Italian Geological Society*; Crescenti, U., D'Offizi, S., Merlino, S., Sacchi, L., Eds.; Società Geologica Italiana: Roma, Italy, 2004, pp. 131–140.

29. Vitale, S.; de Santis, A.; di Mauro, D.; Cafarella, L.; Palangio, P.; Beranzoli, L.; Favali, P. Geostar deep seafloor missions: Magnetic data analysis and 1D geoelectric structure underneath the Southern Tyrrhenian Sea. *Ann. Geophys.* **2009**, *52/1*, 57–63.

30. Turco, E.; Zuppetta, A. A kinematic model for the Plio-Quaternary evolution of the Tyrrhenian-Apenninic system: Implications for rifting processes and volcanism. *J. Volc. Geoth. Res.* **1998**, *82*, 1–18.

31. Pondrelli, S.; Piromallo, C.; Serpelloni, E. Convergence vs retreat in Southern Tyrrhenian Sea: Insights from kinematics. *Geophys. Res. Lett.* **2004**, *31*, doi:10.1029/2003GL019223.

32. Selli, R.; Lucchini, F.; Rossi, P.L.; Savelli, C.; del Monte, M. Dati geologici, petrochimici e radiometrici sui vulcani centro-tirrenici. *Gior. Geol.* **1977**, *42*, 221–246. (in Italian)

33. Savelli, C.; Gasparotto, G. Calc-alkaline magmatism and rifting of the deep-water volcano of Marsili (Aeolian back-arc, Tyrrhenian Sea). *Mar. Geol.* **1994**, *119*, 137–157.

34. Trua, T.; Serri, G.; Marani, M.; Renzulli, A.; Gamberi, F. Volcanological and petrological evolution of Marsili seamount (southern Tyrrhenian Sea). *J. Volc. Geoth. Res.* **2002**, *114*, 441–464.

35. Faggioni, O.; Pinna, E.; Savelli, C.; Schreider, A.A. Geomagnetism and age study of Tyrrhenian seamounts. *Geophys. J. Int.* **1995**, *123*, 915–930.

36. Marani, M.P.; Trua, T. Thermal constriction and slab tearing at the origin of a super-inflated spreading ridge: Marsili volcano (Tyrrhenian Sea*). J. Geophys. Res.* **2002**, *107*, doi:10.1029/2001JB000285.

37. Nicolosi, I.; Speranza, F.; Chiappini, M. Ultrafast oceanic spreading of the Marsili Basin, southern Tyrrhenian Sea: Evidence from magnetic anomaly analysis. *Geology* **2006**, *34/9*, 717–720.

38. Cocchi, L.; Caratori Tontini, F.; Muccini, F.; Marani, M.P.; Bortoluzzi, G.; Carmisciano, C. Chronology of the transition from a spreading ridge to an accretional seamount in the Marsili back-arc basin (Tyrrhenian Sea). *Terra Nova* **2009**, *21*, doi:10.111/j.1365-3121.2009.00891.x.

39. Dekov, V.M.; Savelli, C. Hydrothermal activity in the SE Tyrrhenian Sea: An overview of 30 years of research. *Mar. Geol.* **2004**, *204*, 161–185.

40. Italiano, F.; Caso, C.; Cavallo, A.; Favali, P.; Fu, C.; Iezzi, G.; Martelli, M.; Mollo, S.; Paltrinieri, D.; Paonita, A.; *et al.* Geochemical features of the gas phase extracted from sea-water and rocks of the Marsili seamount (Tyrrhenian sea, Italy): Implications for geothermal exploration projects. In Proceedings of the ICGG11, International Conference on Gas Geochemistry 2011, San Diego, CA, USA, 28 November–4 December 2011; Volume 67–68.

41. Lagmay, A.M.F.A.; Tengociang, A.M.P.; Marcos, H.B.; Pascua, C.S. A structural model for geothermal exploration in Ancestral Mount Bao, Leyte, Philippines. *J. Volc. Geoth. Res.* **2003**, *122*, 133–141.

42. Murton, B.J.; Parson, L.M. Segmentation, volcanism and deformation of oblique spreading centres: A quantitative study of the Reykjanes Ridge. *Tectonophysics* **1993**, *222*, 237–257.

43. McGuire, W.J. Volcano instability: A review of contemporary themes. *Geol. Soc. Lond. Spec. Publ.* **1996**, *110*, 1–23, doi:10.1144/GSL.SP.1996.110.01.01.

44. Blanco-Montenegro, I.; Nicolosi, I.; Pignatelli, A. Magnetic imaging of the feeding system of oceanic volcanic islands: El Hierro (Canary Islands). *Geophys. J. Int.* **2008**, *173*, 339–350.

45. Caso, C.; Signanini, P.; de Santis, A.; Favali, P.; Iezzi, G.; Marani, M.P.; Paltrinieri, D.; Rainone, M.L.; di Sabatino, B. Submarine geothermal systems in Southern Tyrrhenian Sea as future energy resource: The example of Marsili seamount. In Proceedings of the World Geothermal Congress 2010, Bali, Indonesia, 25–30 April 2010; pp. 1–9.

46. Italiano, F.; Bonfanti, P.; Ditta, M.; Petrini, R.; Slejko, F. Helium and carbon isotopes in the dissolved gases of Friuli region (NE Italy): Geochemical evidence of CO_2 production and degassing over a seismically active area. *Chem. Geol.* **2009**, *266*, 76–85, doi:10.1016/j.chemgeo.2009.05.022.

47. Italiano, F.; Sasmaz, A.; Yuce, G.; Okan, O. Thermal fluids along the East Anatolian Fault Zone (EAFZ): Geochemical features and relationships with the tectonic setting. *Chem. Geol.* **2013**, *339*, 103–114, doi:10.1016/j.chemgeo.2012.07.027.

48. Hildenbrand, T.G.; Tosenbaum, J.; Kauahikaua, J. Aeromagnetic study of the island of Hawaii. *J. Geophys. Res.* **1993**, *98*, 4099–4119.

49. Tivey, M. Fine-scale magnetic anomaly field over the southern Juan de Fuca Ridge: Axial magnetisation low and implications for crustal structure. *J. Geophys. Res.* **1994**, *99*, 4833–4855.

50. Irving, E. The Mid-Atlantic Ridge at 45° N. Oxidation and magnetic properties of basalt; Review and discussion. *Can. J. Earth Sci.* **1970**, *7*, 1528–1538.

51. Johnson, H.P.; Karsten, J.L.; Vine, F.J.; Smith, G.C.; Schonharting, G. A low-level magnetic survey over a massive sulfide ore body in the Troodos ophiolite complex, Cyprus. *Mar. Tech. Soc. J.* **1982**, *16*, 76–79.

52. Uchupi, E.; Ballard, R.D. Evidence of hydrothermal activity on Marsili Seamount, Tyrrhenian Basin. *Deep-Sea Res. Part A, Oceanogr. Res. Pap.* **1989**, *36*, 1443–1448.

53. Caratori Tontini, F.; Cocchi, L.; Muccini, F.; Carmisciano, C.; Marani, M.P.; Bonatti, E.; Ligi, M.; Boschi, E. Potential-field modeling of collapse-prone submarine volcanoes in southern Tyrrhenian Sea (Italy). *Geophys. Res. Lett.* **2010**, *37*, doi:10.1029/2009GL041757,2010.

54. Wallace, P.J. Volatiles in subduction zones magmas: Concentrations and fluxed based on melt inclusion and volcanic gas data. *J. Volc. Geoth. Res.* **2005**, *114*, 441–464.

55. Kieffer, S.W. Seismicity at Old Faithful Geyser: An isolated source of geothermal noise and possible analogue of volcanic seismicity. *J. Volc. Geoth. Res.* **1984**, *22*, 59–95.

56. Crone, T.J.; Wilcock, W.S.D.; Barclay, A.; Parsons, J.D. The sound generated by mid-ocean ridge black smoker hydrothermal vents. *PLoS One* **2006**, *1*, doi:10.1371/journal.pone.0000133.

57. Legaz, A.; Revil, A.; Roux, P.; Vandemeulebrouck, J.; Gouèdard, P.; Hurst, T.; Bolève, A. Self-potential and passive seismic monitoring of hydrothermal activity: A case study at Iodine Pool, Waiomangu geothermal valley, New Zealand. *J. Volc. Geoth. Res.* **2009**, *179*, 11–18.

58. Chouet, B. Long-period volcano seismicity: Its source and use in eruption forecasting. *Nature* **1996**, *380*, 309–316.

59. Kumagai, H.; Chouet, B. The complex frequencies of long-period seismic events as probes of fluid composition beneath volcanoes. *Geophys. J. Int.* **1999**, *138*, F7–F12.

60. Sgroi, T.; Montuori, C.; Agrusta, R.; Favali, P. Low-frequency seismic signals recorded by OBS at Stromboli volcano (Southern Tyrrhenian Sea). *Geophys. Res. Lett.* **2009**, *36*, L04305, doi:10.1029/2008GL036477.

61. Iezzi, G.; Caso, C.; Ventura, G.; Vallefuoco, M.; Cavallo, A.; Behrens, H.; Mollo, S.; Paltrinieri, D.; Signanini, P.; Vetere, F. First documented deep submarine explosive eruptions at the Marsili Seamount (Tyrrhenian Sea, Italy): A case of historical volcanism in the Mediterranean Sea. *Gondwana. Res.* **2014**, *25*, 764–774.

62. Eurobuilding, Marsili Project. Available online: http://www.eurobuilding.it/marsiliproject/ (accessed on 10 March 2014).

63. Armani, F.B.; Paltrinieri, D. Perspectives of offshore geothermal energy in Italy. *EPJ Web Conf.* **2013**, *54*, 02001, doi:10.1051/epjconf/20135402001.

64. Beranzoli, L.; Ciafardini, A.; Cianchini, G.; de Caro, M.; de Santis, A.; Favali, P.; Frugoni, F.; Marinaro, G.; Monna, S.; Montuori, C.; *et al.* A first insight in the Marsili volcanic seamount (Tyrrhenian Sea): Results from ORION-GEOSTAR3 experiment. In: *Seafloor Observatories: A New Vision of the Earth from the Abyss*; Favali, P., Beranzoli, L., de Santis, A., Eds.; Springer-Praxis Publishing: Amsterdam, The Netherlands, 2014; ISBN:978-3-642-11373-4 (in press).

65. Favali, P.; Beranzoli, L.; de Santis, A. *Seafloor Observatories: A New Vision of the Earth from the Abyss*; Favali, P., Beranzoli, L., de Santis, A., Eds.; Springer-Praxis Publishing: Amsterdam, The Netherlands, 2014; ISBN:978-3-642-11373-4.

Geothermal Potential Evaluation for Northern Chile and Suggestions for New Energy Plans

Monia Procesi

Abstract: Chile is a country rich in natural resources, and it is the world's largest producer and exporter of copper. Mining is the main industry and is an essential part of the Chilean economy, but the country has limited indigenous fossil fuels—over 90% of the country's fossil fuels must be imported. The electricity market in Chile comprises two main independent systems: the Northern Interconnected Power Grid (SING) and the Central Interconnected Power Grid (SIC). Currently, the primary Chilean energy source is imported fossil fuels, whereas hydropower represents the main indigenous source. Other renewables such as wind, solar, biomass and geothermics are as yet poorly developed. Specifically, geothermal energy has not been exploited in Chile, but among all renewables it has the greatest potential. The transition from thermal power plants to renewable energy power plants is an important target for the Chilean Government in order to reduce dependence on imported fossil fuels. In this framework, the proposed study presents an evaluation of the geothermal potential for northern Chile in terms of power generation. The El Tatio, Surire, Puchuldiza, Orriputunco-Olca and Apacheta geothermal fields are considered for the analysis. The estimated electrical power is approximately 1300 MWe, and the energy supply is 10,200 GWh/year. This means that more than 30% of the SING energy could be provided from geothermal energy, reducing the dependence on imported fossil fuels, saving 8 Mton/year of CO_2 and supplying the mining industry, which is Chile's primary energy user.

Reprinted from *Energies*. Cite as: Procesi, M. Geothermal Potential Evaluation for Northern Chile and Suggestions for New Energy Plans. *Energies* **2014**, *7*, 5444-5459.

1. Introduction

Chile is a republic located in South America bordered by the Andes Mountains and the Pacific Ocean. The country is partitioned into 15 administrative regions. Chile is rich in natural resources such as copper, timber, nitrates, precious metals and molybdenum. It is the world's largest producer and exporter of copper and mining is the primary Chilean industry and an essential component of the Chilean economy. Although Chile has an abundance of copper and other mining resources, it has limited indigenous fossil fuels and over 90% of its fossil fuels needs must be met by imports [1].

In 2012, Chile's energy demand was approximately 60,000 GWh and it is projected to increase by 4.9% annually reaching a consumption of 136,000 GWh/year by 2030 [1,2]. The installed capacity for electric generation is 17,500 MWe, producing approximately 65,600 GWh/year [3]. Of the total installed capacity 62% corresponds to fossil-fuelled power plants, 34% to hydropower plants and only 4% to renewable resources [3]. The majority of the total electricity is consumed by the mining and industry sectors. The electricity market in Chile is comprised of two main independent systems:

SING—the Northern Interconnected Power Grid;

SIC—the Central Interconnected Power Grid.

SING has an installed capacity of approximately 3800 MWe, 100% of which is generated by imported fossil fuel. SIC has an installed capacity of 13,500 MWe, 53% of which is generated by imported fossil fuels, 42% by hydroelectric power and the remaining 4% by renewable resources [3].

Therefore, Chile is heavily dependent on imported fossil fuels and hydropower represent the main indigenous energy source of the country. Other renewable resources such as wind, solar, biomass and geothermal energy are poorly developed, but several corporations, such as the Centre for Renewable Energy (CER), have been working to ensure the optimal participation of renewable energies in Chile's energy matrix to contribute to the sustainable development of the country [4].

In 2011, The Global Energy Network Institute (GENI) suggested that a strategic energy plan for Chile was necessary to ensure both energy autonomy of the country and the transition from thermal power plants to renewable energy power plants [4]. This passage from conventional energy to renewable sources is indispensable both to reduce the amount of greenhouse gases emitted into the atmosphere and to reduce the dependence on imported fossil fuels. Under such circumstances, renewable energy resources in Chile should become more relevant. Geothermal energy is not yet exploited here, but among all renewable resources it has the greatest development potential. Although no geothermal power plants have been installed to date, a vigorous geothermal exploration program is under way [5]. A preliminary evaluation of northern Chile's geothermal power potential is approximately 2000 MWe, whereas the country's central-southern region is estimated to be between 1000 and 1500 MWe [6,7]. This makes Chile one of the most attractive countries for the utilisation of geothermal energy.

The purpose of this work is to review the geothermal potential evaluation for northern Chile in terms of power generation, considering the conventional geothermal resources of Surire, Puchuldiza, Apacheta, El Tatio and Irriputunco-Olca. This evaluation could provide useful information for the development of a new strategic energy plan applicable to the SING. Northern Chile has been chosen for the analysis because more data are available than for the central and southern regions, and because of the country's large energy reliance on imported fossil fuels.

2. Regional Setting of the Study Area

Northern Chile has a relatively homogeneous geological setting, consisting of Lower Miocene–Pleistocene ignimbrite deposits and andesitic–rhyolitic volcanic products overlying Middle Cretaceous–Upper Miocene volcano-sedimentary formations [8–13]. The latter hosts the main hydrothermal reservoirs that predominantly consist of andesitic lava and pyroclastic flows, conglomerates, breccias, sandstones, siltstones, limestones, marls and evaporites [10,14]. The main hydrothermal systems are located within the NS-, NW-trending grabens [8,10,15] on the western side of the Pliocene–Holocene Central Andean Volcanic Zone.

For each studied system (Figure 1), a brief geographical/geological setting with relative descriptions of the geothermal manifestations at the surface is reported. The geothermal fields of

Colpitas and Larima are also described, but it was not possible to evaluate their geothermal potential due to a lack of public data.

Figure 1. Geothermal areas of northern Chile. The red box shows the studied area.

2.1. Colpitas

The Colpitas geothermal field is located in the northernmost part of Chile, in the Arica and Parinacota region, with an elevation from 4000 to 5200 m above sea level (a.s.l.). Geologically, the area is characterised by volcanic rocks and volcanoclastic deposits and volcano-sedimentary sequences ranging from Miocene to Holocene [16]. The stratigraphy is mainly comprised of the Lupica Formation (Upper Oligocene-Miocene), corresponding to the *basement* of the basin. It is constituted by rhyolitic ash flow tuffs, andesitic lavas and subvolcanic plugs as well as epiclastic sandstones. The Lupica Formation underlies, in the western part of the study area, the Huaylas Formation (Upper Miocene). The Huaylas Formation is mainly formed by lacustrine epiclastic sandstones and gravels, with interbedded ash layers and is itself overlain by the Lauca Ignimbrite (Upper Pliocene).The potential reservoir is likely located within Cenozoic volcanoclastic rocks. The thermal springs, located in the northern and southern part of the Colpitas field, have temperatures that range from 28 to 55 °C with a total flow <10 L/s [16]. The North Thermal springs are interpreted as being the most representative of deeply derived thermal waters [16]. Stable isotope composition indicate that thermal spring waters have been subject to evaporation, or more likely, varying amounts of mixing with groundwater brine that underlines the salt deposits in the basin [16]. Na/K ratio of the North Thermal Springs gives the hottest equilibration temperatures of approximately 235 °C. Many of the springs have moderate bubbling of what is likely to be CO_2 and H_2S gas [16]. Recently, a slimhole well was drilled to prove the existence of a geothermal reservoir and to evaluate the potential of the field [3,16].

2.2. Surire

The Surire hydrothermal system is located at an altitude of 4000–4300 m a.s.l., in the southeast part of the Salar de Surire. Many volcanoes rise above 5500 m around the salar, some lavas have flowed into it and are now partially buried by sediments. The volcanoes are dacitic and andesitic; the dacities generally occur as domes and the andesities form stratovolcanoes. In 1972, as many as 133 thermal discharges occurred in an approximately 15 km^2 area [15]. Presently, most of the bubbling pools and thermal springs are located along the southern border of the salar and have temperatures between 20 and 80 °C [15]. The Database of Geothermal Resources in Latin America & the Caribbean [17] indicated a potential temperature of the geothermal reservoir of 110–234 °C and an electric power output of 50–60 MWe.

2.3. Puchuldiza

The Puchuldiza-Tuja hydrothermal system is located at an altitude of approximately 4100–4200 m a.s.l. and is 27 km SW of the active Isluga volcano that is characterised by permanent fumarolic activity [18]. The hydrothermal area of Puchuldiza is limited both to the north and south by several Plio-Pleistocene stratovolcanoes reaching altitudes higher than 5000 m. A Pleistocene-Holocene fracture system characterized by northeast-southwest faults affected the Plio-Pleistocene volcanoes and produced, in part, the surficial hydrothermal activity. The fluid discharges within the Puchuldiza area are controlled by the Churicollo, Puchuldiza and Tuja faults. Several thermal springs with low gas emissions surround the main emission areas [19]. The geothermal reservoir is hosted in the Utayane Ignimbrite and the Puchuldiza Formation. The permeability of the volcanic formations is due to cooling joints and tectonic fractures [6]. Geothermometry evaluations suggest that the fluid reservoir has a relatively high equilibrium temperature, up to 270 °C [18]. The Database of Geothermal Resources in Latin America & the Caribbean [17] indicated a potential electric power output of 190 MWe.

2.4. Lirima

The Lirima geothermal field is located at an altitude of 3900 m a.s.l., 25 km SW of the Sillajguay volcanic chain. The geology of the area is characterized by Mesozoic basement rocks constituted by clastic-carbonate sequences, volcanic and sedimentary rocks of Oligocene-Miocene and middle Miocene to Pleistocene volcanic edifices [20]. In the Lirima area have been recognized bubbling pools, along the western side of the field, and three main sites with thermal springs [21]. The thermal springs are characterized by temperatures between 38 °C and 80 °C, high Cl and B concentration, $\delta^{18}O$ enrichment, and relatively low Mg concentration; consistent with deep circulation from a geothermal reservoir, and low mixing degree. Minimum temperatures from water and gas geothermometers range from 200 to 200 °C.

2.5. Apacheta

Apacheta is located 105 km NE of Calama City and 55 km NW of the El Tatio hydrothermal system. A 180 m deep well (PAE-1) drilled by the Chilean National Mining Company (CODELCO) in 1998 produced steam measured at 88 °C [22]. Fluid discharges emitting superheated steam (up to 118 °C) [22] with high flow rates are found along the eastern flank of the 5150 m high Apacheta volcano. Currently, in the Apacheta geothermal field, a project (Cerro Pabellón) is underway for power production from geothermal resources. Project feasibility studies began in 2005, then four wells were drilled from 2009 to 2010, whose depths reached between 1300 and 2000 m. The results of the production and injection tests showed the presence of a liquid-dominated reservoir with a maximum-measured temperature of 260 °C. ENEL Green Power has planned the drilling of 13 wells to operate a 50 MWe power plant. This project represents the first commercial-scale geothermal plant in the country [7].

2.6. Irriputunco-Olca

The Irriputuncu-Olca field is characterised by the presence of the Irriputuncu and Olca Volcanos, located in the Chilean Altipiano at 4000–5000 m a.s.l., and in the vicinity of the copper mine. Irriputuncu is an active dacitic stratovolcano, with fumaroles at the top crater and one acid-sulphate hot spring at the base of the volcano. Two slim boreholes (800 and 1430 m in depth) measured a bottom hole temperature close to 150 °C and 195 °C (at 3350 and 3000 m a.s.l., respectively) [23]. Time domain electromagnetic (TEM) and Magnetotelluric (MT) data suggests the presence of a potentially deeper reservoir at approximately 220 °C [23]. Olca is an andesitic volcano, of which the TEM-MT data exhibit two conductive layers intercalated with resistive zones. The possible thickness of the reservoir is 2000–3000 m and for the surface area there are three estimates, conservative (7.5 km^2), likely (15 km^2) and optimistic (45 km^2) [23]. Preliminary results, assuming up to 10 MWe/km^2 and a reservoir temperature of 230–300 °C, suggest a potential for electric generation between 75 and 450 MWe [23].

2.7. El Tatio

El Tatio is located 100 km E of Calama City at an altitude of 4300 m a.s.l. Several thermal springs, fumaroles, geysers and boiling and mud pools are present. Hydrogeological models [24,25] indicate that meteoric waters infiltrate in recharge areas 15 km E of the field. The main hydrothermal reservoir is confined within the permeable Puripicar Formation and the Salado Member. An important secondary aquifer occurs in the Tucle Dacite subunit that is capped by the impermeable Tatio Ignimbrite [24]. In Figure 2, the simplified geological profile and circulation conceptual model are shown. The potential geothermal reservoir is hosted in the Puripicar Formation and Salado Member, although a temperature of approximately 170 °C was recorded in the permeable levels hosted in the Grupo volcanic de Tucle [6].

In Figure 3, a profile of the El Tatio Graben is shown. It crosses through the wells numbered 1, 4, 9 and 7, from NW to SE. In the boreholes, three permeable zones were detected. The permeability essentially originated by tectonic fracturing or rapid cooling of the volcanic bodies [9]. The temperatures of the three permeable zones ranged from 170, 230 to 260 °C, moving from the Grupo volcanico de Tucle to Puripicar Formation and Salado Member, respectively [9].

Figure 2. Simplified geological map of the El Tatio geothermal field, geological profile and circulation conceptual model. The dashed box represents the area in Figure 3 (modified from [9]).

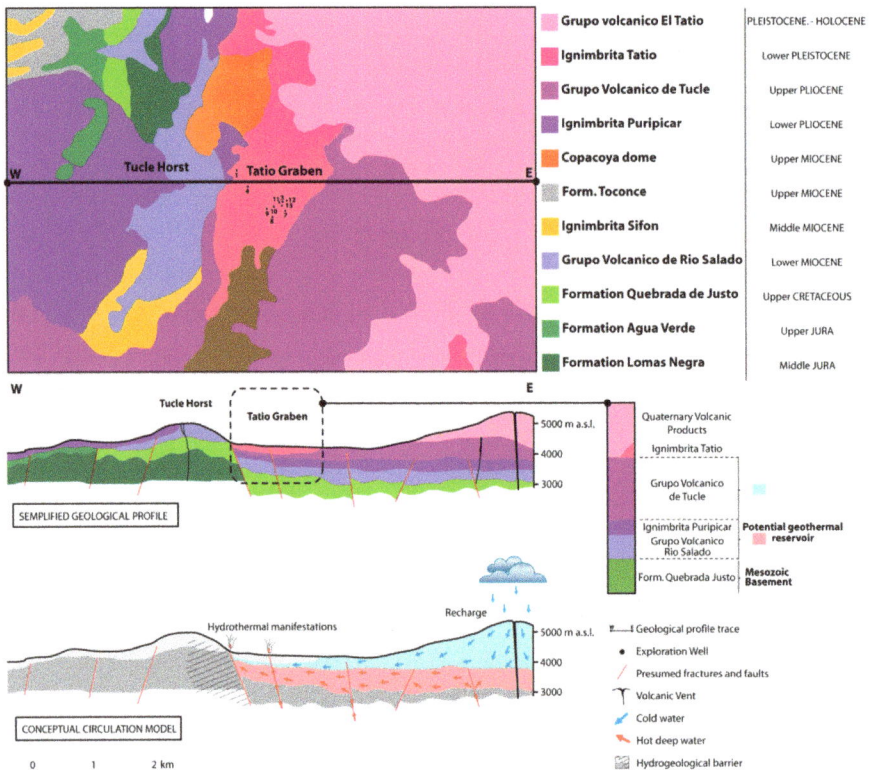

Figure 3. The geological profile from NW-SE of the El Tatio graben through the boreholes numbered 1, 4, 9 and 7 (modified from [9]). For further details see text in the previous page; paragraph relative to El-Tatio.

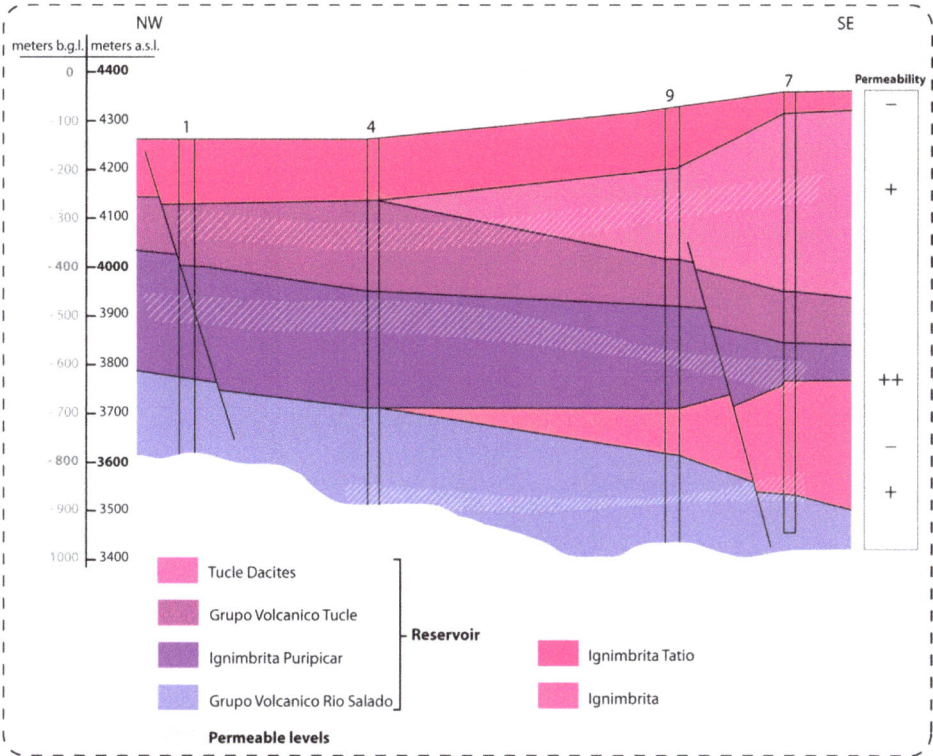

2.8. Geothermal Exploration in Chile

Initial geothermal exploration in the Central Andean Volcanic Zone took place in late 1960s in response to increasing Chilean energy demands. At El Tatio, a pre-feasibility investigation, funded in 1967 by the Corporation for the Promotion of Development and the United Nations Development Program (CORFO/UNDP), was followed by geological, geophysical and geochemical surveys from 1968 to 1980. Six 600 m deep exploration wells, drilled between 1969 and 1971, encountered temperatures up to 250 °C. Seven production wells drilled in 1973 and 1974 disclosed three discrete reservoirs with temperatures up to 260 °C. Three of these wells produced an average of 14.7 kg/s (adequate for 6 MW each); two other wells produced less, but could still be capable of 5 MW each. An electric power output of 100 MWe for the El Tatio geothermal field was estimated [26]. At Puchuldiza, geological, geochemical and geophysical studies were performed by CORFO/UNDP (from 1968 to 1974) and by Japan International Cooperation Agency (JICA) (from 1978 to 1980) to evaluate the geothermal potential. Six wells were drilled in 1976 and a depth of 1200 m was reached. Temperatures up to 175 °C were measured at depth of 900 m [27,28].

In the Surire zone, geological and geochemical [15,29] investigations were performed by CORFO between 1972 and 1979. Reservoir temperatures up to 230 °C were estimated by geothermometric calculations based on the water chemistry of the thermal discharges [29]. Geological and geochemical studies were conducted but no wells were drilled.

Geothermal exploration in the Central Andean Volcanic Zone was abandoned in 1982 because of both the remote location of the hydrothermal systems and economic factors. After almost three decades, private and governmental companies have planned to conduct a new phase of geothermal exploration in the systems investigated between 1969 and 1982 as well as in other areas of northern Chile. New slim holes will be drilled at Puchuldiza, Polloquere, Pampa Lirima, Colpitas and Juncalito by Energía Andina, which plans to have a geothermal plant working by 2015 [30].

In central-southern Chile, the geothermal activity is related to the Pocuro fault system (33°–34° S) and to the Liquiñe-Ofqui Fault Zone (39°–46° S). Several slimholes have been drilled in Tinguiririca, Calabozos, Laguna del Maule, Chillán and Tolhuaca with a potential output estimated at 3–10 MW per well (Figure 4). At Tolhuaca, two holes have been drilled up to depths of 1200 m and a 50 MW geothermal power plant is planned to start production in 2013. The potential is estimated at 600 MWe to 950 MWe in this area [31]. Currently, 79 exploration and 7 exploitation concessions have been given to Chilean and foreign companies [3].

Figure 4. Geothermal areas of southern Chile.

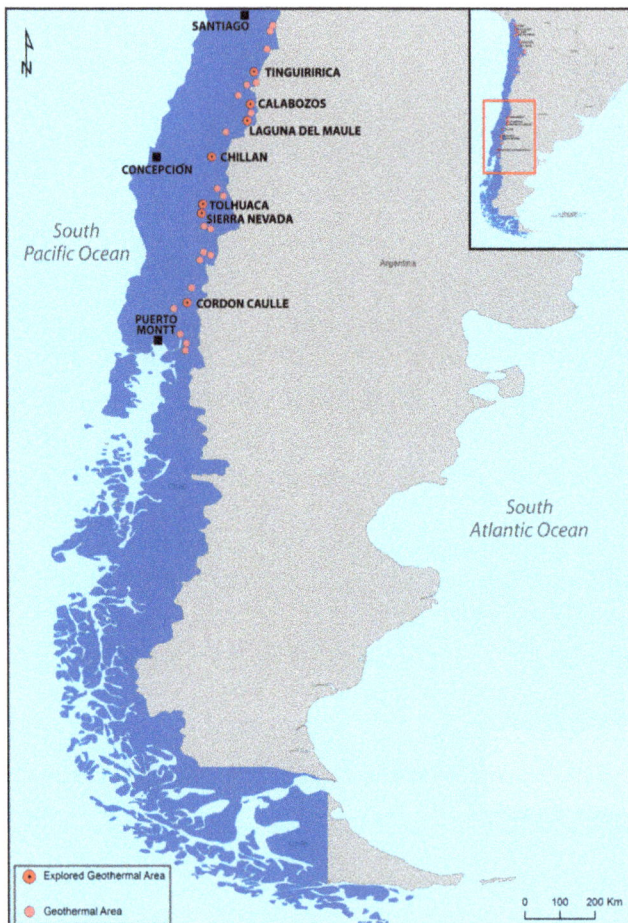

3. Methodology

In the selected potential geothermal fields, the minimum and maximum electric energy supply (E_{su}) was evaluated considering an operation time (OT) of 8000 h/year. To evaluate the E_{su} it was necessary to estimate both heat (Q) and electric power (W_e).

The recoverable Q of examined geothermal reservoir was computed using the following equation:

$$Q = m \cdot C_w \cdot (T - T_0) \tag{1}$$

where m is the mass flow rate of the geothermal system (kg/s), C_w (J/kg·K) is the specific heat capacity of the fluids contained in the geothermal reservoir, T is the reservoir temperature (K) and T_0 is the reinjection temperature. The minimum reservoir temperature T^- and the maximum reservoir temperature T^+ were used to compute the minimum (Q^-) and maximum (Q^+) heat, respectively.

The production rate m of the examined geothermal system is obtained by multiplying the volume of the geothermal reservoir V (km^3) and the specific productivity (m_w) [32] derived from the flow well tests. The W_e was estimated using the following equation:

$$W_e = Q \cdot \eta \tag{2}$$

where the η represents the efficiency of the selected power plant. The electric power was computed considering both conventional and binary cycle power plants as ORC (Organic Rankine Cycle). For conventional and binary plants, a η of 20% was considered, whereas the value of η, relative to geothermal binary power plants, was computed following the methodology proposed by Di Pippo [33]. Generally, the ideal cycle for a binary plant is the Carnot Cycle, but this assumption is inappropriate and can result in misleading conclusions [33–36]. Carnot's ideal cycle produces the highest efficiency with respect to any other cycle operating between a heat source and a sink, but the Carnot Cycle is applicable only to reversible processes. This property means that all heat transfer and work processes must be thermodynamically perfect and these conditions are impossible for a real cycle. A more useful model is the triangular (or trilateral) cycle, which considers the heating medium not as an isothermal source but rather as a fluid that cools as it transfers heat to the cycle working fluid [33]. Therefore, the efficiency (η) was computed considering a triangular cycle and following the equation:

$$\eta_{TR} = (T_H - T_C)/(T_H + T_C) \tag{3}$$

where T_H and T_C represent the heat source and the fixed condensing temperature, respectively. In this work, a T_C of 50 °C was considered. In order to convert from the ideal cycle to the practical it is necessary to apply a relative efficiency [33]. Real binary plants have demonstrated relative efficiencies of about 55% ± 10% [33]. Thus, one may estimate the efficiency of a binary plant using the approximate formula:

$$\eta_{b[TR]} = 0.55(T_H - T_C)/(T_H + T_C) \tag{4}$$

Therefore, the minimum and maximum *electric energy supply* (E_{el}) was evaluated for the selected geothermal systems based on 8000 working hours per year. The total E_{su} from the selected geothermal systems was evaluated. Moreover, the CO_2 emissions saving was computed considering the substitution for the estimated E_{su} of the crude oil with geothermal energy.

The relationship between CO_2 emissions and the different energy resources, such as coal, crude oil, natural gas and geothermal energy, is shown in Table 1.

Table 1. CO_2 emissions for kWh by coal, petroleum, natural gas and geothermal energy [37,38].

CO_2 Emissions (g/kWh)	Coal	Petroleum	Natural Gas	Geothermal Energy
	949	892	598	122
References	[37]	[37]	[37]	[38]

4. Results and Discussion

For the analysis, the geothermal systems of El Tatio, Surire, Puchuldiza, Irriputunco-Olca and Apacheta were considered. A satisfactory dataset is available only for the El Tatio geothermal field, whereas for the other systems partial data are present. The specific productivity (m_w) is available only for the El Tatio geothermal field where the Q and W_e were computed, whereas for the others systems, data for W_e were sourced from previous specific literature [17,22,23]. For the El Tatio geothermal system, the boreholes numbered 7, 10 and 11 provide the main information in terms of specific productivity m_w (kg/s), minimum and maximum temperature and type of fluid (Table 2). This geothermal system is water-dominated. The recorded m_w ranges from 37 to 77 kg/s, the temperatures range from 170 to 260 °C and the C_w was considered to have a value of 0.0042 J/kg·K. The minimum temperatures are recorded in the Tucle Formation, whereas the highest temperatures are in the Puripicar and Rio Salado Formation.

Table 2. Available data for the El Tatio geothermal field.

Well	m_w (kg/s)	C_w (J/kgK)	T^- (°C)	T^+ (°C)	K
7	77	0.0042	170	260	273.15
10	37	0.0042	170	260	273.15
11	74	0.0042	170	260	273.15
Reference	[1]	-	[1]	[1]	-

The estimated minimum and maximum heat (Q^-, Q^+) ranges from 1482 to 2752 MW$_{th}$, for a reservoir volume of approximately 15 km^3 (Table 3). The electric power (W_e^-, W_e^+) was estimated for both conventional [c] and binary geothermal power plants [b]. For temperatures up to 170 °C, an efficiency ($\eta_{b[TR]}$) for a geothermal binary plant of 0.09 was computed considering the Equation (4). The minimum estimated electric power ($W_{e\ [b]}^-$) is 45 MW$_e$, whereas the maximum electric power ($W_{e\ [c]}^+$), which was computed considering a conventional plant with an efficiency (η_c) of 0.2, is 174 MW$_e$. The relative energy supply ($E_{su[b]}$, $E_{su[c]}$) for an operation time of 8000 h/year, is 362 and 1391 GWh/year for binary and conventional plants, respectively.

For the other geothermal systems (*i.e.*, Surire, Puchuldiza, Irriputuncu-Olca and Apacheta), the energy supply has been evaluated for the electric power values provided by scientific literature [17,22,23]. The Database of Geothermal Resources in Latin America & the Caribbean [17] indicated an electric power output of 50–60 and 190 MWe for Surire and Puchuldiza, respectively. Reyes *et al.*, in

2011 [23], estimated the electric power for the Irriputunco-Olca system to be 75–450 MWe, and Urzuà et al., in 2002 [22], evaluated the electric power for the Apacheta geothermal field to be 400 MWe. The computed energy supply (E_{su}) for the systems listed above is reported in Table 4. For the evaluation, an operation time of 8000 h/year was considered. The values range from 400 to 3600 GWh/year. The maximum estimated value of the total energy supply for northern Chile is approximately 10,200 GWh/h whereas the minimum total energy supply is around 3000 GWh/year.

The total CO_2 emissions corresponding to 10,200 GWh/h are 1.2 Mton/year. The CO_2 emissions would be approximately 9 Mton/year for the same energy amount if sourced from petroleum only.

Northern Chile has large geothermal potential. The evaluations performed for the El Tatio, Surire, Puchuldiza, Irriputunco-Olca and Apacheta geothermal fields show that the geothermal energy could provide approximately 10,200 GWh/year, with an installed capacity of approximately 1300 MWe. Currently, the SING has an installed capacity of 3800 MWe, approximately 100% of which is generated by imported fossil fuel. Geothermal electric power could replace more than 30% of the SING installed capacity, decreasing the reliance on foreign fossil fuel providers. Furthermore, substituting 1300 MWe of fossil fuel for geothermal energy means a CO_2 emissions savings of approximately 8 Mton per year.

The geothermal potential of northern Chile could be greater than 1300 MWe for two main reasons: (1) it was not possible to evaluate some explored areas because of the lack of public data (especially Larima and Colpita); (2) there are many unexplored areas characterised by medium temperatures that could be exploited using binary systems such as ORC and/or Kalina cycles.

The proposed approach is reasonable for regional estimates of efficiency and power output and cannot replace detailed heat balance analyses needed for plant design.

Table 3. Evaluations of the thermal and electric power for the El Tatio geothermal field. The letters [b] and [c] means binary cycle and conventional, respectively.

EL Tatio	m * (kg/s)	C_w (J/kg·K)	T^- (°C)	T^+ (°C)	T_0 (°C)	K	V (km³)	Q^- (MW$_{th}$)	Q^+ (MW$_{th}$)	$\eta_{b[TR]}$	η_c	W_e^- [b] (MW$_e$)	W_e^+ [c] (MW$_e$)
-	63	0.0042	170	260	40	273.15	15	1482	2752	0.09	0.2	45	174

* Arithmetic mean computed starting from Q_w values listed in Table 2.

5. Conclusions

Chile is rich in natural resources such as copper, timber, nitrates, precious metals and molybdenum. Although Chile has an abundance of copper and mining resources, it has limited indigenous fossil fuels, and over 90% of the fossil fuels must be imported. The SING has an installed capacity of 3800 MWe, 100% of which is generated by imported fossil fuel.

As suggested by the Global Energy Network Institute (GENI) in 2011, a strategic energy plan for Chile is necessary to ensure the transition from traditional power plants to renewable energy power plants. This transition is necessary both to reduce the amount of greenhouse gases into the atmosphere and to reduce the country's dependence on imported fossil fuels. The development of geothermal energy represents a useful tool for achieving this objective.

Table 4. Evaluations of the thermal and electric power for the northern Chile.

Geothermal System	Elevation	A (km²)	H (km)	m (kg/s)	Hydrothermal Manifestations	Geological Features	T (°C)	T (°C)	Reservoir	W_e^- (MW$_e$)	W_e^+ (MW$_e$)	OT (h/y)	E_{sw}^- [b] (GWh/y)	E_{sw}^+ [c] (GWh/y)	Reference
El Tatio	4,300	30	0.5	63	Geysers, thermal springs, fumaroles, bubbling pools, mud pools, hydrothermal alteration. $T \sim 86$ °C	N-S graben filled by Miocene to Pleistocene ignimbrites and andesitic volcanoes.	170	260	Liquid Dominated	45	174	8,000	362	1391	Table 2; [29]
Surire	4,000	45	-	-	Fumaroles, thermal springs, hydrothermal alteration, bubbling pools. $20\ °C < T < 80\ °C$	Plio-Pleistocene dacitic volcanoes.	110	234	-	50	60	8,000	400	480	[17,30]
Puchuldiza	4,100	50	-	-	Thermal springs, fumaroles, mud pools, hydrothermal alteration, bubbling pools. $20\ °C < T < 90\ °C$	Volcano-tectonic depression surrounded by Plio-Pleistocene volcanoes.	175	200	-	50	190	8,000	400	1,520	[17,30,39]
Iriputunco (OLCA)	4,000	15	2–3	-	Thermal springs, fumaroles. $T > 100\ °C$	Stratovolcano within a NE-SW trending chain of volcanoes constructed within the collapse scarp of a Holocene debris avalanche.	230	300	-	75	450	8,000	600	3,600	[23,30]
Apacheta	4,000	25	-	-	Fumaroles. $T \sim 118\ °C$	Plio-Pleistocene volcanic complex located within a NW trending graben.	200	325	Liquid Dominated	150	400	8,000	1,200	3,200	[22,30,39]
Colpitas	4,000	-	-	-	Thermal springs and bubbling pools, hydrothermal alteration. $28\ °C < T < 55\ °C$	Volcanic rocks, volcanoclastic deposits and volcano-sedimentary sequences from Miocene to Holocene.	135	235	-	-	-	-	-	-	[16]

Table 4. *Cont.*

Geothermal System	Elevation	A (km²)	H (km)	m (kg/s)	Hydrothermal Manifestations	Geological Features	T^- (°C)	T^+ (°C)	Reservoir	W_e^- (MWe)	W_e^+ (MWe)	OT (h/y)	E_{su}^- [b] (GWh/y)	E_{su}^+ [c] (GWh/y)	Reference
Lirima	4,000	-	-	-	Geysers, thermal springs, fumaroles, bubbling pools, hydrothermal alteration. 20 °C < T < 80 °C	Upper tertiary sedimentary and volcanic rocks.	169	211	-	-	-	-	-	-	[17,20,30]
TOT	-	-	-	-	-	-	-	-	-	370	1,274	-	2,962	10,191	-
CO₂ emissions (Mton/year)	-	-	-	-	-	-	-	-	-	-	-	-	0.4	1.2	-

The black dashes indicate No Data availability.

In this paper a new evaluation of northern Chile's geothermal potential was performed, focusing on the El Tatio, Surire, Puchuldiza, Orriputunco-Olca and Apacheta geothermal fields. Thermal electric power and electric energy supply were calculated, although a satisfactory dataset is available only for El Tatio. The total estimated electric power for northern Chile is approximately 1300 MWe with an energy supply of 10,191 GWh/year. This means that more than 30% of SING's energy could substitute fossil fuel for geothermal energy, saving approximately 8 Mton of CO_2 per year. Geothermal energy development could be a useful resource for the mining and industry sectors, which represent Chile's primary energy users. It is important to note that dedicated field analysis for overcoming natural barriers, such as altitude, climate, water scarcity and distances from urban centres, will be necessary to accelerate Chilean geothermal development.

Acknowledgments

A special thanks is due to Alfredo Lahsen for his willingness and for providing many rare and useful papers, and UGI (Unione Geotermica Italiana) for giving the authorisation to publish this work. Albert Genter, François Vuataz and Franco Tassi are warmly thanked for the suggestions they provided in an early version of the manuscript. The author wishes to express her gratitude to the anonymous reviewers for their constructive and useful comments. A final thanks goes to Fedora Quattrocchi and Daniele Cinti, scientific managers of the project which supported this work.

Conflicts of Interest

The author declares no conflict of interest.

References

1. New Zealand Trade and Enterprise. *Geothermal Energy in Chile—Market Profile*; Exporter Guide; New Zealand Trade and Enterprise: Auckland, New Zealand, 2012.
2. National Energy Strategy 2012–2030, Energy for the Future. Available online: http://www.minenergia.cl/estrategia-nacional-de-energia-2012.html (accessed on 21 August 2014).
3. Barría, C. *Geothermal Energy in Chile*; Global Geothermal Development Plan Roundtable: The Hague, The Netherlands, 2013.
4. Woodhouse, S.; Meisen, P. *Renewable Energy Potential of Chile*; Global Energy Network Institute: San Diego, CA, USA, 2011.
5. Bertani, R. Geothermal power generation in the world 2005–2010 update report. *Geothermics* **2012**, *41*, 1–29.
6. Lahsen, A.A. Geothermal Exploration in northern Chile—Summary. Circum-Pacific energy and mineral resources. *Memoir* **1976**, *25*, 169–175.
7. Hodgson, S.F. Focus on Chile. *Geotherm. Resour. Counc. Bull.* **2013**, *42*, 24–37.
8. Francis, P.W.; Rundle, C.C. Rates of production of the main magma types in the Central Andes. *Geol. Soc. Am. Bull.* **1976**, *87*, 474–480.

9. Lahsen, A.; Trujillo, P. El campo geotermico de El Tatio, Chile. In Proceedings of the 2nd Symposium on Development and Use of Geothermal Resources, San Francisco, CA, USA, 20–29 May 1975. (In Spanish)

10. Marinovic, N.; Lahsen, A. Calama map, Antofagasta Region, scale 1:250,000. *Serv. Nac. Geol. Min.* **1984**, *58*, 140.

11. Montgomery, E.L.; Rosko, M.J. Groundwater exploration and wellfield development in the Pampa Lagunillas and Pampa Lirima areas, Provincia de Iquique, Chile. *Andean Geol.* **1996**, *23*, 135–149.

12. Polanco, E.; Gardeweg, M. Preliminary study of the volcanic stratigraphy of Upper Cenozoic at Pampa Lirima and Cancosa, 1st Region highland, Chile (19°45'–20°00' S and 69°00'–68°30' W). In Proceedings of the 9th Chilean Geology Congress, Puerto Varas, Chile, 31 July–4 August 2000; pp. 324–328.

13. Ahumada, S.; Mercado, J.L. Origin and Geological–Structural Evolution of the Pampa Apacheta Sector, 2nd Region, Antofagasta. Unpublished Undergraduate Thesis, Universidad Católica, Norte, Chile, 2008. (In Spanish)

14. García, M.; Gardeweg, M.; Clavero, J.; Hérail, G. Arica map: Tarapacá Region, scale 1:250,000. *Serv. Nac. Geol. Min.* **2004**, *84*, 150.

15. Trujillo, P. *Study of the Thermal Manifestations of Suriri*; Unpublished Report; Committee for Geothermal Energy Resources (CORFO): Santiago, Chile, 1972; p. 15. (In Spanish)

16. Aguirre, I.; Clavero, S.; Simmons, S.; Giavelli, A.; Mayorga, C.; Soffia, J.M. "Colpitas"—A New Geothermal Project in Chile. *Geotherm. Resour. Counc. Trans.* **2011**, *35*, 1141–1146.

17. Battocletti, B.L. *Geothermal Resources in Latin America and the Caribbean*; U.S. Department of Energy, Office of Geothermal Technologies: Albuquerque, NM, USA, 1999.

18. Céspedes, L.; Clavero, J.; Cayupi, J. Hazard management at Isluga volcano, Northern Chile: Preliminary results. In Proceedings of IAVCEI General Assembly, Pucón, Chile, 13–20 November 2004.

19. Letelier, M. *Geochemistry of Thermal Manifestations in Puchuldiza and Surrounding Areas*; Unpublished Report; Committee for Geothermal Energy Resources (CORFO): Santiago, Chile, 1981; p. 60. (In Spanish)

20. Arcos, R.; Clavero, J.; Giavelli, A.; Simmons, S.; Aguirre, I.; Martini, S.; Mayorga, C.; Pineda, J.; Soffia, J. Surface Exploration at Pampa Lirima Geothermal Project, Central Andes of Northern Chile. *Geotherm. Resour. Counc. Trans.* **2011**, *35*, 689–693.

21. Tassi, F.; Aguilera, F.; Darrah, T.; Vaselli, O.; Capaccioni, B.; Poreda, R.J.; Huertas, A.D. Fluid geochemistry of six hydrothermal systems in the Arica-Parinacota, Tarapacà and Atofagasta regions (Northern Chile). *J. Volcanol. Geotherm. Res.* **2010**, *192*, 1–15.

22. Urzúa, L.; Powell, T.; Cumming, W.B.; Dobson, P. Apacheta, a new geothermal prospect in northern Chile. *Geotherm. Resour. Counc. Trans.* **2002**, *26*, 65–69.

23. Reyes, N.; Vidal, A.; Ramirez, E.; Arnason, K.; Richter, B.; Steingrimsson, B.; Acosta O.; Camacho, J. Geothermal Exploration at Irruputuncu and Oca Volcanoes: Pursuing a Sustainable Mining Development in Chile. *Geotherm. Resour. Counc. Trans.* **2011**, *35*, 983–986.

24. Cusicanqui, H.; Mahon, W.A.J.; Ellis, A.J. The geochemistry of the El Tatio geothermal field, Northern Chile. In Proceedings of the 2nd UN Symposium, Development and Utilization of Geothermal Resources, San Francisco, CA, USA, 20–29 May 1975; pp. 703–711.

25. Giggenbach, W.F. The isotope composition of waters from the El Tatio geothermal field, Northern Chile. *Geochim. Cosmochim. Acta* **1978**, *42*, 979–988.

26. Huttrer, G.W. The status of world geothermal power production 1990–1994. *Geothermics* **1996**, *25*, 165–187.

27. Japan International Cooperation Agency (JICA). *Geothermal Power Development Project in Puchuldiza Area*; Unpublished Report; Japan International Cooperation Agency (JICA): Osaka, Japan, 1979; p. 109.

28. Japan International Cooperation Agency (JICA). *Report on Geothermal Power Development Project in Puchuldiza Area*; Unpublished Report; Japan International Cooperation Agency (JICA): Osaka, Japan, 1981; p. 48.

29. Cusicanqui, H. *Geochemical Study of the Suriri Thermal Area, Arica Province, 1st Region Estudio Geoquímico del área Termal de Suriri–Provincia de Arica–I Región*; Unpublished Report; Committee for Geothermal Energy Resources (CORFO): Santiago, Chile, 1979; p. 29. (In Spanish)

30. Sánchez, P.; Morata, D.; Lahsen, A.; Aravena, D.; Parada, M.A. Current Status of Geothermal Exploration in Chile and the Role of the New Andean Geothermal Centre of Excellence (CEGA). *Geotherm. Resour. Counc. Trans.* **2011**, *35*, 1215–1218.

31. Lahsen, A.; Munoz, N.; Parada, M.A. Geothermal Development in Chile. In Proceedings of the World Geothermal Congress, Bali, Indonesia, 25–30 April 2010.

32. Doveri, M.; Lelli, M.; Marini, L.; Raco, B. Revision, calibration, and application of the volume method to evaluate the geothermal potential of some recent volcanic areas of Latium, Italy. *Geothermics* **2010**, *39*, 260–269.

33. DiPippo, R. Ideal thermal efficiency for geothermal binary plants. *Geothermics* **2007**, *36*, 276–285.

34. Muffler, P.; Cataldi, R. Methods for regional assessment of geothermal resources. *Geothermics* **1978**, *7*, 53–89.

35. DiPippo, R. Second Law assessment of binary plants generating power from low-temperature geothermal fluids. *Geothermics* **2004**, *33*, 565–586.

36. Procesi, M.; Cantucci, B.; Buttinelli, M.; Armezzani, G.; Quattrocchi, F.; Boschi, E. Strategic use of the underground in an Energy mix plan: Synergies among CO_2, CH_4 geological storage and geothermal energy. Latium Region case study (Central Italy). *Appl. Energy* **2013**, *110*, 104–131.

37. Bloomfield, K.K.; Moore, J.N.; Neilson, R.M., Jr. Geothermal Energy Reduces Greenhouse Gases. *Geotherm. Resour. Counc. Bull.* **2003**, *32*, 77–79.

38. Armannsson, H.; Fridriksson, T.; Kristjansson, B.R. CO_2 emissions from geothermal power plants and natural geothermal activity in Iceland. *Geothermics* **2005**, *34*, 286–296.

39. Enel Green Power. *Experience as an Investor in Chile*; Enel Green Power: Rome, Italy, 2011.

Implications of Spatial Variability in Heat Flow for Geothermal Resource Evaluation in Large Foreland Basins: The Case of the Western Canada Sedimentary Basin

Simon Weides and Jacek Majorowicz

Abstract: Heat flow and geothermal gradient of the sedimentary succession of the Western Canada Sedimentary Basin (WCSB) are mapped based on a large thermal database. Heat flow in the deep part of the basin varies from 30 mW/m^2 in the south to high 100 mW/m^2 in the north. As permeable strata are required for a successful geothermal application, the most important aquifers are discussed and evaluated. Regional temperature distribution within different aquifers is mapped for the first time, enabling a delineation of the most promising areas based on thermal field and aquifer properties. Results of previous regional studies on the geothermal potential of the WCSB are newly evaluated and discussed. In parts of the WCSB temperatures as high as 100–210 °C exist at depths of 3–5 km. Fluids from deep aquifers in these "hot" regions of the WCSB could be used in geothermal power plants to produce electricity. The geothermal resources of the shallower parts of the WCSB (>2 km) could be used for warm water provision (>50 °C) or district heating (>70 °C) in urban areas.

Reprinted from *Energies*. Cite as: Weides, S.; Majorowicz, J. Implications of Spatial Variability in Heat Flow for Geothermal Resource Evaluation in Large Foreland Basins: The Case of the Western Canada Sedimentary Basin. *Energies* **2014**, *7*, 2573-2594.

1. Introduction

The Western Canada Sedimentary Basin (WCSB) is known for its large reserves of oil, gas and coal. In times of public discussion on climate change and the greenhouse gas emissions that come with burning of fossil fuels, the focus of interests shifts gradually towards renewable energy production such as geothermal energy. In western Canada, geothermal energy could play a role in replacing some fossil-fuel generated heat energy used as an energy source for warm water provision, district heating, industrial processes, or even electric power production. The feasibility of producing geothermal heat is strongly dependent on the thermal and geological conditions of the subsurface. Naturally, sufficient temperature is a primary constraint. However, only in a situation where a significant amount of warm fluid is produced will a geothermal project be successful. Therefore information on reservoir properties, particularly porosity and permeability, are crucial for geothermal exploration.

In this study information on subsurface thermal conditions and geology is combined by mapping the temperature for different stratigraphic depths, and overlaying the distribution of potential geothermal target formations on these maps.

2. Previous Work

2.1. Thermal Field

The study of geothermal heat in the WCSB has a long history (see Majorowicz and Jessop [1] and Majorowicz and Grasby [2] for a review of the early work). The first precise heat flow measurements were done by Garland and Lennox [3] in shallow 300–1000 m deep wells near Leduc (67 mW/m^2) and Redwater (61 mW/m^2) in the vicinity of Edmonton. Majorowicz et al. [4] applied a paleoclimatic correction which increased these values by 12% to 75 mW/m^2 for the Leduc well and 68 mW/m^2 for the Redwater well.

The first regional WCSB basin analysis of geothermal patterns from industrial temperatures was done in 1981 by Majorowicz and Jessop [1] for Alberta, Saskatchewan and the Northwest Territories (NWT). Lam and Jones [5] and Jones and Majorowicz [6] expanded the database available to Majorowicz and Jessop and conducted thermal conductivity, heat generation and heat flow studies of the sedimentary basin and Precambrian basement rocks. They found that heat flow patterns poorly correlate with heat generation of the Precambrian basement rocks from decay of ^{235}U-, ^{232}Th- and ^{40}K- isotopes. This has been recently confirmed by Majorowicz et al. [7]. It is contradictory with Bachu [8] who assumed that heat flow in the basin is controlled by variability of heat generation of the basement and influence of hydrodynamics is marginal (see discussion section for more information).

The first attempt to predict and map temperature at the geological surfaces was done by Jones et al. [9] in 1985 for the Paleozoic erosional surface and the Precambrian surface of the Alberta basin, followed by mapping of temperature at Precambrian surface for the larger area of the whole WCSB by Bachu [8] in 1993.

Majorowicz et al. [10] identified significant overestimation of temperatures from Alberta industrial well logs from shallow depths (<1000 m). This has been determined from high precision temperature logs conducted in shallow wells that have been allowed to reach thermal equilibrium. These findings have been confirmed by recent studies in the northern half of Alberta using tens of thousands of industrial temperature measurements from three independent datasets: Annual Pool Pressure surveys (APP), Drill Stem Tests (DST) and Bottom Hole Temperatures (BHT) constrained by equilibrium high precision logs coupled with 33 Thermal Conductivity wells were used to provide a more accurate prediction of the temperature gradient of the northern Alberta part of the WCSB [7,11]. The results of these recent northern Alberta studies showed the need for this study which covers all of the WCSB.

One of the main reasons for the overestimation of BHT's from shallow wells are seasonal effects on analogue thermometers which were used until the 1980's [11]. These thermometers recorded the maximum temperature in a well, which was assumed to have been recorded at the bottom of the well. However, during the summer months surface air temperatures exceed the BHT and can lead to overestimation of BHT's [11].

No high precision temperature data exists below 1000 m with the exception of one deep well on the outskirts of Fort McMurray in the shallow north eastern part of the basin, drilled 2400 m into the basement granites below 0.5 km of sediments [7].

2.2. Studies on the Geothermal Potential of Deep Aquifers

The first study on the geothermal potential of deep aquifers in the WCSB was published by Lam and Jones [12] in 1985. In their paper the authors examined aquifer porosity, thickness, water chemistry and water recovery in the area of Hinton-Edson in western Alberta, concluding that especially the Mississippian and Upper Devonian carbonate rocks have a good geothermal potential. In a second study Lam and Jones [13] investigated the geothermal potential in the Calgary area. Despite the low geothermal gradient, the authors stated that the Calgary area is an attractive location for geothermal recovery due to the relatively thick sedimentary succession and the substantial population of the city. Similar to the results of their study of the Hinton-Edson area, the largest potential for geothermal purposes in the Calgary area was also found in Upper Devonian and Mississippian carbonate rocks. Jessop and Vigrass [14] published a report on a geothermal well which was drilled in 1979 into the depth of 2214 m at the Campus of the University of Regina (Saskatchewan). Tests showed an excellent geothermal potential, but unfortunately the large sports building that was intended to be the load for the well was not built, so the well has only been used as a research facility.

In 2011 the Geological Survey of Canada released a report [15] which synthesizes previous geothermal studies and delineates the potential of the different geothermal resource types in Canada. A major finding of the report is that the highest geothermal potential (for electricity production) exists in the volcanic belts of the Cordillera and in parts of the WCSB (northeastern British Columbia, northern Alberta and southern Northwest Territories). The report describes the other deeper parts of the WCSB as a very large resource for direct heat use. In 2013 the British Columbia Ministry of Energy and Mines assessed the geothermal resource in the Devonian Carbonates of the Clarke Lake gas field in northeastern British Columbia [16]. In central Alberta Weides *et al.* [17] mapped porosity, permeability and temperature of four Devonian carbonate aquifers and the Cambrian Basal Sandstone Unit [18], concluding that all five formations are potentially useable for geothermal heating applications. Using a similar approach, Weides *et al.* [19] investigated the geothermal potential of the siliciclastic Granite Wash Unit in northwestern Alberta. Ferguson and Grasby [20] examined the deep clastic reservoirs of the Winnipeg and Deadwood formations (Basal Clastics) in Saskatchewan, finding that these formations have "geothermal potential for development of direct use and electricity generation systems". Besides depth, thickness and temperatures of the Basal Clastics, Ferguson and Grasby focused on injection rates from existing disposal wells, most of which operate at flow rates between 30 and 140 L/s.

3. Methods

3.1. Temperature Database

In this paper heat flow and geothermal gradient data for the WCSB were compiled from previous research. The heat flow and geothermal gradient data base from Majorowicz and Grasby for western and Northern Canada [2] has been expanded by additional heat flow and geothermal gradient studies which were conducted as part of Helmholtz-Alberta Initiative for the northern half of the Alberta territory [2,4,7,11]. The recent compilation done by the Geological Survey of Canada for all of Canada includes this dataset [15]. The Majorowicz and Grasby [2] compilation mainly was based on the corrected bottom hole temperatures (BHTs) and drill stem test (DST) temperature records, with few (5) precise temperature depth logs in equilibrium wells. This compilation has been expanded with a dataset containing estimates of the geothermal gradient from temperatures taken by industry and reported to the Alberta Energy Conservation Board. The same dataset has been used by Majorowicz and Moore [21] for their first Canadian evaluation of feasibility of Enhanced Geothermal Systems in the Alberta basin. The resulting dataset used in this study consists of about 70,000 single values (from APP's, DST' and BHT's) from more than 26,400 wells. The data were carefully filtered and corrected for equilibrium conditions. More detailed information on data quality and handling of the dataset is found in Gray *et al.* [11]. The heat flow data used in this article is based on conductivities of the main 13 rock types in the WCSB, which were determined from about 1405 measurements [22].

3.2. Mapping of Geothermal Data

The distribution of geothermal gradient and heat flow were mapped for the whole sedimentary succession deeper than 1 km. For calculation of the maps the ArcGIS 10.1 Geostatistical Analyst extension was used. In a first step, the dataset was checked for outliers. All heat flow values which were unusually high (above 100 mW/m²) or low (below 30 mW/m²) have been removed from the dataset. In total 462 values were identified as outliers, of which the majority showed no spatial consistency. These extreme values probably are the result of measurement or notation errors and do not represent the real thermal conditions. The resulting heat flow dataset includes 74,728 heat flow values from 26,421 wells. For those wells for which more than one heat flow value exists, the arithmetic average was calculated. The heat flow map was calculated using the simple kriging algorithm. The data were declustered to adjust for preferential sampling. A stable omnidirectional semivariogram was modelled, using 25 lags with a length of 10,000 m each, a nugget of 0.11, a range of 165 km and a partial sill of 0.50 (Figure 1a).

A similar approach was applied to map the geothermal gradient. First, all geothermal gradient values which were unusually high (above 80 °C/km) or low (below 10 °C/km) have been removed (37 values), resulting in a dataset of 68,377 gradient values from 26,492 wells. For those wells for which more than one gradient value exists, the arithmetic average was calculated. The geothermal gradient was mapped applying the simple kriging algorithm. A tetraspherical omnidirectional semivariogram was modelled, using 25 lags with a length of 10,000 m each, a nugget of 0.13, a range of 230 km and a partial sill of 0.39 (Figure 1b).

Figure 1. (a) Sample variograms and variogram models for heat flow; **(b)** Geothermal gradient.

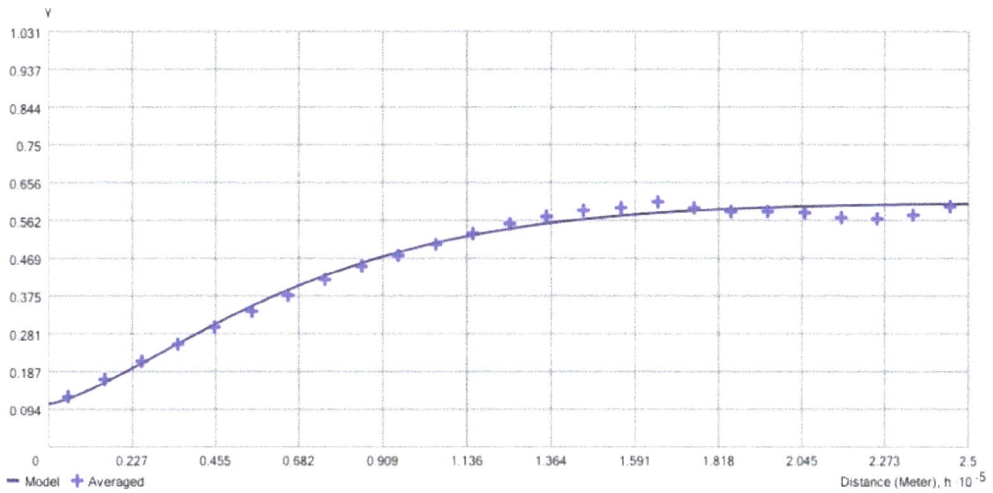

(a)

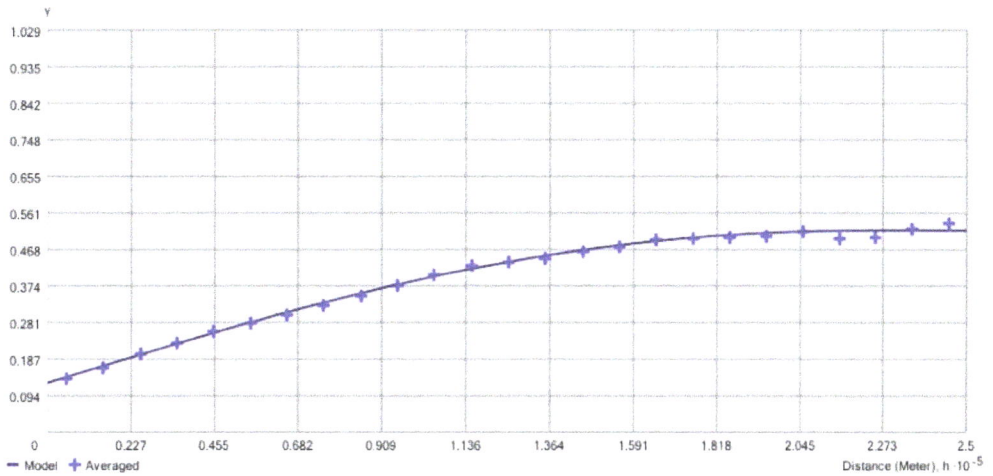

(b)

The geothermal gradient map was then used to calculate the temperature distribution for different stratigraphic units. The benefit of this approach over mapping temperature at a constant depth is that the resulting maps combine two key aspects relevant for geothermal exploration: temperature and geology. Five stratigraphic units where chosen for the maps: the top of the Precambrian basement, the Devonian Beaverhill Lake Group, the Devonian Winterburn Group and the Mississippian succession, and the bottom of the Cretaceous succession (sub-Mannville unconformity). First, structure depth maps were calculated for the five stratigraphic units using the well control data of the Geological Atlas of the Western Canada Sedimentary Basin [23], applying the ordinary kriging algorithm. To obtain the temperature distribution at depth, the raster of the geothermal gradient map was then multiplied with the raster of the depth distribution. Because at shallow depths less than 1 km subsurface temperatures are generally too low for geothermal applications, and as temperature measurements from shallow wells (less than 1 km) tend to be biased [11], the depth range shallower than 1 km is not displayed on the temperature distribution maps.

In addition to the temperature and depth information, the geographical extension of potential geothermal target formations was added to the maps. These formations were either chosen because they have already been in the focus of earlier geothermal exploration studies, or because they have been described in the literature as porous (and permeable) and therefore could host larger amount of warm fluids. It has to be pointed out that the potential geothermal target formations in most cases do not have the same depth as the particular temperature map (and as the depth contours), but rather are located a few hundred meters above or below, because the maps report the temperature at the top or bottom of a specified formation. A brief overview on the potential geothermal target formations is given in Table 1. Figure 2 shows examples of core samples from some of the formations.

Table 1. Potential geothermal target formations in the WCSB (for parameters see Table 2).

Period	Group	Formation	Lithology	References	Figure
Cretaceous	Mannville		sandstone	[12]	9
Cretaceous	Mannville	Cadomin	sandstone & congl.	[12]	9
Mississippian	Rundle		carbonates	[12,13]	8
Mississippian	-	Charles	carbonates	-	8
Mississippian	-	Banff	limestone	-	7
Devonian	Wabamun	Wabamun	dolomite	[12,13,17]	7
Devonian	Winterburn	Nisku	carbonates	[12,13,17]	7
Devonian	Woodbend	Grosmont	dolomite	-	6
Devonian	Woodbend	Leduc	dolomite	[12,17]	6
Devonian	Woodbend	Cooking Lake	reefal carbonates	[17]	6
Devonian	Beaverhill Lake	Slave Point	reefal carbonates	[12,16,19]	6
Devonian	Beaverhill Lake	Swan Hills	reefal carbonates	[19]	6
Devonian	Elk Point	Pine Point	dolostone	-	6
Devonian	-	Granite Wash Unit	sandstone	[19]	5
Ordovician	Basal Clastics	Winnipeg	sandstone	[14,20,24]	5
Cambrian		Deadwood	sandstone		5
Cambrian	-	Basal Sandstone Unit	sandstone	[13,17,18]	5

Figure 2. Core samples from potential geothermal target formations.

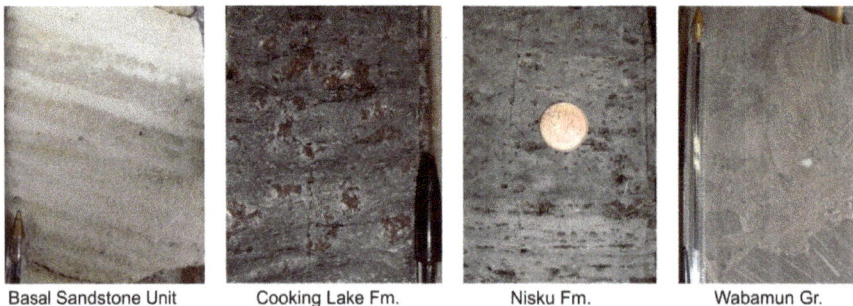

Basal Sandstone Unit   Cooking Lake Fm.   Nisku Fm.   Wabamun Gr.

Table 2. Results from previous regional geothermal studies in the WCSB; See Figure 4 for location of the study areas; data is taken from [13] for Calgary, [16] for Clarke Lake, [17,18] for Edmonton, [12] for Hinton-Edson, [19] for Peace River, and [14,20] for Saskatchewan.

Area & basin depth	Geothermal gradient [°C/km]	Best aquifer	Lithology	Aquifer depth [km]	Thickness [m]	Porosity [%]	Permeability [mD]	Temp. [°C]	Other potential aquifers
Calgary (1) 3.4–4.2 km	23.6	Leduc	reefal carbonate, dolomitized	3.7–4.0	up to 300	-	-	87–94	BSU, Nisku, Wabamun, Elkton
Clarke Lake (2) 2.4–2.6 km	up to 50–55	Slave Point & Keg River	reefal carbonate, dolomitized	2.0–2.1	up to 200	up to 25	-	110–123	-
Edmonton (3) 1.8–3.5 km	34.6	BSU	sandstone	1.8–3.5	28–45	7–19	1–>1000 (avg. ~1)	62–122	Cooking Lake, Leduc, Nisku, Wabamun
Hinton–Edson (4) 4–6 km	29.2	Leduc	reefal carbonate, dolomitized	3.4–5.4	up to 250	6–12	-	99–158	Slave Point, Nisku, Wabamun, Elkton, Belloy
Peace River (5) 1.7–2.3 km	33	Granite Wash Unit	sandstone	1.7–2.4	<30	2–19	1–>200 (avg ~1–10)	50–75	Slave Point
Saskatchewan (6) 2.2 km	28.1	Basal Clastics	sandstone	0.4–3.0	50–550	11–17	100–200	40–100	-

4. Results

4.1. Heat Flow and Geothermal Gradient

The heat flow in the WCSB generally ranges from 30 to 100 mW/m^2, being 60.4 mW/m^2 on average (Figure 3a,b). The highest heat flow is found in the northern part of the WCSB in the Northwest Territories, and adjacent northeastern British Columbia (B.C.) and northwestern Alberta. Other larger positive anomalies exist at the southeastern margin of the WCSB in the area of Regina (Saskatchewan) and Brandon (Manitoba), and in the western part of central Alberta (Figure 3b). Larger negative heat flow anomalies are found in northeastern Alberta (south of Fort McMurray) and in southern Alberta in the area of Calgary. Generally a northerly trend of increasing heat flow exists.

Figure 3. (**a**) Data points used for mapping of heat flow; dataset consists of on 74,728 values at 26,421 locations; (**b**) Heat flow of the WCSB; map was calculated using the simple kriging algorithm.

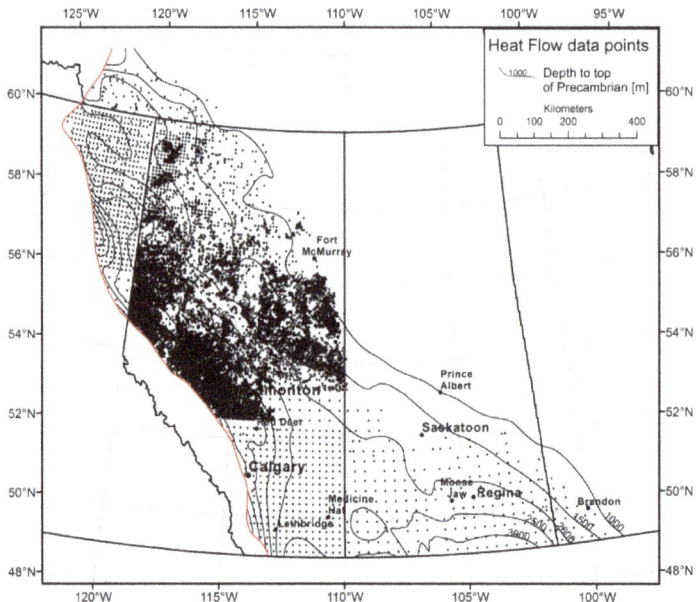

(**a**)

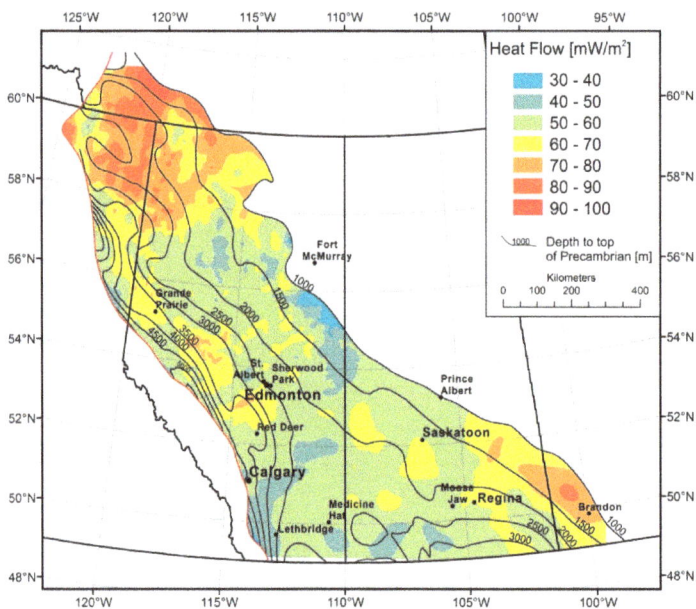

(**b**)

The geothermal gradient in the WCSB ranges from 20 to 55 °C/km, with an average value of 33.2 °C/km (Figure 4). The distribution of the thermal gradient follows the same trend of increasing values towards the northern WCSB.

Figure 4. Geothermal gradient of the WCSB based on 68,377 gradient values from 26,492 wells; map was calculated using the simple kriging algorithm. Black boxes represent the location of previous geothermal studies (see Table 2).

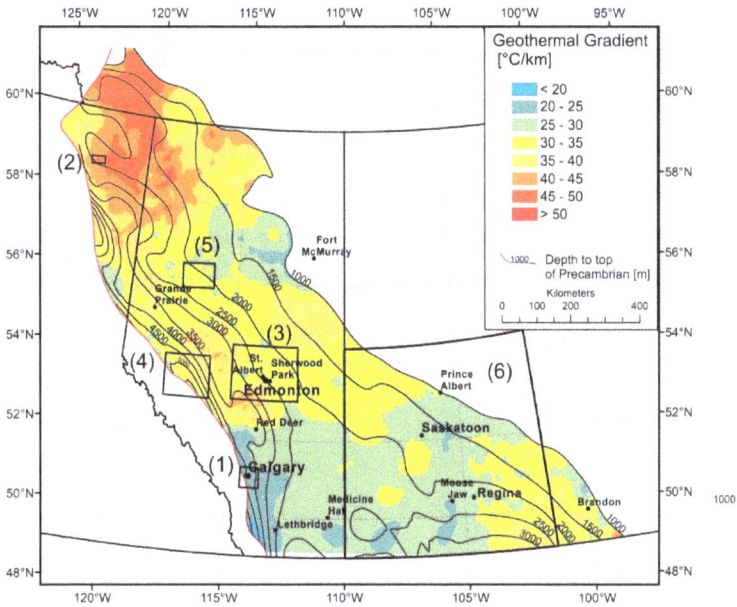

4.2. Temperature at Depth and Distribution of Potential Geothermal Target Formations

At the base of the sedimentary column the highest temperatures are found in the deepest parts of the basin close to the Cordillera, reaching values above 180 °C at a depth of 4.5 km and more (Figure 5). In the deeper half of the WCSB, at depth below 2–2.5 km, temperatures are above 70 °C, thus sufficient for district heating (see also Discussion Section). Potential geothermal target formations at the basal part of the basin fill are the siliciclastic deposits of the Cambrian Basal Sandstone Unit in central Alberta and western Saskatchewan, the Cambro–Ordovician Basal Clastics in eastern Saskatchewan, and the Devonian Granite Wash Unit in northwestern Alberta.

At the stratigraphic depth of the Devonian, the porous deposits from the carbonate platforms and reefal buildups in the Alberta and B.C. part of the WCSB are the potential targets formations (see Figures 6 and 7). Porous carbonate formations from the Carboniferous succession are deposited throughout the major part of the deeper WCSB (Figure 8). The temperature distribution at the sub-Mannville unconformity is the shallowest map presented in this study (Figure 9). Potential geothermal target formations are the sandstones and conglomerates of the Mannville Group above the unconformity, which reach temperatures above 60 °C in the cities of Red Deer and Great Prairie at a depth of about 2 km, and are naturally warmer at greater depths closer to the Cordillera.

Figure 5. Temperature at the top of the Precambrian basement with potential geothermal target formations; formations outline from Trotter [25] (Granite Wash Unit), from Slind *et al.* [26] and Dixon [24] (Basal Clastics), and from Pugh [27,28] (Cambrian Basal Sandstone Unit BSU).

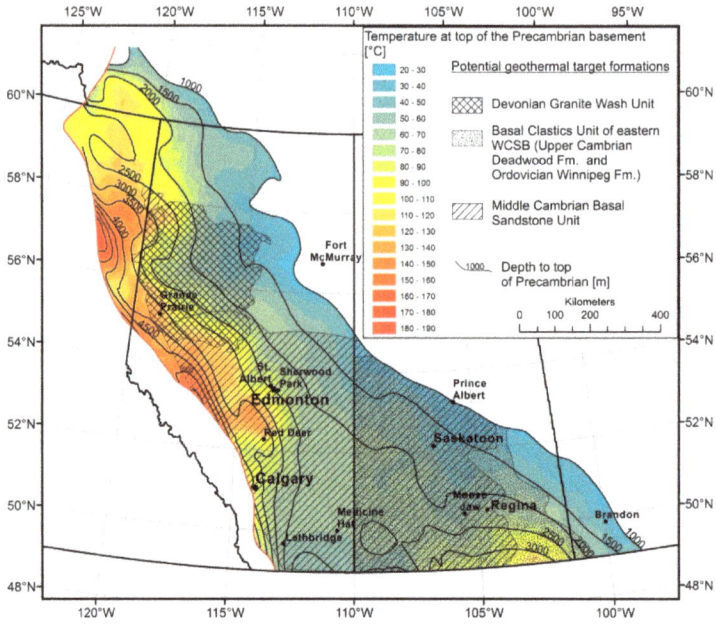

Figure 6. Temperature at the base of Beaverhill Lake Group; formations outline from Switzer *et al.* [29] (Woodbend Group), from Oldale and Munday [30] (Beaverhill Lake Group) and from Meijer Drees [31] (Elk Point Group).

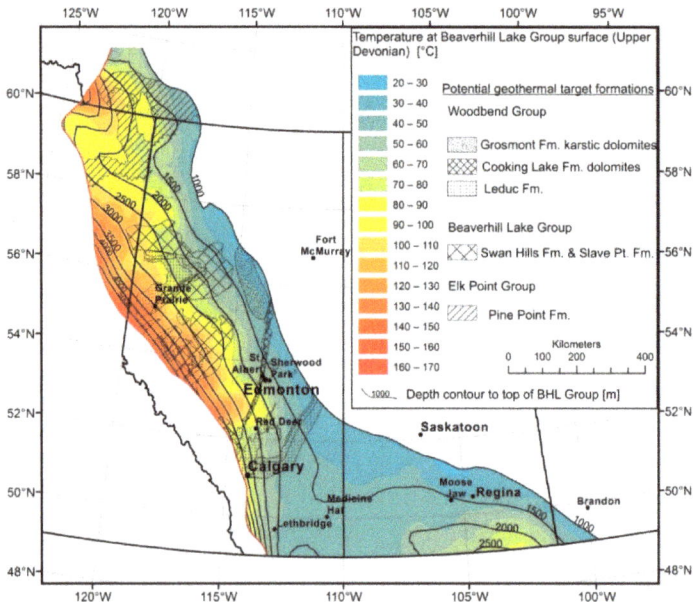

Figure 7. Temperature at the top of the Winterburn Group; formations outline from Switzer *et al.* [29] (Winterburn Group) and from Halbertsma [32] (Wabamun Group).

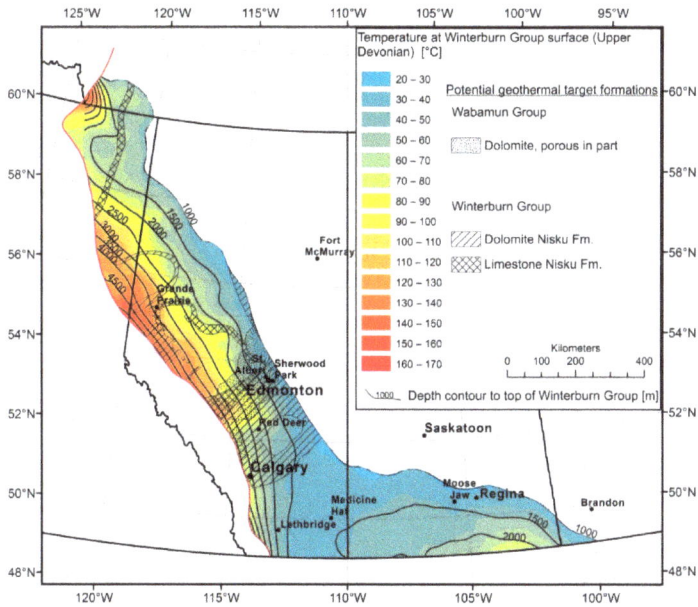

Figure 8. Temperature at the top of the Mississippian; formations outline from Richards *et al.* [33].

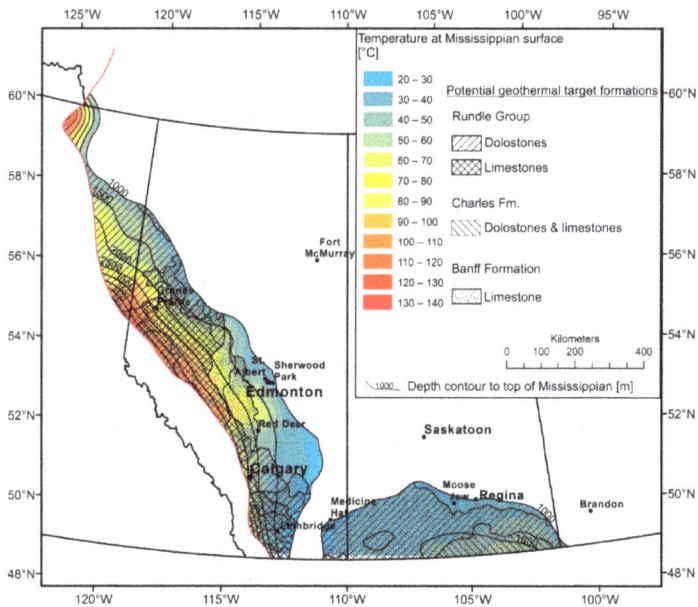

Figure 9. Temperature at the sub-Mannville unconformity; formations outline from Hayes *et al.* [34].

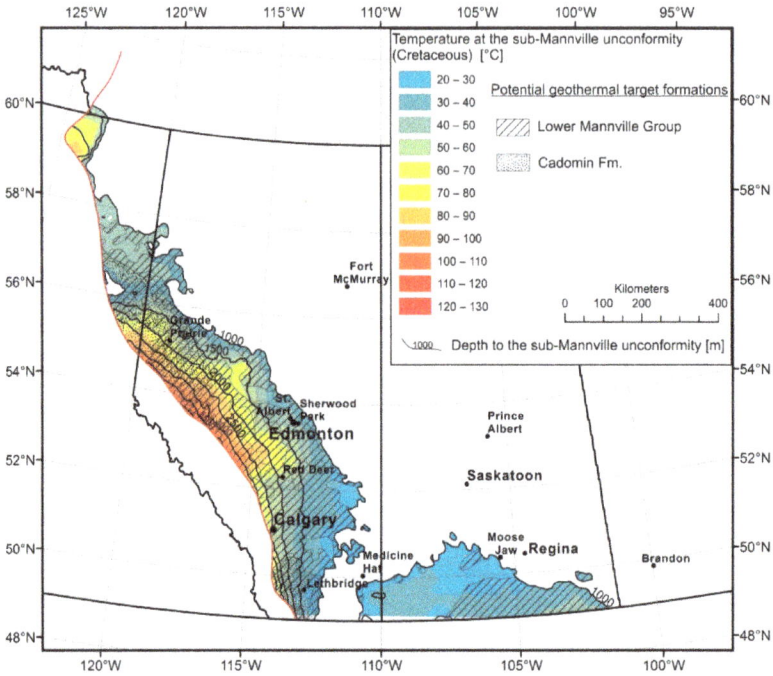

5. Discussion

5.1. Controls upon Thermal Field

It is noticed that the thermal field in the WCSB is highly variable. Heat flow in the deep part of the basin varies from 30 mW/m² in the south to 100 mW/m² in the north; the geothermal gradient varies from as low as 20 °C/km to over 55 °C/km. While values in the range of 30–60 mW/m² and 20–30 °C/km are typical for the Precambrian basement platform filled with sediment, values of 70–100 mW/m² and 40–55 °C/km can be considered as high respectively anomalous.

There are several controlling factors for geothermal gradient:

1. Thermal conductivity;
2. Heat flow;
3. Gravity driven convectional heat transport.

Thermal conductivity k controls the geothermal gradient at constant heat flow Q. The Q/k relationship for any depth along the vertical z axis of the well follows Fourier's law:

$$Q/k = dT/dz \tag{1}$$

where: T is the temperature at depth; z is the vertical depth; dT/dz is the temperature gradient and Q is the heat flow.

Thermal conductivity k for the crust for Canada is given by Jessop [35] and Beach, Jones and Majorowicz [22]. Beach *et al.* [22] based their statistic on 1405 values measured on core samples from Alberta basin rocks with use of the divided bar method. Typically, thermal conductivity is 3 W/mK for crystalline rocks, and 2 W/mK for sediments, which serve as a thermal blanket over the top of the crystalline crust. Figure 10 shows a good example of the control of k upon heat flow for a constant $Q = 70$ mW/m^2 and a mean thermal gradient approximation for the sedimentary succession in this paper.

Figure 10. Temperature depth (gray continuous profile) and thermal conductivity k (step line) control based on an example of a location in the deep foreland basin in British Columbia part (123°W 57°N; see Figure 3 for location). Approximation of the mean thermal gradient is also shown by a broken line.

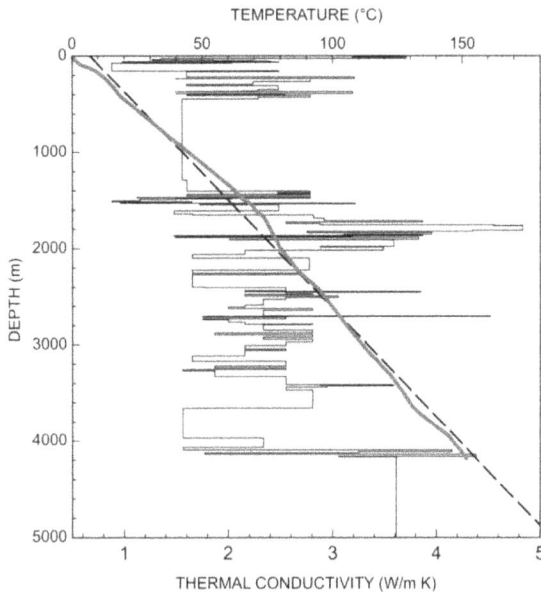

Heat flow at the surface is composed of the heat generation in the sediments (10^{-1} μW/m^3), in the granitic upper crust (1–10 μW/m^3), in the basaltic crust (10^{-2} μW/m^3), and of the contribution from below the crust, which consists of input from transient sources and radiogenic heat production at a very low rate (10^{-3} μW/m^3) [7]. While for several regions in the world a heat flow—heat generation relationship has been established (mainly for the measurements taken in the granitic batholiths [35]), it is difficult to find one for the heat flow estimates *vs.* heat generation of the basement of the WCSB [6]. In case of WCSB estimate of contribution from the upper crust can be based on ^{235}U, ^{232}Th and ^{40}K radiogenic elements contribution [6,36] and lower crust and mantle contribution [36]. This shows that the so called "reduced heat flow" from the mantle and the lower crust is 37 mW/m^2 (S.D. = 2 mW/m^2) [36]. The upper crustal contribution varies in much wider range due to much larger variability of heat production of the "granitic" crust [6]. Its contribution will depend on the thickness the upper crustal high heat generating ("granitic") part and the mean heat generation which differs between 1.1 μW/m^3 (Precambrian shield) and 2.4 μW/m^3 (WCSB) [36].

In the WCSB Burwash and Burwash [37] have provided data on uranium and thorium concentrations for 182 samples from the Precambrian basement in Alberta, Saskatchewan, Manitoba and British Columbia, and for the southern part of the Mackenzie Corridor of the southern Northwestern Territories. The measurements were made by the delayed neutron activation method. Jones and Majorowicz [6] included additional data from the Peace River area in north western Alberta (total of 229 samples analyzed in a nuclear reactor facility at the University of McMaster in Ontario). First analysis and mapping of the heat generation trends across the WCSB was reported by Jones and Majorowicz [6] who delineated three major high heat generation trends across the basement underlying the basin and concluded that these do not correlate with heat flow for the same study area (based on their heat flow data). Burwash and Burwash [37] and Bachu and Burwash [38] also did further analysis and mapping for Alberta and the WCSB, respectively. It has been noticed that the mean heat production for the WCSB is 2.4 $\mu W/m^3$ [6,36] which is more than two times higher that of the eastern Canada for the shield 1.1 $\mu W/m^3$ [36]. This can to some extent explain the elevated high heat flow in the WCSB based on the data which was used for the heat flow map (Figure 3b). The average heat flow for the WCSB is 60 mW/m^2, with a standard deviation of 9 mW/m^2, calculated from 74,728 determinations. If the average heat flow is calculated from the geostatistical interpolation grid presented in Figure 3b, the result is almost the same (61 mW/m^2). This is much higher than the heat flow examined for the Precambrian shield which is closer to 42 mW/m^2 (S.D. 9 mW/m^2) [36]. There is a difference of approx. 18–19 mW/m^2 which cannot be explained by the contribution of radiogenic elements in the sediments. It can be explained by the difference in mean heat generation between the shield and the WCSB, which differs from 1.1 $\mu W/m^3$ to 2.4 $\mu W/m^3$ respectively [36]. If the upper high heat productive "granitic" crust is about 15 km thick, the difference of 1.3 $\mu W/m^3$ in heat generation will explain the difference of about 20 mW/m^2.

A study based on gamma spectroscopy and API gamma logs from a 2.4 km deep well in the NE Alberta part of the WCSB [7] shows a large contrast in the contribution of radiogenic elements to heat production in the sedimentary succession (0.6 $\mu W/m^3$) and in the Precambrian granites of the upper crust (3.2 $\mu W/m^3$).

Temperatures in sedimentary rocks of the foreland basin can be influenced to some extent by non-conductive mechanisms, such as fluid flow. This occurs mainly through flow through porous aquifer conduits in the sedimentary succession above the westward deepening basement (Figure 11), however, flow through faults in the basement cannot be excluded. It was shown by previous research that in these porous sedimentary rocks the calculated surface Q values are significantly different (up to 50%) from the conductive Q, depending upon the nature of the hydrogeological system and its geometry which has been changing over time [1,10]. This was later questioned by Bachu and Burwash [38] who speculated on the relation of heat flow and heat generation as the main factor controlling distribution of thermal field in the WCSB. They argued that Darcy flow rates are too small to make an impact on regional-scale heat flow. Also hydraulic heads and Darcy fluid flow rates with reducing hydrodynamic influence upon heat flow have been diminishing over time due to the erosional change in topography. In the area towards the deep basin foothills of the

Rocky Mountains about 2 km of erosion has taken place since the uplift during the Laramide orogeny [39].

Majorowicz *et al.* [10] numerically tested the extent of hydrodynamic influence across the basin using a 2D numerical model constrained by revised thermal data. For this model a finite element mesh was generated which rebuilds the geometry of the cross section shown in Figure 11 (model is shown in lower panel of Figure 12). For the major fluid conduits like the Devonian carbonates or the Cambrian Basal Sandstone Unit the range of hydraulic conductivities was estimated. The Tertiary and Cretaceous shale units were assumed to have minimal permeability. Topography controls gravity driven flow patterns. Analysis shown in Figure 12 demonstrates that Darcy velocities of 0.01 to 1 m/yr can explain only 10–15 °C/km of thermal gradient elevation, and consequently cannot alone explain observations of temperature gradients elevated 30–40 °C/km above typical values for the basin. From the thermal gradient map (Figure 4) some reduction of gradients can be observed in high topography areas in the western part of the foreland basin, while some positive anomalies are located further east at a distance of 100 km and more, as predicted by the simple model which was made along the cross section through the central foreland basin (Figure 12a,b).

Figure 11. Geological cross section used for the thermal model (see Figure 12); modified from Wright *et al.* [40].

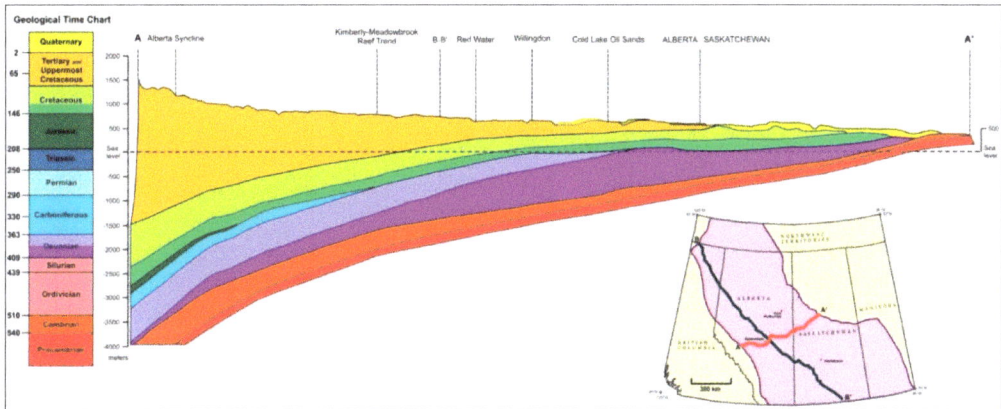

Figure 12. (a) Geothermal gradient across the WCSB profile from numerical modeling [10] for a base heat flow of 70 mW/m², a thermal conductivity model of the sedimentary cover and a surface temperature constrained by 0 °C for two scenarios of gravity driven regional fluid flow (upper panel-velocity of 10^{-2} m/year and lower panel-10^2 m/year. The surface temperature of 0 °C was chosen because the thermal field in the deeper sediments is still in equilibrium with this temperature [4]. The red curve shows the smoothed thermal gradient; (b) Assumed flow paths (modified from [10]).

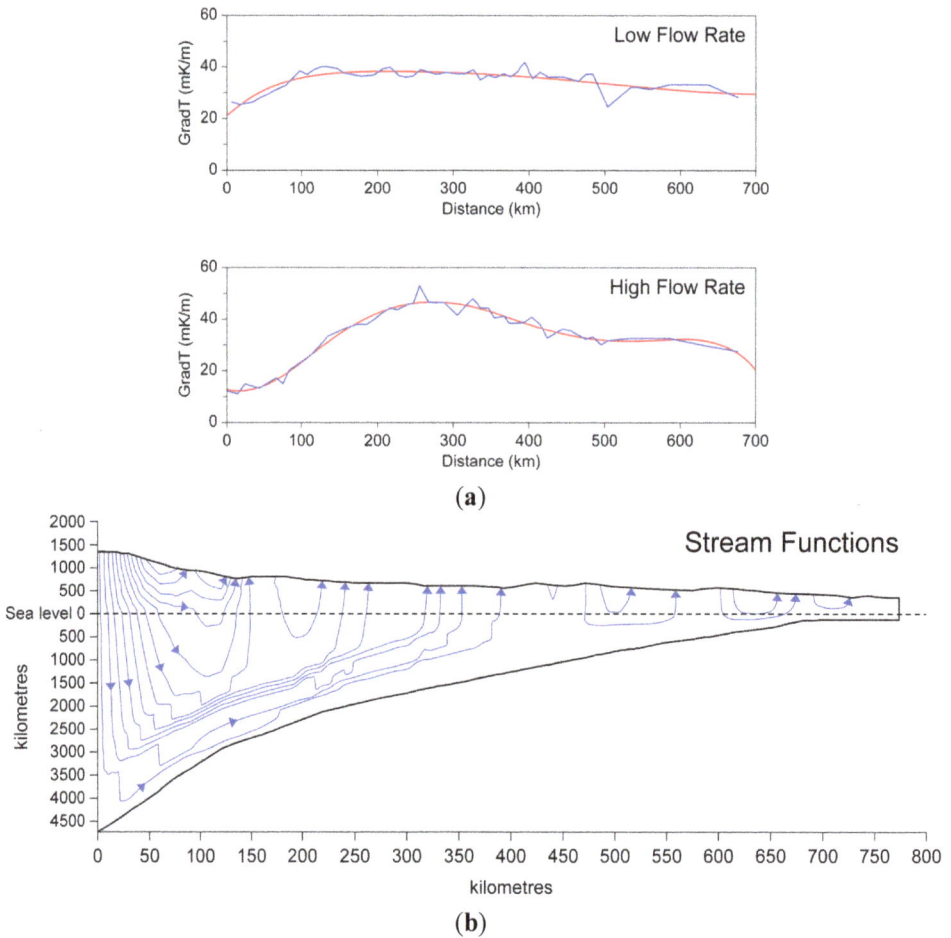

(a)

(b)

5.2. Geothermal Potential Zones

Other than previous studies on the geothermal potential of deep aquifers which all focused on a scale of several 10 km to few 100 km, this paper investigates temperatures and extension of potential geothermal target formations on the scale of the whole WCSB. With help of the maps presented in this paper the best locations for geothermal energy utilization can be identified both laterally and in the rock column on the WCSB scale. However, besides temperature, an appropriate (porous and permeable) reservoir is mandatory for a successful geothermal project. Though the extension of geological formations can be mapped over large distances with manageable effort, the facies of the formations, which controls the distribution of porosity and permeability, must always be investigated on a smaller scale. It is possible to map facies or reservoir properties such as porosity on a large scale, using well logs and core analysis data for example (see [17,19]). However, to obtain a reliable facies map for the scale of the WCSB, an enormous amount of well data would need to be collected, interpreted, classified and mapped, which is beyond the scope of this paper. Previous studies on geothermal reservoir parameters, though they are fragmented throughout the WCSB, can help to increase our knowledge on formation properties at a regional scale. Table 2 summarizes the major findings of these previous studies, and Figure 4 gives the location of these studies projected on the geothermal gradient map. Table 2 gives information on porosity and permeability which are important properties for reservoir evaluation. However, it must be emphasized that high porosities and permeabilities do not necessarily result in high flow rates, which are crucial for the success of a geothermal well. As a result of the large scale of this study, it was not possible to estimate of flow rates from single well tests for all formations presented. Lam and Jones [12,13] calculated flow rates from DST`s for some aquifers in their geothermal exploration studies. In the Leduc Fm. in the central part of the Hinton-Edson area (area 4 in Figure 4; for formation properties see Table 2) flow rates of more than 400 m^3/h are reached [12]. This value is high and can be compared to the wells at the geothermal power plants of Landau and Unterhaching in Germany, which produce at rates of 180–540 m^3/h from carbonatic aquifers at a depth of 3–3.4 km [41,42].

Depending on subsurface temperature and the heat demand at the surface, different applications for using geothermal resources are possible. In Figure 13 different geothermal potential zones are presented for the WCSB, depending on the Precambrian surface temperature, after a classification of Líndal [43]: (1) potential for warm water provision (>40 °C); (2) potential for domestic heating (>70 °C); (3) marginal potential for electrical power production (>100 °C); and (4) good potential for electrical power production (>150 °C). For the major part of the WCSB the temperatures at depths below 1.3 km are high enough to be used for warm water provision or balneological use. Underneath the large urban areas of Edmonton and Calgary, fluid temperatures are sufficient to be used for district heating purposes. Here geothermal heat production appears as a feasible option for a green, sustainable and economic way to reduce dependency on fossil fuels and decrease greenhouse gas emissions. For the southern part of Saskatchewan, Ferguson and Grasby [20] found that direct use of geothermal energy could be quite successful due to the high injection rates and sufficiently high temperatures.

Replacement of gas heating with geothermal systems could form part of a long range target for industrial emissions reduction. Based on the calculations from Majorowicz and Moore [44] 1000 heat generating systems (with 2 wells each) across Alberta drawing 100 °C from deep wells in deep sedimentary basin can save about 30 MT CO_2 per year. For a comparison, the oil sands industry generates some 34.7 MT CO_2 and other greenhouse gases [45]. 1000 wells is a small number compared to >300,000 oil and gas wells drilled in Alberta.

142

Figure 13. Possible geothermal applications based on the temperature at the top of the Precambrian basement.

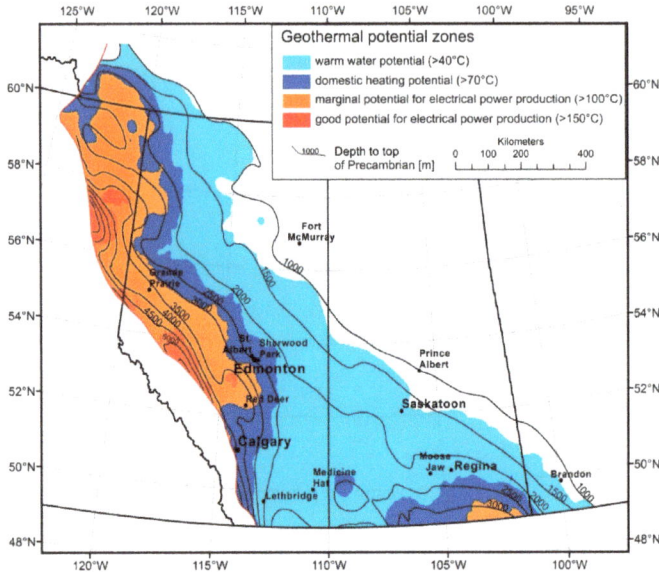

Electrical power production from geothermal heat is generally possible in the deepest part of the basin in vicinity of the Cordillera. A suitable spot for a geothermal power plant would be the geothermal anomaly around the hamlet of Winfield, located 100 km southwest of Edmonton and in direct vicinity of the Altalink transmission line. Here temperatures above 150 °C are found in the Basal Cambrian Sandstone Unit at a depth of 3.7 km. Another good location for a geothermal power plant is found in the area near Hinton in western Alberta, where temperatures in the Leduc Fm. at a depth of 5 km are above 150 °C [12]. Marginal potential for electrical power production exists at the Clarke Lake gas field near Ft. Nelson in northeastern B.C., where temperatures above 110 °C are found in 2.1 km deep Middle Devonian porous reefal carbonates [16], and in southeastern Saskatchewan in the highly permeable Basal Clastics aquifer, where temperatures are around 100 °C at depth of 3 km [20]. Generally it has to be emphasized that all locations presented here as favorable for geothermal utilizations represent locations with technical geothermal potential, based on the distribution of temperatures and potentially permeable formations. However, besides temperature the critical point in the development of a geothermal project is to achieve high flow rates. Hence, in the first phase of a local scale exploration study flow rates from DST`s should be analyzed to evaluate whether a site has an economic geothermal potential. In some cases, depending on the geological and economic situation, stimulation techniques like massive waterfrac treatments or acid injection could be applied to increase permeability of the reservoir.

While exploitation of geothermal resources generally can help to significantly reduce Western Canada's CO_2 emissions, geothermal power production could also lower the power costs for remote communities and reduce their dependency on diesel fuel transports. Electricity costs in remote areas of Canada range from 0.40 to 1.3 $/kWh [46]. Compared to the feed in tariffs for electricity from geothermal power plants in Germany of 0.20–0.28 $/kWh [47], or to the electricity generation costs from low temperature binary developments provided by the International Energy Agency [48], which range from 0.08 to 0.22 $/kWh, geothermal energy production could be economically in remote areas of Canada.

6. Conclusions

The thermal field of the WCSB is highly variable. The heat flow ranges from 30 mW/m^2 in the south to high 100 mW/m^2 in the north, while the geothermal gradient varies from as low as 20 °C/km to over 55 °C/km. The controlling factors of the thermal field in WCSB are poorly understood, and a heat flow—heat generation relationship cannot be established for the entire WCSB. Convective heat transport through fluid flow across the basin can partly explain observed thermal gradient variations.

For most of the WCSB potential geothermal target formations are present at sufficient depth. Especially the deep foreland basin clastic and carbonate plays offer potential for geothermal applications. In the large urban areas of Edmonton and Calgary, fluid temperatures are in the range of 80–90 °C and could be used for district heating, warm water provision, and for industrial applications. In the deepest basin, potential for electricity production by applying EGS technology exists.

Acknowledgments

The authors would like to acknowledge the helpful reviews by three anonymous reviewers. This study is part of the Helmholtz–Alberta Initiative (HAI), which is a research collaboration between the Helmholtz Association of German Research Centres and the University of Alberta.

We would like to thank HAI, and especially Theme 4 "Geothermal Energy" research coordinators Martyn Unsworth (UofA Edmonton) and Ernst Huenges (GFZ Potsdam), for enabling such an in depth study which increased our knowledge on the geothermal energy potential in the area.

We would like to thank Inga Moeck (UofA) for encouraging us to further conduct and deepen this study, which at an earlier point was presented during the "Reservoirs in foreland basin" session at the "Sedimentary Basins Jena 2013" conference in Jena, Germany.

Conflicts of Interest

The authors declare no conflict of interest.

References

1. Majorowicz, J.; Jessop, A. Regional heat flow patterns in the western Canadian sedimentary basin. *Tectonophysics* **1981**, *74*, 209–238.
2. Majorowicz, J.; Grasby, S.E. Heat flow, depth–temperature variations and stored thermal energy for enhanced geothermal systems in Canada. *J. Geophys. Eng.* **2010**, *7*, 232, doi:10.1088/1742-2132/7/3/002.
3. Garland, G.; Lennox, D. Heat flow in western Canada. *Geophys. J. Int.* **1962**, *6*, 245–262.

4. Majorowicz, J.; Gosnold, W.; Gray, A.; Safanda, J.; Klenner, R.; Unsworth, M. Implications of post-glacial warming for northern Alberta heat flow- correcting for the underestimate of the geothermal potential. *GRC Trans.* **2012**, *36*, 693–698.

5. Lam, H.; Jones, F. Geothermal gradients of Alberta in western Canada. *Geothermics* **1984**, *13*, 181–192.

6. Jones, F.; Majorowicz, J. Regional trends in radiogenic heat generation in the Precambrian basement of the Western Canadian Basin. *Geophys. Res. Lett.* **1987**, *14*, 268–271.

7. Majorowicz, J.; Unsworth, M.; Chacko, T.; Gray, A.; Heaman, L.; Potter, D.; Schmitt, D. Geothermal energy as a source of heat for oil sands processing in northern Alberta, Canada. In *Heavy Oil/Bitumen Petroleum Systems in Alberta and beyond*; Hein, F.J., Leckie, D.A., Suter, J., Larter, S., Eds.; American Association of Petroleum Geologists (AAPG): Tulsa, OK, USA, 2013; Volume 64, pp. 725–746.

8. Bachu, S. Basement heat flow in the Western Canada sedimentary basin. *Tectonophysics* **1993**, *222*, 119–133.

9. Jones, F.; Lam, H.-L.; Majorowicz, J. Temperature distributions at the Paleozoic and Precambrian surfaces and their implications for geothermal energy recovery in Alberta. *Can. J. Earth Sci.* **1985**, *22*, 1774–1780.

10. Majorowicz, J.A.; Garven, G.; Jessop, A.; Jessop, C. Present heat flow along a profile across the Western Canada Sedimentary Basin: The extent of hydrodynamic influence. In *Geothermics in Basin Analysis*; Springer: Heidelberg, Germany, 1999; pp. 61–79.

11. Gray, D.A.; Majorowicz, J.; Unsworth, M. Investigation of the geothermal state of sedimentary basins using oil industry thermal data: Case study from Northern Alberta exhibiting the need to systematically remove biased data. *J. Geophys. Eng.* **2012**, *9*, 534–548.

12. Lam, H.-L.; Jones, F. Geothermal energy potential in the Hinton-Edson area of west-central Alberta. *Can. J. Earth Sci.* **1985**, *22*, 369–383.

13. Lam, H.; Jones, F. An investigation of the potential for geothermal energy recovery in the Calgary area in southern Alberta, Canada, using petroleum exploration data. *Geophysics* **1986**, *51*, 1661–1670.

14. Jessop, A.M.; Vigrass, L.W. Geothermal measurements in a deep well at Regina, Saskatchewan. *J. Volcanol. Geotherm. Res.* **1989**, *37*, 151–166.

15. Grasby, S.E.; Allen, D.M.; Chen, Z.; Ferguson, G.; Jessop, A.; Kelman, M.; Majorowicz, J.; Moore, M.; Raymond, J.; Therrien, R. *Geothermal Energy Resource Potential of Canada*; Geological Survey of Canada: Calgary, AB, Canada, 2011; p. 322.

16. Walsh, W. Geothermal resource assessment of the Clarke Lake Gas Field, Fort Nelson, British Columbia. *Bull. Can. Pet. Geol.* **2013**, *61*, 241–251.

17. Weides, S.; Moeck, I.; Majorowicz, J.; Palombi, D.; Grobe, M. Geothermal exploration of Paleozoic formations in Central Alberta. *Can. J. Earth Sci.* **2013**, *50*, 519–534.

18. Weides, S.; Moeck, I.; Majorowicz, J.; Grobe, M. The Cambrian Basal Sandstone Unit in Central Alberta—An investigation of temperature distribution, petrography and hydraulic and geomechanical properties of a deep saline aquifer. *Can. J. Earth Sci.* **2014**, submitted.

19. Weides, S.; Moeck, I.; Schmitt, D.; Majorowicz, J. An integrative geothermal resource assessment study for the siliciclastic Granite Wash Unit, north western Alberta (Canada). *Environ. Earth Sci.* **2014**, doi:10.1007/s12665-014-3309-3.

20. Ferguson, G.; Grasby, S.E. The geothermal potential of the basal clastics of Saskatchewan, Canada. *Hydrogeol. J.* **2014**, *22*, 143–150.

21. Majorowicz, J.; Moore, M.C. *Enhanced Geothermal Systems (EGS) Potential in the Alberta Basin*; University of Calgary, Institute for Sustainability, Energy, Environment and Economy: Calgary, AB, Canada, 2008; p. 34.

22. Beach, R.; Jones, F.; Majorowicz, J. Heat flow and heat generation estimates for the Churchill basement of the Western Canadian Basin in Alberta, Canada. *Geothermics* **1987**, *16*, 1–16.

23. Mossop, G.D.; Shetsen, I. *Geological Atlas of the Western Canada Sedimentary Basin*; Canadian Society of Petroleum Geologists and Alberta Research Council: Edmonton, AB, Canada, 1994.

24. Dixon, J. Stratigraphy and facies of Cambrian to lower Ordovician strata in Saskatchewan. *Bull. Can. Pet. Geol.* **2008**, *56*, 93–117.

25. Trotter, R.D. *Sedimentology and Depositional Setting of the Granite Wash of the Utikuma and Red Earth areas, North-Central Alberta*; Dalhousie University: Dalhousie, NS, Canada, 1989.

26. Slind, O.; Andrews, G.; Murray, D.; Norford, B.; Paterson, D.; Salas, C.; Tawadros, E. Middle Cambrian to Lower Ordovician Strata of the Western Canada Sedimentary Basin. In *Geological Atlas of the Western Canada Sedimentary Basin*; Special Report; Mossop, G., Shetsen, I., Eds.; Canadian Society of Petroleum Geologists and Alberta Research Council: Edmonton, AB, Canada, 1994; Volume 4, pp. 87–108.

27. Pugh, D.C. *Subsurface Lower Paleozoic Stratigraphy in Northern and Central Alberta*; Geological Survey of Canada, Department of Energy, Mines and Resources: Ottawa, ON, Canada, 1973; p. 54.

28. Pugh, D.C. *Subsurface Cambrian Stratigraphy in Southern and Central Alberta*; Geological Survey of Canada, Department of Energy, Mines and Resources: Ottawa, ON, Canada, 1971; p. 33.

29. Switzer, S.; Holland, W.; Christie, D.; Graf, G.; Hedinger, A.; McAuley, R.; Wierzbicki, R.; Packard, J. Devonian Woodbend-Winterburn Strata of the Western Canada Sedimentary Basin. In *Geological Atlas of the Western Canada Sedimentary Basin*; Special Report; Mossop, G., Shetsen, I., Eds.; Canadian Society of Petroleum Geologists and Alberta Research Council: Edmonton, AB, Canada, 1994; Volume 4, pp. 165–202.

30. Oldale, H.; Munday, R. Devonian Beaverhill Lake Group of the Western Canada Sedimentary Basin. In *Geological Atlas of the Western Canada Sedimentary Basin*; Special Report; Mossop, G., Shetsen, I., Eds.; Canadian Society of Petroleum Geologists and Alberta Research Council: Edmonton, AB, Canada, 1994; pp. 149–163.

31. Meijer Drees, N.C. Devonian Elk Point Group of the Western Canada Sedimentary Basin. In *Geological Atlas of the Western Canada Sedimentary Basin*; Special Report; Mossop, G., Shetsen, I., Eds.; Canadian Society of Petroleum Geologists and Alberta Research Council: Edmonton, AB, Canada, 1994; Volume 4, pp. 129–138.

32. Halbertsma, H. Devonian Wabamun Group of the Western Canada Sedimentary Basin. In *Geological Atlas of the Western Canada Sedimentary Basin*; Special Report; Mossop, G., Shetsen, I., Eds.; Canadian Society of Petroleum Geologists and Alberta Research Council: Edmonton, AB, Canada, 1994; Volume 4, pp. 203–220.

33. Richards, B.; Barclay, J.; Bryan, D.; Hartling, A.; Henderson, C.; Hinds, R. Carboniferous Strata of the Western Canada Sedimentary Basin. In *Geological Atlas of the Western Canada Sedimentary Basin*; Special Report; Mossop, G., Shetsen, I., Eds.; Canadian Society of Petroleum Geologists and Alberta Research Council: Edmonton, AB, Canada, 1994; Volume 4, pp. 221–250.

34. Hayes, B.; Christopher, J.; Rosenthal, L.; Los, G.; McKercher, B.; Minken, D.; Tremblay, Y.; Fennell, J.; Smith, D. Cretaceous Mannville Group of the Western Canada Sedimentary Basin. In *Geological Atlas of the Western Canada Sedimentary Basin*; Special Report; Mossop, G., Shetsen, I., Eds.; Canadian Society of Petroleum Geologists and Alberta Research Council: Edmonton, AB, Canada, 1994; Volume 4, pp. 317–334.

35. Jessop, A.M. *Thermal Geophysics*; Elsevier: Amsterdam, The Netherlands, 1990; p. 316.

36. Jessop, A.M. Thermal input from the basement of the Western Canada Sedimentary Basin. *Bull. Can. Pet. Geol.* **1992**, *40*, 198–206.

37. Burwash, R.; Burwash, R. A radioactive heat generation map for the subsurface Precambrian of Alberta. *Geol. Surv. Can. Pap.* **1989**, *89*, 363–368.

38. Bachu, S.; Burwash, R.A. Regional-scale analysis of the geothermal regime in the Western Canada Sedimentary Basin. *Geothermics* **1991**, *20*, 387–407.

39. Majorowicz, J.; Jones, F.; Ertman, M.; Osadetz, K.; Stasiuk, L. Relationship between thermal maturation gradients, geothermal gradients and estimates of the thickness of the eroded foreland section, southern Alberta Plains, Canada. *Mar. Pet. Geol.* **1990**, *7*, 138–152.

40. Wright, G.; McMechan, M.; Potter, D. Structure and Architecture of the Western Canada Sedimentary Basin. In *Geological Atlas of the Western Canada Sedimentary Basin*; Special Report; Mossop, G., Shetsen, I., Eds.; Canadian Society of Petroleum Geologists and Alberta Research Council: Edmonton, AB, Canada; 1994; Volume 4, pp. 25–40.

41. Geothermie Neubrandenburg GmbH. Größtes geothermisches Kraftwerk in Süddeutschland: Wärmegeführtes Kalina-Kraftwerk in Unterhaching. Available online: http://www.gtn-online.de/Projekte/TiefeGeothermie/Projektbeispiel/grotesgeothermischeskraftwerkinsuddeutschland (accessed on 26 March 2014).

42. Geox GmbH. Geothermische Stromerzeugung in Landau. Available online: http://www.geox-gmbh.de/media/Downloadbereich/projekt_1407internet-x.pdf (accessed on 26 March 2014).

43. Líndal, B. Industrial and Other Applications of Geothermal Energy: (Except Power Production and District Heating). In *Geothermal Energy: Review of Research and Development*; United Nations Educational, Scientific and Cultural Organization (UNESCO): Paris, France, 1973; pp. 135–148.

44. Majorowicz, J.; Moore, M. The feasibility and potential of geothermal heat in the deep Alberta foreland basin-Canada for CO2 savings. *Renew. Energy* **2014**, *66*, 541–549.

45. Biello, D. How much will tar sands oil add to global warming? Available online: http://www.scientificamerican.com/article/tar-sands-and-keystone-xl-pipeline-impact-on-global-warming (accessed on 17 April 2014).

46. Arriaga, M.; Cañizares, C.A.; Kazerani, M. Renewable energy alternatives for remote communities in Northern Ontario, Canada. *IEEE Trans. Sust. Energy* **2013**, *4,* 661–670.

47. Einspeiseverguetung.info. Einspeiseverguetung Geothermie. Available online: http://www.einspeisevergütung.info/?page_id=41 (accessed on 26 March 2014).

48. International Energy Agency. Renewable Energy Essentials: Geothermal. Available online: https://www.iea.org/publications/freepublications/publication/Geothermal_Essentials.pdf (accessed on 26 March 2014).

Estimating Limits for the Geothermal Energy Potential of Abandoned Underground Coal Mines: A Simple Methodology

Rafael Rodríguez Díez and María B. Díaz-Aguado

Abstract: Flooded mine workings have good potential as low-enthalpy geothermal resources, which could be used for heating and cooling purposes, thus making use of the mines long after mining activity itself ceases. It would be useful to estimate the scale of the geothermal potential represented by abandoned and flooded underground mines in Europe. From a few practical considerations, a procedure has been developed for assessing the geothermal energy potential of abandoned underground coal mines, as well as for quantifying the reduction in CO_2 emissions associated with using the mines instead of conventional heating/cooling technologies. On this basis the authors have been able to estimate that the geothermal energy available from underground coal mines in Europe is on the order of several thousand megawatts thermal. Although this is a gross value, it can be considered a minimum, which in itself vindicates all efforts to investigate harnessing it.

Reprinted from *Energies*. Cite as: Díez, R.R.; Díaz-Aguado, M.B. Estimating Limits for the Geothermal Energy Potential of Abandoned Underground Coal Mines: A Simple Methodology. *Energies* **2014**, *7*, 4241-4260.

1. Introduction

Large mining areas in Europe are currently being affected by closure processes, which are mainly due to the progress in mining works and changes in mining activities. Mine closure creates negative social, economic, urban and environmental effects on the affected areas. Although mines present high potential for geothermal utilization of low-temperature water, which could be used for heating and cooling purposes, only a few cases have been reported in Europe, Canada, USA or China where this potential geothermal energy from underground mines has been actually detected and used.

In Europe, some cases from The Netherlands, Germany, Poland, United Kingdom, Norway and Spain have been reported by several authors [1–3]. Power obtained from mine water can only be a few kilowatts thermal (kWt) from small installations, like Freiberg (Germany) or Shettleston and Lumphinnans (Scotland, UK); but there are also large installations which extract several megawatts thermal (MWt) from mine waters like those in Heerlen (Netherland), Mszczonow (Poland), and Mieres (Spain). Therefore, a vast geothermal potential is not being exploited; nevertheless, there is always a problem before the start of any project using the geothermal power of a mine. The procedure begins when there is an abandoned mine, as in the two cases shown in Figure 1 (one is an old gallery, and the other is an old shaft). These photos illustrate the reality of the initial stage of these projects: just evidence of the existence of two underground mines which were in operation in the past.

Figure 1. (**A**) Abandoned underground mine, Mariana mine (Asturias); (**B**) Abandoned underground mine, Olloniego colliery (Asturias).

(**A**)   (**B**)

Without doubt, there are strong reasons related to sustainability and ecology which make this kind of projects worthy of research; nevertheless, economic viability is always a strong point helping support the project. Thus, the first question that should be solved is: could the use of geothermal power from these mines be profitable? When revising the specialized literature, not many cases have been described. Relevant research has been carried out mainly in Canada [4–10] (the Springhill case is of special interest because it has been operating since 1988) and Europe [11–18]. Another few more cases of geothermal use of mines have also been reported concerning mines in other countries such as the USA [19,20] and China [21–23]. An interested reader can find more complete information about these cases in [1–3] and [16]. In many of these works, hydrogeological models are described. In any cases, a great deal of information is needed to conduct these studies successfully. When the mine is in operation, to find this information is easy. On the contrary, it is difficult to obtain such information when the mine was closed many years ago and the company has disappeared. It is often necessary to study old public administration documents or even visit historic libraries or registries, making this an archaeological and/or industrial patrimony task rather than technical research. Such an amount of work would be only justified in those cases where a significant geothermal energy extraction is expected. The same question can be asked at another level. For example, in countries having a long mining tradition, is it reasonable to carry out actions related to geothermal use from abandoned mines which imply a large investment? How much geothermal energy is expected to be extracted from abandoned mines? In this sense, the main objective of this research work is to develop a method which allows estimating lower and upper limits easily for the geothermal potential of an abandoned underground coal mine. This is important in order to evaluate the possibility of an actual geothermal use of an abandoned mine or of a future use for a mine near closure. The method would help to decide under which conditions it becomes interesting to start a project for using this energy from mines.

In this work, the "geothermal potential" of a mine is the total amount of geothermal energy (or geothermal power) which can be obtained from this mine. It is easy to understand that one of the factors influencing geothermal potential is the volume and characteristics of the voids created by mining activity (*i.e.*, these voids can be stable in the long run and remain open, or they can cave in immediately and be filled with rock debris). This value is directly related to coal output. Our goal is to relate the present geothermal potential of an underground coal mine with the total saleable coal production yielded by the mine through its operation history. The main advantage of this proposal is that the coal output of a mine is a well-known parameter, which is always easy to find because it has been recorded over the years by public administration. On the other hand, wide coal mining experience helps us to establish an easy-to-use method.

Thus, a simple formula is proposed:

$$W_t \approx k \times P_T \tag{1}$$

where W_t is the value for geothermal power of the mine in MW thermal (MWt); P_T is the total saleable coal production in millions of tonnes (Mt); and k is the factor of proportionality which has to be estimated empirically.

Before starting a project for the use of geothermal power of an abandoned mine, it would be very interesting to make a "reasonable estimation" of the minimum and maximum thermal energy which can be recovered from mine water. If the minimum quantity is sufficiently high, a project for its use could be proposed. Nevertheless, under some unfavorable conditions (as for example a mine located far away from inhabited areas) the project could be rejected if it were not economically feasible, even assuming maximum heat recovery.

A "reasonable" maximum and minimum value for the ratio W_t/P_T will be determined in the following. Regarding our mines, the authors have determined that $k_{min} \approx 0.25$ and $k_{max} \approx 1.0$. It is not possible to make accurate predictions using this methodology; nevertheless, if technicians were able to predict a minimum value (or lower limit) for the geothermal potential of a mine and this value is high enough, the development of a project can be justified. On the other hand, if a maximum value of geothermal power is calculated and it is not sufficiently high to support a project, it is clear that it is better not to spend resources in this project.

In order to define this empirical method, coal output and two other parameters of the mine also have to be known: maximum water pumping and average quantity of air flow. Two different mines (different coalfields, history, exploitation methods, hydrogeological conditions...) are described here below. An analysis of these mines helps to understand the method and the value of the characteristic parameters, k_{min} and k_{max}.

It has to be pointed out that the method is only proposed for underground coal mines. As is well known, due to its sedimentary origin, coal usually appears in Nature as coal seams. This fact makes underground coal mines have a more topologically arranged structure than base-metals mines. Consequently, it is easy to find relationships between different parameters which allow the final relationship between geothermal power and total coal ouput to be found. For example, in a given coal field, the total length of galleries is approximately proportional to coal production. In base-metals mines it is more difficult to establish these kinds of relationships due to variable metal concentration in the rockmass. This does not mean that in this kind of mines it could not occur; in fact, a similar formula could probably be established for base-metals mines, but since no data is available a similar empirical law cannot be defined.

2. Brief Description of the Proposed Method

As previously mentioned, the method is reduced to a simple formula (Equation (1)). This formula allows estimation of the geothermal power of the mine W_t (in MW thermal or MWt), from a well-known parameter which is the total saleable coal output P_T (in millions of tonnes or Mt) produced by this mine during operation period. In effect, P_T is easy to find, since it has always been recorded over the years by different administrations. In order to define the method, the parameter k has to be defined or estimated, which implies knowing both the geothermal power and coal output of a mine at a given moment.

In any active mine, there is always a constant flow of two fluids which interchange heat with the rockmass: water and ventilation air. An estimation of the geothermal power which could be supplied by the mine can be deduced from the total heat extracted by these two fluids from the mine.

Assuming that a certain quantity of water flow Q_w (m³/s) is pumped from the mine and assuming that the temperature of the water has increased in ΔT (°C) in its flow through the rockmass, the heating power W_w (Watts) which heats the water is:

$$W_w = Q_w \times d_w \times s_w \times \Delta T_w \tag{2}$$

d_w and s_w are respectively the density (kg/m³) and the specific heat (J/kg·°C) of water.

In the same way, if the air flow rate Q_a (m³/s) is extracted from the mine by the main exhaust fans and its temperature has increased by ΔT_a (°C), the heating power W_a which heats the air is:

$$W_a = Q_a \times d_a \times s_a \times \Delta T_a \tag{3}$$

where d_a and s_a are respectively the density and the specific heat of air.

Nevertheless, the increase of water and air temperature can be produced by other heat sources in the mine which are not related to the heating capacity of the ground. In underground coal mines, the most important artificial heating source is the electrical equipment which also contributes to increasing the water and air temperature.

Assuming a total electrical power E (MW) installed in the mine and an electrical performance of r (%), the total power transferred to the air/water would be:

$$W_e = (100 - r) \times E \tag{4}$$

Under these conditions, the total thermal power effectively released from the rockmass or transferred from the rockmass to water and air is:

$$W_{tmin} = W_w + W_a - W_e \tag{5}$$

This is a real value, since it has directly obtained from experience and, without doubt, the available geothermal power of this mine is at less equal to it.

In Asturias, the average temperature of water pumped from mines is about 18 °C [24,25] whereas the average temperature of water at the surface is around 12 °C, and consequently, $\Delta T_w \approx 6$ °C. On the other hand, the water density and specific heat are $d_w = 1000$ kg/m³ and $s_w = 4186$ J/kg·°C respectively. The heating power which heats the water, which is a part of geothermal power of the mine, can be estimated by:

$$W_w = 25.1 \times Q_w \tag{6}$$

where W_w is in megawatts thermal (MWt) when Q_w is in cubic meters per second (m³/s).

In winter, the average temperature of mine air is also about 18 °C and the temperature of the air at the surface is about 7 °C. On the other hand, taking into account that the air humidity within the mine is almost constant and close to 90%, its density and specific heat are 1.18 kg/m³ and 1020 J/kg·°C respectively; the thermal power necessary to heat the air is thus:

$$W_a = 0.013 \times Q_a \tag{7}$$

W_a is in megawatts thermal (MWt) when Q_a is in cubic meters per second (m³/s).

On the other hand, assuming an electrical performance of 90%, the total power is:

$$W_e = 0.10 \times E \tag{8}$$

Under these conditions, the total thermal power released from the rockmass or transferred from the rockmass to water and air is:

$$W_t = W_w + W_a - W_e = 25.1 \times Q_w + 0.013 \times Q_a - 0.10 \times E \tag{9}$$

Taking into account the previous explanation, the ratio can be easily calculated as:

$$k = W_t/P_T \tag{10}$$

The analysis of two mines where this ratio reaches low and high values respectively allows us to estimate a minimum and maximum value for k.

The method is useful to perform geothermal resource estimates for given mining regions where coal extraction data are available; nevertheless, it is important to point out that it should not be used to design a geothermal system at a mine site.

The most important factor when using geothermal energy from flooded mines is that there must be a customer for the energy nearby. However, villages or even towns and cities have typically grown due to a mine having started its mining activity nearby. Consequently, a lot of mines in Europe are near populated areas and the geothermal energy can be used directly in district heating or similar systems.

3. Empirical Estimation of Limit Values for Parameter k

3.1. Case History 1: La Camocha Colliery

This mine has exploited an independent coal field in the past. Coal seams were mainly very steep (70° dips) and of low-medium thickness (1.5 m in average). The mining method used initially was the traditional inverted steps method (with backfilling) when mining was manual by means of pick hammers. More recently, sublevel caving with explosives has been used successfully.

This is an example of estimating the geothermal power of abandoned coal mine from historic coal output data, air flow rate and quantity of water pumped. As the quantity of water pumped out of this mine can be considered rather low, data will be used for the estimation of a minimum value k_{min}.

In this case, coal output data are known from the first year of the mine until the last year of exploitation. Annual coal production of the mine during its history is represented in Figure 2A.

The exploitation of the mine started during the 1930's. During the Second World War in Europe, the price of coal increased and this caused an increase in coal production. The maximum was reached in 1960 and then, output decreased quickly mainly due to emigration of miners to other coal fields in Europe which offered better working conditions (in this case, to Germany and Belgium). This tendency continued until 1970. However, the energy crisis in the 70s made the production of coal interesting and coal output increased again. Such an increase continued until about 1995. Then, changes in the world market and in European politics caused coal output to decrease drastically until mining ceased, in 2007. Until then, total accumulated production was about 16 million tonnes of saleable coal (Figure 2B).

Figure 2. Coal production at La Camocha Colliery (Asturias, Spain). (**A**) Coal output, in t/year; (**B**) Cumulative total output from approximately 1932 to mine closure.

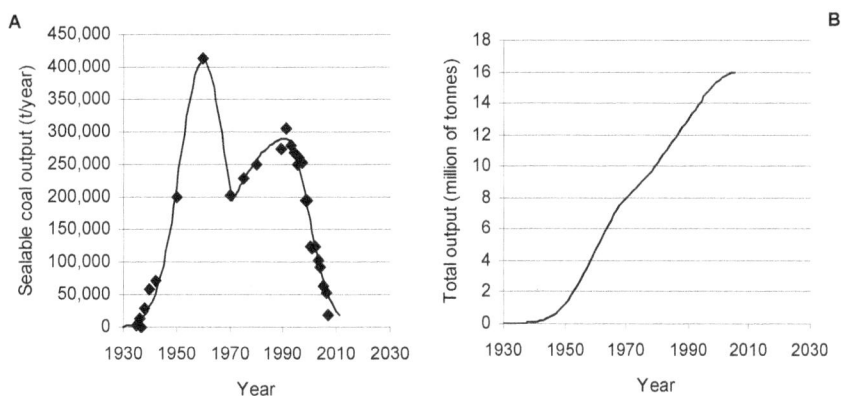

The use of this method implies knowing not only coal output but also another two parameters of the mine: water pumping and quantity of air flow. It is not always possible to obtain the necessary information; nevertheless, in this case, a record for the water pumping over several years has been obtained (Figure 3A).

This is a characteristic curve which decreases over time. There is a initial period of transient regime when the water originally contained in rockmass flows towards the mine. Afterwards, a stationary regime is reached (the water inflow into the mine is equal to the water inflow within the rockmass) and the quantity of water flow remains more constant during the years. The greatest water flow rate pumped was 200 m^3/h.

As can be seen from Figure 3B, water pumping is related to coal output because underground voids created by mining works become channels for water flow. On the other hand, ground movements caused by caving and land subsidence generate an increasing rock mass permeability. But, the relationship between water and coal production is a gross estimation because other significant factors, such as the rainfall, influence the mine water inflow. In many cases, water inflow increases more with the extension of the mine than—for a given extension—with depth.

Figure 3. (**A**) Water pumped at La Camocha colliery per year; (**B**) Water pumped at La Camocha colliery *versus* annual coal output.

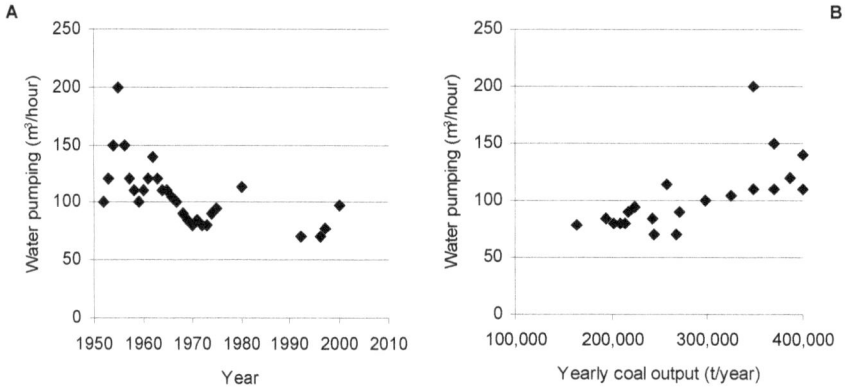

The second factor required is the air flow rate in the mine. This parameter does not vary greatly year by year; so, to develop a simple procedure, a unique representative value will be chosen. Thus, the average quantity of air flow extracted from mine by the main fans determined from measurements taken during the last years of the history of the mine is a representative value. Here, air flow rate supplied to the mine by two main exhaust fans was aproximately 60 m³/s, and was practically constant over the years. When the value of this parameter is not known, it can be deduced from mining experience. Figure 4A,B represent the specific methane emissions in m³ of methane per raw ton and the average methane flow in m³ of gas per day for a typical underground mine in Asturias [26–28]. In this mine, the average output in the last 50 years is 300,000 tonnes of saleable coal per year (about 2500 of raw tonnes per day). This mine was not very gassy so, for this output level, the methane flow is of about 10,000 m³ of gas per day or 0.115 m³/s. Methane concentration in ventilation return in our coal mines is usually about 0.20%; consequently, the fresh airflow rate is about 60 m³/s.

The last factor is electrical power. Traditionally, these mines have low mechanisation. We can assume that electrical power of mining equipment in the mine is lower than E = 1.0 MW.

It now becomes easy to estimate k_{min}. The maximum value of recorded Q_w has to be selected. In this case, for 200 m³/h, Q_w= 0.055 m³/s. On the other hand, Q_a = 60 m³/s.

By replacing these values in Equation (9):

$$W_{tmin} = 25.1 \times Q_w + 0.013 \times Q_a - 0.10 \times E = 1.38 + 0.78 - 0.10 \approx 2.1 \text{ MWt} \qquad (11)$$

This is a real value for geothermal power that can be extracted from La Camocha mine and, although Q_w is a maximum, W_{tmin} is a minimum because the water inflow in this mine can be considered low.

Figure 4. Typical methane emissions parameters from underground coal mines in Asturias (**A**) Specific emissions; (**B**) Average methane flow.

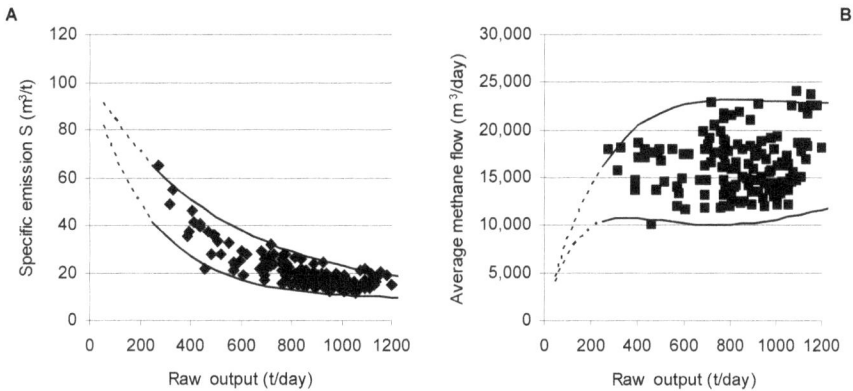

In order to estimate a geothermal power for a given total accumulated coal output, the relationship between W_{tmin} and P_T is calculated for all the years for which P_T is known (Figure 5A and Equation (10)):

$$k_{min}(t) = W_{tmin}/P_T(t) = 2.1/P_T(t) \tag{12}$$

In order to better understand the results, this relationship is represented against the accumulated output P_T in Figure 5B.

Figure 5. (**A**) Relationship between geothermal power and output per year (La Camocha colliery); (**B**) Relationship between geothermal power and output *versus* accumulated output (La Camocha colliery).

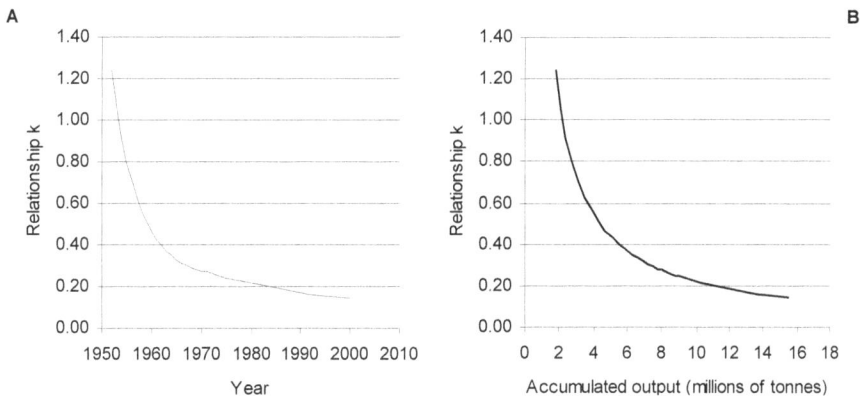

3.2. Case History 2: Figaredo Colliery

The second example is similar to the previous one in some aspects; for example, in the latter also vertical coal seams were also mined. However, there are several factors which significantly influence the water inflow into the mine, making it greater than in the former case. One factor is that the longwall method with caving was used more extensively thereby increasing the permeability of the rockmass. The second aspect is that there is a river above the mine and it has been demonstrated that a stream of water had flown from the river into the mine. The last issue is that this mine is connected to three other collieries and could receive water from them. For all these reasons, this practical case will be used to estimate a maximum value of k.

The history of coal production in this colliery is shown in Figure 6. The exploitation of coal started at the end of the XIX century (Figure 6A). It had increased during the second half of the 20th century due to the Second World War and the subsequent petroleum crisis. Peak production was reached in the decade of the 80s and afterwards production felt until the closure of the mine in 2007. Total accumulated production was about 10 million tonnes of saleable coal over more than 100 years (Figure 6B).

Figure 6. Coal production at Figaredo colliery (Asturias, Spain) **(A)** Coal output, in t/year; **(B)** Cumulative total output from 1910 to mine closure.

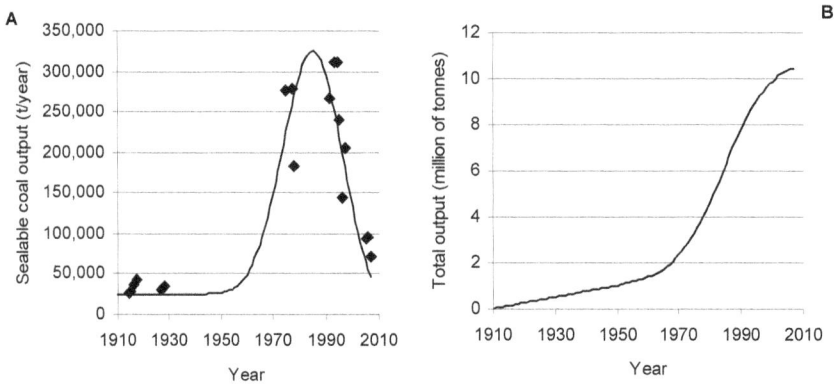

Figure 7A shows the water pumped from the mine for 20 years and, in Figure 7B, water flow rate pumped from the mine is related to the coal output. As can be inferred from Figure 7, water inflow in this mine is greater than in the mine previously analysed, in particular it is as much as five times higher. The maximum quantity of pumped water reached 1000 m^3/h or 0.277 m^3/s. On the other hand, the air flow rate recorded in the last year of the mine was 90 m^3/s and the electrical power of mining equipment is lower than E = 1.0 MW.

Proceeding as in the previous case, geothermal power could be obtained from Equation (9):

$$W_{tmax} = 25.1 \times Q_w + 0.013 \times Q_a - 0.10 \times E = 6.97 + 1.17 - 0.10 = 8.1 \text{ MWt} \qquad (13)$$

This is a real value for geothermal power which can be extracted from the mine and it can be considered as a maximum by comparing it with other mines in Asturias. In order to estimate geothermal power for a given total accumulated coal output, the relationship W_{tma}/P_T is calculated for all the years in which P_T is known (Figure 8A and Equation (10)):

$$k_{max}(t) = W_{tmax}/P_T(t) = 8.1/P_T(t) \qquad (14)$$

This relationship is represented as a function of the accumulated output P_T in Figure 8B, in order to better understand the results.

Figure 7. (**A**) Water pumped at Figaredo colliery per year; (**B**) Water pumped at Figaredo colliery *versus* yearly coal output.

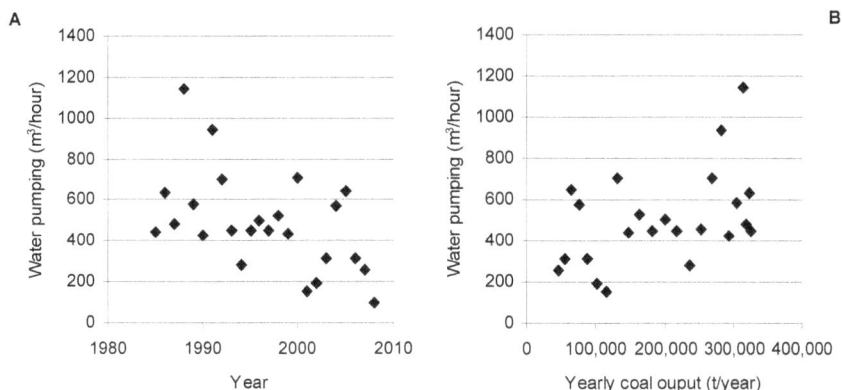

Figure 8. (**A**) Relationship between geothermal power and output per year (Figaredo colliery); (**B**) Relationship between geothermal power and output *versus* accumulated output (Figaredo colliery).

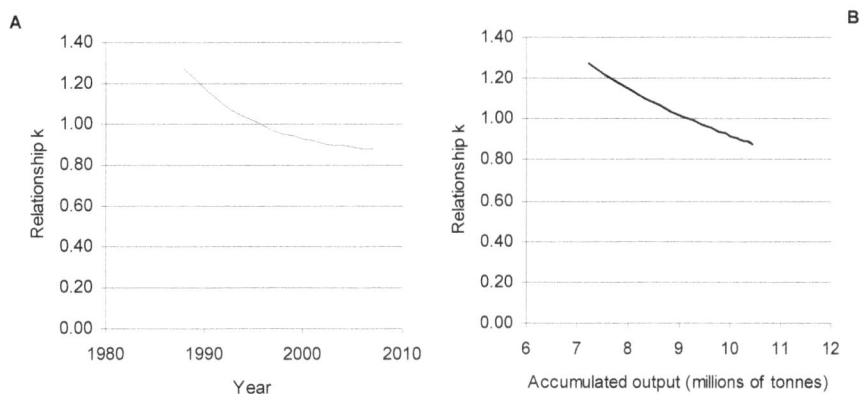

3.3. Determination of Parameter k Based on Experience

In Figure 9, the value of k has been represented as a function of the accumulated coal output for the period of activity of these two representative mines. This method for selecting the proper value of k would be useful for mines producing, at least, 5 million tonnes of saleable coal.

As it can be deduced directly from experience (Figure 9), a reasonable minimum value for k would be $k_{min} = 0.2 - 0.4$ while, a reasonable maximum for the parameter would be $k_{max} = 0.90 - 1.20$.

This means that, with regard to the assessing geothermal power in Asturias, the value k_{min} would be a conservative value and this geothermal power could be actually reached, while k_{max} is an optimistic value and it would be hard (even impossible) to reach the corresponding estimated geothermal power.

Figure 9. (**A**) Minimum relationship between geothermal power and output *versus* accumulated output; (**B**) Maximum relationship between geothermal power and output *versus* accumulated output.

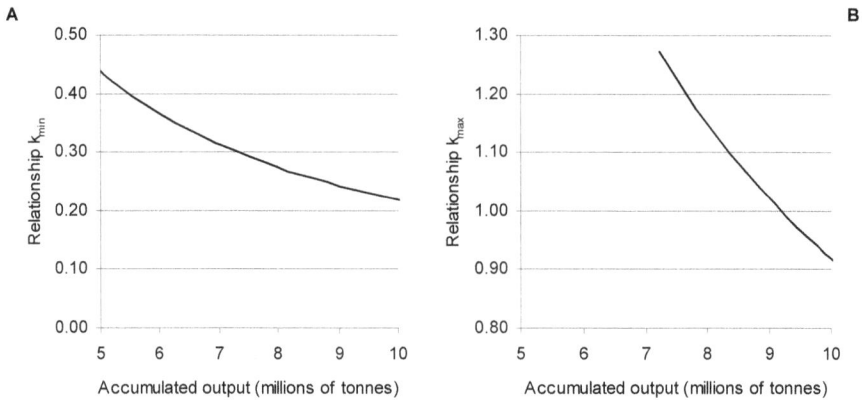

4. Analytical and Semi-Empirical Estimation of Limit Values for Parameter K

It is not easy to estimate "theoretically" reasonable maximum and minimum values for geothermal power. But a semi-empirical approach could be used if the total length of galleries in the mine is known. If this length parameter is unknown it could be either obtained from historical data or it could also be deduced from coal output as explained below.

After a study carried out in Spain in 1990 [29], the length of galleries excavated in rock in Spanish coal mines varies from 4 to 9.5 km per million of saleable tonnes. In the case of gateroads or galleries advanced in coal seams this value ranges from 6 to 12 km. The report gave data from a total of about 25 underground mines in Spain (pit-coal, anthracite and lignite). In this period, a number of large-scale mines were in operation and these values can be taken as representative for any mine. In the following, a minimum value for the gallery length excavated in rock and a maximum value for the total length of excavated galleries (rock + coal) will be necessary. Taking into account the above, limits of 5 km and 20 km per million of tones have been chosen.

Figure 10 shows the total yearly length of galleries excavated in rock and gateroads excavated in coal seams for several mines in Asturias. The value is related to the yearly coal output and it is given in mm per ton (equivalent to kilometre per million tonnes). These Figures illustrate that these values, 5 and 20 km, could be accepted for mines in Asturias rather than show data from which these values could be deduced mathematically.

In order to define a procedure, a typical mine in Asturias with a total output of about 10 million of saleable coal has been assumed. The total necessary gallery length would be about 200 km. This value is in agreement with real data, since the total length of galleries excavated in Figaredo Colliery has been about 254 km.

Figure 10. (A) Yearly ratio of length of new galleries excavated to coal output; **(B)** Yearly ratio of length of new gateroads excavated to coal output.

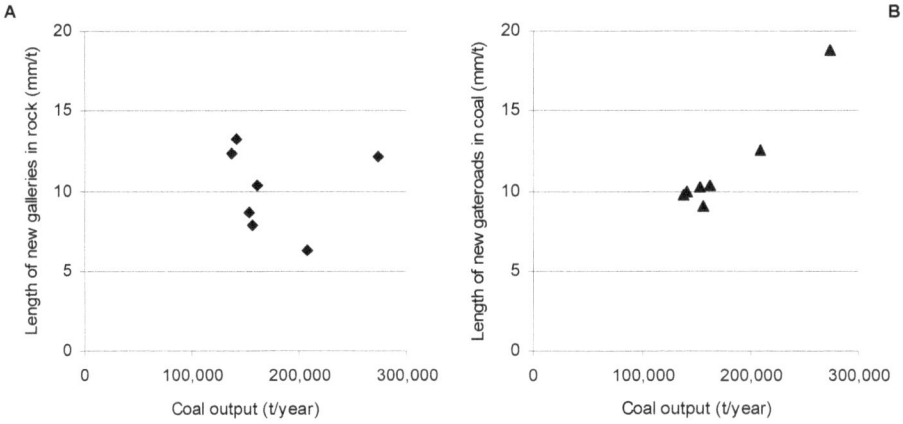

After research in a typical mine in Asturias [30], it has been found that the potential geothermal power of a 1 km gallery is approximately 50 kWt = 0.050 MWt. This means that the total geothermal power of a mine which has produced P_T tonnes along its life would be:

$$W_{tmax} = 0.050 \times 20 \times P_T = 1.0 \times P_T \tag{15}$$

Therefore the value of k which can be taken as a maximum would be a constant:

$$k_{max}(t) = W_{tmax}/P_T(t) = 1.0 \tag{16}$$

This value has to be considered a "maximum" because the factor 50 kW/km was deduced from data of galleries at a depth of 500 m, where rockmass temperature was about 28 °C. It is clear that galleries at a lower depth would have smaller geothermal potential. On the other hand, it is assumable that most of the galleries excavated in rock maintain their section, with no significant convergences. Another assumed factor is that the distance between galleries is enough to allow extraction of the maximum heat from the rockmass, which is not always realistic. Finally, in order to recover this amount of heat, large quantities of water should be used (which is not always possible).

A more conservative value can be estimated if it is assumed that gateroads (galleries following coal seams) would collapse and water could not flow through them. In this case, only galleries excavated in rock are stable in the long term and only these galleries could behave as paths for water flow. Moreover, the smaller ratio of galleries' length excavated in rock to coal output is chosen. Consequently, only 5 km of galleries are useful per million of coal tonnes:

$$W_{tmax} = 0.050 \times 5 \times P_T = 0.25 \times P_T \tag{17}$$

Consequently the value of k which can be taken as a minimum is constant and its value is:

$$k_{min}(t) = W_{tmin}/P_T(t) = 0.25 \tag{18}$$

These theoretical results are represented in Figure 11, in a graphical output similar to the one that shows the more experimental results previously deduced (in Figure 9).

Figure 11. (**A**) Minimum relationship between geothermal power and output *versus* accumulated output; (**B**) Maximum relationship between geothermal power and output *versus* accumulated output.

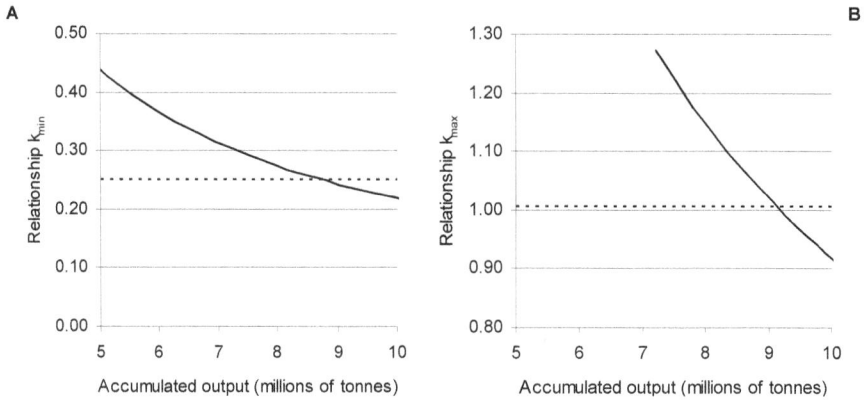

5. Using the model

5.1. Estimation of the Geothermal Power of a Mine

It now becomes easy to estimate the geothermal power potential of any mine in Asturias, by applying the above-described method.

The case example selected to validate the method is a coal mine having mainly low, steep coal seams at a moderate depth of 400 m. The coal is anthracite, without methane and the exploitation method is longwall with caving. This coal field did not have easy access to the rest of the region, so its mining history starts about the middle of the 20th century, when a power station was built near the coalfield (Figure 12). Production drastically increased in 2000, due to the mechanisation of the works in order to mine a 4 m thick coal seam by the longwall method [31].

Figure 12. Coal production of an underground mine in Asturias (**A**) Coal output, in t/year; (**B**) Cumulative total output from approximately 1954 to present time.

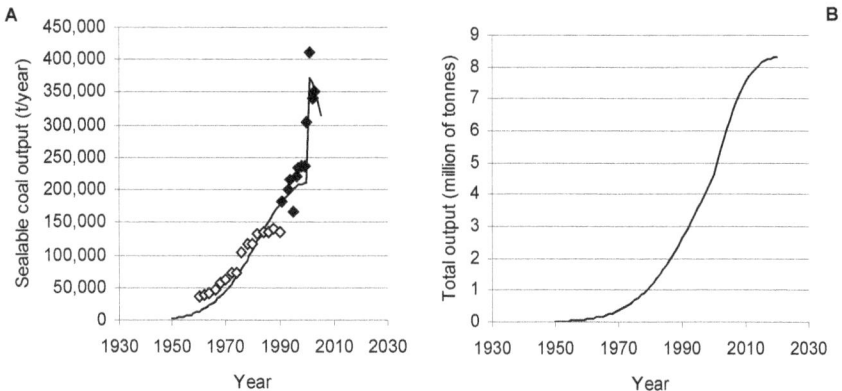

Studies conducted during 2003 have provided some valuable information, which can be used to validate the approach. Up to this date, the mine had produced about $P_T = 5 \times 10^6$ tonnes of saleable coal. Taking the value $k_{min} \approx 0.25$, the minimum expected geothermal power of the mine would be:

$$W_{tmin} = 0.25 \times 5 = 1.25 \text{ MWt} \tag{19}$$

This minimum value can be verified from real data obtained from the mine. In 2003, the total water inflow into the mine was $Q_w \approx 162 \text{ m}^3/\text{h} = 0.045 \text{ m}^3/\text{s}$ while the air flow rate was $Q_a \approx 25 \text{ m}^3/\text{s}$ [24,32]. In this case, it can also be assumed that the electrical power of mining equipment is about $E = 1.0$ MW.

Consequently, a realistic value for geothermal power would be:

$$W_t = 25.1 \times Q_w + 0.013 \times Q_a - 0.10 \times E = 1.13 + 0.32 - 0.10 = 1.35 \text{ MWt} \tag{20}$$

This value is greater than the previously calculated minimum value.

Taking $k_{max} \approx 1.0$, the upper limit for geothermal power (which is not expected to be reached) would thus be:

$$W_{tmax} = 1.0 \times 5 = 5.0 \text{ MWt} \tag{21}$$

Nevertheless, in opposition to estimation of the minimum parameter, the assessment of the maximum value cannot be proved.

5.2. Estimation the Geothermal Power for Several Mines in the Same Coalfield

This section shows the typical problem of estimating the geothermal power potential of many abandoned mines for a given coalfield in Europe, applying it to Asturian mines. The total underground coal output in Asturias during the last 200 years is shown in Figure 13. It is a fairy moderate production of only 110 million tonnes. Actually, the output is about 1 million tonnes per year. Data is only referred to pit-coal produced in the Central Coal Basin.

Actual data from 1980 and 2004 allow verification of the simplified method. In 1980, total coal ouput of mines in Asturias was about 89 million tonnes of saleable coal ($P_T = 8.9 \times 10^7$ tonnes). This production was obtained mainly from 25 collieries, so the average saleable production reached was about 3.5 Mt per mine facility. So, at the present time these mines could be considered "old mines" and the approach could be used.

For the following minimum and maximum values, $k_{min} \approx 0.25$ and $k_{max} \approx 1.0$, the minimum and maximum expected geothermal power of the mines would thus be:

$$W_{tmin} = 0.25 \times 89 = 22.2 \text{ MWt} \tag{22}$$

$$W_{tmax} = 1.00 \times 89 = 89.0 \text{ MWt} \tag{23}$$

It can be estimated that the total mine water evacuated from underground mines in Asturias was more $4.0 \times 10^7 \text{ m}^3$ per year or $Q_w = 1.2 \text{ m}^3/\text{s}$, as a study about mine water carried out in 1980 reports [33]. On the other hand, following the procedure explained above, coal output for this year was 5.5×10^6 tonnes, and the total air flow rate supplied to the mines would be more than 1500 m^3/s. With this input data, the actual geothermal power could then be estimated as:

$$W_t = 25.1 \times Q_w + 0.013 \times Q_a = 30.1 + 19.5 = 49.6 \text{ MWt} \qquad (24)$$

Figure 13. Total coal production of underground mines in Asturias (**A**) Coal output, in thousand t/year; (**B**) Cumulative total output from 1850 to present time, in million tonnes.

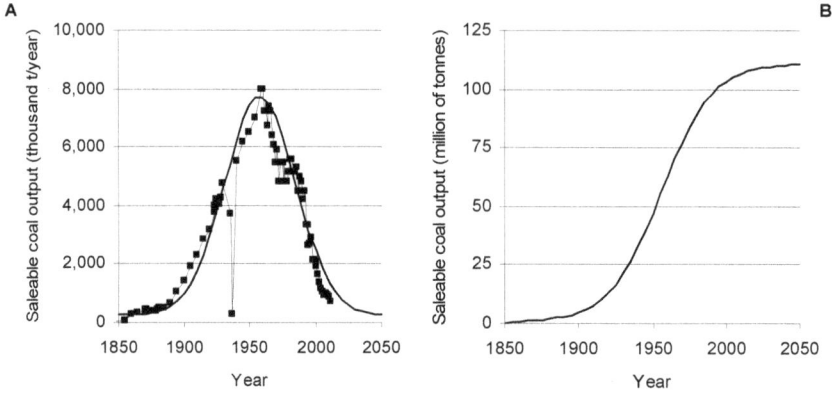

This value is higher than the minimum geothermal power estimated before and, it is obviously lower than the maximum one. In 2004, the total coal output of mines in Asturias was about 104 million tonnes of saleable coal ($P_T = 1.04 \times 10^8$ tonnes). Taking the values $k_{min} \approx 0.25$ and $k_{max} \approx 1.0$, the minimum and maximum expected geothermal power would be:

$$W_{tmin} = 0.25 \times 104 = 26.0 \text{ MWt} \qquad (25)$$

$$W_{tmax} = 1.00 \times 104 = 104.0 \text{ MWt} \qquad (26)$$

A study carried out in 2004 [25] demonstrated that the total mine water pumped from underground mines in the Central Coal Basin in Asturias was more than 36×10^6 m^3 per year or $Q_w = 1.1$ m^3/s. On the other hand, following the procedure developed in this research, for a yearly coal output of approximately 1.8×10^6 tonnes, and a total air flow rate supplied to the mines of more than 500 m^3/s in 2004, the estimated geothermal power is:

$$W_t = 25.1 \times Q_w + 0.013 \times Q_a = 27.6 + 6.5 = 34.1 \text{ MWt} \qquad (27)$$

A value which is also between the minimum and maximum values previously estimated.

It is important to point out that, as recorded in [25], the population of villages and towns close to these mines reaches 500,000 inhabitants which could directly use this geothermal power.

5.3. Could the Total Geothermal Power of Abandoned Mines in Europe be Estimated?

An accurate quantification of the geothermal power of abandoned mines in Europe would not only contribute to making the right decisions but also help to find proper uses for existing funds. This is obviously an interesting problem which cannot be solved at this stage of research; the main but not the only reason, is that at the present time, thousands of mines remain abandoned in Europe with no information available and without reported data; nevertheless, at least, an attempt to estimate the potential of abandoned coal mines could be made by applying the proposed method.

The graphical output in Figure 14A shows a gross estimation of the total coal production in the European Union for the last 150 years [34,35]. The accumulated coal output could reach the value of 11,000 million tonnes (Figure 14B).

Figure 14. Estimation of total coal production of underground mines in Europe (**A**) Coal output, in million t/year; (**B**) Cumulative total output from 1850 to 2000, in million tonnes.

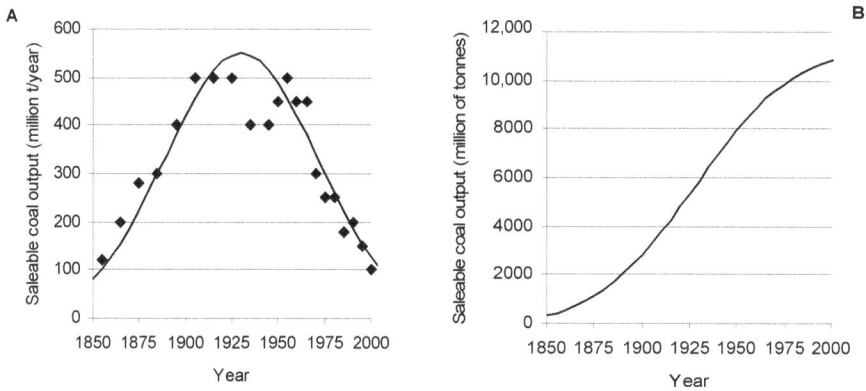

Taking the value $k_{min} \approx 0.25$, the total geothermal power potential in Europe could be assessed:

$$W_{tmin} \approx 0.25 \times P_T = 0.25 \times 11000 = 2750 \text{ MWt} \tag{28}$$

This is, about 3000 MWt could be extracted only from abandoned underground coal mines. The conclusion is that, as stated above, a vast geothermal potential from abandoned mines is not being exploited in Europe.

Furthermore, at this point, it would be interesting to make a gross estimation of hypothetical reduction in CO_2 emissions due to the use of this unexploited geothermal power.

Assuming that $W_t = 3000$ MWt, and for a coefficient of performance COP $= 4$, the useful thermal power is:

$$W_u = W_t \times COP/(COP - 1) = 3000 \times 1.33 \approx 4000 \text{ MW} \tag{29}$$

Assuming that the power is used h $= 24$ h/day and d $=30$ days/month during m $= 6$ months/year, the total energy would thus be:

$$E_u = W_u \times (12 - m) \times d \times h = 4000 \times 6 \times 30 \times 24 \approx 17,280,000 \text{ MWh/year} = 17.3 \text{ TWh/year} \tag{30}$$

The ratio between tonnes of CO_2 emissions and MWh produced depends on the source. In order to produce 1 MWh of thermal energy, it is necessary to emit 0.850, 0.450 or 0.200 tonnes of CO_2 to the atmosphere depending on wether electrical, fuel or natural gas has been used as a primary energy source [30]. Assuming an average ratio of 500 t/MWh the production of 17.3 TWh/year would imply a total emission of more than 8.5 million tonnes per year.

The value of this ratio is 0.170 for geothermal power by means of heat pumps. Consequently, in this case, CO_2 emissions would be only 3 million tonnes per year, thus yielding a reduction of CO_2 emissions of about 5 million of tonnes/year.

Finally, it is important to point out that this 3000 MWt could be extracted only from abandoned coal mines. It is unquestionable that a quantity of similar magnitude could be extracted from base-metals mines. So, the total amount of geothermal energy which could be recovered from underground mines in Europe could be as much as 6000 MWt. This value is equivalent to the energy supplied by 6000 eolic generators or equivalently, to the energy supplied by a wind power park with more than 150 generators for each country in the European Union. For this reason, promoting the widespread use of this source of renewable energy is of the most importance.

6. Conclusions

The following conclusions can be deduced from research carried out so far:

- Although mines present a high potential for geothermal utilization, there are only few cases known in Europe where this potential has been detected, and accurately used.
- A method has been developed to allow a non-complex estimation of the limits for the geothermal potential of an abandoned underground coal mine, from the value of its total production.
- The method is useful for making geothermal resource estimates for given mining regions where coal extraction data are available; it should not be used to design a geothermal system at a mine site.
- The specific maximum and minimum values, $k_{min} = 0.25$ and $k_{max} = 1.0$, could also be applied in coal regions similar to Asturias. Many parameters can influence these values, as for example thermal properties and hydrogeological characteristics of the rockmass, average temperature of virgin rock and gradient of temperature with depth, climate and average temperatures of the air and the river water and mining methods...*etc.* Consequently, values of k_{min} and k_{max} could be different in other regions.
- Assuming that the application of the formula has a high level of uncertainty, it has been estimated that an underground coal mine has a geothermal power of approximately 2.5 MWt per each 10,000,000 of tonnes produced.
- At least approximately 3000 MWt could be used from underground coal mines in the European Union, without including base-metals mines; the potential for coal mines is equivalent to 3,000 eolic generators or thereabouts, to the energy supplied by a wind power park with 90 generators for each country in the European Union.
- If this energy potential were used, an important reduction in CO_2 emissions of approximately 5 million tonnes of CO_2 per year could be reached.
- A good practice in mining management would be to make some mine-measurements, such as recording air flow rates, quantity of water actually pumped or air and water temperatures; this data would be of the most interest for future studies, especially when approaching the mine closure date.

Author Contributions

Rafael Rodríguez conceived and developed the idea behind the present research and he proposed the methodology and analytic procedure. Rafael Rodriguez and María B. Díaz have carried out the literature review and manuscript preparation including the searching and study of historical data of mines used in the definition of k_{min} and k_{max}. Final review, including final manuscript corrections, was done by Rafael Rodríguez.

Conflicts of Interest

The authors declare no conflict of interest.

References

1. Hall, A.; Scott, J.A.; Shanga, H. Geothermal energy recovery from underground mines. *Renew. Sustain. Energy Rev.* **2011**, *15*, 916–924.
2. Klinger, C.; Charmoille, A.; Bueno, J.; Gzyl, G.; Garzon Súcar, B. Strategies for follow-up care and utilisation of closing and flooding in European hard coal mining areas. *Int. J. Coal Geol.* **2012**, *89*, 51–61.
3. Younger, P.L. Hydrogeological challenges in a low-carbon economy. *Q. J. Eng. Geol. Hydrogeol.* **2014**, *47*, 7–27.
4. Jessop, A.M.; MacDonald, J.K.; Spence, H. Clean energy from abandoned mines at Springhill, Nova Scotia. *Energy Sources* **1995**, *17*, 93–106.
5. Jessop, A.M. Geothermal energy from old mines at Springhill, Nova Scotia, Canada. In Proceedings of the World Geothermal Congress, Florence, Italy, 18–31 May 1995; pp. 463–468.
6. Raymond, J.; Therrien, R. Low-temperature geothermal potential of the flooded Gaspé Mines, Québec, Canada. *Geothermics* **2008**, *37*, 189–210.
7. Ghoreishi, S.A.; Ghomshei, M.M.; Hassani, P.; Abbasy, F. Sustainable heat extraction from abandoned mine tunnels: A numerical model. *J. Renew. Sustain. Energy* **2012**, *4*, doi:10.1063/1.4712055.
8. Grasby, S.E.; Allen, D.M.; Bell, S.; Chen, Z.; Ferguson, G.; Jessop, A.; Kelman, M.; Ko, M.; Majorowicz, J.; Moore, M.; *et al. Geothermal Energy Resource Potential of Canada*; Open File 6914; Geological Survey of Canada: Ottawa, ON, Canada, 2012.
9. Renz, A.; Ruhaak, W.; Schatzl, P.; Diersch, H.-J.G. Numerical modelling of geothermal use of mine water: Challenges and examples. *Mine Water Environ.* **2009**, *28*, 2–14.
10. Raymond, J.; Therrien, R. Optimizing the design of a geothermal district heating and cooling system located at a flooded mine in Canada. *Hydrogeol. J.* **2014**, *22*, 217–231.
11. Malolepszy, Z. Man-made, low-temperature geothermal reservoirs in abandoned workings of underground mines on example of Nowa Ruda coal mine, Poland. In Proceedings of the International Geothermal Conference, Reykjavík, Iceland, 14–17 September 2003; pp. 23–29.

12. Malolepszy, Z. Low temperature, man-made geothermal reservoirs in abandoned workings of underground mines. In Proceedings of the 28th Workshop on Geothermal Reservoir Engineering, Stanford, CA, USA, 27–29 January 2003; Stanford University: Stanford, CA, USA, 2003; pp. 259–265.

13. Malolepszy, Z.; Demollin-Schneiders, E.; Bowers, D. Potential use of geothermal mine waters in Europe. In Proceedings of the World Geothermal Congress 2005, Antalya, Turkey, 24–29 April 2005; International Geothermal Association: Antalya, Turkey, 2005; pp. 1–3.

14. Demollin-Schneiders, E.; Malolepszy, Z.; Bowers, D. Potential use of geothermal energy from mine water in Europe for cooling and heating. In Proceedings of the International Conference Passive and Low Energy Cooling for the Built Environment, Santorini, Greece, 19–21 May 2005; Santamouris, M., Ed.; Heliotopos Conferences: Santorini, Greece, 2005; pp. 683–685.

15. Tóth, A.; Bobok, E. A prospect geothermal potential of an abandoned copper mine. In Proceedings of the 32nd Workshop on Geothermal Reservoir Engineering, Stanford, CA, USA, 22–24 January 2007; pp. 1–3.

16. Hamm, V.; Bazargan Sabet, B. Modelling of fluid flow and heat transfer to assess the geothermal potential of a flooded coal mine in Lorraine, France. *Geothermics* **2010**, *39*, 177–186.

17. Ferket, H.L.W.; Laenen, B.J.M.; Van Tongeren, P.C.H. Transforming flooded coal mines to large-scale geothermal and heat storage reservoirs: What can we expect? In *Mine Water—Managing the Challenges*, Proceedings of the International Mine Water Association Congress 2011, Aachen, Germany, 4–11 September 2011; Rüde, T.R., Freund, A., Wolkersdorfer, C., Eds.; International Mine Water Association: Wendelstein, Germany, 2011; pp. 171–176.

18. Uhlík, J.; Baier, J. Model evaluation of thermal energy potential of hydrogeological structures with flooded mines. *Mine Water Environ.* **2012**, *31*, 179–191.

19. *Municipal building, Park Hills, Missouri*; Ghpc #CS-064; Geothermal Heat Pump Consortium: Washington, DC, USA, 1997.

20. Watzlaf, G.R.; Ackman, T.E. Underground mine water for heating and cooling using geothermal heat pump systems. *Mine Water Environ.* **2006**, *25*, 1–14.

21. Guo, P.; He, M.; Yang, Q.; Chen, C. Wellhead anti-frost technology using deep mine geothermal energy. *Min. Sci. Technol. China* **2011**, *21*, 525–530.

22. He, M.; Zhang, Y.; Guo, D.; Qian, Z. Numerical analysis of doublet wells for cold energy storage on heat damage treatment in deep mines. *J. China Univ. Min. Technol.* **2006**, *16*, 278–282.

23. Zhang, Y.; Guo, D. Effect of cold energy storage of doublet-wells aquifer thermal energy storage in Sanhejian coal mine. *Energy Procedia* **2012**, *14*, 1730–1734.

24. Toraño, J.; Rodríguez, R.; Rivas, J.M. Application of numerical methods in the analysis of the hydrogeology of an area affected by underground mining. In Proceedings of the Congresso de Métodos Computacionais em Engenharia, Lisboa, Portugal, 31 May–2 June 2004; Asociación Portuguesa de Mecánica Teórica Aplicada y Computacional (AMPTAC); Servicio de publicaciones del CIMNE: Barcelona, Spain, 2004; pp. 1–15.

25. Jardón, J.; Pendas, F.; Ordóñez, A.; Cordero, C.; Álvarez, C.; Garzón, B. Aprovechamiento de las aguas de mina en la Cuenca Central Asturiana como recurso hídrico y energético. In Proceedings of the XII International Congress on Energy and Mineral Resources, Oviedo, Spain, 7–11 October 2007; Consejo Superior de Colegios de Ingenieros de Minas: Madrid, Spain, 2007; pp. 1–7. (In Spanish)

26. Luque, V.C.; Pedeca, S.C.L.E. *Manual de ventilación de minas*; Pedeca S. Coop. Ltda: Madrid, Spain, 1988; p. 732. (In Spanish).

27. Díaz Aguado, M.B.; González Nicieza, C. Control and prevention of gas outbursts in coal mines, Riosa-Olloniego coalfield, Spain. *Int. J. Coal Geol.* **2007**, *69*, 253–266.

28. Rodríguez, R.; Lombardía, C. Analysis of methane emissions in a tunnel excavated through Carboniferous strata based on underground coal mining experience. *Tunn. Undergr. Space Technol.* **2010**, *25*, 456–468.

29. Rambaud, C. Situación actual y tendencias futuras en las labores de preparación. In Proceedings of the Jornadas Técnicas sobre Labores de Preparación, Madrid, Spain, 21–22 November 1990; Instituto Tecnológico Geominero de España: Madrid, Spain, 1990; pp. 23–25. (In Spanish)

30. Rodríguez, R.; Díaz, M.B. Analysis of the utilization of mine galleries as geothermal heat exchangers by means a semi-empirical prediction method. *Renew. Energy* **2009**, *34*, 1716–1725.

31. Alvarez, T.J.; Rodríguez, R.; Rivas, J.M.; Casal, M.D. Economic and technical results mining a 4 m thick coal seam in Spanish Carbonar Colliery. *Glückauf* **2003**, *139*, 323–328.

32. Toraño, J.; Rodríguez, R.; Rivas, J.M. New techniques *versus* experience: A real case of mine ventilation analysis. In Proceedings of the 10th U.S./North American mine ventilation symposium, Anchorage, AK, USA, 16–19 May 2004; Bandopadhyay, S., Ganguli, R., Eds.; Taylor & Francis: London, UK, 2004; pp. 487–492.

33. Pendás, F.; García, M.P. Caracterización hidrogeológica de la minería de la Cuenca Central Asturiana. In Proceedings of the VIII Congreso Internacional de Minería y Metalurgia, Oviedo, Spain, 16–22 October 1988; Asociación Nacional de Ingenieros de Minas de España: Oviedo, Spain, 1988; pp. 279–298. (In Spanish)

34. Del Rosal, I. La productividad en la minería española del carbón. *Revista de Minas* **1999**, *19–20*, 213–218. (In Spanish)

35. Del Rosal, I. La reconversión del carbón, una dependencia plena de la decisión pública. *La Empresa Pública* **2000**, *1*, 156–166. (In Spanish)

Chapter 2:
Uptake of Geothermal Energy

Geothermal Power Growth 1995–2013—A Comparison with Other Renewables

Ladislaus Rybach

Abstract: Based on global statistical data the current status of deep geothermal resource utilization for electricity generation is presented. Particular attention is paid to growth rates. The rates are compared with those of other renewable energies (biomass, hydro, solar photovoltaic (PV), wind). Whereas wind and solar PV exhibit annual growth rates of 25%–30% since 2004, geothermal growth is only about 5% per year. Geothermal electricity production (in TW·h/yr) was higher until 2011 than from solar PV, but is now clearly falling behind. So far the global geothermal electricity generation is provided nearly entirely by hydrothermal resources, which exist only under specific geologic conditions. Further development (=increasing production capacity) based on this resource type alone will therefore hardly accelerate to two-digit (>10% per year) growth rates. Faster growth can only be achieved by using the ubiquitous petrothermal resources, provided that the key problem will be solved: establishing a universally applicable technology. This would enable to create, at any requested site, feasible and efficient deep heat exchangers for enhanced geothermal systems (EGS) power plants—irrespective of the local subsurface conditions. Goals and challenges of this technology are addressed.

Reprinted from *Energies*. Cite as: Rybach, L. Geothermal Power Growth 1995–2013—A Comparison with Other Renewables. *Energies* **2014**, *7*, 4802-4812.

1. Introduction

Renewable sources of electricity are generating increasing interest and having a corresponding impact on the energy scene. Geothermal energy sources have the advantage of providing base-load electricity, *i.e.*, independent of daily, seasonal or climatic variations and thus can complement other, intermittently producing renewable sources like wind or solar. Whereas wind and solar energy sources are abundant on the surface, for geothermal sources one has to go deep, usually a few kilometers. In the following, geothermal electricity is addressed and the global electricity supply from various renewable sources will be presented and compared. The growth rate in renewable power generation is a decisive factor on the electricity market.

2. Geothermal Power Generation

There are two main types of deep geothermal resources from which electricity can be produced: hydrothermal and petrothermal. Hydrothermal resources have naturally occurring geothermal fluids at depth, often originating from surface infiltration of precipitation. The fluids can be used as heat carriers and taken out from the ground through boreholes. Such hydrothermal resources like deep aquifers exist only when specific geologic/hydrogeologic conditions prevail, which makes them rather rare. Petrothermal resources on the other hand are more or less ubiquitous and immense; they consist basically of the "heat in place" in deep rock formations. The heat must be therefore extracted, e.g., by establishing a fluid circulation through a special, man-made heat exchanger at depth (see below for details). So far 99.99% of all existing geothermal power plants use hydrothermal resources. Figure 1 shows schematically the two resource types.

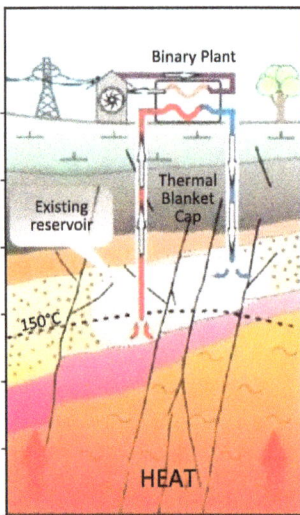

Figure 1. The two types of geothermal resources capable to generate electric power. The hydrothermal type on a natural reservoir that can feed, besides binary power plants, also geothermal steam condensing turbines. For the petrothermal type the reservoir needs to be created (details see text). Modified from Figure 3 in Geothermal Electricity (GEOELEC) Resource Assessment Protocol [1].

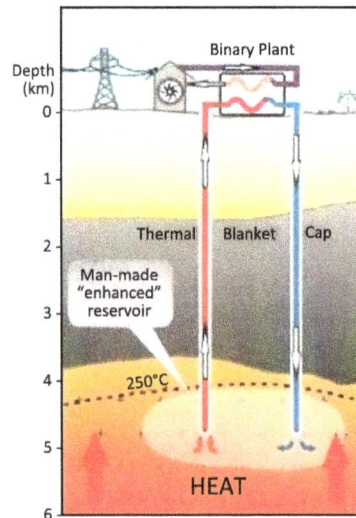

Geothermal power plants provide base-load electricity. Currently, the total globally installed capacity amounts to about 12 GWe, in 24 countries, with a total production of 76 TW·h/yr [2]. So far, practically all power plants use hydrothermal resources. Geothermal power generation started in 1904 in Larderello, Italy. In earlier days, reservoirs with dry steam have been tapped, later also those with steam/water mixtures. Such high-temperature fields (>200 °C in less than 2 km depth) are mostly located in volcanic areas and are correspondingly rare. The average power plant size is about 50 MWe. The largest hydrothermal plant to date, at Toanga (previously called Nga Awa Purua) in New Zealand operates with a single 140 MWe turbine unit and is fed by only six production wells [3].

With advanced technology such as binary power plants it is now possible to convert heat to power also with lower fluid temperatures (100–120 °C). But the conversion efficiency is correspondingly low (a few percentage points only) and the plant size is also limited (only a few MWe).

Below a global comparison is presented between geothermal and the other renewable energies, in terms of both potential and power generation. Development growth is presented for wind, solar photovoltaic (PV) and geothermal power and compared for the time period 1995–2013. In addition, a comparison is made of the annual geothermal production in 2013 with the renewables hydropower, biomass, solar PV and wind.

3. Large Geothermal Potential

A highly respected source (World Energy Assessment (WEA)—a collaborative effort between United Nations Development Programme (UNDP), United Nations Development of Economic and Social Affairs (UN DESA) and the World Energy Council (WEC)) attests the largest potential value to geothermal energy among all forms of renewable energy sources. The comparison is given in Table 1.

Table 1. Potential of renewable energy sources, from World Energy Assessment (WEA) [4].

Energy source	Capacity (EJ/yr)
Geothermal	5000
Solar	1575
Wind	640
Biomass	276
Hydro	50
Total	7541

The values are given in capacity units, *i.e.*, energy per unit time. It is obvious that geothermal energy has the largest capacity, although the accuracy of the reported number is limited. This potential is so far only marginally developed.

4. Growth Comparison over the Time Period 1995–2013

Geothermal power development data is available for the time period 1960–2013 according to Geothermal Energy Association (GEA) 2012 [5] and 2000–2013 from GEA 2014 [6]. The growth is practically linear, with only small increase rate changes lately, see Figure 2.

New data on the development of power generation from renewable sources is given in REN21 2014 [2]. The installed capacity of wind power shows a clearly accelerating trend of an exponential nature (Figure 3), with an annual growth rate of about 25%.

A similar trend of exponential growth is reported for solar PV power, both grid-connected and off-grid production already for 1995–2008. In Figure 4, the geothermal power growth in the same period—from [5]—is plotted for comparison. It is evident that geothermal had the lead over solar PV in the time before year 2007. Afterwards solar PV clearly took over.

Figure 2. Growth of installed geothermal power (MWe) worldwide over the years 2000–2014 (from Geothermal Energy Association (GEA) 2014 [6]). Global growth 2004–2012 ~4%.

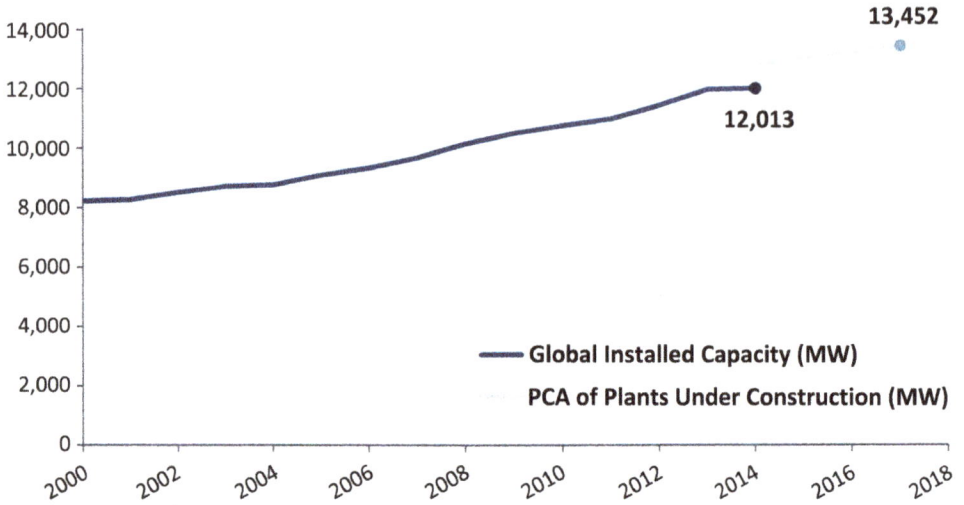

Figure 3. Growth in global wind power development (GWe) over the years 2000–2013 from REN21 [2].

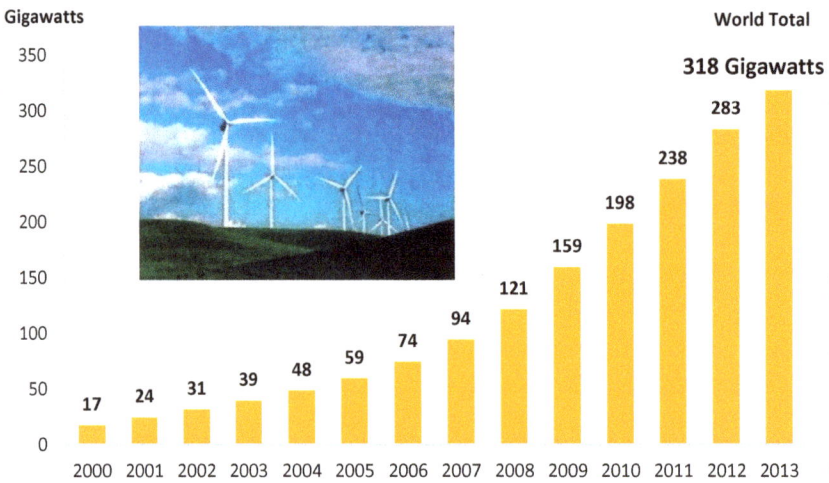

Figure 4. Global growth in solar photovoltaic (PV) development (MWe) 1995–2008 from REN21 2009 [7]. Geothermal power growth is plotted for comparison (data from [5]). Until about 2007 geothermal power was far ahead of solar PV.

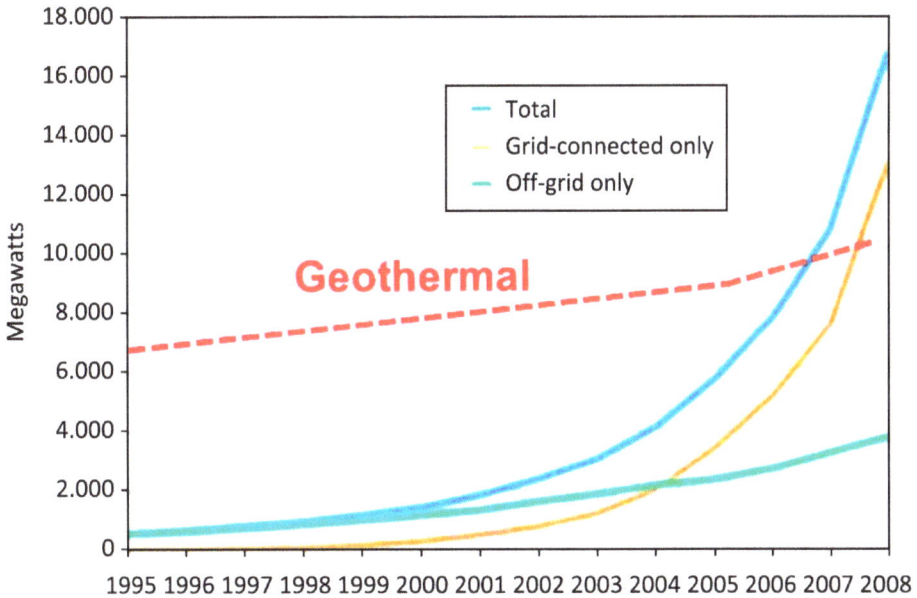

Now new solar PV data are available. Figure 5 shows the situation by end of 2013. For comparison, the geothermal data are plotted again. It is clear that geothermal is now left far behind. Here it must be noted that practically all geothermal power originates so far from hydrothermal resources.

Figure 5. Global growth in solar PV development (MWe) until 2013 from REN21 2014 [2]. Geothermal power growth (dashed line) is plotted for comparison—from Figure 2. Whereas solar growth is annually about 40%, geothermal growth remains low at 4% per year.

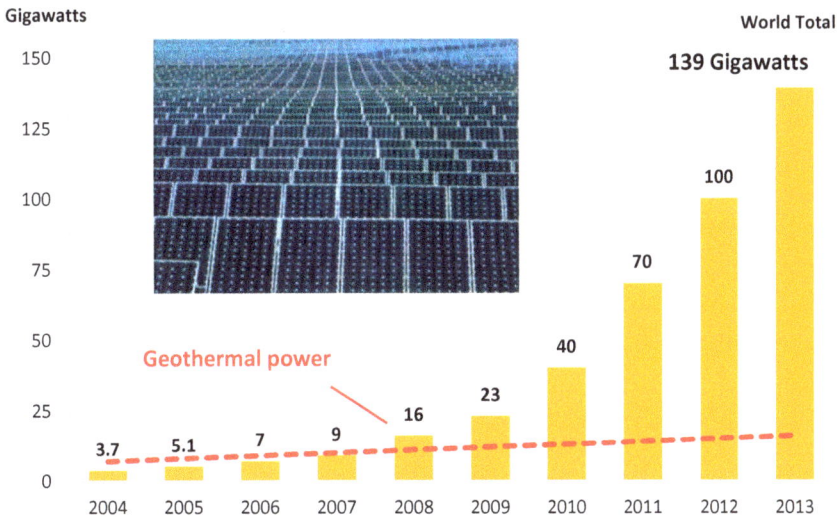

Here it must be emphasized that the Figures 2–5 refer to installed capacity, not to actual power production. What counts is the produced amount of electricity. Annual production data (in TW·h) are assembled in Table 2 for various renewable sources. Wind is not blowing at all times; the sun is shining only during daytime whereas geothermal production can go on at practically all times (except for production stops, for example during maintenance operations). This is reflected by the capacity factor (basically the percentage of yearly operating hours), given in Table 2. Sometime in 2011 solar PV took also over geothermal electricity in terms of global annual production; and the solar-geothermal gap is thus further increasing.

Table 2. Comparison of global electricity production by renewable technologies in 2013 (data from REN21 [2]).

Technology	Installed capacity		Annual production		Availability
	GWe	%	TW h/yr	%	%
Hydropower	1000	64.2	3680	74.9	42
Biomass	88	5.7	405	8.2	53
Wind	318	20.4	585	11.9	21
Geothermal	12	0.8	76	1.5	72
Solar PV	139	8.9	170	3.5	14
Total	1557	100	4916	100	-

From the above comparison it is evident that currently geothermal power development is left behind wind and solar PV. Whereas geothermal development growth is more or less linear (steady but slow growth—just a few percent increase per year), wind and solar PV exhibit accelerating growth with a clearly exponential tendency. To keep pace geothermal growth needs to be speeded up too; in the following some possible ways and means to accomplish this are addressed, primarily for power generation.

5. How to Achieve Accelerated Geothermal Power Growth?

Until today the growth in installed geothermal power capacity originated entirely from "conventional", hydrothermal resources. Such resources are found in numerous but special places, with high-temperature geothermal fluids present in the subsurface at relatively shallow depths (2–4 km). Such "anomalous" places can mainly be found in volcanic terranes or in other regions, depending on their plate tectonic settings (details see e.g., in [8]). It can be expected that geothermal power development based on conventional high-enthalpy resources will remain more or less linear in the future; therefore some new technology is needed to provide the exponential growth component. In the following the case is made that enhanced geothermal systems (EGS) technology could play this role.

In a study commissioned by the Intergovernmental Panel on Climate Change (IPCC) a team of authors [8] estimated the growth curve in geothermal power development from the present to year 2050. Figure 6 shows the result (installed capacity as well as power production). The curves in Figure 6 also exhibit exponential character.

Figure 6. Installed geothermal capacity and electricity production since 1995 and forecasts for 2010–2050. From [8].

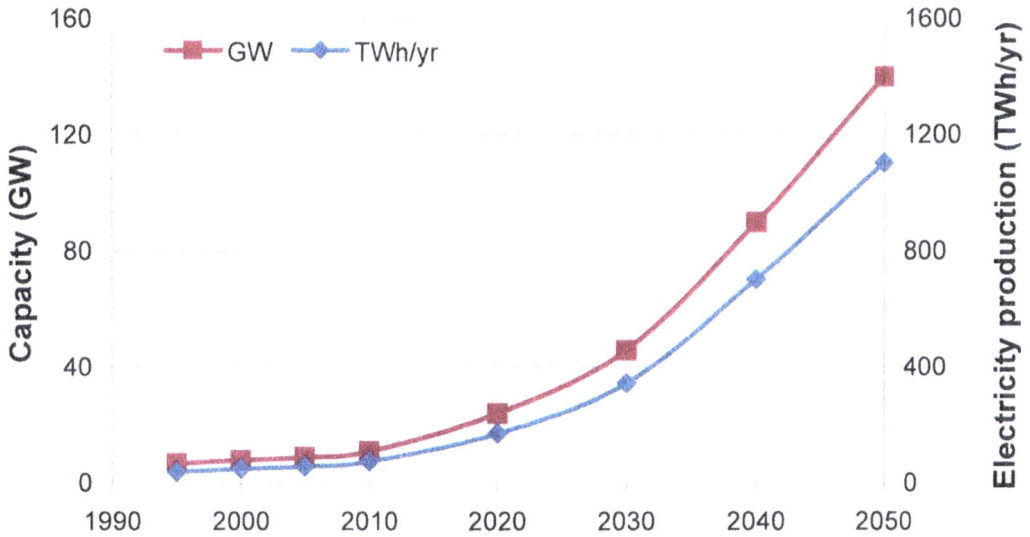

6. EGS Technology: Goals and Open Questions

The renowned Massachusetts Institute of Technology (MIT) [9] study "The Future of Geothermal Energy—Impact of EGS on the United States in the 21th Century" suggests that EGS will be the future of geothermal energy utilization. EGS is an umbrella term for various other denotations such as Hot Dry Rock, Hot Wet Rock, and Hot Fractured Rock. The MIT study determined EGS resources >200,000 EJ alone for the USA, corresponding to 2000 times the annual primary energy demand.

The EGS principle is simple: in the deep subsurface where temperatures are high enough for power generation (150–200 °C) an extended, well distributed fracture network is created and/or enlarged to act as new fluid pathways and at the same time as a heat exchanger ("reservoir"). Water from the surface is pumped through this deep reservoir using injection wells and recovered by production wells as steam/hot water. The extracted heat can be used for district heating and/or for power generation.

The core piece of an EGS installation is the heat exchanger at depth. It is generally accepted that it must have a number of properties in order to be technically feasible and economically viable. These refer to the total volume, the total heat exchange surface, the flow impedance, and the thermal and stress-field properties. The key properties are summarized in Table 3.

Table 3. Required properties for an enhanced geothermal systems (EGS) reservoir (after RHC Platform [10]).

Fluid production rate	50–100 kg/s
Fluid temperature at wellhead	150–200 °C
Total effective heat exchange surface	>2 ×106 m^2
Rock volume	>2 ×108 m^3
Flow impedance	<0.1 MPa/(kg/s)
Water loss	<10%

Although the minimum requirements for an economically viable EGS reservoir are herewith set, their realization in a custom-made manner to comply with differing site conditions is not yet demonstrated. The key issue is the development of a technology to produce electricity and/or heat from a basically ubiquitous resource, in a manner relatively independent of local subsurface conditions, *i.e.*, to develop a technology for the creation of EGS downhole heat exchangers—wherever needed—with the properties quantified above. Therefore, several questions about establishing and operating EGS heat exchangers that are still open need to be addressed and answered. Here are some of the key issues:

- Development of a technology to produce electricity and/or heat from a basically ubiquitous resource, in a manner more or less independent of site conditions.
- Site exploration must clarify the local temperature and stress field, lithology, kind and degree of already existing fracturation, natural seismicity.
- In creating EGS heat exchangers at several kilometers depth, questions of rock mechanics like the role of anisotropy degree, stress change propagation/transmission—fast/"dry"? slow/"wet"? (under different site conditions)—need to be answered.
- EGS induced seismicity (during stimulation in establishing the EGS heat exchanger but also during production) becomes a real issue, and thus needs to be controlled. Magnitudes need to be limited since public acceptance will be decisive [11].
- Uniform connectivity throughout a planned reservoir cannot yet be engineered. There is no experience with possible changes of an EGS heat exchanger over time; permeability enhancement (e.g., new fractures generated by cooling cracks) could increase the recovery factor while permeability reduction (e.g., by mineral reactions) or short-circuiting could reduce recovery.
- This leads to the question of production sustainability. The production level needs to be set in order to guarantee longevity of the system (details in [12]).

7. Increasing EGS Power Plant Size

In order to play a significant role on the electricity scene, geothermal power plants should have the size of at least some tens of MWe. So far, EGS plants (Soultz sous Forêts in France, Landau and Insheim in Germany) have just a few MWe installed capacity. Wind generators nowadays come with at least 2 MWe and can be installed, especially offshore, in large numbers.

One of the main future R&D goals will be to work out how and to what extent could the EGS power plant size be upscaled. So far, there are only some theoretical calculations available; see e.g., [13]. In this publication an EGS scheme with 24 injection and 19 production wells is modelled, providing a net power output of around 60 MWe. Of course such ideas need to become substantiated by field evidence.

8. European Geothermal Growth Perspectives, Financial Aspects

The European Union has ambitious goals in term of renewable energy growth; the "20-20-20 goal" (20% share of renewable energies, 20% energy savings and 20% CO_2 emission reduction until 2020) clearly calls also for more geothermal electricity. In the Union, a goal of 3 GWe EGS capacity has been proposed for the year 2020 and further substantial EGS growth by European Geothermal Energy Council (EGEC) [14], see Figure 7.

Figure 7. Vision of European Geothermal Energy Council (EGEC) about geothermal electricity growth until 2050, from EGEC 2012 [14]. The largest share should come from EGS.

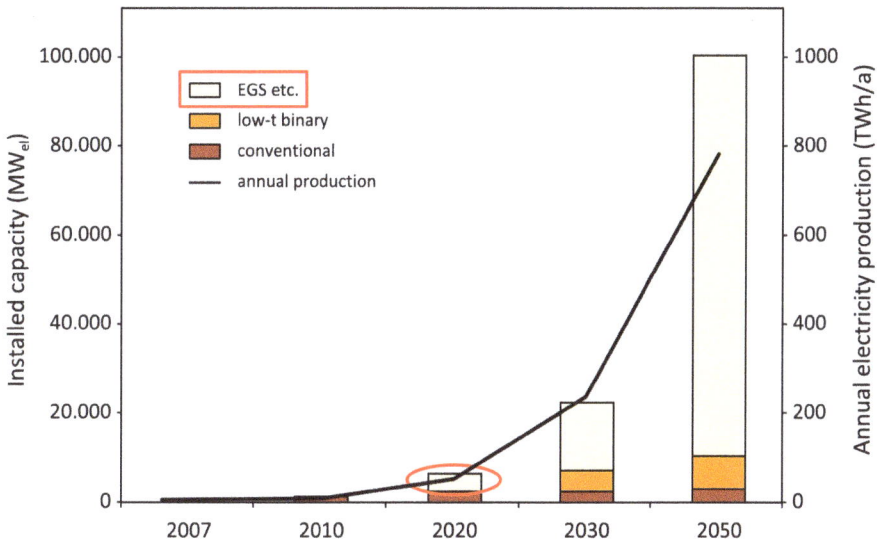

What financial sources would be needed for the realization of this vision? Currently the cost of establishing a generating capacity of 1 MWe from EGS (including exploration, drilling, stimulation, power plant, *etc.*) is estimated to be around 16 million € according to Geothermal Electricity (GEOELEC) [15]. Thus the 3 GWe EGS capacity foreseen for 2020 in Europe (circled in Figure 7) would require an investment of about 50 billion €. Today it is unclear where such a funding would come from.

It is obvious from the above-described knowledge gaps that very substantial R&D efforts are still needed to make EGS become the future of geothermal energy. Whereas some of the problems could be tackled by broad-based international cooperation, national R&D programs have to provide additional means for the challenge. Public funding, mainly by governmental agencies, will be indispensable.

Although envisaged for conventional geothermal resources, Ibrahim [16] describes five steps to expedite development; four of them are based on fund allocations from national or regional governments. In any case it will be crucial to make rapid progress in tackling and solving the above-mentioned, still open financing problems.

9. Conclusions and Outlook

Geothermal power develops steadily world-wide, albeit with modest growth rates (a few % per year). In some countries, like in Iceland, New Zealand or Turkey, the growth is remarkable. At the same time, wind and solar PV develop exponentially, with 30%–40% annual growth. In other words: globally, geothermal power falls increasingly back behind electricity from wind and solar PV.

Therefore, geothermal power growth should be accelerated. Since the development of hydrothermal resources cannot be accelerated much, mainly because such resources are limited; the only option that remains are petrothermal resources. The only problem: how to get out the heat in place? In particular, the following questions need to be addressed:

- Where? (favorable site conditions → exploration)
- How? (sufficient, deep heat exchanger realization → proper, site dependent stimulation—without significant induced seismicity)
- With what efficiency? (recovery factor → enhancement of heat extraction, production sustainability). *Recovery factor, R (%) = extractable heat/heat in place*

Besides, upscaling EGS power plant size will be decisive. EGS pilot plants are badly needed, as is long-term experience. In addition, the financing of all the R&D needs should also get solved. All these open questions need to be answered—and rather quickly so.

The future of geothermal power will strongly depend on to what extent can be the power plant deployment accelerated. Other sources of renewable energy are developing rapidly, especially wind and solar PV: wind energy recently accomplished to install 35 GWe additional capacity per year; solar PV reached 39 GWe/yr, whereas geothermal power growth remains below 2 GWe/yr. Even when one takes into account the higher geothermal capacity factor the need for speeding-up geothermal development is obvious. Accelerating EGS development could provide a break-through, under the condition that the necessary significant funding needs can be met. This, in turn, will require heavy engagement of both the public and the private sector.

Conflicts of Interest

The author declares no conflict of interest.

References

1. Van Wees, J.D.; Calcagno, P.; Dezayes, C.; Lacasse, C. *A Methodology for Resource Assessment and Applications to Core Countries*; Geothermal Electricity (GEOELEC): Brussels, Belgium, 2013. Available online: http://www.geoelec.eu/concep/library/ (accessed on 22 July 2014).

2. *Renewables 2014: Global Status Report 2014*; Renewable Energy Policy Network for the 21st Century (REN21): Paris, France, 2014.

3. Tadao, H.; Muto, T. Technical Features of Nga Awa Purua Geothermal Power Station New Zealand. In Proceedings of Renewable Energy 2010, Yokohama, Japan, 27 June–2 July 2010; Abstract O-Ge-4-4. Available online: http://www.re2010.org/ (accessed on 22 July 2014).

4. *World Energy Assessment (WEA) Report: Energy and the Challenge of Sustainability*; United Nations Development Programme (UNDP), Bureau for Development Policy, One United Nations Plaza: New York, NY, USA, 2000.

5. *Geothermal: International Market Overview Report*; Geothermal Energy Association (GEA): Washington, DC, USA, 2012.

6. *Annual U.S. & Global Geothermal Power Production Report*; Geothermal Energy Association (GEA): Washington, DC, USA, 2014.

7. *Renewables 2009: Global Status Report 2009*; Renewable Energy Policy Network for the 21st Century (REN21): Paris, France, 2009.

8. Fridleifsson, I.B.; Bertani, R.; Huenges, E.; Lund, J.W.; Ragnarsson, A.; Rybach, L. The Possible Role and Contribution of Geothermal Energy to the Mitigation of Climate Change. In Proceedings of the IPCC Scoping Meeting on Renewable Energy Sources, Luebeck, Germany, 20–25 January 2008; Volume 20, pp. 59–80.

9. Tester, J.W.; Anderson, B.J.; Batchelor, A.S.; Blackwell, D.D.; DiPippo, R.; Drake, E.M.; Garnish, J.; Livesay, B.; Moore, M.C.; Nichols, K.; *et al. The Future of Geothermal Energy—Impact of Enhanced Geothermal Systems (EGS) on the United States in the 21st Century*; Massachusetts Institute of Technology (MIT): Cambridge, MA, USA, 2006; p. 358. Available online: http://www1.eere.energy.gov/library/default.aspx?page=4 (accessed on 22 July 2014).

10. *Strategic Research Priorities for Geothermal Technology*; European Technology Platform on Renewable Heating and Cooling (RHC): Brussels, Belgium, 2012; p. 65. Available online: http://www.rhc-platform.org/publications/ (accessed on 22 July 2014).

11. Majer, E.; Baria, R.; Stark, M. *Protocol for Induced Seismicity Associated with Enhanced Geothermal Systems*; Report Produced in Task D Annex I; International Energy Agency—Geothermal Implementing Agreement: Taupo, New Zealand, 2008.

12. Rybach, L.; Mongillo, M. Geothermal sustainability—A review with identified research needs. *GRC Trans.* **2006**, *30*, 1083–1090.

13. Vörös, R.; Weidler, R.; de Graaf, L.; Wyborn, D. Thermal Modelling of Long Term Circulation of Multi-Well Development at the Cooper Basin Hot Fractured Rock (HFR) Project and Current Proposed Scale-Up Program. In Proceedings of the Thirty-Second Workshop on Geothermal Reservoir Engineering, Stanford, CA, USA, 22–24 January 2007.

14. European Geothermal Energy Council (EGEC). EGEC Vision for 2050 on Geothermal Power in Europe (Vision of ETP GEOELEC), 2012. Available online: http://egec.info/policy/tp-geoelec/ (accessed on 22 July 2014).

15. *Geothermal Investment Guide*; Geothermal Electricity (GEOELEC): Brussels, Belgium, 2011; p. 32. Available online: http://www.geoelec.eu/concep/library/ (accessed on 22 July 2014).

16. Ibrahim, H.D. Why is Geothermal Development Slow—How to Accelerate It? *Petrominer Monthly Magazine*, 1 July 2009; Volume 7, pp. 28–29.

National Deployment of Domestic Geothermal Heat Pump Technology: Observations on the UK Experience 1995–2013

Simon Rees and Robin Curtis

Abstract: Uptake of geothermal heat pump technology in the UK and corresponding development of a domestic installation industry has progressed significantly in the last decade. This paper summarizes the growth process and reviews the research that has been specifically concerned with conditions in the UK. We discuss the driving forces behind these developments and some of the supporting policy initiatives that have been implemented. Publically funded national trials were completed to assess the performance and acceptance of the technology and validate design and installation standards. We comment on both the technical and non-technical findings of the trials and the related academic research and their relevance to standards development. A number of technical issues can be identified—some of which may be particular to the UK—and we suggest a number of research and development questions that need to be addressed further. Current national support for the technology relies solely on a tariff mechanism and it is uncertain that this will be effective enough to ensure sufficient growth to meet the national renewable heat target in 2020. A broader package of support that includes mandatory measures applied to future housing development and retrofit may be necessary to ensure long-term plans for national deployment and decarbonization of heat are achieved. Industry needs to demonstrate that efficiency standards can be assured, capital costs reduced in the medium-term and that national training schemes are effective.

Reprinted from *Energies*. Cite as: Rees, S.; Curtis, R. National Deployment of Domestic Geothermal Heat Pump Technology: Observations on the UK Experience 1995–2013. *Energies* **2014**, *7*, 5460-5499.

1. Introduction

Exploitation of geothermal energy at shallow depths (<500 m) by the application of heat pump technology has grown rapidly in a number of countries since the 1990s so that more than 2.7 million world-wide installations have been reported [1]. The most common form of geothermal or ground-source heat pump (GSHP) technology uses anti-freeze fluid circulated in a closed loop heat exchanger and the vapor-compression refrigeration cycle to deliver thermal energy for space heating and hot water production. In the UK, the form of closed loop heat exchanger used in domestic systems is usually a single U-tube vertical borehole (up to 100 m deep), horizontal parallel loop or horizontal "slinky" device (1–2 m deep). Space heating in UK domestic properties using this technology is nearly always hydronic with heat emitters that are radiator/convectors or under-floor heating and domestic hot water production is often supplemented by electric resistance heating.

The diffusion of geothermal heat pump technology into the UK domestic building market has lagged that found in several northern European and North American countries by more than a decade. The national scene has changed from only a few installations known in the 1990s to several thousand being installed annually and a young installation industry being established. Domestic geothermal heat pumps now feature in the government's national carbon reduction strategies, renewable energy production targets and related market incentive schemes. These developments have been accompanied by national field trials to support technical evaluation and standards development. A national industry body has been established that promotes the technology and provides support for training and standards development. Although early diffusion of the technology has been arguably successful, the UK GSHP industry has recently faced very challenging economic circumstances and changing regulatory and market incentive frameworks.

In this paper, we review the time-line of the deployment of geothermal heat pump technology in the UK domestic market and attempt to identify some of the key growth factors. Three phases of national field trials of the technology and both the technical and socio-technological conclusions that have been drawn are summarized. We furthermore identify a number of technical issues that may be particular to the UK situation that would benefit from further research. We finally reflect on the recent changes in policy and the regulatory framework and prospects for further market growth.

1.1. The National Context

Understanding of the apparently reluctant uptake of geothermal heat pump systems in the UK requires some appreciation of the technological context and the national energy scene. The UK has a maritime climate such that winter temperatures are mild compared to Scandinavian and Central European climates and moderate in summer such that domestic cooling is unnecessary. The mild climate and abundance of fossil fuels has meant that relatively poor insulation and airtightness standards have been tolerated and are reflected in much of the historic building stock. The predominant domestic heating technology is natural gas fueled hydronic heating (more than one million gas boilers are sold each year). This predominance is a natural reflection of the abundance of natural gas resources available in the last few decades and a well-developed gas distribution network. Mains gas heating is used in 84.2% of households with 9% using electric heating (1.9 million households). The third most common fuel is heating oil which is used in 3.9% of all households and is much more common in rural areas [2]. Fuels such as LPG, biomass and coal are currently used only in small quantities. Hydronic heating systems are typically designed for high temperature operation with wall mounted radiator heat emitters. Under-floor heating is not common, even in new housing.

The predominance of hydronic heating systems using natural gas boilers is also reflected in the national skills and knowledge base. The absence of demand for domestic air conditioning has meant that skills and training in small-scale refrigeration systems is very limited. This is similarly reflected in the education and training that has been demanded by, and delivered to, the domestic heating industry.

This historic abundance of coal resources, and more recently natural gas, has meant that much of the UK electrical energy supply is derived from fossil fuel sources. Only approximately 20% of power is generated by nuclear sources and the renewable contribution has been insignificant until very recently. This has meant that the carbon emissions associated with electrical energy are relatively high compared with other energy sources. Consequently, although the carbon emissions reduction benefits of using geothermal heat pumps can be demonstrated, these benefits are more marginal than in some other countries [3]. A technical feature of the national electrical power distribution system is that domestic buildings have single-phase supplies that have limited ability to tolerate compressor start-up currents [4]. This has placed a limit of approximately 8–12 kW on the maximum capacity of most domestic heat pump systems in the existing housing stock.

The broad situation regarding accepted technologies, skills base, and energy supply can be contrasted with that in a number of other countries that have demonstrated significant uptake of geothermal heat pump technology. In the USA, for example, the prevalence of domestic air conditioning means that there is a good refrigeration skills base and consumers are willing to accept central air systems with little difficulty. In many parts of the USA there are also established land drilling industries able to offer geothermal borehole drilling at acceptable cost. In countries such as Sweden, Austria and Switzerland, more prevalent hydro electricity generation makes electric heating more economic and also advantageous in terms of carbon emissions compared to fossil fuels than in the UK. The mild climate in the UK also means that the efficiency advantages of geothermal heat pumps are not as great compared to air-source heat pumps (ASHP) as in countries with more severe winters such as those in Central Europe, Canada and Sweden.

1.2. Market Development

Worldwide surveys of geothermal heat pump developments have been reported at five-year intervals since 1975 (e.g., [5]) and have included reports from the UK since 1985 [6]. Although such international data highlights significant growth in application of heat pumps to domestic buildings in the 1990s in several parts of Europe and the USA, little activity was reported for the UK in early surveys. Although there is evidence of the application of domestic ground source heat pumps in the UK since 1960 [7], it is thought that the first system of what may now be considered conventional, borehole based, closed-loop configuration was installed in 1994 [8]. A total of 40 UK installations were reported in 2000 in contrast to an estimated 500,000 installations worldwide and a growth of 59% in terms of global installed capacity in the 1995–2000 period [9].

It is not until the first decade of the current century that developments beyond single properties and establishment of viable installation industry in the UK can be identified. In 2005 [5] it was reported that, "*the country now understands that ground-source heat pumps, connected to the electricity grid, offer very substantial reductions in overall carbon emissions compared to conventional fossil-fueled systems*". It was also reported that approximately 500 domestic systems were known [10] and estimated to have a combined capacity of 10.2MWt and an annual energy use of 45.6 TJ/year. The 2000–2005 period, therefore, represents a significant change in the deployment of geothermal heat pump technology in the UK.

Geothermal heat pump market data reporting annual sales in the UK has been available since 2007 [11] and reflects annual installations at a rate of 2400 in that year rising to 3980 in 2009. Similar installation rates were reported in the international survey published in 2010 [1]. Data for 2012/2013 indicates installation rates have fallen to approximately 3000 per annum following the broader downturn in the economy. Following the introduction of the EU Renewable Energy Sources directive, cumulative data reporting the uptake of geothermal heat pump technology along with estimates of thermal energy production within EU Member states has been recorded by EurObserv'ER [12,13]. These data, in the case of the UK submissions, are based on market reports and so included a figure of 3980 annual installations in 2009 [12] and an estimated cumulative uptake of 14,330 systems. This cumulative figure had risen to 17,760 in 2012 [13]. The latter data is consistent with the figure of approximately 16,000 being installed by the end of 2012 reported to the European Geothermal Energy Congress [14]. Installations were projected to grow by a further 3000 in 2013 [11].

Figure 1. (a) Total geothermal heat pump installations in the UK ranked with other EU Member States as of 2012; **(b)** total installations per million capita. Data from [12,13].

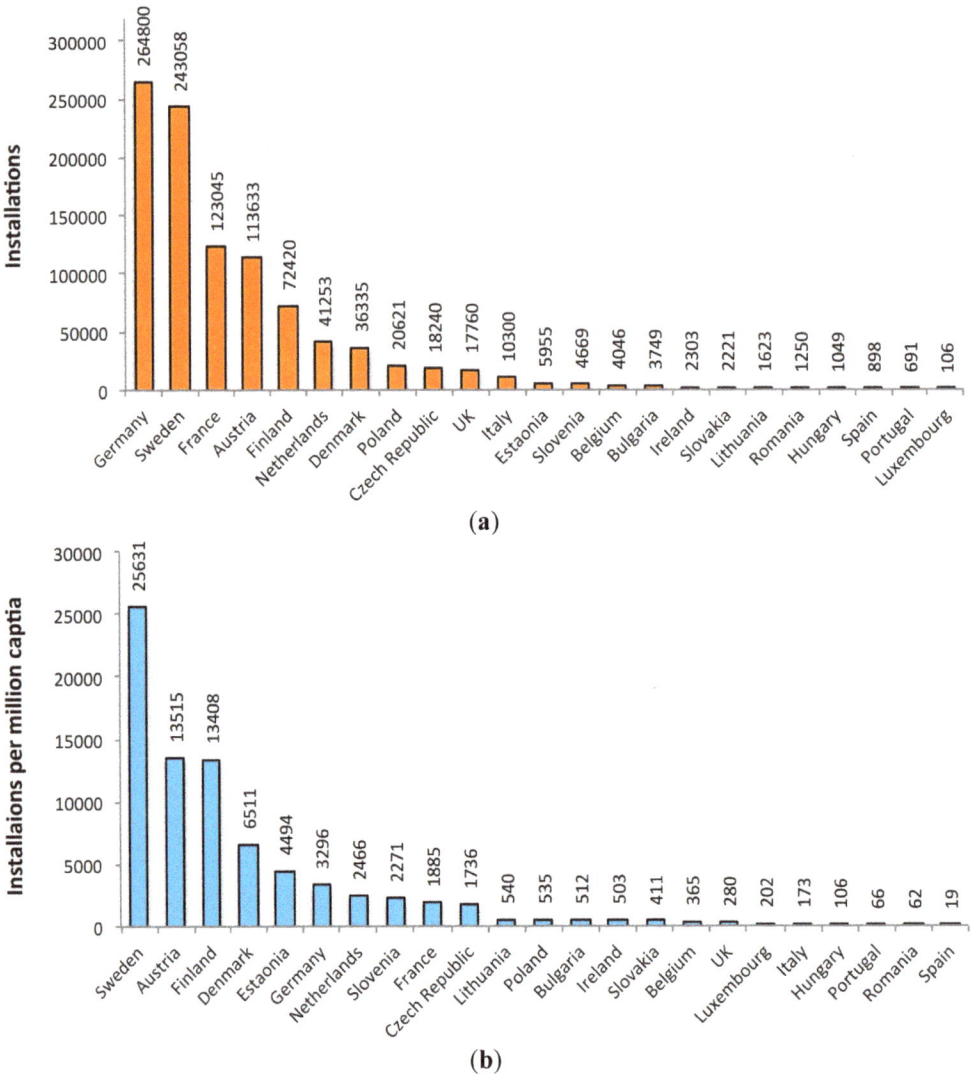

(a)

(b)

These data have been put into a European context in Figure 1 along with data formulated on a per capita basis. The significant deployment of the technology in Sweden, Germany and Austria has been demonstrated for some time and is unsurprising [1,15,16]. Uptake in France has been rapid in recent years [17]. The data shown in Figure 1 suggests that the UK installation industry is in the early stages of growth but small relative to other EU Member States. Penetration into the domestic heating market is broadly demonstrated by the data compared on a per-capita basis (Figure 1b). Although these data suggest penetration is currently poor in the UK, comparison with smaller countries but with similar climates (Netherlands and Denmark) suggests there is considerable potential for further growth.

1.3. Policy and Support Programme Development

1.3.1. Capital Grant Programmes

The development of UK energy efficiency policy can be traced from the oil crisis of 1973 and has varied in its emphasis on interventionist or market led measures through several governments of differing political persuasion [18]. It is not until the 1990s that carbon emissions reduction is explicitly addressed in policy documents [19] and even more recently that renewable energy policy and support programmes have emerged. The move towards an integrated approach to significant carbon emissions reduction that embraces energy efficiency and renewable energy is most clearly marked by the policy statements in the 2002 Energy Review [20] and the legislation in the 2003 Energy White Paper "Our Energy Future—Creating a Low Carbon Economy" [21]. These policy statements also sought to address the issues of energy security and fuel poverty and placed a noticeable emphasis on community scale measures. One of the first actions arising from the Energy Review was the release of £100 million for renewable energy support of which £31 million went into a PV demonstration programme, £10 million into the "Clear Skies" programme and £3.1 million into the corresponding 'Scottish Community and Householders Renewables Initiative' (SCHRI) to support small-scale renewable energy schemes.

The Clear Skies programme [22] offered capital grant support for a range of small-scale "microgeneration technologies" that included many recognized renewable technologies. In the few years leading up to the development stage of the Clear Skies programme (during planning in 2002 it had been known as the "Community and Household Capital Grant Scheme") there was both inter-departmental debate and external consultation as to what technologies might be classified as "microgeneration" and furthermore receive grants. Geothermal heat pumps were incorporated into the definition of microgeneration technologies but not without presentation of the case for support from industry stakeholders. It proved valuable to be able to point to some of the early UK examples that had shown promising levels of performance (e.g., [23]) as well as a larger body of international good practice (e.g., [24]). The argument that geothermal heat pumps could, in the long run, make significant contributions to carbon reduction—given minimum performance levels and also in view of long-term plans for decarbonization of the grid—was accepted. The technologies supported by the Clear Skies programme were then: solar thermal, micro-wind, small-scale hydro, biomass boilers and geothermal heat pumps (photovoltaics being funded under a separate £31m Major Photovoltaics Demonstration programme). The legal definition of the term microgeneration was later formalized in the 2004 Energy Act [25] and is a broader definition that includes fuel cells, biofuels, micro-CHP, wave, tide and "*other sources of energy…which would, in the opinion of the Secretary of State, cut emissions…*". This definition has provided the flexibility to include air-source heat pumps (ASHP) in the microgeneration classification more recently.

The Clear Skies initiative was later transformed into the "Low Carbon Buildings Programme" (LCBP) that ran between 2006 and 2010 with a total funding commitment of £137 million [26]. This brought together capital grant support for photovoltaic systems into the same programme as the other microgeneration technologies. The majority of the funds went to householders and non-profit organizations providing housing (termed Registered Social Landlords or RSLs in the UK). Separate funding streams within the LCBP were initiated for medium and large-scale systems with capacities exceeding the 50 kWe/45 kWth limits imposed in the Clear Skies programme. A second phase of the LCBP extended funding to community non-domestic buildings after 2007. Of the various microgeneration technologies supported, more than 75% of the funding went to support solar thermal and PV systems and only 8.2% to support geothermal heat pumps [26].

The end of the LCBP programme (2010) coincided with a change in UK government and a pause in the funding of capital grants for householders. Development of a novel tariff-based support programme for renewable heat [27] was initiated following the Climate Change Act 2008 and the development of a national renewable energy strategy [28] that recognized increasing adoption of renewable sources of domestic heating would be an essential element of the national commitment to carbon reduction. These plans evolved into the non-domestic Renewable Heat Incentive (RHI) programme in 2011 and a domestic RHI scheme that was to start in summer 2013 [29]. Although this had been planned before the change in government, the new government continued to develop this approach but raised questions as to lack of budgetary control [30]. Consultation was extended and finalization of tariff rates delayed so that payments for domestic installations could not start until April 2014. An interim domestic capital grant (voucher) scheme was eventually introduced to boost the uptake of renewable heating technologies in the run-up to the introduction of RHI tariff payments. This programme was the Renewable Heat Premium Payment (RHPP) programme [31] and ran in 2012–2013. This programme offered £1250 capital grants for domestic GSHP installations but with the condition that the money would be reclaimed if RHI payments were claimed later.

1.3.2. Energy Supplier Obligation Programmes

Privatization of the UK electricity industry following the Electricity Act 1989 and the gas industry following the Gas Act 1994, also saw the establishment of respective independent regulatory authorities, OFFER and OFFGAS. The legislation gave the regulators powers to set binding standards of performance on the energy suppliers. In particular, a series of Energy Efficiency Supplier Obligation Programmes (EESOP 1–3) were established that obligated the suppliers to implement measures—usually working with third-party contractors—that could be demonstrated to save energy [32] and that were assessed with respect to an annual target set by the regulator. The three EESOP schemes ran between 1994 and 2002 with the savings targets ultimately set at 4.9 TWh (electricity) and 6.1 TWh (gas) and programme costs of £55million per annum. The regulations allowed the costs to be passed on to the consumer directly—initially amounting to £1 per annum per customer. These programmes achieved their savings targets for the most part by funding retrofits of insulation in homes but also boiler replacements and (for a limited period) heavily subsidized distribution of compact florescent lamps.

After 2002 (*i.e.*, around the time of the initial growth of the GSHP industry noted earlier and concurrently with Clear Skies) the supplier obligations took the form of the Energy Efficiency Commitment programme (EEC1) [33] followed by EEC2 in the 2005–2008 periods [34]. These were significantly more ambitious programmes with target savings set at 62 TWh and 130 TWh and programme spending of £500 million and £1.2 billion respectively. These targets (following revised legislation in the Utilities Bill 2000) were set by the Secretary of State for Energy rather than the independent regulators [32]. There were three changes between the EESOP and EEC programs that are significant in this context. In the EEC there was firstly, a specific intent to address fuel poverty (Fuel poverty is defined in this context as the situation where consumers spend more than 10% of their income on energy), secondly application to domestic properties alone and, thirdly the application of microgeneration technology alongside conventional energy saving measures [32].

In the run-up to the start of the EEC1 programme, it was proposed that GSHP systems could make a contribution to reduction of fuel poverty if they could be successfully incorporated into small social housing properties. This was appealing to RSLs in that they had become under obligations to improve housing standards after the publication of new government policy in the Housing Green Paper of 2000 [35]. GSHPs were, in principle, a very effective way of reducing running costs and improving thermal comfort in properties using coal or oil. Social housing projects also offered the potential for economically scaling-up installation into projects that could be managed in conjunction with energy suppliers. This concept was taken up by a consortium led by John Parker at Earth Energy Engineering, the energy supplier Powergen (later Eon), a GSHP installer, and a UK heat pump manufacturer, Calorex [36]. The regulators were persuaded that sufficient energy and carbon savings could be achieved over the lifetime of the system and so new GSHP installations could count towards the EEC energy saving targets. The energy supplier was satisfied this could be done at an acceptable cost. A new heat pump was developed that was optimized for smaller UK properties that needed both heating and hot water generation and could operate with radiator heat emitters and simple controls [37]. This form of GSHP system (denoted the Powergen "heatplant") was first installed at ten new Metropolitan Housing Trust properties in Nottingham in 2001 [38,39] followed by a project in Cornwall with Penwith Housing Association (Figure 2) involving retrofit to fourteen small properties [40]. The system was deployed at other social housing projects (mostly retrofit to off-gas-grid properties) during the EEC1 programme [41] and in growing numbers during EEC2. We describe the technical development of the heat pump later in the paper.

The Supplier Obligations took the form of a programme known as the Carbon Emissions Reduction Target (CERT) in the period 2008–2012 [42]. In this programme the effectiveness of the measures implemented by the supply companies was assessed against a lifetime carbon emission reduction rather than an energy reduction target. The Department of Energy and Climate Change set this target at a total of 293 million tonnes of CO_2. The programme expenditure amounted to approximately £1.2 billion per annum and this corresponded to approximately £51 per annum added to consumer bills [32]. In this programme, minimum levels of performance were imposed for certain measures (e.g., 68% for insulation) but not minimum levels of support for GSHP or other microgeneration technology installation. However, as measures in the CERT programme were assessed in terms of carbon emission savings, there was additional motivation to implement GSHP retrofits where the original form of heating had been electric resistance, or even coal fired heating.

Figure 2. (a) Drilling operations at a social housing project in 2002 [40]; **(b)** an example of a "heatplant" installation in a small house retrofit project [36].

(a)   (b)

The most recent form of the UK supplier obligation policy (2012–2017) has been the Energy Company Obligation (ECO) programme [43] that had an initial budget of £1.3 bn. The budget was split between three sub-programmes. These were: the Affordable Warmth Obligation (targeted at households at risk of fuel poverty); the Carbon Saving Obligation (targeted at insulation of "hard-to-treat" houses), and; the Carbon Saving Communities Obligation (targeted at specific economically deprived areas). The ECO programme has differed from the CERT programme in having a distinct emphasis on insulation measures. The "innovation" measure that encouraged a modest level of GSHP uptake in the CERT scheme was removed. The programme budget and scope was changed in 2014 and we comment on this further in the final discussion section of this paper.

1.3.3. Support Programme Outcomes

Some measure of the significance of these support programmes can be made by comparing the total GSHP installations reported by the industry noted above, with the numbers of installations supported by grants that have been disclosed. We suggested earlier that the 2000–2005 period was significant in terms of the establishment of the UK GSHP installation industry as installations grew from 40 at the start of this period to approximately 500 by 2005. This period corresponds largely to the implementation of the Clear Skies programme in the UK. In 2005 it was reported that 500 GSHP installations had received grants under this programme [44]. Although the EEC1 supplier obligation programme was in effect in this period, it provided support for only 40 installations [33]. The LCBP that followed on from the Clear Skies initiative, and ran until 2010, supported the installation of 1573 GSHPs. These were nearly all domestic systems with 843 grants given to individual households and the remainder of the GSHP systems being delivered via grants to RSLs and other non-profit organizations [26].

Although the portion of the Clear Skies funding that was directed into GSHP installations was relatively small, an important aim of the programme was dissemination of information about microgeneration technologies to householders and other stakeholders—lack of information being a recognized barrier to uptake of renewable technologies and energy efficiency measures [45,46]. We suggest that this was particularly important in view of the fact that this was the UK's first comprehensive consumer focused renewable energy support programme. Although some microgeneration technologies in the programme, such as solar thermal, had some record of success in earlier decades and something of a public profile, very little information relating to geothermal heat pump technology had been made available to the public in the UK. We suggest there was some benefit, in terms of promotion of information, from GSHP technology being presented alongside other renewable technologies as part of a broad programme of microgeneration deployment.

A criticism of the first phase of the LCBP programme, based on responses to questionnaires completed by individual householders, has been that a significant number of grant recipients would have purchased a system even if they had not received a grant [26]. It is not clear to what extent this was true of GSHP installations as compared to the other technologies in the programme (80% of the funding went to solar thermal and PV installations). This may reflect two other factors: firstly, that the grants provided a mean of only 10.6% of the GSHP installation cost; secondly, the demographic data showed that most of the householder grants went to owners of relatively large properties (4 bedrooms or more) and so these respondents may have had good access to other funds.

The administrators of the LCBP programme have acknowledged this criticism [26] but suggest that the programme had broader value in developing a quality assurance framework (the MCS discussed below) that was important to establishing good market conditions and incubating the industry. The same criticism was not made of the second phase of the LCBP that had provided grants to non-profit community organizations. This funding differed in that the grants were larger (a mean of 46.6% of the total cost) and for systems with higher capacity (a mean of 40.3 kW per scheme). We suggest that having a national programme like Clear Skies and LCBP that offered some assurance of quality and independent consumer advice also played an important role in encouraging new SMEs into the GSHP installation industry at what was an embryonic stage in its development.

The three supplier obligation schemes that have run through the period of interest (EEC1, EEC2 and CERT) have seen only a small proportion of programme funds spent on GSHP installations. However, as these schemes have grown to become substantial streams of funds, the number of domestic GSHP systems that have been funded has amounted to a significant proportion of the total installed in the UK. The EEC1 scheme was established when interest in GSHP was embryonic and saw only 40 GSHP systems installed before the advent of EEC2 [33]. The EEC2 programme ran concurrently with much of the LCBP and funded the installation of 1500 GSHP systems [34]. The most substantial supplier obligation scheme, CERT, provided funding for 4497 installations between 2008 and early 2011 [42].

The supplier obligation programmes did not have the same public information dissemination and independent advice brief as Clear Skies and LCBP. However, their impact on the development of the GSHP installation industry is probably greater in view of the overall number of systems they funded and the fact that the level of support was greater. Many of the projects funded consisted of groups of properties—both new and retrofit—and this allowed some economies of scale to be gained. For example, as the mobilization costs associated with drilling operations are substantial, it is more efficient to drill at groups of properties than at individual houses in different locations. This was to the benefit of the SMEs entering the industry (including drilling contractors) and to clients in reducing costs. There is evidence that in social housing projects funded by these programmes, there have been fuel poverty benefits for tenants as a result of the GSHP installations, particularly where the heating fuel was previously oil or coal, or the heating system was electric resistance heating—addressing fuel poverty has been one of the aims of the supplier obligation programmes.

Taken together, the Clear Skies, LCBP and RHPP programmes provided grants for 4022 installations. The supplier obligation schemes provided funding for a total of 6037 installations. The annual installation data is shown over the 2000–2013 period in Figure 3 and highlights the significance of the number of systems supported by both types of programme in relation to the total number of systems installed. Altogether, the support programmes have funded 10,059 installations representing 57% of the total installations reported [13,47]. Prior to 2006 a high proportion of the installations have been supported by the Clear Skies programme. These data show 2006–2009 was a period of rapid expansion of the industry during which the number of installations supported by private funding has also grown significantly. In the 2009–2012 period much of the growth appears to have come from the CERT programme. In the April 2012 to December 2013 period, the growth is almost entirely attributable to the RHPP programme.

Figure 3. Growth of ground-source heat pump (GSHP) installations in the UK part funded by public capital grant programmes (Clear Skies and "Low Carbon Buildings Programme" (LCBP)), energy supplier obligation schemes (EEC1, EEC2, CERT) and, the most recent Renewable Heat Premium Payment (RHPP) grant programme. Data for total installations is a combination of data reported by WREC [1], BSRIA [48] and EurObserv'ER [13]. Installations funded without grant support are categorized as "other funding" in this figure.

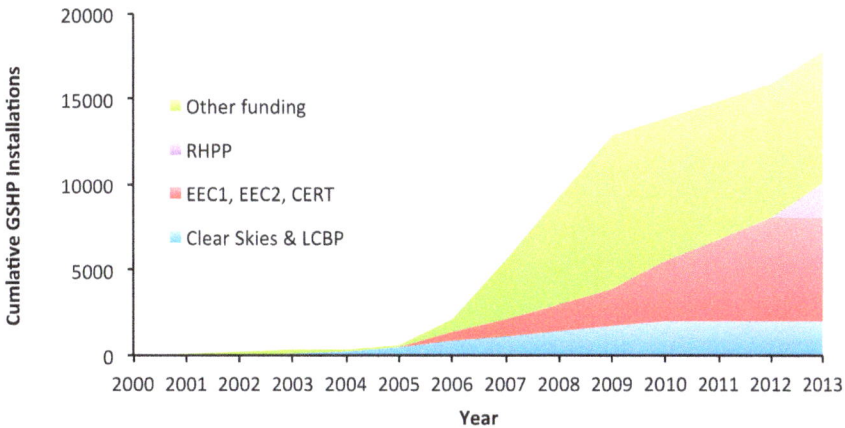

1.4. Other Supporting Measures

The importance of accessible consumer information, development of standards and skills to the acceptability and uptake of domestic renewable energy technologies and development of installation industries has been widely acknowledged [44,45]. National support programmes such as Clear Skies and the LCBP have played an important role in the dissemination of information (e.g., case studies [49]) together with related quasi-governmental organizations (e.g., the Energy Saving Trust, EST) in the provision of consumer advice. Although the UK GSHP installation industry broadly acknowledged the need for coordination with regard to promotion, research, publication, training and standards there was no obvious professional or trade body in existence with which it could align (the Heat Pump Association and the British Drilling Association probably are the most closely related).

Progress towards an industry body was made by the formation of the "Ground Source Heat Pump Club" in 2004. This organization was hosted by the National Energy Foundation (NEF)—a non-profit entity that had a record of promotion and support for energy efficiency and renewable technology industries. This club later became a more formally organized trade body in the form of the Ground Source Heat Pump Association and has become independent of NEF [50]. The body has been able to make representation to government departments on behalf of the industry with some success [51] and has taken something of a lead in developing industry standards and training.

Where there is rapid development by entry of new installers into a market for a relatively novel form of renewable technology, there must be some risk of adverse consequences to the health of that market if system design, products and installations fall short of good practice. Development of the Clear Skies programme (2002) was accompanied by governmental recognition of the value of both product labeling and installer quality assurance measures as part of the development of a national microgeneration support programme (product labeling for boilers and certification of installers was already well established). To this end, the programme initiated a requirement for grants only to be made for installations completed by registered installers and for registered products. This aspect of the programme later became the Microgeneration Certification Scheme (MCS) [52] which formalized the requirements for installer registration and started to introduce new installation standards for each of the technologies supported. An installer standard for domestic GSHP systems was first introduced in 2008 and has been developed significantly in light of the monitoring and performance evaluation programmes discussed below. There is evidence that development of the standard for heat pumps has had an important role to play in seeing that the lessons learned from the field trials are translated into practice—as we discuss later.

The technical complexities of GSHP systems are reflected in the blend of skills and knowledge required for implementation and consequently reflected in the training needs of the installation industry. At the professional level, some geological, geotechnical and HVAC engineering competence is required. At technician level, skills that are normally divided between drilling, heating systems and refrigeration competences need to be brought together. The dominance of fossil fueled hydronic heating in the UK has meant that heat pumps have never featured in heating technician training and only appeared in refrigeration and air conditioning training programmes. The need for appropriate training programmes that give some assurance of quality implementation and, at the same time, ensure the supply of suitably trained personnel is not a constraint on growth was recognized in the articles of the EU RES directive. The directive required all member states to have training programmes for renewable technologies like GSHP in place by 2012. Progress towards this end has been slow in most Member States.

Some effort has been made in the EU to develop coordinated heat pump technician training in the form of the EU-HPCERT programme [53]. The GEOTRAINET project has sought to develop training materials and programme frameworks for drilling experts and building design professionals. These programmes are yet to be implemented in a nationally coordinated manner in any EU Member States. In the UK progress is being made on the part of the drilling industry [54] and the GSHPA is leading efforts to engage the qualification accrediting and awarding bodies that have oversight of technician training and qualifications [55]. Although many colleges have taken initiatives to include renewable heating technologies in technician training there is still not an obvious place for this to fit with recognized qualifications. The primary mechanism for assuring personnel are competent remains the MCS scheme. This does not require particular levels of recognized qualification and allows more than one route to registration. Assurance mostly comes from a need for contractors to be audited [56]. Training in the requirements of the MCS installation standards is currently provided by third-party training bodies and heat pump manufacturers. The need for a nationally recognized system of training and qualifications that is integrated with existing accredited courses remains.

2. System Performance Evaluation

2.1. Performance Metrics and Benchmarks

Whether one wants to compare theoretical and actual thermodynamic performance, energy efficiency, running costs or carbon emissions, some form of metric that is a ratio of useful heat output to electrical energy input is required. The thermodynamic metric Coefficient of Performance (COP) is useful to define rates of heat transfer in relation to power at a particular steady-state operating condition. This is useful in the context of product specification (catalogue data), labeling and standards (e.g., EN 14511-2 [57]) to enable heat pump devices to be compared at the design stage and with reference to the theoretical Carnot efficiency. However, if one is concerned with realistic operating conditions, then an integrated measure (*i.e.*, based on energy rather than power ratios) that recognizes the heat pump device as part of a larger heat delivery system is required. Such metrics are commonly termed Seasonal Performance Factor (SPF). Their definition is highly significant if one wants to address the question of what forms of heating system should be supported as part of a national carbon emission reduction strategy, expected running costs and what constitutes the renewable energy output of heat pump systems.

Precise definition of SPF that is applicable to a range of system configurations is not easy to establish. The EU standard EN15316-4-2 [58] defines SPF as *"the ratio of the total annual energy delivered to the distribution subsystem for space heating and/or domestic hot water to the total annual input of driving energy ... plus the total annual input of auxiliary energy."* Gleeson and Lowe, in an analysis of available data and metrics [59], identified thirteen different variations in the definition of SPF in several heat pump trials so that it appears the definition in EN15316-4-2 is not sufficiently precise. The complications arise as different numbers of circulation pumps, buffer/storage tanks, and supplementary electric resistance heater can be found in practical heat pump systems—many of which provide both space heating and domestic hot water. This complicates comparisons between different field trials and between different heating technologies significantly. It furthermore complicates what are the most appropriate monitoring arrangements and what can be deduced from the results [60].

This issue has received attention during the period we have reviewed (e.g., [61]) such that there has been some movement towards a consensus following the SEPEMO-Build project which sought to derive well defined forms of SPF metric that could be practically measured and form a useful basis of performance comparison [62]. This methodology defines four metrics—SPF_{H1}, SPF_{H2}, SPF_{H3} and SPF_{H4}—that have an increasing number of electrical energy inputs included respectively. SPF_{H1} includes only the heat pump compressor and auxiliary energy (controls *etc.*) and will have the highest value of these metrics (closest to the steady-state COP). SPF_{H2} also includes the ground loop circulation pump electrical energy. SPF_{H3} includes any electrical energy associated with a boost resistance heater in the heat pump package. SPF_{H4} further includes the electrical energy associated with the heating or hot water distribution system. The EU has accepted these definitions for defining the renewable energy delivered by heat pump systems [63] and are discussed in detail in UK field trial documentation [64]. The system boundaries are indicated in Figure 4.

Figure 4. System boundaries and their relation to Seasonal Performance Factor (SPF) metrics 1–4. The system efficiency recorded in the first phase of the UK national field trials (SEFF) can be considered SPF_{H5} and includes the output of the hot water system.

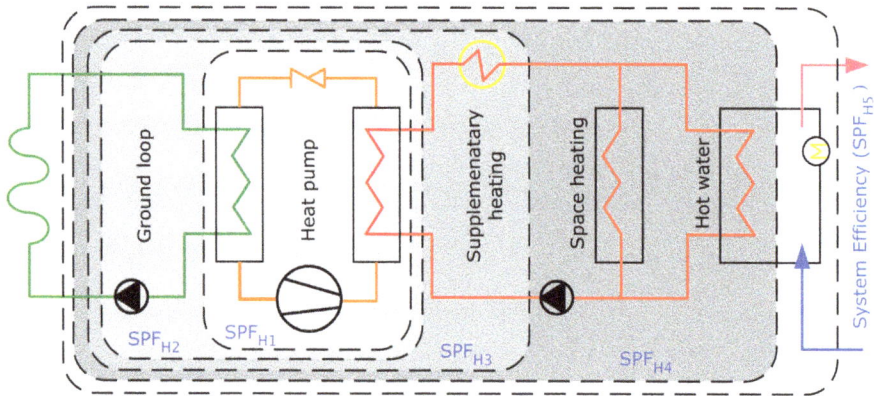

If the aim is to compare the seasonal performance of heat pump systems with other forms of heating and hot water production (gas fired boilers being the most relevant alternative in the UK) then either SPF_{H2} or SPF_{H4} could be appropriate metrics. SPF_{H2} is often easier to measure as it only includes the main equipment in the heat pump cabinet (excluding any built-in boost electric element) and the ground-loop circulating pump. The equipment included (and monitoring points required) to measure SPF_{H4} varies considerably according to the configuration of equipment outside the heat pump cabinet—principally whether buffer and/or hot water tanks are separate components or are not required and the number of circulation pumps. The metric SPF_{H4} is arguably the most appropriate if one is to consider likely energy costs. In the recent UK evaluations of heat pump performance [64,65] SPF_{H4} has been adopted for most comparisons. In evaluating both the minimum performance standard and the renewable energy contribution made by heat pumps SPF_{H2} has been adopted within the EU [63].

What, then, might be regarded as a minimum level of performance that should be expected? The question has been asked for a number of reasons—both with a view to householder expectations and broader performance of the housing stock with GSHP—and might be answered from an energy, carbon emissions or cost perspective. Broader assessments such as Life Cycle Analysis [66] and Carbon Footprinting [67] have been put forward in academic studies. In both these broader forms of assessment operating efficiency (*i.e.*, SPF) is also the most significant parameter [67].

The approach taken in the EU (with respect to the Renewable Energy Sources (RES) directive [68]) is based on consideration of primary energy *i.e.*, input at the power station. This is reasonable if one seeks to make a simplified comparison with systems that rely on local combustion of fuel such as gas, oil or biomass boilers in terms of overall energy efficiency. The minimum performance in order for a system to be counted as a renewable energy source is defined by an SPF_{H2} value greater than $1.15/\eta$, where η is the ratio of (at national grid level) electrical energy delivery to primary energy input and 15% distribution losses are assumed. The value of η varies from country to country and over time (gradually increasing) but for the sake of uniformity and application over the 2010–2020 timeframe, a conservative value of 0.455 has been agreed [63]. The means, after rounding, a minimum SPF_{H2} value of 2.5 is required.

With respect to seasonal efficiency levels in the UK, and comparison with other forms of heating in the context of a national carbon emissions reduction strategy, the metric SPF_{H4} can be used to define benchmark efficiencies along with fuel carbon factors and the efficiency of the alternative system. This form of analysis helps answer the question as to what level of performance is required for a GSHP to have an advantage over other forms of heating in terms of carbon emissions rates. This is a simple calculation and such results have recently been reported [47]. If the comparison is made between a natural gas fired systems (carbon factor 0.185 $kgCO_2/kWh$) and a heat pump using grid electricity (carbon factor 0.480 $kgCO_2/kWh$), the minimum SPF_{H4} in order to show lower emissions is 2.21. This assumes the seasonal efficiency of modern gas boilers is 85%—a value reported in recent boiler national trials using similar methodology to the heat pump field trials [69]—and that room conditions are comparable. When compared to other fuels the minimum value is lower: compared with oil the minimum would be 1.65; compared to Liquified Petroleum Gas (LPG) the value would be 1.9.

Similar calculations can be made to find the minimum SPF required for the GSHP system to have lower running costs than other forms of heating. Using nationally reported unit fuel costs (2013 values) the minimum SPF_{H4} required when compared with the annual energy cost of a natural gas fueled system would be 2.49 according to the Department of Energy and Climate Change (DECC) [47]. Compared to an oil fired system this value would be 1.82 and compared to a low-tariff electric night-storage resistance heating system, the value would be 1.5 [47].

Given the benchmark minimum SPF_{H4} values noted above, and that the differential between SPF_{H4} and SPF_{H2} is fractional [47] (in the approximate range 0.1–0.3), the EU benchmark of 2.5 for SPF_{H2} seems quite appropriate for the UK at present.

2.2. National Field Trials and Monitoring Programmes

The Energy Saving Trust (EST) has been responsible for coordinating national-scale field trials of a number of microgeneration technologies during the Clear Skies and Low Carbon Buildings Programmes. In 2008 a new trial was organized to evaluate the "real world" performance of domestic heat pump systems—both air and ground source [70]. This field trail was funded by a number of energy suppliers (who had already installed a significant number of systems by 2008) and several heat pump manufacturers and installers. The broad aim was to evaluate the differences between stated (lab-based) performance and the seasonal performance when installed and operated in typical households—in a similar manner to earlier national trials of solar thermal and photovoltaic (PV) domestic systems. The methodology included technical monitoring but also analysis of user surveys. This was the first large-scale trial of domestic GSHP systems in the UK. The results of this trial and the follow-on monitoring programmes, have received a good deal of scrutiny and proved a useful source of evidence to those interested in grant support policy, national carbon reduction strategies, product development, installation standards and the broader debate as to the future role of GSHPs in the UK.

The trial sought to collect evidence from a broad sample of systems that had already been installed. The systems included in the trial were accordingly widely distributed around the UK. The population of 83 sites (54 GSHP, 29 ASHP) included a range of: building forms, type of heat emitter, new and retrofit installations, tenant and owner-occupied properties and hot water provision. A few systems were integrated with solar thermal systems. In addition to the system electrical demands and heat outputs, other data collected included ground loop temperatures, internal and external air temperatures, energy costs, installation configuration and sizing data. A total of 14 different heat pump manufacturers' equipment were included in the sample. This 'phase 1' trial data was collected between Spring 2009 to Spring 2010.

2.2.1. National Field Trial Results (Phase 1)

The practicalities of instrumenting a wide variety of different heat pump products and system configurations seem to have proved more complex than expected so that energy flows could not be resolved to the system boundaries defined as SPF_{H2} or SPF_{H4} for the first phase data. The efficiency metric reported was denoted "system efficiency" and is essentially expanded from the SPF_{H4} boundary so that domestic hot water energy was measured at the outlet of the storage cylinder rather than the inlet (see Figure 4). Values can, accordingly, be expected to be lower than SPF_{H4} values. The difficulty with this definition is that, particularly in systems with separate storage cylinders, the measured hot water energy is sensitive to tank losses; cold water feed temperatures and usage pattern [60]. (It should also be said that at the time of the first phase of the field trials there was no consensus as to monitoring standards and the work emerging from the SEPEMO project was yet to be published). In DECC's first RHPP data analysis [47] the system efficiency data from the first phase EST trial was redefined as SPF_{H5} (see Figure 4).

The system efficiency data derived from the phase 1 measurements from GSHP installations is summarized in Figure 5 in the form of a histogram. The sample size finally reported was reduced to 49 GSHP and 22 ASHP after some data was rejected for quality control reasons. The mean GSHP system efficiency was reported as 2.39 and the range as 1.55–3.47. The corresponding results for ASHP were a mean of 1.83 and range of 1.2–2.2.

Of the ten key findings stated in the final report of this first UK national field trial [70] those that received the most attention were *"The system efficiency figures for the sample of ground source heat pumps were lower than those monitored in similar European field trials"* and *"Heat pump performance is sensitive to installation and commissioning practices."* [70]. The first of these findings is evident from comparisons with trials in Germany, Switzerland and Denmark [59]. The second finding is also clear if one considers the differences between the equipment rated COP values stated by manufacturers and the measured efficiencies. Installed seasonal efficiencies were expected to be lower but not by so much or with such a large range. Although the performance of some systems was good, and comparable with other European reports, a mean value of 2.31 is low in relation to the EU benchmark for renewable systems of 2.5. An independent study [60] that collected data in the following year reported similarly disappointing system efficiencies (along with slightly higher SPF_{H4} values) from a sample of 10 small RSL properties in Harrogate that used the same heat pump: mean 2.21; range 2.12–2.33.

Figure 5. GSHP system efficiency (SPF$_{H5}$) data reported from Phase 1 of the national heat pump field trials. Data from [70].

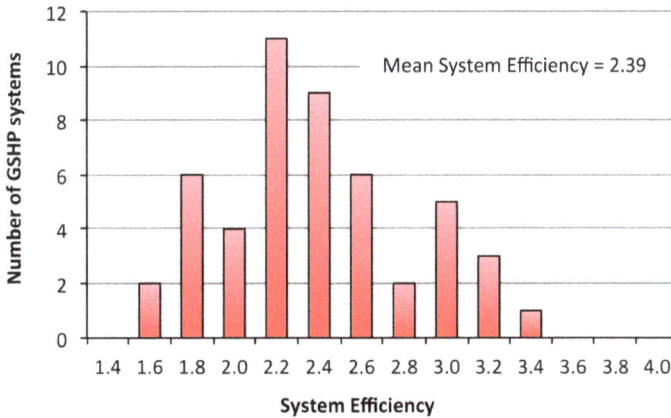

2.2.2. National Field Trial Results (Phase 2)

Some of the systems in the phase one trial could arguably be regarded as faulty or even failures and unduly skewing the results and so should have been excluded from the final results presented. The counter-argument is that the sample represented industry practice and what users were experiencing. In any case, the large range of system efficiencies indicated poor design and installation practice in all but the upper quartile of installations. The Department of Energy and Climate Change (DECC) initiated a programme of detailed data analysis in cooperation with the EST and heat pump installers in 2010 to identify likely technical factors in the poor performing systems [71]. The main consequences of this were further development of the Microgeneration Installation Standards [72] that we discuss later in this paper and, concurrently, initiation of a second national field trial.

The second phase of the national field trials sought to take advantage of what had been learnt from the detailed technical analysis [71] and saw implementation of a set of intervention measures at 32 of the original sites identified as poorly performing. Some well-performing sites from phase 1 were also included in the sample. The sample comprised 21 GSHP installations and 15 ASHP. The level of instrumentation was increased so that it was possible to separate out SPF$_{H2}$, SPF$_{H4}$ and system efficiency (SPF$_{H5}$) values for nearly all the systems. The intervention measures were classified as major (12 sites), medium (9 sites) or minor (11 sites). Major interventions included replacement of a heat pump and repairs to a ground loop. Medium interventions included installation of a new hot water tank, new radiators and circulating pumps. Minor interventions included additional insulation and modified control settings [65].

Results from a further year of monitoring were presented in the form of SPF$_{H2}$ and SPF$_{H4}$ (DECC having adopted the latter metric for comparisons with other heating systems) and these are summarized for the GSHP systems in Figure 6 [64]. The mean SPF$_{H2}$ value for the GSHP sites was 3.1 and the mean SPF$_{H4}$ value (20 of 21 sites) was 2.82. When monitored SPF$_{H2}$ values were compared with the EU benchmark (2.5) 20 of the 21 GSHP sites was found to exceed this. Of the sites that had either major or medium interventions made, 17 of the 20 showed noticeable improvements in system efficiency. Three showed small deterioration in system efficiency (the results are not broken down into GSHP or ASHP and so it is difficult to comment further). Of the sites where minor or no interventions were made there were small changes in system efficiency but a similar number showed improvement as showed deterioration [65]. The range of results remained significant and suggests that installation and design practice remained variable. User behavior was found to be significant in some cases (related to hot water usage). Although it was only possible to upgrade a relatively small number of the monitored systems to the new requirements, the measures addressed in the revised installer standards were judged to be validated by the improved results [64].

Figure 6. Seasonal Performance Factor data (SPF$_{H2}$ and SPF$_{H4}$) reported from Phase 2 of the national field trial (after implementation of improvement measures). Data from [64].

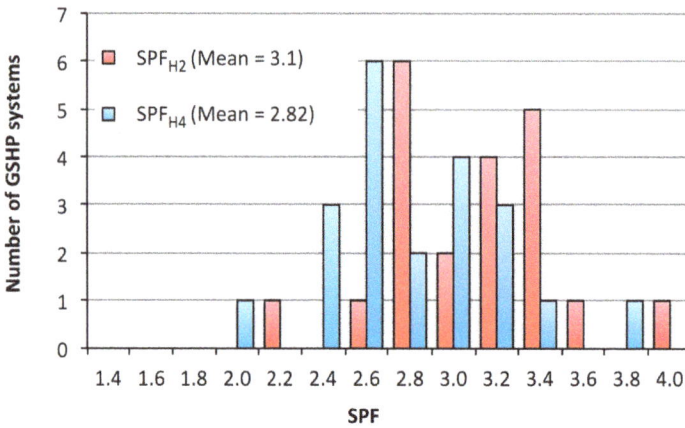

2.2.3. Initial RHPP Performance Data

The RHPP programme was a means of providing capital grants to householders and RSLs for renewable heating equipment in anticipation of the RHI tariff scheme. Monitoring of heat pump installations was incorporated into the programme and built on the methodology developed in the earlier national field trials. Householders were incentivized to participate in the monitoring exercise by modest additional grant payments. The result has been that a significantly larger sample has been included in the monitoring exercise (124 GSHP in the sample after data quality control). Performance data from December (Testing showed December was reasonably representative of heating season behavior without temperature correction [47]) of the first phase of the programme (August 2011–March 2012) was presented later in 2013 [47]. System efficiencies (*i.e.*, SPF$_{H5}$, calculated with some assumptions about hot water losses) from this report are shown along with the EST trial phase 1 data in Figure 7. The SPF$_{H4}$ data is shown in histogram form in Figure 8 along with data from the national field trial phase 2.

There are some differences in the approach taken to monitoring and data analysis in the RHPP data compared to the earlier monitoring programmes besides the fact that the sample size is larger. Detailed diagnoses of poorly performing systems were not attempted in the same way as Phase 2 of the field trials. Hence, although installers were active in correcting faults, these were not prompted by interventions recommended following analysis of monitoring data. The seasonal performance data such as that presented in Figures 7 and 8 is consequently representative of industry design and installation practice in 2012–2013.

Figure 7. Seasonal Performance Factor data (SPF$_{H5}$) reported from December 2013 RHPP programme data [47] and compared with the Energy Saving Trust (EST) national field trial Phase 1 data [71]. Data from [47,71].

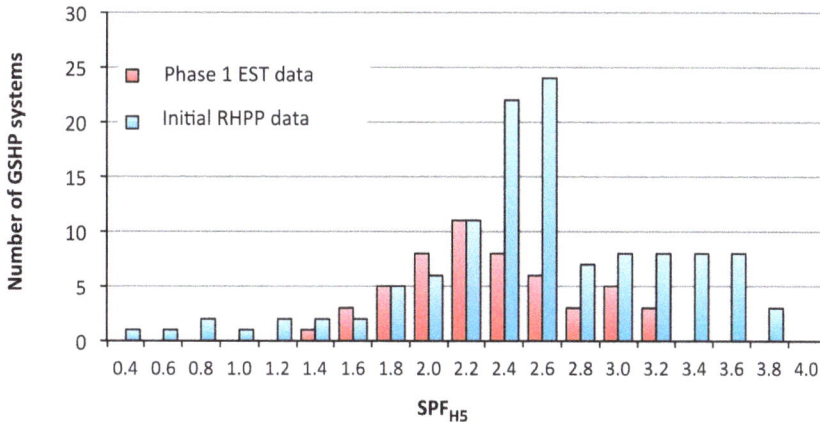

Figure 8. Seasonal Performance Factor data (SPF$_{H4}$) reported from December 2013 RHPP programme data [47] and compared with the EST national field trial Phase 2 data [64]. Data from [47,64].

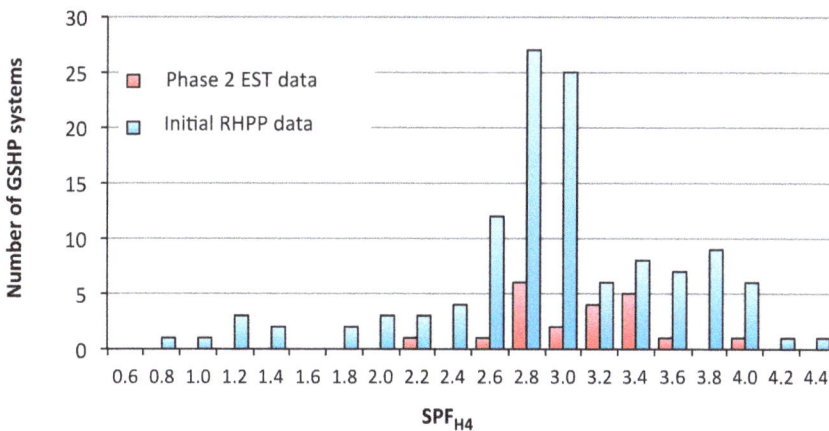

The December 2013 RHPP data [47] is indicative of some improvement in seasonal performance in that both the mean SPF$_{H4}$ and SPF$_{H5}$ efficiencies are increased relative to those found in both phases of the earlier EST national field trials. The mean SPF$_{H4}$ was found to be 2.92 (2.82 in phase 2) and the mean SPF$_{H5}$ to be 2.74 (2.39 in phase 1). However, the ranges of the efficiencies continue to be significant—a long tail with relatively few values being evident at the lower end of the range—although this may reduce when a whole season of data is included in the analysis and further quality checks are made. Comparisons with the EU renewable energy benchmark ($SPF_{H2} > 2.5$) and estimates of carbon emissions and energy cost reduction relative to other heating systems were also presented. Of the GSHP installations, 84% were shown to be above the EU threshold. Using analysis similar to that noted in Section 2.1, 64% of the GSHP systems would have shown reduced fuel costs relative to natural gas heating systems. The percentage of systems resulting in cost reductions was increased for other fossil fuels. The percentage of systems showing reduced carbon emissions relative to natural gas heating was 88%—and higher when compared to other fossil fuels [47].

2.3. Grant Recipient Characteristics, Experiences and Behavior

All the support programmes providing grants to individual householders (Clear Skies, LCBP and RHPP) have collected user data from grant recipients. These data have included feedback about the operation of the programme and satisfaction with the system along with information that characterizes the property, type of system and location [26]. In the first national field trials the EST and Open University researchers (Caird *et al.* [73,74]) sought to investigate the relationships between operating efficiency and the behavior and characteristics of the users by employing questionnaires and carrying out in-depth interviews with householders. The population consisted of 78 users of which 48 were private households and 30 were social housing tenants. Fifty of the properties had GSHP systems and the remainder used ASHPs. The study of ten social housing installations in the later Harrogate study [75] included user surveys using a similar questionnaire and interview approach.

Overall levels of satisfaction with the heat pump systems in the EST field trial (phase 1) were good: 83% of users agreed or strongly agreed that the system made their home warm and comfortable; 86% of users said the system met their domestic hot water requirements [73]. Although these levels of overall satisfaction were high, responses to detailed questions showed only 63% of users were satisfied with the level of support they received from the suppliers of the system and only 62% were satisfied with the running cost savings. This dissatisfaction is to some degree a reflection of the technical problems and poor efficiencies measured during the first phase of the trial [70].

There were several questions in the survey used by Caird *et al.* [cite references] where responses noticeably differed according to whether the users were social housing tenants or private householders. Ownership status, in itself, was not necessarily an indicator of cause as ownership also correlated with property and system characteristics—Social housing properties were smaller, had a higher proportion of radiators and lower building fabric standards. A greater number of social housing properties were retrofit with heat pumps rather than being new properties. Private houses tended to be newer, better insulated and a higher proportion had under-floor heating. Private householders also interacted with the purchasing, installation and commissioning processes in a different manner to social housing tenants. Private householders tended to be involved with information collection and decision-making earlier in the process whereas, in the case of social housing installations, the landlord procured the system.

One aim of Caird *et al.* [73] was to try and establish any correlations between user characteristics, behavior and system performance. To this end, they categorized installations according to whether their system efficiency fell below 2, in the interval 2–2.5, or above 2.5 and examined the correlation with type of heat pump and ownership status. Their most significant finding was that 95% of the best performing systems were in private households. This was partly a reflection the characteristics of the systems in that a higher proportion of private households had GSHP and under-floor heating. However, Caird *et al.* [73] also point out that there was a higher level of knowledge and understanding of the systems among private households: 82% of users of the higher performing systems stated they had either "a fair amount" or "a lot" of knowledge and understanding of the heat pump system and only 4% of social housing tenants stated they had a "lot" of knowledge [73].

A significant cause of dissatisfaction amongst users in the EST field trial was the difficulties in understanding the system operating instructions and uncertainty in how best to operate the controls [73]. Forty four percent of all heat pump users said they were uncertain how best to operate the controls and this was expressed almost as equally by private householders (17 of 32) as social housing tenants (15 of 32). Responses to these questions where one element of the survey that differentiated GSHP negatively was in that 22 of those dissatisfied were users of GSHP systems and 10 were users of ASHP.

Other behavioral factors that were investigated included user choices about operating temperatures, window opening and system time control. Users were typically advised to leave the system on (enabled) at all times. This is firstly reflected in the fact that 76% of all users left the system on all day and night. This proportion was higher among private householders (85%) and GSHP users (85%). This practice is reflected in the operation of the systems with efficiency greater than 2.5 in that all of these systems were operated (enabled) continuously. Practice was more varied amongst social housing tenants and ASHP users. Although 71% of social housing tenants said they left the heating on for long periods of the day only 55% left it on all night and 59% when out of the house compared to 82% and 87% of private householders respectively. Open user responses noted by Caird *et al.* [73] suggest housing managers sought to advise tenants to leave the system on but this operating pattern was not universally accepted—most likely as it was contrary to experience with previous heating systems. This issue seems conflated with the reportedly poor written instructions that users received. As a consequence of the survey responses the EST field trial report concluded that user behavior did have an impact of performance levels and that there was a need for clearer and simpler user advice [70]. Similar dissatisfaction with operating instructions and variability in operating behavior and window operation were observed in the Harrogate study [76].

The RHPP programme collected feedback from participants through online questionnaires [77]. The body of user data was substantial as a consequence of the fact that all recipients had to complete the questionnaire online in order to claim the grant payment. This resulted in 804 sets of GSHP user data collected soon after installation of the system (phase 1 of the RHPP program) and placed some emphasis on the motivation for the purchase, experience of installation, technical support and operating instructions. A follow-up online questionnaire was completed by 544 GSHP users after the completion of the first heating season with an alternative set of questions more concerned with operating experience and behavior [77].

The levels of overall satisfaction with renewable heating systems in the programme (GSHP, ASHP and biomass boilers) were high. User satisfaction with GSHP systems was 90% in the follow-up responses. This indicates some improvement over the user experiences in the first EST field trial. However, the responses to questions about initial faults and failures and difficulties in understanding how to get the best from the systems indicated there was further room for improvement on the part of installers and manufacturers. Of the grant recipients with GSHP systems, 61% reported that they needed additional advice after installation. Twenty-two percent of GSHP system users reported manufacturing or installation faults [77] (levels very similar to those with ASHP or biomass boiler installations). Although user satisfaction overall was high and no similar data for more conventional heating systems is available, these levels of difficulty of operation and initial faults seem high.

Although overall satisfaction with the temperatures achieved was high (95% with GSHP systems) there are some interesting trends in reported uncomfortably cold or hot periods. Air-source heat pump users reported slightly more hours being too cold at night: 13% compared to 9% for GSHP users. This may reflect some drop-off in ASHP heating capacity in the particularly cold 2013 winter [77]. Only 1% of ASHP or biomass boiler users reported being hot. Of GSHP users, 3% reported being too hot on the coldest nights. Some responses indicated GSHP users behaved differently to users of other systems. Users were asked to indicate up to three types of action they took when they felt too hot. These responses are shown in Figure 9. When comparing behavior between users of different heating system types, relatively few GSHP users changed the timing of system operation or turned the system off (15% and 9% respectively). Users of biomass boilers changed timing and system-off periods more noticeably (44% and 40% respectively). More GSHP users respond by opening windows and doors during periods when they are too hot (30% and 15%) than users of biomass boilers (11% and 7% respectively). Although the number of users in the sample affected is small, overheating suggests energy demand could be further reduced by better space temperature control.

Figure 9. User responses to overheating reported from online follow-up questionnaires in the first phase of the RHPP programme [77]. Adapted from [77].

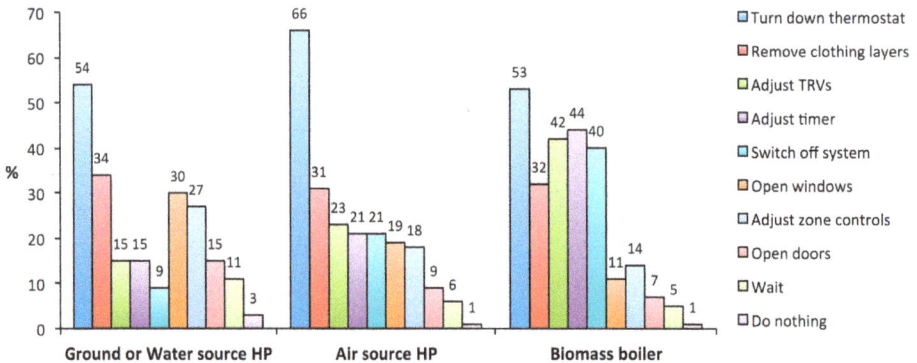

Other user data collected in the first phase of the RHPP programme [77] included information about the type of property, user income level and regional location. These data give some indication of the nature of the GSHP market at that time. The RHPP programme was conceived as a precursor to the RHI tariff programme and so RHPP user data is probably representative of initial RHI programme participants. Although some of the programme funds were ring-fenced for RSLs (who showed some preference for ASHP rather than GSHP systems [78]) the number of systems in private ownership was very much higher in the RHPP programme than previous programmes such as LCBP [26]. This is reflected in the property size and income level data: both being higher than reported in the LCBP. Regression analysis showed that both property size and income levels were furthermore higher amongst GSHP users than either ASHP or biomass users [77]. The average number of rooms in houses with GSHP systems was 4.1 and the average household income was £61,500 (UK median number of bedrooms was 3 [79] and average earnings in this period were £27,000 [80]). The average household income of ASHP users in the programme was £10,000 lower. Compared to the other renewable heating technologies it was found significant that more GSHP installations were in new properties built for older householders planning to stay in the property for many years [77]. The characteristics of such users and their properties is in contrast to those of the many social housing properties and users that received GSHPs in larger numbers in earlier programmes and in the streams of funding within the RHPP available to RSLs. This suggests something of a bi-polar nature to the market at present with relatively few systems being taken up by middle-income homeowners in average size properties.

The RHPP user data shows some regional variation in the uptake of the renewable heating technologies supported by the programme. To be eligible to receive a grant, users had to be off the gas grid. This firstly means that most installations are further from the central axis of England, further from the central urban belt in Scotland and more frequently occur in the southwest and eastern England, Wales and the border and highland regions of Scotland. Air-source heat pumps installations have been geographically widely dispersed. However, the mapping data [77] shows something of an inverse relationship between biomass boiler and GSHP adoption. For example, in parts of Scotland there is good availability of biomass fuel and GSHP installations are much less common than biomass boilers. Conversely, in the east of England there is very little forestry and so a higher density of GSHP installations and many fewer biomass boiler installations were reported.

3. Technology Adaptations and Development

One of the key findings of the first phase EST national field trial was that "the system efficiency figures of the sample of ground source heat pumps were lower than those monitored in similar European field trials" [70]. This finding led to questions as to whether this was related to poor design and installation practice, or whether there were peculiarities of the UK situation—such as geological conditions, housing and heating system design or climate—that meant the technology would not perform as well as reported elsewhere [81]. Although behavioral factors had some impact this was not sufficient to explain the wide range in performance. A further detailed technical study was carried out by DECC and the monitoring contractors in the following year with the aim of identifying faults and sensitivity to variations in design and installation practice [70]. Technical issues were identified with particular sites that were the most likely causes of poor performance. Some of these issues were factors in multiple sites. The technical issues identified can be grouped—in roughly descending order of likely impact on efficiency—as follows [71]:

(1)   Under-sizing of the heat pump;
(2)   Under-sizing of the ground heat exchanger;
(3)   Poor insulation standards (pipes and tanks);
(4)   Flow temperature unnecessarily high;
(5)   Excessive pump usage (time control or number of pumps);
(6)   Poor control.

The most noticeable and recurring consequence of under-sizing of either the heat pump in relation to space heating demand or under-sizing the ground loop was excessive use (without the user being aware) of supplementary electric resistance heating—either built into the heat pump equipment or a separate tank immersion heater. The similar but smaller-scale study in Harrogate highlighted poor control as the most likely technical cause of poor performance [76].

Some of the problems enumerated above had been largely avoided in the first projects in small social housing projects with the "heatplant" package discussed earlier [37]. In these systems, the heat pump capacity was well matched to the design heating loads of the houses in question. This meant developing a new heat pump with a capacity that was noticeably lower (3.5 kW and later 5 kW) than equipment available from other parts of Europe at the time. The ground loop in these installations was sized rather conservatively. Excessive use of electric heating was avoided as no "cassette" electric heat was built into the heat pump or included elsewhere in the heating circuit. The heat pump had been designed to operate in dual mode and deliver hot water at 65 °C so that resistance heating of the hot water tank was not usually necessary. Although these systems used radiators, retrofit of insulation in many cases meant that they could be generously sized. Controls were very similar to the two-channel devices familiar in UK gas fired boiler systems and, by relying on a room sensor, achieved closed-loop control and so some minimization of flow temperatures according to variations in both climate and user behavior.

3.1. Technical Performance

Gleeson and Lowe [59] attempted to systematically compare the first phase EST field trial data with efficiencies reported in other European countries but faced some difficulty due to the differing system boundaries adopted in different trials. (The UK data was inevitably lower due to the large system boundary than adopted elsewhere). One set of Swedish data [82] from systems retrofit in older homes reported SPF_{H4} values in a similar range to those in the UK trials (a mean of 2.6 and range 2.4–2.9) but most trials, even after allowances for differences in system boundaries were considered, indicated noticeably better efficiencies were being achieved elsewhere in western Europe. One comparable Danish field trial of 138 GSHP in a mix of new and retrofit installations [83], reported a mean SPF_{H4} value of 3.03 being achieved. A 2010 Swedish trial [84] reported results in terms of SPF_{H3} with a mean value of 3.26 and range of 2.6–3.6. Higher values have been reported for a survey of retrofit installations in 36 German homes [85]: a mean SPF_{H3} of 3.88 and range 3.1–5.1. The highest SPF_{H4} values were reported in a trial of installations in 56 new German homes [86] where a mean of 3.75 was found. The low performance reported in the first phase EST trials led to

Gleeson and Lowe [59] suggesting the outlook for carbon reduction from heat pump installation in the UK was uncertain.

Taken together, the European field trials considered by Gleeson and Lowe [59] showed that features found to broadly correlate with good performance were: (i) low temperature heat emitters; (ii) capacity chosen to minimize resistance heating, and; (iii) a higher proportion of space heating demand compared to hot water demand. This has been largely confirmed in the detailed analysis of the UK field trial data [71] that led to new rules being established for system capacity and heat emitter selection [72] (see Section 3.2). Improvements due to these revised standards should be reflected (not all systems were implemented after the change in standards) in the later field trial and RHPP data and some progress towards European performance levels is evident. For example, the mean SPF_{H4} value of approximately 3 shown in the RHPP data (see Figure 8) is similar to that in the Danish trials (3.03) [59,83]. Nevertheless, the higher seasonal performances reported in the German trials suggests there is room for further improvement in the UK context.

Some authors have suggested that "poor quality" installation and building construction (in relation to the best European standards) have been to blame [81]. It is clear that building fabric insulation levels and airtightness have a direct relationship to absolute capacity requirements. However, it is less clear how this affects the ratio of heat output to power input. Gleeson and Lowe examined this issue briefly and compared home heating demands in the different trials on a unit area basis. In the German trials of new housing installations heating demands varied between 85 kWh/m^2/year and 340 kWh/m^2/year [59]. This range overlaps with that typical in UK housing (mean of 90 kWh/m^2/year in 2004). This is a reflection of the fact that, although UK insulation standards are lower, the climate is also milder than either Scandinavia or Germany. Consequently, it is hard to see "poorer insulation" as a significant issue in itself.

3.2. Standards Development

One of the main outcomes of the first phase EST field trial [71] was the identification of a need for better heat pump installer training, installation standards and design guidance adapted for the UK context. DECC established a working committee, drawn from the heat pump industry, to address many of the issues surrounding design and installation of domestic heat pump systems. This led to revision of the MCS GSHP installation standard MICS 3005 [72] and production of supplementary standards and guidance on ground heat exchanger sizing [87], heat emitter selection [88] and hydraulic design [89]. The standards relating to heat pump sizing and heat emitter selection are common to both ASHPs and GSHPs. Development of the standards was supported by separate technical studies of the effects of thermostatic radiator valves [90], buffer tanks and cycling behavior [91]. The installation standards seek to address the problems of excessive use of electric backup heating, poor ground heat exchanger design, unnecessarily high temperatures and excessive pump energy use that were identified in the detailed analysis of the field trial data and enumerated above.

Revision of the MCS heat pump installation standard (MIS 3005 version 3.1) [72] introduced a requirement to match the heat pump capacity to 100% of the house design heat loss (for mono-energetic systems). This is a rather different approach to that taken in other parts of Europe where common practice is to size to allow the final 10%–20% of peak load to be met by electrical resistance heating. The latter practice can be more optimal in terms of capital cost and not very damaging to running costs if the number of hours in the year when peak loads are approached is small. The relationship between heat pump capacity and design load depends on how loads are calculated and the design temperatures chosen. In the revised UK standard 99th percentile coldest temperatures from typical year climate data are defined and so, arguably, there may be short periods where heat pump capacity is exceeded. The argument for taking this conservative approach is that it is simply defined and also robust given the uncertainties in heating load calculation and given that any resistance heating use is very detrimental to carbon efficiency because of the UK's relatively high electricity carbon factor.

Evidence that ground heat exchangers were too small also emerged from the national field trials. For some time there had been a concern that there may not have been any (domestic) sizing tools available to GSHP installers that had been specifically evolved for the UK. Whilst generic sizing software such as EED [92], GLHEPro [93] and GLD [94] had been available and quite capable of generating suitable design data given appropriate input of local ground properties, these were largely beyond the reach of many small domestic installers operating under the UK MCS scheme, whether in terms of cost, or complexity. Several of the European heat pump manufacturers were offering their own dedicated software, but it was not clear whether this had been checked or modified for UK conditions. The other European standard that was in existence is VDI-4640 [95], but there were also concerns as to whether this was applicable to all locations in the UK, or met with newly specified UK design requirements.

For ground heat exchanger sizing it was decided to develop a paper based (*i.e.*, non-software) methodology that would cover boreholes, horizontal EU style collectors and Slinkies, for sizes up to 45 kW [96]. The adoption of a non-software approach avoids the issue of trying to support different operating systems on different proprietary software, all at different version levels. The methodology was designed to cover all UK climate and geological conditions, ranging from the Scottish off-islands to as far south as the Isles of Scilly. The philosophy that has been adopted is that, provided an installer follows this sizing methodology, a conservative GSHP design should result. For installers that have better "knowledge" and/or access to other recognized tools, it is possible that more cost-effective designs can be developed. The tables were developed using GLHEPro [93] and EED [92] for boreholes, and CLGS [97] and GLD for horizontal and Slinky systems. The outcomes were cross-checked with VDI-4640 [95], and a few cases of heat pump manufacturers' software. In the final manifestation, three sets of tables are available (boreholes, horizontal pipe, Slinkies) for the range 1200–3600 of full load equivalent operating hours. Each graph in the standard provides ground extraction thermal capacity plotted against ground equilibrium temperature, and ground thermal conductivity. For the three different types of ground heat exchanger, the configuration is defined in some detail, e.g., borehole spacing, pipe diameters, trench spacing *etc*.

Excessive circulating pump energy demands were identified, in certain monitored systems, of significantly undermining energy performance [64,71]. This was a combination of excessive pump size, poor or non-existent hydraulic design of the ground loop array and also poor control of the running period. The problem of excessive pump power was partly addressed by the requirement introduced into the MCS installation standard [72] that pump power should not exceed 3% of the heat pump's thermal capacity. This was intended to encourage better hydraulic design and selection of efficient pumps. Design guidance on hydraulic performance was also incorporated into the standards by providing engineering data in graphical form for a range of pipe sizes and different anti-freeze fluids. A flow chart based approach to hydraulic design is provided, based around the pressure drop values given in the tables. By simple iteration it is possible to arrive at a ground loop layout that achieves turbulent flow in the active ground loop elements whilst minimizing circulating pump power, and also meeting the MCS target for the additional parasitic energy. The design methodology and pressure drop tables are also provided [89] as an addendum to the MCS Heat Installation Standard [72].

Pump energy demands are also a function of the length of time they are made to run. The RHPP data showed that faulty circulating pump operation (or poor control design) continued to be an issue. DECC engineers accordingly devoted some effort to automatically detecting pump operation and identifying problem systems by analyzing the streams of monitoring data. They took a similar approach to analyzing hot water production demands. Although the fault detection algorithms no doubt need further development, making intelligent use of performance and energy data could address some of the variations in performance levels that persist in the recent data (e.g., sites shown in the lower quartile of Figure 7). Carrying out fault detection and diagnosis by remote data collection is very challenging as installation of additional monitoring equipment has shown to be highly error prone (a high proportion of monitoring equipment in the RHPP was found not to comply with the specification and resulted in rejection of the data) but also requires handling and processing of large amounts of data [47]. We suggest that a useful approach would be for heat pump equipment to include factory-installed monitoring equipment and to have local data processing and fault diagnosis capabilities. Such approaches (along with automated commissioning) are taken in some non-domestic heating and cooling equipment [98] and have been shown to be applicable to domestic heat pumps [99] but with an emphasis on running faults rather than energy performance. This functionality may also enable better after-installation care on the part of installers/suppliers—something user surveys have highlighted as needing improvement—and may also improve user confidence.

3.3. System Dynamics and Control

One of the noticeable differences between the sample of GSHP in German new houses [86] noted earlier and discussed by Gleeson and Lowe [59], and those in the UK field trials are the total floor areas. The floor areas in the German sample ranged from 90 m^2 to 360 m^2 with a mean value of 189 m^2. This is noticeably larger than the 91 m^2 average of the UK properties. As the UK field trials included a number of relatively small social housing properties that had been retrofitted with GSHP, a significant number of properties had floor areas smaller than the national average. Small floor area has a number of implications in terms of system design and building thermal behavior. Small property size (either in terms of floor area or perimeter) tends to mean that domestic hot water demand is a higher proportion of the whole *i.e.*, the heat pump will operate for a large proportion of running hours with high delivery temperatures. It also means that heat emitters tend to be radiators as under-floor heating has a limited specific output and heat losses are more proportional to perimeter length rather than floor area. Having said this, it should be noted that there were some properties in the UK field trials that were small retrofit projects with radiators that performed relatively well.

A further point of difference between some of the UK properties in the heat pump trials and those in other parts of Europe was the relatively high thermal mass of the traditional wall construction. Boait *et al.* [76] pointed out, in their study of small social housing properties in Harrogate, that the masonry construction results in a noticeably higher thermal time constant and this has implications for the design of the heating system controls. This generally makes the control problem more challenging but also gives rise to energy savings opportunities where night set-back operation is introduced [76]. These energy saving opportunities are missed if users are told to operate the system continuously.

A further consequence of property size being relatively small is that casual heat gains associated with occupant activity (e.g., operating appliances) are more significant in relation to the system capacity. Similarly, the action of opening a door or window introduces instantaneous heat losses that are more significant than in a larger property—the most modern and efficient of which are likely to have mechanical ventilation and heat recovery and so lower and more constant ventilation heat loses. The control systems in the houses of the Harrogate study used an open-loop principle where heating output was adjusted in response to the return heating water temperature and did not have any form of room sensor (the particular Swedish heat pump manufacturer's normal recommendation). In situations where a casual heat gain causes local and short term overheating, where users have been instructed to leave the heat pump running continuously, and where the control system is not responsive, users must resort to opening windows to reduce the temperature but at the expense of unnecessary heat loss. This form of response to overheating seems to be reflected in the RHPP programme user behavior noted above in relation to Figure 8, in that opening doors and windows in response to overheating was much more common among users of GSHP systems than users of biomass boilers. It should be noted that not all the GSHP systems in the trials had the same form of open-loop control as in the Harrogate study. Some did incorporate room sensors and some of these are known by the authors to be among the systems that performed well, but, as control system type or strategy was not systematically reported or studied in the trials, it is difficult to draw firm conclusions. The issue of variation in technical performance of the control systems is also conflated with the dissatisfaction amongst a significant number of users with the operating instructions/training provided [73]. We consequently suggest that there is scope for further research into the optimal form of heat pump system control in the UK context—possibly relying on more intelligent algorithms but also offering interfaces that are more intuitive to users.

A further technical issue related to system control is that of dynamic cycling behavior of heat pumps and the application of buffer tanks [91]. In domestic heat pumps the predominant capacity control mechanism is simple cyclic switching of the compressor. When the heat pump is at rest, refrigerant pressures tend to equalize and the lubricating oil tends to settle. Consequently, in the dynamic start-up phases of operation the system operates inefficiently until the proper pressures are established and the compressor runs with less than ideal lubricating conditions. If cycling is excessive (in terms of cycle duration or a high number of cycles per hour) this results in a deterioration of SPF but also some concern as to reduced compressor life. One approach to maximizing cycle times is to include a buffer tank at the outlet of the heat pump from which the heating system draws warm fluid as demanded and to which the heat pump adds heat according to tank temperature. It is evident from the UK trials that industry practice as to the inclusion of buffer tanks varies considerably. The value of configuring the heating system in this way, and the relationship to TRV operation, was investigated following the first phase of the UK field trials in a combined modeling and experimental study [100]. This study showed that operation of TRVs in systems with small numbers of radiators could exacerbate frequent cycling in some situations and that a small buffer tank could be beneficial. Higher frequency cycling was shown to result in modest reductions in system efficiency. A further benefit of buffer tanks is that it allows more flexible design of the heating system control strategy in that the output of the heat pump is separated from the operating temperature and flow (e.g., circulating pump operation) of the heating distribution system. The use of buffer tanks is therefore an issue that bares further investigation in the context of optimal control.

An alternative approach to capacity control is to vary the speed of the compressor—sometimes referred to as inverter control. This approach is more expensive (in terms of the heat pump package) than simple cyclic controls but in other unitary refrigeration equipment has met with successful efficiency improvements. However, this technology has not penetrated the UK domestic GSHP heat pump to any noticeable extent, unlike the ASHP market where inverter models are more prevalent. Wider exploitation of inverter drive technology could also help in a UK context in that it should enable lower start-up currents [4] but may require further development to ensure presentation of better harmonic characteristics to the power grid [101].

4. Discussion

Policy measures supporting the uptake of renewable energy technologies usually take the form of packages that include a balance of capital grants, tariff mechanisms, quotas, public procurement and mandatory regulation [102]. A particular feature of the UK microgeneration support strategy is the various forms of supplier obligation programmes that have operated through energy supply companies. Although only a small portion of the programme funds have been directed to GSHP installations they have, nevertheless, represented a significant source of funding for the industry. These programmes have been particularly beneficial in funding larger-scale social housing projects. Registered Social Landlord clients could continue to be an important sector of the domestic GSHP market but, as current support programmes are very market oriented, the industry faces increasing competition from ASHPs in this sector.

The most recent form of supplier obligation scheme is the Energy Company Obligation (ECO). This scheme has not, to date, resulted in the same opportunities for GSHP installers or promoted renewable technologies to the same extent as the earlier CERT programme. The scheme, in view of the fact that the costs are visibly added to consumer energy bills, has been subject to political pressures and has recently been cut back by one third [103]. The scheme currently focuses on domestic insulation provision (with emphasis on social housing) and has introduced funding for district heating connections. Although this emphasis is arguably well founded in terms of marginal cost and marginal carbon reduction it can be seen, nevertheless, as a missed opportunity for support of geothermal heat pump uptake. Funding for larger-scale social housing projects that include GSHP may require RSLs to resort to more innovative funding mechanisms.

The current UK government has introduced a marked swing from historic energy efficiency policy towards a strongly market-based approach [18]. Since the end of the RHPP programme, capital grants have stopped and support will rely on a renewable heat tariff mechanism (the RHI). Capital support for GSHP and other renewable technologies will be available to householders through the UKs "Green Deal" programme which provides up-front capital loans rather than grants and these are repaid through the homeowners' energy bills [104]. In this scheme, the technical suitability of a particular energy efficiency or renewable technology measure is ranked according to the results of an individual property energy assessment and the costs are offset by renewable heat or feed-in tariff payments over the life of the system. No significant data is currently available to indicate the success or failure of this novel approach. This approach is clearly highly sensitive to individual attitudes, for example, attitudes to debt, likely energy costs and the individual's long-term property ownership plan. At the time of writing we must conclude that the level of support for GSHP installations that may come from these programmes is highly uncertain.

It is, perhaps, surprising that there is little evidence that the UK Building Regulations [105] that determine allowable levels of energy efficiency and carbon emission rates have not had any impact on the uptake of domestic GSHP technology—in spite of the higher standards that have been prompted by the EU Energy Performance of Buildings Directive (EPBD). Essentially domestic building designers are still able to show compliance by suitable choice of insulation and inclusion of a condensing gas boiler. There has been limited uptake of solar thermal technology in the new building stock provided by large-scale suppliers but no other renewable technologies. Small developments and individually commissioned properties seem to be the only exceptions and are relatively few in number. That new houses be "zero carbon" (zero carbon is defined with respect to installed heating, hot water and lighting demands in this context.) by 2016 has been a stated UK policy for some time. Two sets of standards have been developed by government in the interest of moving energy performance in this direction: the mandatory Building Regulations—which are to be revised again in 2016—and a set of voluntary standards known as the Code for Sustainable Homes (CSH) that are aspirational and go beyond the mandatory standards [106]. These latter standards could be adopted by individual contractors voluntarily but also specified as an additional planning requirement by local authorities—a power granted under the Planning and Energy Act 2008.

The higher-level CSH standards (code levels 5 and 6) require, in practice, that renewable technology be deployed in some form [107]. If the mandatory Building Regulations were to enforce the energy/carbon efficiency standards set out in the higher levels of the CSH when they are revised in 2016, this may force an increased rate of uptake of a range of renewable heat technologies including GSHP. However, recent UK government consultation (led by the Department of Communities and Local Government, DCLG rather than the DECC) that has been notionally motivated by a desire for simplification of regulation has changed expectations. It is consequently proposed to have a regulatory regime focused entirely on the building regulations and eliminate the CSH standards [108]. It is further proposed to repeal the provisions of the Planning and Energy Act 2008 so that local authorities are no longer able to insist on higher energy performance standards or levels of renewable energy deployment (a change of possibly greater significance in the non-domestic GSHP sector). Housing developers may furthermore be given the option to invest in off-site renewable energy sources rather than integrate renewable technologies into the house design to show compliance with the "zero carbon" requirement. At this point it is unclear what these proposals may amount to in house design trends but, if such proposals are adopted, it could be a missed opportunity to drive further uptake of renewable technologies.

Renewable energy policy in the UK has firstly been set out with a 2020 milestone in mind [28]. This has been mostly driven by the timeline set out in the EU RES directive that has prompted policies concerned specifically with renewable heating for the first time. Following the national binding targets for long term carbon emissions reduction introduced by the Climate Change Act 2008, policy and planning (for all sectors) has focused on 2030 and 2050 milestones. Policy advice and setting and monitoring of targets have become the responsibility of the independent Committee on Climate Change (CCC) who initially set out four carbon budgets for the years up to 2027 [109]. Recent carbon reduction and energy supply planning has also turned to 2030 and 2050 timeframes [110]. Much of this planning and the underlying modeling is based on the electricity grid being steadily decarbonized and provision of heat moving from fossil fuels to electrically driven heat pumps. The current CCC "medium abatement" scenario is for 6.8 million heat pumps to be in use by 2030 [109]. This is predicated on heat pumps being a competitive and consumer-appealing technology after 2030 so that the majority of new homes have heat pumps.

The optimism of this view of long-term heat pump deployment was reflected in the national renewable energy strategy published in 2009 but with respect to the 2020 milestone. These proposals formed the basis of the UK National Renewable Energy Action Plan (NREAP) submitted to the European Commission in response to the requirements of the RES directive in 2010 [111]. The UK NREAP makes a commitment to deployment of GSHPs that is derived from a projection of 330,000 units being installed by 2020 [112]. This estimated deployment appears to be based on an installation growth rate of slightly more than 50% per year leading up to 2020. Domestic GSHP installations were expected to mostly occur in the off-gas-grid and new housing sectors and financial support was expected to be at a similar level to that of the CERT programme. Since the NREAP was submitted in 2010, circumstances have changed considerably in that the economic downturn has severely limited new house building and the nascent UK heat pump industry capacity has faltered. Furthermore, it is clear that the ECO supplier obligation scheme will not provide the support for GSHP installations that CERT did [42,43]. Consequently it is hard to see that either the new housing market will have developed or industry capacity grown to the levels reflected in the UK NREAP. Poor progress along the trajectory set out in the plan was pointed out in the most recent report on the EurObserve'ER data [13]. The renewable thermal energy contribution from geothermal heat in 2012 was reported to be 23 ktoe, which is significantly lower than the 174 ktoe target for 2012 included in the UKs NREAP [111].

Deployment of GSHP systems under the RHPP programme focused on off-gas-grid retrofit applications. Where existing electrical resistance heating was replaced by heat pumps (either ASHP or GSHP) there has been no evidence of detrimental impact on local power distribution networks. However, in most of the UK's long term carbon/energy transition plans (2030 and 2050 timeframes) electrification of heating by large scale uptake of heat pumps is seen as a key element [113] and so impact on the national power grid and demand profiles is a genuine concern [114]. Incorporating large numbers of heat pumps into the UK power distribution system has to be seen in the context of many factors that are driving change and innovation in the UK power generation and distribution industry and alongside wider scale distributed generation (e.g., domestic PV) and adoption of electric vehicles [114]. Whole system national modeling studies [110] have shown that heat pumps can play an important role in facilitating demand management in that they offer opportunities for energy storage [110]. It therefore seems important that the heat pump industry is responsive to this requirement and enables intelligent interaction between the grid and heat pump installations and designs heating systems that include appropriate levels of thermal storage [115].

5. Conclusions

Deployment of domestic geothermal heat pumps in the UK has grown from a handful of systems installed in the late 1990s to approaching 18,000 at the time of writing. The industry has passed through an embryonic stage until 2003 and has grown approximately linearly in the 2003–2013 period with the support of a variety of grant programmes. The UK's adoption of domestic GSHP technology has lagged that of many other EU countries which is mostly a reflection of the well developed gas grid and consumer preference for high temperature hydronic heating and hot water generation. The dominance of gas fueled heating is reflected in the national skills base and consumer awareness.

Some of the barriers to deployment of GSHPs have been addressed by the technology being promoted through a national microgeneration grant programme that has disseminated information and initiated the development of installation standards and product certification. The industry has been able to organize and represent itself through the formation of a national trade association that is active in standards and training development as well as dissemination of information.

Support programmes that have been either focused on householders or have operated on a larger scale through energy supplier obligations have provided funding for over half of the domestic GSHP installations to date. However, funding of installations outside of the support programmes has dropped significantly during the recent economic downturn. We suggest, therefore, that the industry has yet to reach a stage of firm and sustainable growth and remains highly dependent on support programmes. In the near future (2014–2017) this will only be the tariff-based Renewable Heat Incentive and associated Green Deal programmes, the success of which—in that this market-based approach is sensitive to consumer attitudes, market conditions and fuel price changes—is highly uncertain.

Our review of data produced by earlier energy supplier obligation programmes shows that they have been very effective in providing support for GSHP installations and assisting in growth of the industry. GSHPs installed in the social housing projects funded through the supplier obligation programmes have made a contribution to alleviating fuel poverty. Supplier obligation programmes have supported the installation of approximately twice the number of heat pumps as the householder grant programmes although the latter have probably been more successful in promoting the technology and providing consumer information.

The UK has operated a series of relatively large-scale national heat pump field trials. These initially indicated the industry was delivering too many installations that were performing poorly relative to efficiency levels reported in other European countries. The trials have provided good evidence for the development of improved installation standards and given some insight into the technical issues that were not being addressed consistently. The most recent monitoring data shows that performance has been improved such that there can be high confidence that GSHP can contribute real carbon emission savings and running cost reductions. Average seasonal efficiencies are still short of the best European standards. We have identified a number of areas where technical improvements could be made and highlighted the need for further research into control of heat pumps in small houses with high thermal mass that are typical of the UK retrofit market. Research that has drawn on UK user responses has shown there is a need for users to receive better information about system operation and for control systems that are more intuitive.

UK carbon emissions reduction and renewable energy plans are based on electrification of the domestic heat sector through the adoption of heat pumps. The long-term prospects for GSHP adoption are, on the face of it, very favorable. However, progress with extensive adoption of heat pumps through the 2020s, when the electricity grid should have made some progress towards decarbonization, is dependent on having a sustainable installation industry that is considerably larger than now exists. It will correspondingly be important that the installation industry is sufficiently supported in the years approaching 2020 that both consumer and investor confidence grows. The absence of a supportive supplier obligation scheme makes growth in the run-up to 2020 very difficult to predict. The failure to show year-on-year increases in the rate of installation means that matching the levels of deployment envisaged in the UK NREAP will make the obligations of the RES directive very difficult to meet. We suggest that progress towards sustainable growth in the GSHP industry before 2020 may not be possible without further regulatory measures implemented through vehicles such as the Building Regulations or the type of explicit directives that have ensured uptake of condensing boilers and fluorescent domestic lighting in the recent past. This need for regulatory measures to provide sufficient assurance of long-term growth was highlighted in the detailed bottom-up modeling prepared for the CCC [113]. One regulatory measure that could be considered would be to follow the example of the Swedish building regulation implemented in 1984 that required all new heating systems to be designed to operate at low temperature (55 °C) [116]. This opened a path for later deployment of heat pumps in Sweden when the technology became more consumer-appealing and cost effective. This measure would have little cost impact in well-insulated homes in the UK, where radiator sizes are small in any case, and would also benefit condensing boiler operation in the medium term.

Although the new RHI programme payments for renewable heat from GSHPs will go some way to leveling the playing field in terms of financial payback, and there are opportunities in the off-gas-grid sector, GSHP technology will face strong competition from ASHPs. Although social housing development has played a significant role in the growth of the GSHP industry, recent evidence suggests there may be an increasing preference for ASHPs amongst such developers [78]. Recent grant uptake also suggests that larger new up-market housing may play a more significant role in the domestic GSHP market. For GSHP technology to be appealing to home owners and developers the industry will need to demonstrate that higher efficiencies can be consistently assured and make progress in reducing the capital costs and complexities (both technical and contractual) associated with ground heat exchanger installation. Technology such as automated monitoring and fault detection may assist in ensuring consistent performance. However, achieving higher levels of efficiency, lower fault levels and higher consumer satisfaction at the same time as industry growth, will require more comprehensive and better integrated industry training to be fully established and standards to be consistently met.

Acknowledgments

The authors wish to thank Chris Wickins of the Department of Energy and Climate Change for his helpful comments during the preparation of this paper. We would also like to thank Tom Pine of GeoScience for his assistance with the installation data.

Author Contributions

Simon Rees was responsible for the primary analysis of the market and support programme data. He carried out the review of the national field trials and non-technical data. Robin Curtis provided background information regarding historical development of the GSHP installation industry and was responsible for part of the content relating to technical developments. Robin Curtis provided material relating to installation standards development.

Conflicts of Interest

The authors declare no conflict of interest

References

1. Lund, J.; Freeston, D.; Boyd, T. Direct utilization of geothermal energy 2010 worldwide review. *Geothermics* **2011**, *40*, 159–180.
2. Baker, W. *Off-Gas Consumers Information on Households Without Mains Gas Heating*; Consumer Focus: London, UK, 2013; p. 54.
3. Bayer, P.; Saner, D.; Bolay, S.; Rybach, L.; Blum, P. Greenhouse gas emission savings of ground source heat pump systems in Europe: A review. *Renew. Sustain. Energy Rev.* **2012**, *16*, 1256–1267.
4. Singh, H.; Muetze, A.; Eames, P. Factors influencing the uptake of heat pump technology by the UK domestic sector. *Renew. Energy* **2010**, *35*, 873–878.
5. Lund, J.; Freeston, D.; Boyd, T. Direct application of geothermal energy: 2005 worldwide review. *Geothermics* **2005**, *34*, 691–727.
6. Garnish, J.D. Geothermal Resources of the UK—Country Update Report. *Trans. Geotherm. Resour. Counc.* **1985**, *9 International*, 217–222.
7. Sumner, J.A. *Domestic Heat Pumps*; Prism Press: London, UK, 1976.
8. Curtis, R. Earth Energy in the UK. *Geo Heat Cent. Bull.* **2001**, 23–30.

9. Lund, J.; Freeston, D. World-wide direct uses of geothermal energy 2000. *Geothermics* **2001**, *30*, 29–68.

10. Batchelor, T.; Curtis, R.; Ledingham, P. Country update for the United Kingdom. In Proceedings of World Geothermal Congress 2005, Antalya, Turkey, 24–29 April 2005; pp. 1–5.

11. Karpathy, B.Z. *Heat Pumps United Kingdom World Renewables 2013 A Multi Client Study*; BSRIA: Bracknell, UK, 2013.

12. EurObserv'ER. Ground-Source Heat Pump Barometer. *J. Des Énergies Renouv.* **2011**, *205*, 82–101.

13. EurObserv'ER. *Heat Pumps Barometer*; EurObsv'ER: Brussels, Belgium, 2013; p. 17.

14. Curtis, R.; Ledingham, P.; Law, R.; Bennett, T. Geothermal Energy Use, Country Update for United Kingdom. In Proceedings of European Geothermal Congress 2013, Pisa, Italy, 3–7 June 2013; pp. 1–9.

15. Fawcett, T. The future role of heat pumps in the domestic sector. In Proceedings of the ECEEE 2011 Summer Study, Energy Efficiency First: The Foundation of a Low-Carbon Society, Belambra Presqu'île de Giens, City, France, 6–11 June 2011; pp. 1547–1557.

16. Le Feuvre, P. Le An Investigation into Ground Source Heat Pump Technology, its UK Market and Best Practice in System Design. Master's Thesis, Univeristy of Strathclyde, Glasgow, UK, 2007, p. 180.

17. *Outlook 2009 European Heat Pump Statistics*; European Heat Pump Association: Brussel, Belgium, 2009; p. 39.

18. Mallaburn, P.; Eyre, N. Lessons from energy efficiency policy and programmesin the UK from 1973 to 2013. *Energy Effic.* **2014**, *7*, 23–41.

19. Department of the Environment. *This Common Inheritance: Britain's Environmental Strategy*; HMSO: London, UK, 1990; p. 291.

20. Performance Inovation Unit. *The Energy Review*; Cabinet Office Performance and Innovation Unit: London, UK, 2002; p. 216.

21. Department of Trade and Industry (DTI). *Our Energy Future—Creating a Low Carbon Economy*; DTI: London, UK, 2003; p. 142.

22. BRE. Renewable energy grants top £5.5 million. Availabe online: http://www.bre.co.uk/news/Renewable-energy-grants-top-55-million-230.html (accessed on 12 June 2014).

23. BRECSU. *Heat Pumps in the UK—A Monitoring Report (GIR 72)*; BRECSU: Watford, UK, 2000; p. 16.

24. Van de Ven, H. Ground-source heat pump systems, An international overview. *IEA Heat Pump Cent. Newsl.* **1999**, *17*, 10–12.

25. HMG. *Energy Act 2004*; HM Government: London, UK, 2004; p. 296.

26. Gardiner, M.; White, H.; Munziger, M.; Ray, W. *Low Carbon Building Programme 2006–2011 Final Report*; Department of Energy and Climate Change: London, UK, 2011; p. 169.

27. Radov, D.; Klevnäs, P.; Lindovska, M. Design of the Renewable Heat Incentive Study for the Department of Energy & Climate Change. NERA Econdomic Consulting: London, UK, 2010; p. 69.

28. HMG. *The UK Renewable Energy Strategy*; HM Government: London, UK, 2009; p. 236.

29. OFGEM. Domestic Renewable Heat Incentive. Available online: https://www.ofgem.gov.uk/environmental-programmes/domestic-renewable-heat-incentive (accessed on 12 June 2014).

30. Barker, G. *Departmental Note: Support for Renewable Heat Technologies in the Domestic and non-Domestic Sectors*; Department of Energy and Climate Change: London, UK, 2013; p. 1.

31. Department of Energy and Climate Change (DECC). *Renewable Heat Incentive, Consultation on Proposals for a Domestic Scheme*; DECC: London, UK, 2012; p. 32.

32. Rosenow, J. Energy savings obligations in the UK—A history of change. *Energy Policy* **2012**, *49*, 373–382.

33. Office of Gas and Electricity Markets (OFGEM). *A Review of the Energy Efficiency Commitment 2002–2005*; OFGEM: London, UK, 2005; p. 83.

34. Office of Gas and Electricity Markets (OFGEM). *A Review of the Energy Efficiency Commitment 2005–2008*; OFGEM: London, UK, 2008; p. 62.

35. Department of the Environment, Transport and the Regions (DETR). *Quality and Choice: A Decent Home for All: The Housing Green Paper*; DETR: London, UK, 2000; p. 132.

36. Dewdney, P. Powergen Enters New Era of Low-Cost, Low-Carbon Heating. Available online: http://www.heatpumps.org.uk/CaseExDomestic.html (accessed on 18 August 2014).

37. Calorex. *Case Study, Ground Source Heat Pumps Cutting Carbon Emissions*; Calorex Ltd.: Maldon, Essex, UK, 2007; pp. 16–17.

38. BBC. Hot Earth to Heat Homes. Available online: http://news.bbc.co.uk/1/hi/england/1670092.stm (accessed on 12 June 2014).

39. Hill, D. Ground Sourced Heat Pumps, Our experiences and where to next. In *GSHPA Seminar, 2005*; GSHPA: Milton Keynes, UK, 2005; pp. 1–3.

40. BRE. *Ground Source Heat Pumps Case Study*; BRE: Watford, UK, 2004; pp. 1–4.

41. Calorex. *Calorex Case Study—Housing*; Calorex Ltd.: Maldon, Essex, UK, 2006; p. 1.

42. Office of Gas and Electricity Markets (OFGEM). *A Review of the Fourth Year of the Carbon Emissions Reduction Target*; OFGEM: London, UK, 2012; p. 55.

43. Office of Gas and Electricity Markets (OFGEM). Energy Companies Obligation (ECO). Available online: https://www.ofgem.gov.uk/environmental-programmes/energy-companies-obligation-eco (accessed on 12 June 2014).

44. Energy Savings Trust (EST). *Element Energy Potential for Microgeneration, Study and Analysis Final Report*; EST: London, UK, 2005.

45. Caird, S.; Potter, S.; Herring, H. *Consumer Adoption and Use of Household Renewable Energy Technologies, Report DIG-10*; The Open University: Milton Keynes, UK, 2007; p. 57.

46. Balcombe, P.; Rigby, D.; Azapagic, A. Motivations and barriers associated with adopting microgeneration energy technologies in the UK. *Renew. Sustain. Energy Rev.* **2013**, *22*, 655–666.

47. Wickins, C. *Preliminary Data from the RHPP Heat Pump Metering Programme*; Department of Energy and Climate Change: London, UK, 2014; p. 61.

48. Karpathy, Z. *Heat Pump Markets UK in Europe*; BSRIA: Bracknell, UK, 2012.

49. DECC. *Low Carbon Building Programme Householder's Project Case Study—Ground Source Heat Pump*; Department for Communities and Local Government: London, UK, 2010; p. 4.

50. NEF. Ground Source Heat Pump Association Comes of Age. Available online: http://www.nef.org.uk/about-us/press-releases/ground-source-heat-pump-association-comes-of-age (accessed on 9 May 2014).

51. GSHPA. GSHPA press release—Ground Source Heat Pump RHI tariff expected to double. Available online: http://www.gshp.org.uk/pdf/GSHPA_Press_Release_24_January_2013.pdf (accessed on 12 May 2014).

52. Microgeneration Certification Scheme (MCS). *History of the Microgeneration Certification Scheme*; MCS: Watford, UK, 2013; p. 5.

53. European Certified Heat Pump Installer. Available online: http://www.heatpumpcentre.org/en/newsletter/previous/Documents/HPC-news_2_2004.htm (accessed on 12 June 2014)

54. BDA. BDA awarded ConstructionSkills grant for Geothermal Drilling. Available online: http://www.britishdrillingassociation.co.uk/news/BDA-awarded-ConstructionSkills-grant-for-Geothermal-Drilling (accessed on 12 June 2014).

55. Wincott, N. GeoTrainet and European Wide Training Developments. In *GSHPA Technical Seminar 2013*; GSHPA: Milton Keynes, UK, 2013; p. 24.

56. Brooke, S.B.; Matthews, D.; Willson, C. *UK Literature Review for International Energy Agency (IEA) Annex 36 on Investigating the Effect of Quality of Installation and Maintenance on Heat Pump Performance*; Department of Energy and Climate Change: London, UK, 2013.

57. CEN. *EN 14511-2, Air Conditioners, Liquid Chilling Packages and Heat Pumps with Electrically Driven Compressors for Space Heating and Cooling—Part 2: Test Conditions*; CEN: Brussels, Belgium, 2007.

58. CEN. *EN 15316-4-2, Heating Systems in Buildings—Method for Calculation of System Energy Requirements and System Efficiencies—Part 4-2: Space Heating Generation Systems, Heat Pump Systems*; CEN: Brussels, Belgium, 2008.

59. Gleeson, C.; Lowe, R. Meta-analysis of European heat pump field trial efficiencies. *Energy Build.* **2013**, *66*, 637–647.

60. Stafford, A. Long-term monitoring and performance of ground source heat pumps. *Build. Res. Inf.* **2011**, *39*, 566–573.

61. Wemhorner, C.; Dott, R.; Afjei, T.; Huber, H.; Helfenfinger, D.; Keller, P.; Furter, R. *Calculation Method for the Seasonal Performance of Heat Pump Compact Units and Validation*; Swiss Federal Office of Energy: Bern, Switzerland, 2007.

62. Zottl, A.; Nordman, R.; Rivière, P. SEPEMO—Common Monitoring Methodology. *Eur. Heat Pump News* **2011**, *3*, 5–6.

63. European Commission. *Decision (2013/114/EU) Establishing the Guidelines for Member States on Calculating Renewable Energy from Heat Pumps from Different Heat Pump Technologies pursuant to Article 5 of Directive 2009/28/EC of the European Parliament an dof the Council*; Official Journal of the European Union: Brussels, Belgium, 2013; pp. 27–35.

64. Dunbabin, P.; Charlick, H.; Green, R. *Detailed Analysis from the Second Phase of the Energy Saving Trust's Heat Pump Field Trial*; Department of Energy and Climate Change: London, UK, 2013; p. 123.

65. Energy Saving Trust (EST). *The Heat is on: Heat Pump Field Trials Phase 2*; EST: London, UK, 2013; p. 40.

66. Greening, B.; Azapagic, A. Domestic heat pumps: Life cycle environmental impacts and potential implications for the UK. *Energy* **2012**, *39*, 205–217.

67. Johnson, E. Air-source heat pump carbon footprints: HFC impacts and comparison to other heat sources. *Energy Policy* **2011**, *39*, 1369–1381.

68. European Commission. *Directive 2009/28/EC of the European Parliament and of the Council of 23 April 2009 on the Promotion of the Use of Energy from Renewable Sources and Amending Subsequently Replealing Directives 2001/77/EC and 2003/30/EC*; Official Journal of the European Union: Brussels, Belgium, 2009; pp. 16–62.

69. Orr, G.; Lelyveld, T.; Burton, S. *Final Report : In-situ Monitoring of Efficiencies of Condensing Boilers and Use of Secondary Heating*; Gastec at CRE: Cheltenham, UK, 2009; p. 44.

70. Energy Saving Trust (EST). *Getting Warmer: A Field Trial of Heat Pumps*; EST: London, UK, 2010.

71. Dunbabin, P.; Wickins, C. *Detailed Analysis from the First Phase of the Energy Saving Trust's Heat Pump Field Trial*; Department of Energy and Climate Change: London, UK, 2012.

72. Microgeneration Certification Scheme (MCS). *Microgeneration Installation Standard: MIS 3005, Requirements for Contractors Undertaking the Supply, Design, Installation, Set to Work Commissioning and Handover of Microgeneration Heat Pump Systems, Version 3.1*; MCS: Watford, UK, 2008; pp. 1–47.

73. Caird, S.; Roy, R.; Potter, S. Domestic heat pumps in the UK: User behaviour, satisfaction and performance. *Energy Effic.* **2012**, *5*, 283–301.

74. Roy, R.; Caird, S. Diffusion, User Experiences and Performance of UK Domestic Heat Pumps. *Energy Sci. Technol.* **2013**, *6*, 14–23.

75. Stafford, A.; Lilley, D. Predicting in situ heat pump performance: An investigation into a single ground-source heat pump system in the context of 10 similar systems. *Energy Build.* **2012**, *49*, 536–541.

76. Boait, P.; Fan, D.; Stafford, A. Performance and control of domestic ground-source heat pumps in retrofit installations. *Energy Build.* **2011**, *43*, 1968–1976.

77. DECC AECOM. *Analysis of Customer Data from Phase One of the Renewable Heat Premium Payments (RHPP) Scheme*; Department of Energy and Climate Change: London, UK, 2013; p. 110.

78. Energy Saving Trust. RHPP RSL Fast Track Competition Winners 2013–2014. Available online: http://www.energysavingtrust.org.uk/Generating-energy/Getting-money-back/Renewable-Heat-Premium-Payment-RHPP-Social-Landlords-Competition-Phase-Two-Extensions-2013-14/RHPP-RSL-Fast-Track-Competition-Winners-2013-2014 (accessed on 19 May 2014).

79. Office for National Statistics. Home ownership and renting in England and Wales—Detailed Characteristics. Available online: http://www.ons.gov.uk/ons/rel/census/2011-census/detailed-characteristics-on-housing-for-local-authorities-in-england-and-wales/short-story-on-detailed-characteristics.html (accessed on 22 June 2014).

80. Office for National Statistics. *Annual Survey of Hours and Earnings, 2013 Provisional Results*; Office for National Statistics: London, UK, 2013; pp. 1–45.

81. Delta Energy & Environment. *Heat Pumps in the UK: How Hot Can They Get?* Delta Energy and Environment: Edinburgh, UK, 2011; p. 8.

82. Stenlund, M.; Axell, M. *Residential Ground-Source Heat Pump Systems—Results from a Field Study in Sweden*; SP Swedish Technical Research Institute: Borås, Sweden, 2007; pp. 1–10.

83. Pederson, S.; Jacobsen, E. *Approval of Systems Entitled to Subsidies. Measurements Data Collection and Dissemination*; Danish Technological Institute: Taastrup, Denmark, 2011; p. 12.

84. Nordman, R.; Andersson, K.; Axell, M.; Lindahl, M. *Calculation Methods for SPF for Heat Pump Systems for Comparison, System Choice and Dimensioning*; SP Swedish Technical Research Institute: Borås, Sweden, 2010; p. 82.

85. Russ, C.; Miara, M.; Platt, M.; Günther, D.; Kramer, T.; Dittmer, H.; Lechner, T.; Kurz, C. *Feldmessung Wärmepumpen im Gebäudebestand*; Fraunhofer ISE: Freiburg, Germany, 2010; pp. 1–21. (In German)

86. Miara, M.; Gunther, D.; Kramer, T.; Oltersdorf, T.; Wapler, J. *Messtechnische Untersuchung von Wärmepumpenanlagen zur Analyse und Bewertung der Effizienz im realen Betrieb*; Fraunhofer ISE: Freiburg, Germany, 2011; p. 154. (In German)

87. Microgeneration Certification Scheme (MCS). *MCS 022: Ground Heat Exchanger Look-up Tables, Supplementary Material to MIS 3005, Issue 1.0*; MCS: Watford, UK, 2011; pp. 1–22.

88. Microgeneration Certification Scheme (MCS). *MCS 021: Heat Emitter Guide for Domestic Heat Pumps, Issue 1.0*; MCS: Watford, UK, 2013; p. 9.

89. GeoEnergy. *MCS Procedure and Charts for Designing the Hydraulics and Associated Pumping Power of Closed Loop GSHP Systems under MCS*; MCS: Watford, UK, 2012; p. 39.

90. Green, R.; Knowles, T. *The Effect of Thermostatic Radiator Valves on Heat Pump Performance*; EA Technology Consulting: Chester, UK, 2011; p. 21.

91. Charlick, H. *Investigation of the Interaction between Hot Water Cylinders, Buffer Tanks and Heat Pumps*; Kiewa GASTEC at CRE: Cheltenham, UK, 2013.

92. Hellström, G.; Sanner, B. Software for dimensioning of deep boreholes for heat extraction. *Proc. CALORSTOCK* **1994**, *94*, 195–202.

93. Spitler, J.D. GLHEPRO—A design tool for commercial ground loop heat exchangers. In Proceedings of the Fourth International Heat Pumps in Cold Climates Conference, Aylmer, QC, Canada, 17–18 August 2000; p. 16.

94. Gaia Geothermal LLC. Ground Loop Design User's Guide. Available online: http://www.groundloopdesign.com/downloads/GLD_2.x/manual.pdf (accessed on 21 August 2014).

95. Verein Deutscher Ingenieure. *VDI-Richtlinie 4640: Thermische Nutzung des Untergrundes–Blatt 2: Erdgekoppelte Wärmepumpenanlagen*; Beuth Verlag: Berlin, Germany, 2001.

96. Curtis, R.; Pine, T.; Wickins, C. Development of new ground loop sizing tools for domestic GSHP installations in the UK. In The Proceedings of European Geothermal Congress 2013, Pisa, Italy, 3–7 June 2013; pp. 1–10.

97. International Ground Source Heat Pump Association (IGSHPA). *CLGS Ground Heat Exchanger Design Program*; IGSHPA: Stillwater, OK, USA, 2005.

98. Schein, J.; Bushby, S.T. A heirarchical rule-based fault detection and diagnostic method for HVAC systems. *HVAC&R Res.* **2006**, *12*, 115–126.

99. Zogg, D. Fault Diagnosis for Heat Pump Systems. Ph.D.Thesis, Swiss Federal Insitute of Technology (ETH), Zurich, Switzerland, 2002.

100. Curtis, R.; Pine, T. *Effects of Cycling on Domestic GSHPs Supporting Analysis to EA Technology Simulation/Modelling*; Mimer Geoenergy: Falmouth, UK, 2012; p. 28.

101. Heffernan, W.; Watson, N. Harmonic performance of heat-pumps. *J. Eng.* **2013**, *25*, 1–14.

102. Connor, P.; Bürger, V.; Beurskens, L.; Ericsson, K.; Egger, C. Devising renewable heat policy: Overview of support options. *Energy Policy* **2013**, *59*, 3–16.

103. Department of Energy and Climate Change (DECC). *The Future of the Energy Company Obligation*; DECC: London, UK, 2014; p. 63.

104. Department of Energy and Climate Change (DECC). *The Green Deal, A Summary of the Government's Proposals*; DECC: London, UK, 2010; p. 22.

105. HMG. *The Building Regulations, Conservation of Fuel and Power Approved Document Part L1A (2013 version)*; HM Government: London, UK, 2013; p. 48.

106. Department for Communities and Local Government (DCLG). *Code for Sustainable Homes, A Step-Change in Sustainable Home Building Practice*; DCLG: London, UK, 2006; p. 31.

107. Department for Communities and Local Government (DCLG). *The Code for Sustainable Homes, Case Studies Volume 2*; DCLG: London, UK, 2010; Volume 2, p. 50.

108. Department for Communities and Local Government (DCLG). *Housing Standards Review*; DCLG: London, UK, 2013; p. 88.

109. The Committee on Climate Change (CCC). *The Fourth Carbon Budget, Reducing Emissions through the 2020s*; CCC: London, UK, 2010.

110. Strbac, G.; Gan, C.K.; Aunedi, M.; Stanojevic, V.; Djapic, P.; Dejvises, J.; Mancarella, P.; Hawkes, A.; Pudjianto, D.; Openshaw, D.; *et al. Benefits of Advanced Smart Metering for Demand Response based Control of Distribution Networks*; Imperial College London: London, UK, 2010; p. 49.

111. Department for Communities and Local Government (DCLG). *National Renewable Energy Action Plan for the United Kingdom*; DCLG: London, UK, 2009; p. 160.

112. Radov, D.; Klevnäs, P.; Lindovska, M.; Abu-Ebid, M.; Barker, N.; Stabaugh, J. *The UK Supply Curve for Renewable Heat*; NERA Economic Consulting: London, UK, 2009; Volume 689, p. 155.

113. Frontier Economics Element Energy. *Pathways to High Penetration of Heat Pumps*; Frontier Economics Ltd.: London, UK, 2013; p. 147.

114. Guzeleva, D. Challenges in the UK network. In *IET Seminar, Integrating Renewable Energy to the Grid: Optimising and Securing the Network*; IET: London, UK, 2014.

115. Sugden, L. Smart Grids create new opportunities for heat pumps. *REHVA J.* **2012**, 20–22.

116. Karlsson, F.; Axell, M.; Fahlén, P. *Heat Pump Systems in Sweden—Country Report for IEA HPP Annex 28*; SP Swedish Technical Research Institute: Borås, Sweden, 2003; pp. 1–29.

Deep Geothermal Energy Production in Germany

Thorsten Agemar, Josef Weber and Rüdiger Schulz

Abstract: Germany uses its low enthalpy hydrothermal resources predominantly for balneological applications, space and district heating, but also for power production. The German Federal government supports the development of geothermal energy in terms of project funding, market incentives and credit offers, as well as a feed-in tariff for geothermal electricity. Although new projects for district heating take on average six years, geothermal energy utilisation is growing rapidly, especially in southern Germany. From 2003 to 2013, the annual production of geothermal district heating stations increased from 60 GWh to 530 GWh. In the same time, the annual power production increased from 0 GWh to 36 GWh. Currently, almost 200 geothermal facilities are in operation or under construction in Germany. A feasibility study including detailed geological site assessment is still essential when planning a new geothermal facility. As part of this assessment, a lot of geological data, hydraulic data, and subsurface temperatures can be retrieved from the geothermal information system GeotIS, which can be accessed online [1].

Reprinted from *Energies*. Cite as: Agemar, T.; Weber, J.; Schulz, R. Deep Geothermal Energy Production in Germany. *Energies* **2014**, *7*, 4397-4416.

1. Introduction

The importance of renewable energy is growing since it is evident that fossil fuel supplies are finite, politically vulnerable and responsible for climate change. Geothermal energy enjoys a special position amongst the renewable energy sources because it is available all year round, at any time of the day, and can therefore be used for base load energy, for both heat and power production. This paper focuses on deep geothermal energy resources in Germany. In contrast to shallow geothermal energy, deep geothermal energy can be used directly—with a much higher energy output and without the need to install a heat pump to raise the temperature.

In Germany, geothermal energy is defined as energy stored beneath the surface of the solid earth in the form of heat [2]. Although the last volcanic activities ceased in Germany approximately 10,000 years ago, there are many places where deep geothermal energy is available as an environmentally friendly alternative to fossil fuels. Its use helps to reduce the emission of greenhouse gases and to diversify the overall energy supply mix of Germany. The development of geothermal energy is supported by the German Federal Government in terms of project funding, market incentives and credit offers, as well as a feed-in tariff for geothermal electricity.

Unfortunately, the national geothermal resources are limited to intermediate and low enthalpy resources. All present-day geothermal installations use fluids with temperatures evidently below 180 °C. Nevertheless, an economic sector is developing rapidly to exploit the geothermal potential of these low enthalpy resources, which are typically located in deep sedimentary basin and graben structures. The most important geological settings for geothermal energy in Germany are deep Mesozoic sediments, which can be found in the North German Basin, the Upper Rhine Graben, and the South German Molasse Basin (reference [3] and references therein). Extensive, permeable and water-bearing sediments (aquifers) of at least 20 m thickness are referred to as geothermal reservoirs.

The average geothermal gradient, the increase in temperature per depth unit, is 32 K/km [4]. In some areas the gradient can reach values as high as 100 K/km. Aquifers at depths less than 400 m and temperatures below 20 °C are not considered as deep in the context of geothermal energy resources. Aquifers with thermal waters below 60 °C are usually used for spas in Germany. The usage of thermal waters in spas is generally realised with a single production well (singleton) and flow rates rarely above 15 L/s. Hence, these systems rely on the natural recharge of the aquifer. District heating stations usually require temperatures above 60 °C and much higher flow rates. Geothermal district heating is realised with two or more geothermal wells with at least one production well and one injection well (Figure 1). The re-injection of cooled geothermal fluids is necessary to maintain the pressure in the reservoir and in order to avoid the contamination of surface waters or shallow aquifers with high salt loads or even toxic fluid constituents. The flow rates for the primary cycle range from 35 L/s to 150 L/s. Combined geothermal heat and power production requires similar flow rates and additionally temperatures of at least 110 °C for economical operation. Modern conversion techniques such as the Organic Rankine Cycle or the Kalina cycle make it possible to generate power from fluids even below 100 °C. However, geothermal electricity production is still more expensive due to high investment costs than production by conventional coal fired or nuclear power plants in Germany. Geothermal power production is therefore not economical viable without the German Renewable Energy Sources Act (EEG) and the guaranteed feed-in-tariffs for renewable electricity producers. Currently, geothermal power plant operators receive 0.25 €/kWh. Petrothermally produced electricity is even worth 0.30 €/kWh, but no petrothermal-only project has been realised yet in Germany. The government plans a revision of the EEG for 1 August 2014. It is intended to abandon the petrothermal bonus of 0.05 €/kWh and to introduce obligatory direct-marketing for operators. The beginning of direct-marketing depends on the installed capacity. Longer adaption periods will be arranged for small facilities with less than 500 kW output. The additional costs for direct-marketing of geothermal electricity shall be compensated by increasing the feed-in-tariff from 0.25 €/kWh to 0.252 €/kWh. The direct-marketing concept increases financial risks but offers the chance to increase revenues if operators can raise power production at times when electricity prices at the European Power Exchange (EPEX SPOT SE) are higher than the monthly average.

Today, combined heat and power production takes place in the alpine foreland (South German Molasse Basin) and along the upper Rhine valley (Upper Rhine Graben) due to the high geothermal potentials in these areas (Figure 2). Factors which have a major influence on the revenue include the achievable flow rates per well, the temperature and the attainable prices on the heat market. Factors which have a major influence on the costs include the exploration and drilling of reservoirs. Like other renewable energy technologies, geothermal projects also face elevated levels of financing risk due to high upfront costs. However, there are additional risks specific to geothermal. The drilling phase is much more capital intensive than all the previous phases, while still fraught with uncertainty. If the target formation of geothermal drilling turns out to be unsuitable for the project aims, a significant amount of investment is lost. Therefore, each geothermal project needs to be planned individually and must consider the great variation in the geological conditions of the deep subsurface.

Figure 1. Diagram of a doublet for hydrothermal exploitation.

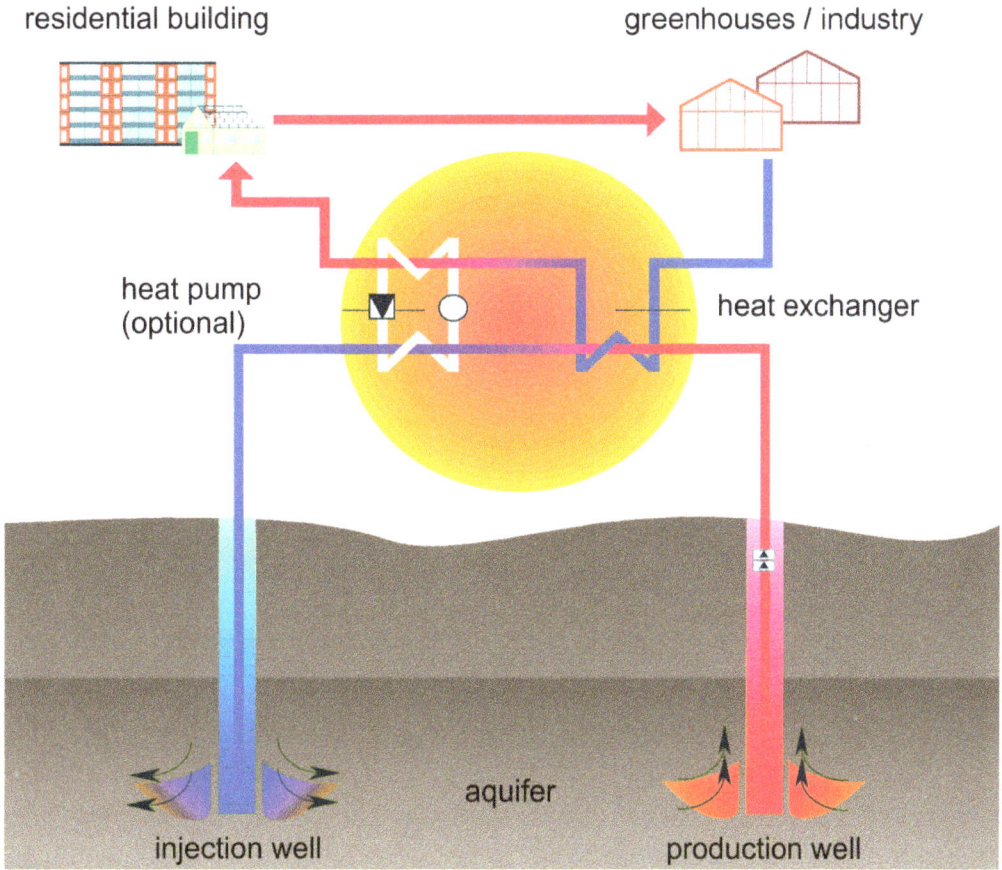

Figure 2. German regions with hydrothermal resources (proven and assumed) and associated temperature ranges. Map adapted from Suchi *et al.* [5] (copyright 2014 Leibniz Institute for Applied Geophysics (LIAG)).

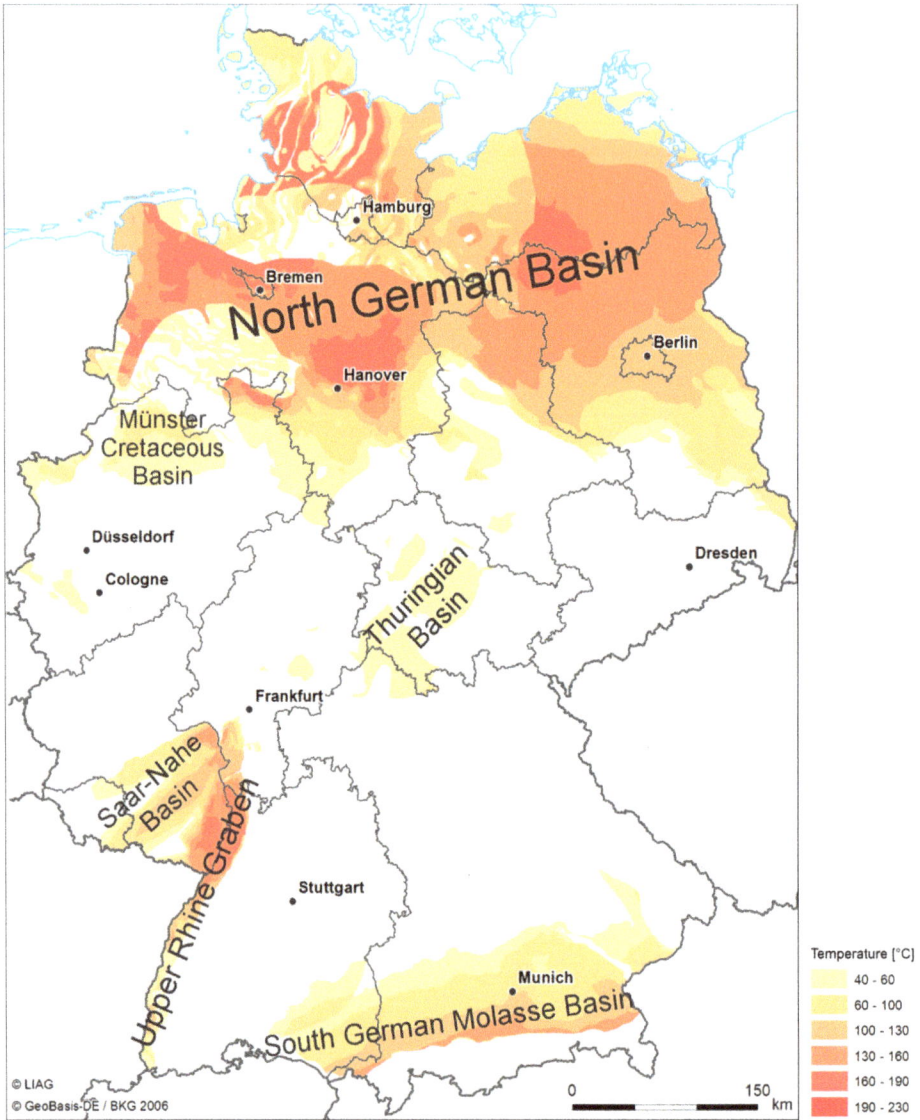

2. Concepts of Geothermal Energy Production

2.1. Hydrothermal Systems

Hydrothermal systems make use of hot water in deep aquifers to produce geothermal energy. The conventional exploitation technique requires a production well to tap the hot water and a re-injection well for water disposal (Figure 1). The water is pumped to the surface where the stored heat is directly used or transferred to a secondary cycle (e.g., a district heating system) via a heat exchanger. In cases where the thermal water is characterised by a high level of salinity or gas concentration it is necessary to maintain the primary circuit under pressure in order to avoid the precipitation of minerals and resultant scaling.

Through a second well the cooled down water is then re-injected back into the same aquifer but at a specific distance to the source in order to keep up the productivity of the aquifer and to prevent a thermal breakthrough. This configuration, consisting of one production and one injection well, is called a doublet, however, there is also the possibility to extend the system by further wells of both types in order to enhance the performance of the installation.

Hydrothermal systems include aquifers which are classified on the basis of the water temperature: hot (above 100 °C), warm (between 60 °C and 100 °C) and thermal (at least 20 °C). Besides the temperature of the groundwater the productivity and injectivity of the wells are important parameters for the economic operation of geothermal plants.

Faults or fault zones can be attributed to hydrothermal systems as well. These geological structures promise a high geothermal potential [6] due to the fact that faults can represent connections to deep and hot rock formations and allow deep groundwater to migrate to shallower depths. In some cases faults and associated rising fluids even reach earth`s surface and can be exploited for balneological applications like for example in the city of Aachen where over 30 hot springs with water temperatures of up to about 70 °C can be found along two lineaments. In order to use these geological features as geothermal reservoirs in a safe and economic way further research has to be carried out to gain knowledge of the characteristics of different fault types like predominant stress and geological conditions, geochemical processes or the degree of permeability.

2.2. Petrothermal Systems

In the case of petrothermal systems thermal energy stored in deep rocks itself is exploited without any need of natural water-bearing rock formations. The most important parameter for petrothermal systems is the temperature at depth. Therefore regions with positive temperature anomalies are of particular interest for the exploitation of geothermal energy because sufficient temperatures can be achieved in shallower depths resulting in lower drilling costs.

In applications like Hot-Dry-Rock-Systems (HDR) or Enhanced-Geothermal-Systems (EGS) deep hot horizons (usually the crystalline basement) at depths of more than 3000 m with temperatures in the order of 150–200 °C are used as a heat exchanger to extract energy from the rock mass. Although the upper parts of the crystalline basement are thought to be fractured and show a certain level of permeability, it is necessary to enhance the network of fractures by stimulation procedures to create a heat exchanger of adequate size for an economic operation. This is usually accomplished by injecting large volumes of water after drilling the first well to increase permeability and migration paths by expanding natural and creating additional fractures (hydraulic stimulation). After successful stimulation a second well is drilled into the fractured zone in order to enable a circular flow: water pumped down the injection well absorbs heat from the hot rock while flowing through the fracture network and is produced from the production well. Like already mentioned in the section on hydrothermal systems precipitation of minerals in the usually oversaturated water can be limited by maintaining the cycle at elevated pressure. These systems are usually designed for the generation of electricity. The Leibniz Institute for Applied Geophysics (LIAG) is currently investigating the petrothermal potential of a 10 km × 12 km area in the Erzgebirge (Saxony) [7]. The idea is to develop the first petrothermal project in Germany.

Petrothermal systems can also be exploited by deep borehole heat exchangers (BHEs) installed in wells with depths ranging from 400 m to about 3000 m. Deep BHEs are closed systems in which a heat transfer medium (e.g., ammonia) is pumped down the annular space of a coaxial pipe system and gets heated by thermal conduction from the surrounding rock on its way down to the bottom of the heat exchanger. Via an insulated inner pipe the heated medium is brought back to surface where the thermal energy is extracted and used for supplying heat.

Operating deep BHEs in Germany exist in Arnsberg (North Rhine-Westphalia) with a total depth of 2835 m heating a spa, Prenzlau (Brandenburg, 2786 m, used for district heating) and Heubach (Hesse) providing heat for industry (773 m).

3. Key Parameters

A profound geothermal resource assessment requires a detailed knowledge of the geologic setting as well as thermal and hydraulic properties of the geothermal reservoir. The most relevant thermal parameter is temperature. Although the vertical temperature profile is approximately linear in many sedimentary settings, there are many locations where the vertical temperature increase is not linear due to strongly varying thermal conductivities within the sedimentary column or due to groundwater advection. The product of the geothermal gradient and the thermal conductivity relates to the heat flow density according to the Fourier equation of conductive heat transfer:

$$q = \lambda \cdot \text{grad}\, T \tag{1}$$

The heat flow density q quantifies the amount of heat per surface area and time. Its value is approximately 70 mW/m^2 on average in Germany [8]. Without utilisation of geothermal energy, this amount of energy is lost to space. The largest contribution to the terrestrial heat flux is related to the decay of radioactive isotopes in crustal rocks. Another major contribution is the remaining heat from the formation of the Earth.

Records of subsurface temperatures exist from approximately 11,000 wells. Equilibrium temperature logs and reservoir temperatures are considered to be the best available data, which require no corrections. Because of the periodic monitoring of some production wells over many years, reservoir temperatures are available in time series; the fluctuation of these temperatures is mainly less than 1 K. Bottom-hole temperature (BHT) data are recorded in almost all industrial boreholes at the deepest point of the well immediately after drilling has stopped. The temperature field around a borehole is usually disturbed by mud circulation related to the drilling process. A number of methods to extrapolate from BHT to the undisturbed temperature have therefore been developed based on various assumptions about the cooling effect of the circulating mud and the thermal behaviour of the borehole and the surrounding rock [4,9,10].

The average surface temperature in Germany is 8.2 °C. The highest temperature ever measured in a German borehole is 253 °C at a depth of 9063 m (KTB, Upper Palatinate). Figure 3 shows the subsurface temperature at 2000 m below sea level (mbsl) and 4000 mbsl. The maps represent horizontal sections of a geostatistical 3D temperature estimate based on measured data [4].

Figure 3. Maps of subsurface temperature distribution (mbsl: meters below sea level).

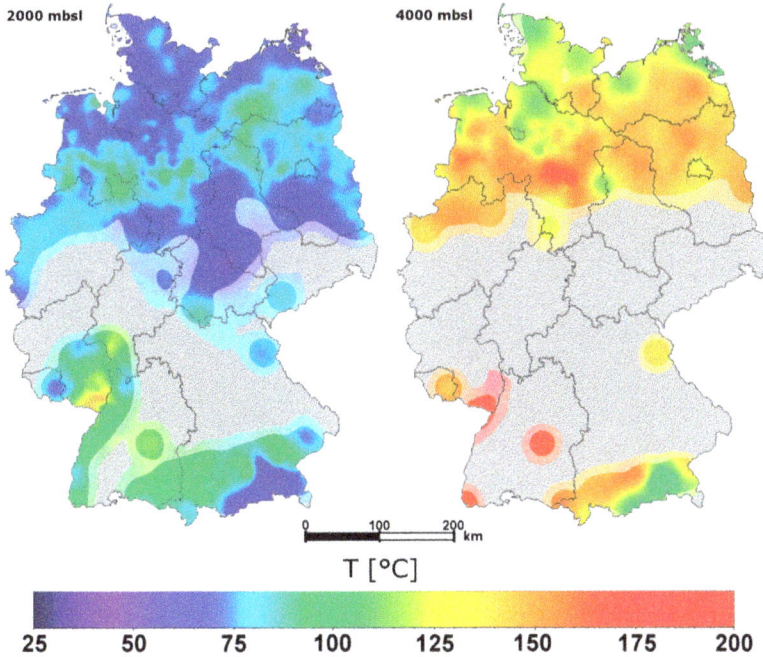

Permeability and hydraulic conductivity describe the ability of a porous medium to let a viscous fluid pass through. The permeability characterizes the conductive properties of the rock matrix, only. The hydraulic conductivity characterizes the conductive properties of the system, including the specific weight and viscosity of the fluid. Hydraulic conductivity K and permeability k are related to each other by following equation:

$$K = \frac{\rho \cdot g}{\mu} k \qquad (2)$$

where g is the acceleration due to gravity, ρ is the fluid density, and μ is its dynamic viscosity. The viscosity has a strong influence on the hydraulic conductivity. The viscosity of water essentially depends on the temperature while the influence of the pressure is rather small. For example, the hydraulic conductivity of an aquifer halves if temperature drops from 100 °C to 45 °C. Direct use applications generally require a hydraulic conductivity of at least 1×10^{-6} m/s.

Porosity and permeability data are available from drill core samples. Hydraulic tests estimate the transmissivity T, which is by definition equal to the integration of the hydraulic conductivities across the aquifer thickness H:

$$T = \sum_{i}^{n} K_i \cdot H_i \qquad (3)$$

T is measured in m²/s. It is possible to derive the transmissivity of porous aquifers from permeability data if the density and dynamic viscosity of the fluids are known [3]. A transmissivity of 5×10^{-5} m²/s, a porosity of 20% and an aquifer thickness of 20 m are considered as minimum values in order to achieve sufficient flow rates.

The productivity index (PI) is also frequently used to describe the hydraulic properties at a geothermal site. The PI comprises hydraulic properties of the tapped aquifer and well specific properties. The production index characterizes the production rate in relation to the pressure drawdown. A pressure drawdown of 1–3 MPa is typical for geothermal facilities [11]. In the case of injection wells, the injection index II is the analogue to the production index. It describes the injection rate in relation to the rise in pressure.

4. Project Planning

The successful development of a geothermal project requires detailed planning, a competent project management, and an effective cooperation of experts from many different fields. In Germany, a geothermal project for district heating takes six years on average. However, development time can vary, depending on the federal regulatory requirements, availability of a drilling rig and other factors. It can be divided into a series of development phases before the actual operation begins:

(1) Preliminary Study (6–12 months)

- Definition of project objectives
- Data compilation
- Technical draft

(2) Feasibility Study (12–24 months)

- Data acquisition
- Quantification of exploration risk
- Financial analysis

(3) Exploration (18–24 months)

- Obtaining permits
- Seismic exploration
- First drilling
- Hydraulic tests and well stimulation
- Decision on strike

(4) Field Development (18–24 months)

- Second drilling
- Hydraulic tests
- Construction of surface facilities
- Securing licence area at the local mining authority

(5) Start-up and Commissioning (3–6 months)

4.1. Preliminary Study

The preliminary study of a geothermal project must point out the main objectives and possible barriers and risks. A compilation and analysis of available data on the geological setting is the starting point of each preliminary study.

Because it is always possible that the prospectivity of a geothermal reservoir penetrated by a borehole is worse than expected or even unsuitable, investors and project operators try to assess the exploration risk and insure against it if possible. The exploration risk therefore has to be precisely defined and used to quantify the probability of the economic success of a project. The exploration risk is defined as the risk of penetrating a geothermal reservoir with one (or more) borehole(s) with inadequate quantity or quality.

The quantity is defined here by the installed capacity of a geothermal plant:

$$P = \rho_F \cdot c_F \cdot Q \cdot (T_i - T_o) \qquad (4)$$

where:

P	Capacity	W
ρ_F	Fluid density	$kg \cdot m^{-3}$
c_F	Specific heat capacity at constant pressure	$J \cdot kg^{-1} \cdot K^{-1}$
Q	Flow rate, production rate	$m^3 \cdot s^{-1}$
T_i and T_o	Input and output temperature of the geothermal plant	K

The key parameters for the estimation of the capacity are production rate Q and temperature T_i (the temperature at the wellhead), which crucially depends on temperature T_A in the aquifer. The production rate depends on the hydraulic conductivity of the tapped aquifer, the depth of the water table, the nominal output of the downhole pump, and the well design. While the first two parameters depend on the local geological setting, the latter two parameters are subject to technical and economic limits.

The quality of a geothermal reservoir mainly concerns the chemical composition of the fluid. Fluids may contain high content of dissolved gas and high salt loads. Materials exposed to hydrothermal fluids could be affected by corrosion. Depending on technology and selected materials, most fluid compositions encountered in Germany are manageable for geothermal exploitation. However, very strong saline fluids could pose a serious problem to the process of heat extraction due to precipitation of solids. Very high salt loads up to 330 g/L have been encountered in deep Mesozoic sandstones of the North German Basin. Fluids of deep aquifers in the Upper Rhine Graben exhibit generally lower salt concentrations. Here, highest salinity values observed are below 150 g/L. Most waters of the Upper Jurassic karst aquifer of the South German Molasse Basin have salt loads below 1 g/L [12].

For assessing the exploration risk the project operator defines limits for T_A and Q, above which a well can be considered a success. For a geothermal well, the probability of success can be estimated by determining the probability of each risk separately and by multiplying the single risks for flow rate and temperature [13]. The experience from other geothermal facilities in the same region also provides valuable information on the feasibility of a geothermal project. A major source of public data relevant for geothermal exploration in Germany can be found in the national geothermal information system GeotIS, which can be accessed online [1,3]. Basically, GeotIS is the digital version of a geothermal atlas. It offers a compilation of data and information about deep aquifers in Germany for possible geothermal use. Extent, depth and temperature of relevant geologic formations are presented for those regions of Germany most relevant to geothermal exploration. Surface and subsurface temperatures are provided where a sufficient amount of data is given. Both, temperature and geologic data have been compiled to state-of-the-art 3D-models. Views of the 3D content can be retrieved interactively by generating temperature plots on stratigraphic surfaces (Figure 4) or geologic cross sections (Figure 5).

Figure 4. Screenshot of geothermal information system (GeotIS) showing an example of an interactive cross section through the eastern part of the North German Basin. Well locations (red) visible on the map strip are projected onto the cross section. Filled circles indicate wells with temperature data. A mouse-over-feature gives additional details of the well on the map strip and temperature values on the cross section. Colours of stratigraphic units, vertical exaggeration and z-axis can be adjusted by the user.

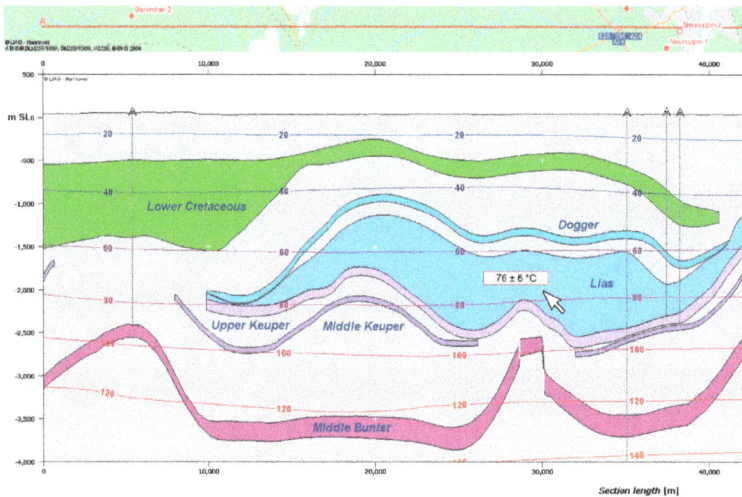

The mouse pointer can be used to obtain a temperature prediction including an uncertainty range of any point in space of the subsurface. Maps of formation permeability are available as part of the information system for many regions of geothermal interest. Furthermore, data such as the locations of wells or seismic profiles can be displayed. In Germany, geothermal resources belong to the mining estate and the legal standard is the Federal Mining Law. The land owner is not the presumptive owner of deep geothermal. Exploration claims and exploitation permits (in German: "Erlaubnisfeld" and "Bewilligung") are granted by the State mining authorities. A query about existing claims, permits, and special federal laws relevant for geothermal projects is therefore one of the first things to do.

Figure 5. Screenshot of GeotIS with a map of the subsurface temperature at the top of the Upper Jurassic formation (Malm) south of Munich (Bavaria). Bold black lines mark the location of faults in the Molasse Basin. The temperature scale adjusts automatically to the existing range of values in the selected area. A mouse-over-feature gives temperature values or the name of a well (not shown in this example). The superimposition of temperature, depth contours, fault lines and topographic features simplifies the first assessment of potential geothermal project locations.

Based on the expected geothermal potential and the general objectives, the preliminary study also provides a rough technical concept that outlines how geothermal energy is brought to the surface and distributed to the users. If it is planned to supply geothermal heat to residential or commercial buildings then existing and new district heating networks are as important to consider as a realistic estimate of the future demand and revenue. Peak demands in periods of extreme cold weather usually need to be backed by gas boilers.

The installation of a power plant results in additional costs but may help to compensate the seasonal fluctuations of the heat demand. Since the EEG guarantees feed-in-tariffs for geothermal electricity of currently 0.25 €/kWh, many geothermal facilities with reservoir temperatures above 110 °C combine heat and power production in Germany. Besides the potential revenues for electricity and heat, estimates of the costs for drilling and surface installations, access to further governmental incentives as well as interest rates on loans are also important aspects within the scope of the financial concept of a preliminary study. Both, prospective costs and revenues depend on site- and project-specific conditions.

4.2. Feasibility Study

The feasibility study continues and deepens the analyses of the preliminary study as it addresses the same questions but provides much more details. Seismic data, well profiles and borehole logs could be purchased in order to refine subsurface models. The depth level, size and hydraulic properties of the target formation need to be estimated and the uncertainty of this estimate quantised. An assessment of possible environmental impacts is also part of the feasibility study. A solid financial concept provides figures on estimated revenues, capital expenditures and operational expenditures. Capital expenditures include exploration, insurance to cover exploration risk, drilling, hydraulic testing/stimulation, and surface installations. Operational expenditures include operating staff, monitoring, maintenance, consumables, electricity, and provisions for depreciation. The decision to realise a project involves the general cost structure and an analysis of the levelised cost of energy of a geothermal facility. The calculation of the levelised cost of energy is based on the total costs throughout the economic lifetime of a geothermal plant in relation to the provided energy. It is therefore a commonly applied approach to compare different energy provision technologies with each other and with the prices that are paid on the energy markets.

Finally, the feasibility study outlines the exploration phase and provides a general project schedule. Mobilising capital and securing governmental support are important steps at this stage. Geothermal projects are usually developed by municipal or private companies in Germany. Within the frame of the market incentive programme (MAP) the German government offers low-interest loans (soft loans) and redemption grants to companies and municipalities. Since 2006 private insurance companies provide policies covering the exploration risk. The first private insurance worldwide covering the exploration risk related to geothermal drilling has been procured for the facility at Unterhaching [13]. Such policies mitigate the risk for investors. The project planner proposes to insure against failure of the drilling and stimulation activities, and defines lower limits for reservoir temperature or flow rate (or both). It is also possible to arrange for partial compensation in case of sufficient but lower than expected flow rates or temperatures. The insurance rate depends on geological setting, reliability of available data, involved risk, coverage extent, insurance sum, project location, and capital costs. The approval of a geothermal drilling plan relies on independent expertise. Insurance companies frequently ask the LIAG to assess the probability of success of geothermal projects. The LIAG provided more than 50 expert reports for investors and insurance companies since 2002.

4.3. Exploration

When the feasibility study turned out to be promising, then the next step is the commissioning of a company for the project management. In Germany, it is highly recommended to organise information events for the local public beforehand in order to mitigate initial concerns about new infrastructure. Before exploration can begin, the project planner needs to apply for an exploration claim at the State mining authority. In order to obtain the exploration claim, it is required to prepare an operating plan and to inform the State mining authority on all intended exploration methods. In Germany, 3D seismic surveys are common practice for the exploration of geothermal resources in sedimentary settings. These 3D seismic surveys are conducted by specialised firms. Seismic data and lithologic well profiles are the best basis for the development of stratigraphic 3D models. Even the reprocessing of older 2D seismic data can provide valuable information. Seismic surveys may also reveal structural information on permeable joints in rock masses without bedding. The results of the seismic survey are also used to find the optimal drilling site and drill paths. The next step is the invitation to bid for the first wellbore. After its completion, hydraulic tests reveal if sufficient high yields can be achieved with a reasonable drawdown. Stimulation measures may help to improve the hydraulic properties of the well. Temperature measurements and a detailed analysis of the fluid chemistry provide further important data. They are needed for decisions on the right technology of the power plant, the dimensioning of the surface installations and the materials used. If the first borehole fails to meet the criteria outlined in the project objectives, the investors may stop the project or drill a side-track. They might also find a substitute geothermal project with lower requirements. If the project continues the project planner develops a work plan and applies for an exploitation permit at the State mining authority. The term is usually for 50 years but a prolongation is possible.

4.4. Field Development

The development phase starts with the invitation to bid for the second wellbore. The technique of directional drilling enables drilling from a single site, intersecting near vertical faults, penetrating the target formation in low angles, and achieving the optimal distance of the landing points of the production and injection well. If the distance in the reservoir is too small the fluid temperature cannot be maintained over the entire lifetime of the system. On the other hand, the distance in the reservoir should not be too large in order to maintain a sustainable yield of the production well.

Productions tests, stimulation measures if needed, and interference tests between the two boreholes complete the drilling and testing activities. The decision on which well to use for production and injection is based on temperature and hydraulic data. A higher temperature at the production well increases the energy output. A greater transmissibility is advantageous for the injection well because cooling a geothermal fluid increases its viscosity. The planning of production and reinjection strategies of a geothermal plant is supported by numerical reservoir models.

The next step comprises the construction of the geothermal plant, the installation of pipelines from the well heads to the geothermal plant, and the connection to a district heating network. The development of geothermal district heating networks is often accomplished in subsequent stages.

The commissioning is the final phase before the geothermal plant starts to supply heat and electricity to the clients on a regular basis. It can take several months to check all components, resolve technical problems, and to optimise operation.

5. Environmental Impact

There will be an initial environmental impact as the geothermal plant is set up. The construction of roads, pipelines and buildings requires energy related to transport and manufacturing of materials. The drilling, testing and stimulation of wellbores also requires energy and causes noise. Since geothermal wells are often drilled in urban areas, efficient noise protection walls are mandatory. The impact on the environment caused by drilling and construction works normally ends once all facilities are built.

The development of enhanced geothermal systems involves hydraulic stimulation which could cause seismicity. The high-pressure injection of water during stimulation lowers the threshold value for rupture by reducing the effective frictional resistance along fault planes. Within sedimentary settings, this induced seismicity is rather low and rarely causes irritation. However, hydraulic stimulation in crystalline bedrocks is more likely to cause sensible seismic events. The extent of surface effects and possible damages of a seismic event can be related to the ground velocity at the surface. The German standard DIN4150 considers a peak ground velocity of 5 mm/s as non-hazardous for almost all building categories.

Induced seismicity can also occur during operation of a geothermal plant. The largest seismic events have been reported for the geothermal plant at Landau. After two years of stable operation, two seismic events with magnitudes of 2.4 and 2.7 in August 2009 resulted in the obligation to reduce primary circulation although no significant damage occurred. Smaller induced seismicity has been reported for the geothermal plant at Unterhaching (Bavaria). Here, seismicity was probably related to the injection of water into a fault system. Induced seismicity resulting from thermal stress generated by re-injection of cooled water is subject to current research.

The amount of land necessary for the construction of a geothermal plant is generally small compared to other renewable energy technologies. Drilling geothermal wells requires space for a drilling pad large enough to allow the manoeuvring of the drilling rig and to store drill pipe, casing, and other equipment. Typically, drilling pads for deep geothermal projects require a total area of 4000–5000 m^2. After completion of drilling and site restoration the land use footprint depends on the type of geothermal facilities installed. A district heating station requires much less space than a combined heat and power production unit with air-cooled condensers.

6. Geothermal Installations

6.1. Overview of Geothermal Installations in Germany

Although only low and intermediate enthalpy resources are available in Germany, some regions are quite suitable for generating geothermal heat for direct use or even for power production. Germany uses its low enthalpy resources predominantly for spas, space heating, and district heating. Intermediate enthalpy resources are mainly used for district heating and in some cases even for power production. Figure 6 gives an overview on the location of geothermal facilities in Germany.

There are 23 facilities with an installed capacity above 1 MW which supply heat for district heating networks and/or produce electricity for the national grid (see Table 1). The map and the table have been derived from the GeotIS [1], which contains more detailed information and data on geothermal power plants, heating stations and spas in Germany that are in operation or under construction. For each installation, details such as installed capacity or mean power production are provided. Statistics on geothermal power and heat production are available on an annual basis.

Figure 6. Geothermal installations in Germany.

Legend:
- Installations with secondary use
- Installations without secondary use
- Electricity generation
- District heating
- Space heating
- Thermal spa
- Aquifer storage
- Other use

Currently, seven of the plants listed in Table 1 produce about 35.5 GWh/a of geothermal power with a total installed capacity of 27.1 MW (Figure 7). Compared to 2012, the capacity for power generation has more than doubled. In the case of heat production 18 of the plants mentioned above with an installed capacity of approximately 200 MW provide 530 GWh/a for district heating networks (in some cases combined with power generation). Together with 167 spas located predominantly in Southern Germany the total annual geothermal heat production in Germany sums up to about 920 GWh/a at an installed capacity of nearly 250 MW (Figure 8).

Table 1. Selected operating parameters of major geothermal installations in Germany. TVD: true vertical depth.

Location	Installed capacity (MW)		Production (GWh/a)		Wellhead temperature (°C)	Production rate (L/s)	Depth (m TVD)	Stratigraphy	Reporting year
	Power	Heat	Power	Heat					
North German Basin									
Neubrandenburg *	-	-	-	-	65–78	11–28	1268	Rhaetian	2011
Neustadt-Glewe	-	4.0	-	17.6	97	11–35	2450	Rhaetian	2012
Waren	-	1.3	-	3.1	63	17	1565	Rhaetian	2012
Upper Rhine Graben									
Bruchsal	0.4	-	1.2	-	123	24	2542	Middle Bunter	2013
Insheim	4.3	-	14.2	-	165	65	3800	Basement	2013
Landau (Palatinate)	3.0	5.0	13.2	3.0	159	40	3291	Bunter, Basement	2012
South German Molasse Basin									
Aschheim	-	9.8	-	48.7	85.4	39–75	2630	Malm	2013
Dürrnhaar **	5.5	-	-	-	ca. 135	ca. 130	4114	Malm	-
Erding	-	10.2	-	37.6	62–63	6–36	2359	Malm	2013
Garching	-	8.0	-	10.2	74	100	2226.3	Malm	2012
Kirchstockach **	5.5	-	-	-	138.8	ca. 120	3881.6	Malm	-
München (Riem)	-	12.0	-	48.0	94.5	35–85	2747	Malm	2011
Oberhaching	-	38.0	-	48.8	127.5	137	3755.2	Malm	2013
Poing	-	9.0	-	37.0	76.2	100	3049	Malm	2013
Pullach ***	-	11.5	-	45.7	104 and 80	55 and 23	3505	Malm	2013
Sauerlach **.***	5.0	4.0	-	-	ca. 140	110	4480	Malm	-
Simbach-Braunau	-	7.0	-	48.0	80.5	61.1	1941	Malm	2013
Straubing	-	2.1	-	2.9	36.5	17.5	824.8	Malm	2011
Traunreut **	-	7.0	-	-	ca. 108	50	4645.8	Malm	-
Unterföhring	-	9.5	-	29.0	86	ca. 75	1986	Malm	2012
Unterhaching	3.4	38.0	6.9	107.6	123.3	120	3590	Malm	2013
Unterschleißheim	-	8.0	-	40.1	78	65–93.3	1960	Malm	2013
Waldkraiburg	-	16.4	-	2.8	109	20	2718	Malm	2012
Total	27.1	200.7	35.5	530.1	-	-	-	-	-

* heat storage facility; ** recent commissioning; and *** triplet.

Figure 7. Geothermal power production in Germany.

Power production

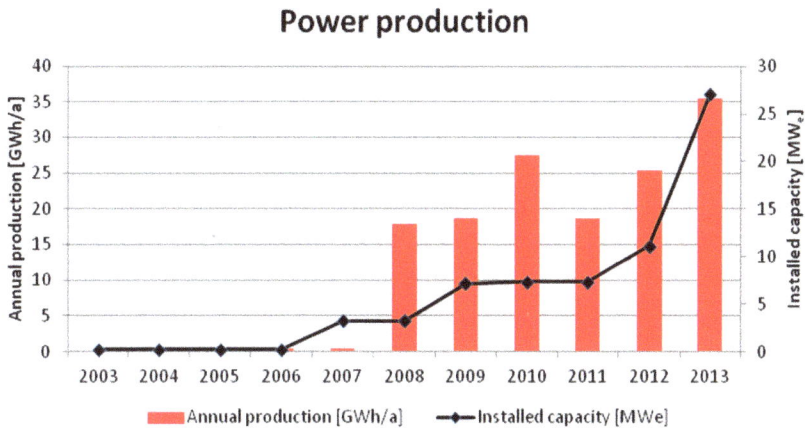

Figure 8. Geothermal direct use (spas and district heating) in Germany.

Direct heat use

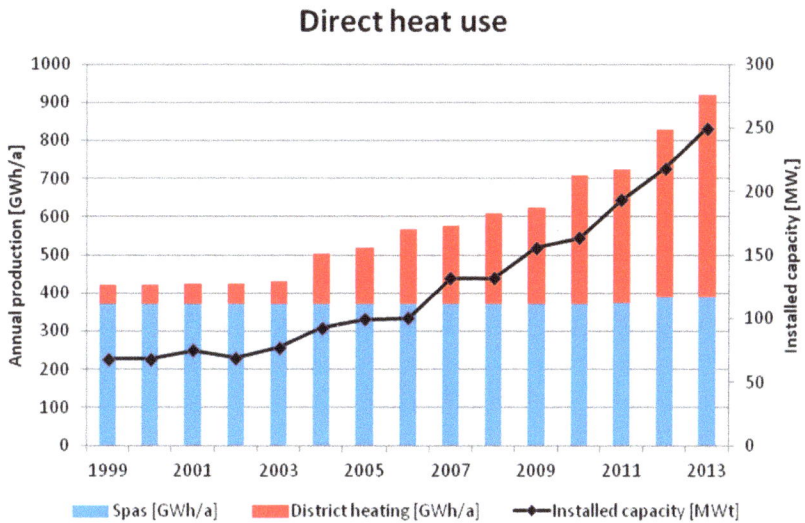

6.2. Geothermal Installations in Northern Germany

In this part of Germany thick Mesozoic sandstone formations represent the most important aquifers for geothermal utilisation. These include sandstones of the Lower Cretaceous, Middle Jurassic as well as reservoirs in the Rhaetian and Liassic sandstones and the Middle Bunter. These formations are widespread and show, especially in north-eastern Germany, favourable properties regarding porosity, permeability and temperature. Besides, the Rotliegend sandstones and the Schilfsandstein of the Middle Keuper are considered as locally promising target horizons for geothermal exploitation [14–17].

The most important geothermal projects in north-eastern Germany include the heating plants in Waren and Neustadt-Glewe as well as the heat storage facility in Neubrandenburg which exploit the sandstones of the Rhaetian/Liassic aquifer complex.

6.3. Geothermal Installations in the Upper Rhine Graben

The geological formations of the Upper Muschelkalk and the Bunter show the highest potential for hydrogeothermal utilisation in the Upper Rhine Graben. In addition, the Hauptrogenstein (Middle Jurassic) located in the southern part of the graben and a Tertiary sequence in the northern part provide suitable conditions for geothermal exploitation. A relatively high temperature gradient indicates the upwelling of deep waters (with high salinity) along fault zones and fissures. This advantageous condition allows the generation of power at all of the larger geothermal facilities located in the Upper Rhine Graben (Bruchsal, Insheim, and Landau (Palatinate); compare Table 1).

6.4. Geothermal Installations in the South German Molasse Basin

The Malm (Upper Jurassic) represents by far the most important water-bearing horizon in southern Germany. Due to the fact that the high permeability of the partially karstified Malm limestones and dolomites allows high production rates (>100 L/s) from depths exceeding 4000 m this geological formation is today the main target for geothermal exploitation in the South German Molasse Basin. The temperature of the usually low mineralised waters is bound to the depth of the Malm horizon which crops out in the north in the Swabian and Franconian Alb but declines to depths below 5000 m when approaching the Alps [18]. For this reason geothermal power generation using thermal water with temperatures up to 140 °C is only possible to the south of Munich whereas farther to the north the main utilisation is aimed at supplying energy for district heating networks. Because of the geological conditions and the proximity to a large number of potential consumers, the area around Munich is one of the best developing regions of Germany in terms of geothermal utilisation. The Pullach geothermal plant has been extended from a doublet to a triplet with a reversal of an injection well to a production well. With the recent commissioning of Dürrnhaar, Kirchstockach, Sauerlach (triplet) and Traunreut the number of operating geothermal facilities in Bavaria amounts to 17 plants whereof four are generating geothermal power (see Table 1). Further projects are already in the stage of construction.

Despite the high density of geothermal installations around Munich there is almost no mutual interference between neighbouring installations. Dussel *et al.* [19] simulated hydraulic and thermal effects in the reservoir for the proposed lifetime of 50 years on the basis of a numeric thermo-hydraulic 3D model. They predict temperature changes after 50 years of operation in the vicinity of injection wells, only. Hydraulic interferences are generally low. The maximum effect between to installations has been estimated with up to 0.3 MPa.

7. Summary and Outlook

Presently, there are 180 geothermal direct-use installations in operation. The installations comprise district heating, space heating in some cases combined with greenhouses and thermal spas. Most of the district heating plants are located in the Bavarian part of the Molasse Basin. From 2003 to 2013, the annual power production increased from 0 GWh to 36 GWh. At the end of 2013, geothermal power generation in Germany reached an installed capacity of 27.1 MW$_e$. However, most geothermal energy is used for heating. From 2003 to 2013, the annual production of geothermal district heating stations increased from 60 GWh to 530 GWh. In 2013, the total installed capacity for geothermal heat production reached 250 MW. Buildings are responsible for about 40% of final energy consumption in Germany. There is still an enormous potential for geothermal direct use installations. Deep geothermal energy accounts for 0.62‰ of total heat supply and 6.92‰ of heat supply from renewable energy sources. Geothermal power production is growing rapidly but on a very small level. It merely accounts for 0.06‰ of total power production and 0.23‰ of green power production in Germany.

Ganz *et al.* [20] expect that due to an exciting new project development in Southern Germany, the installed capacity of geothermal power production will exceed 50 MW$_e$ by 2015 and the installed capacity of deep geothermal heat use may reach 300 MW$_t$ in 2015 with an annual heat production of about 1075 GWh. Information on geothermal energy use in other countries can be found in Ganz and Schulz [21] and Ganz *et al.* [22].

The GeotIS provides up-to-date figures on geothermal facilities and annual geothermal energy statistics for Germany, which are used for national energy statistics on renewable energy, as well as in international reports [22].

Petrothermal energy is of high potential in Germany and could play an important role in German energy production in the long term. A German petrothermal system is currently developed in Saxony. The LIAG has performed a 3D seismic survey in a 10 km × 12 km area in the year 2012 to explore a petrothermal reservoir in a granitic body located in the Erzgebirge. The objective is to develop a detailed 3D model of the subsurface down to a depth of 6 km and to identify and characterise potentially permeable fracture zones as targets for a geothermal borehole using a petrothermal system.

A detailed geological site assessment is still essential when planning a new geothermal facility. As part of this assessment, a lot of geological data, hydraulic data, and subsurface temperatures can be retrieved from GeotIS. The public system offers a wide range of tools to visualise key parameters in regional contexts. It helps to identify promising locations for geothermal projects and provides relevant information and data for preliminary and feasibility studies. However, it is important to note that GeotIS does not replace local feasibility studies.

Acknowledgments

This paper has greatly benefited from data and information provided by GeotIS, which was funded by the Federal Ministry for the Environment, Nature Conservation and Nuclear Safety and is currently funded by the Federal Ministry for Economic Affairs and Energy under project number 0325623A. The authors are grateful to Peter Bayer and two anonymous reviewers for useful comments and suggestions which helped to improve this paper.

Author Contributions

Thorsten Agemar outlined the content and structure of the paper. He prepared Sections 1, 4 and 5. He also contributed to Section 3. Josef Weber is responsible for the energy statistics and provided Sections 2 and 6. Rüdiger Schulz had the initial idea for this manuscript and contributed to all sections, especially to Section 3. Thorsten Agemar and Josef Weber have carried out literature review and finalized the manuscript.

Conflicts of Interest

The authors declare no conflict of interest.

References

1. Geothermal Information System for Germany. Available online: http://www.geotis.de (accessed on 1 July 2014).
2. *VDI-Richtlinie 4640 (Blatt 1): Thermal Use of the Underground Fundamentals, Approvals, Environmental Aspects*; Gesellschaft Energie und Umwelt (VDI): Berlin, Germany, 2010; p. 33.
3. Agemar, T.; Alten, J.-A.; Ganz, B.; Kuder, J.; Kühne, K.; Schumacher, S.; Schulz, R. The geothermal information system for Germany—GeoTIS. *Z. Dt. Ges. Geowiss.* **2014**, *165*, 129–144.
4. Agemar, T.; Schellschmidt, R.; Schulz, R. Subsurface temperature distribution in Germany. *Geothermics* **2012**, *44*, 65–77.
5. Suchi, E.; Dittmann, J.; Knopf, S.; Müller, C.; Schulz, R. Geothermal Atlas to visualise potential conflicts of interest between CO_2 storage (CCS) and deep geothermal energy in Germany. *Z. Dt. Ges. Geowiss.* **2014**, in press.
6. Paschen, H.; Oertel, D.; Grünwald, R. *Möglichkeiten der Geothermischen Stromerzeugung in Deutschland*; Sachstandsbericht, Arbeitsbericht 84; Büro für Technikfolgen-Abschätzung beim Deutschen Bundestag (TAB): Berlin, Germany, 2004. (in German)
7. Lüschen, E.; Schulz, R. 3-D seismic surveys explore German petrothermal reserves. *EOS Trans. Am. Geophys. Union* **2014**, *95*, 237–244.
8. The Global Heat Flow Database of the International Heat Flow Commission. Available online: http://www.heatflow.und.edu/data.html (accessed on 24 April 2014).
9. Hermanrud, C.; Cao, S.; Lerche, I. Estimates of virgin rock temperature derived from BHT measurements: Bias and error. *Geophysics* **1990**, *55*, 924–931.
10. Förster, A. Analysis of borehole temperature data in the Northeast German Basin: Continuous logs *versus* bottom-hole temperatures. *Pet. Geosci.* **2001**, *7*, 241–254.
11. Wolfgramm, M.; Franz, M.; Agemar, T. Explorationsstrategie Tiefer Geothermischer Ressourcen am Beispiel des Norddeutschen Beckens. In *Handbuch Tiefe Geothermie*; Bauer, M., Freeden, W., Jacobi, H., Neu, T., Eds.; Springer Spektrum: Heidelberg, Germany, 2014; pp. 451–493. (in German)
12. Wolfgramm, M.; Seibt, A. Zusammensetzung von Tiefenwässern in Deutschland und ihre Relevanz für Geothermische Anlagen. In Proceedings of the Der Geothermiekongress 2008 Conference, Karlsruhe, Germany, 11–13 November 2008; pp. 503–516. (in German)
13. Schulz, R.; Pester, S.; Schellschmidt, R.; Rüdiger, T. Quantification of Exploration Risks as Basis for Insurance Contracts. In Proceedings of the World Geothermal Congress, Bali, Indonesia, 25–30 April 2010.

14. Katzung, G. *Geothermie-Atlas der Deutschen Demokratischen Republik*; Zentrales Geologisches Institut (ZGI): Berlin, Germany, 1984. (in German)

15. Feldrappe, H.; Obst, K.; Wolfgramm, M. Die mesozoischen Sandstein-Aquifere des Norddeutschen Beckens und ihr Potential für die geothermische Nutzung (in German). *Geol. Wiss.* **2008**, *36*, 199–222.

16. Obst, K.; Brandes, J.; Feldrappe, H.; Iffland, J. *Geologische Karte von Mecklenburg-Vorpommern, Übersichtskarte 1:500.000—Nutzhorizonte des Rhät/Lias-Aquiferkomplexes*; LUNG Mecklenburg-Vorpommern: Güstrow, Germany, 2009. (in German)

17. Obst, K.; Brandes, J. *Geologische Karte von Mecklenburg-Vorpommern, Übersichtskarte 1:500,000—Nutzhorizonte im Mittleren Buntsandstein*; LUNG Mecklenburg-Vorpommern: Güstrow, Germany, 2011. (in German)

18. Stober, I.; Fritzer, T.; Obst, K.; Schulz, R. *Tiefe Geothermie—Nutzungsmöglichkeiten in Deutschland*, 3rd ed.; Federal German Ministry for the Environment, Nature Conservation and Nuclear Safety: Berlin, Germany, 2011.

19. Dussel, M.; Lüschen, E.; Thomas, R.; Agemar, T.; Fritzer, T.; Sieblitz, S.; Huber, B.; Bartels, J.; Wolfgramm, M.; Wenderoth, F.; *et al.* Development of a Thermo-Hydraulic 3D Model of the Deep Carbonatic Malm Aquifer in the Munich Region (Germany) with Special Emphasis on a 3D Seismic Survey. In Proceedings of the European Geothermal Congress, Pisa, Italy, 3–7 June 2013.

20. Ganz, B.; Schellschmidt, R.; Schulz, R.; Sanner, B. Geothermal Energy Use in Germany. In Proceedings of the European Geothermal Congress, Pisa, Italy, 3–7 June 2013.

21. Ganz, B.; Schulz, R. The GIA Trend Report, a New Survey Report about Geothermal Applications and Developments with Trends in Power Generation and Heat Use in IEA-GIA Member Countries. In Proceedings of the European Geothermal Congress, Pisa, Italy, 3–7 June 2013.

22. Ganz, B.; Bendall, B.; Bromley, C.; Busby, J.; de Gregorio, M.; Ketilsson, J.; Kumazaki, N.; López Ocón, C.; Minder, R.; Muller, J.; *et al. Trends in Geothermal Applications*; Survey Report on Geothermal Utilization and Development in IEA-GIA Member Countries in 2011, with Trends in Geothermal Power Generation and Heat Use 2000–2011; International Energy Agency (IEA)-Geothermal Implementing Agreement: Taupo, New Zealand, 2013.

Differences in Public Perceptions of Geothermal Energy Technology in Australia

Simone Carr-Cornish and Lygia Romanach

Abstract: In Australia, geothermal energy technology is still considered an emerging technology for energy generation. Like other emerging energy technologies, how the public perceive the technology and under what conditions they are likely to accept or oppose the technology, remains relatively unknown. In response, this exploratory research utilised online focus groups to identify: (1) the extent of agreement with geothermal technology before and after information, including media reports focusing on a range of the technology's attributes; and (2) how the characteristics of individuals with different levels of agreement vary. After information, within the sample of 101 participants, fewer reported being unsure, the minority disagreed and the majority agreed. Overall, the preference was for projects to be located away from communities. Participants that disagreed or were unsure, were more likely to report lower subjective knowledge of the technology, lower perceived benefits and higher risks, and were less likely to believe people in their community would have the opportunity to participate in consultation. These characteristics suggest there are advances to be made by analyzing what contributes to different levels of acceptance. The findings also suggest that the location of projects will be an important consideration and that the conditions of acceptance are likely to vary amongst community members.

Reprinted from *Energies*. Cite as: Carr-Cornish, S.; Romanach, L. Differences in Public Perceptions of Geothermal Energy Technology in Australia. *Energies* **2014**, *7*, 1555-1575.

1. Geothermal as an Emerging Energy Technology in Australia

Over the last two decades, there has been increasing interest in the mitigation of climate change through the use of low emission energy technologies. However, like many new and emerging technologies, there has also been considerable public opposition to many of the low emission energy technologies being developed [1]. Apart from well known opposition to nuclear power plants [2], substantial opposition has been expressed in relation to wind farms [3], carbon capture and storage technology [4] and geothermal energy projects [5]. Furthermore, there has been increasing recognition by policy-makers and technology developers that not only a lack of opposition but societal acceptance [6] and support at the local level [1] are essential for successful deployment. Geothermal energy technology is an emerging energy technology in Australia and therefore it is relatively unknown how the Australian public is likely to respond to the proposed development of this technology in Australia.

Additionally, Australia has substantial geothermal resources however, there has been limited demonstration of how these resources can be used for power generation [7]. To date, Australia's only geothermal power plant is one that has serviced the remote town of Birdsville, Queensland since 1992, providing approximately 25% of the town's electricity (80 kW) [8]. However, Australia also has some relatively unknown direct-use applications. For example, Perth, Western Australia, has several school and community pools heated by the Perth Basin resource. Tourist spas have also been established, including a facility accessing the Peninsula Hot Springs in Victoria, and another in Queensland that draws from the Great Artesian Basin. More recently, Australian companies have been developing Enhanced Geothermal Systems for larger-scale power generation. These systems consist of hot rocks and the addition of water and/or permeability to extract the heat. With this development, the federal government's Bureau of Resources and Energy Economics forecasts that the use of geothermal resources for energy generation may increase to 8% of total electricity generation in Australia by 2050 [9].

Due to its limited use in Australia there has been very little research to understand the extent to which Australians agree with the use of geothermal energy technology and under what conditions the technology would be accepted. In 2011, the Commonwealth Scientific and Industrial Research Organisation conducted a survey of 1907 Australians which found that just over one-quarter of respondents (27%) reported no knowledge of geothermal, with only 38% of respondents reporting their knowledge as moderate to high [10]. Although self-reported knowledge was low, over half of respondents (57%) agreed with the use of geothermal energy technology in Australia, though 31% reported to be unsure. To contrast, in 2011 a survey of Canadians ($n = 1548$) found 65% had heard of geothermal [11], whereas in a 2007 Eurobarometer survey found that only 44% of Europeans were aware of geothermal as an alternative energy source [12].

A recent report published on the public acceptance of geothermal electricity production in Europe [13] offered practical insights. The insights were drawn from the German media and six case studies of perceptions of projects (one from France, one from Italy and four from Germany). The analysis of media suggested four critical sources of social resistance: environmental issues, missing-involvement (engagement) issues, financial issues and the NIMBYism-syndrome. The reactions of stakeholders reported on by each case study were mostly favourable. However, the majority of the cases demonstrated that the citizens of nearby communities often had limited knowledge of geothermal technology or the specific project. Depending on the project this was due to different factors, for example, limited communication efforts or that communications failed to reach the majority of the citizens. In such cases the lack of involvement contributed to uncertainty about the project, negative perceptions and even opposition. Additional factors of influence included local economic and political factors and experiencing seismicity.

Also emerging from CSIRO's research program were responses collected from workshops with members of the public in Australian capital cities during 2008 to 2009 [14]. Again, participants rated themselves as having low levels of knowledge of geothermal energy technology. Despite the low levels of knowledge and concerns about water usage and seismic activity, participants were overall supportive of using geothermal energy technology in Australia, partly due to identifying geothermal as a renewable energy source. Similarly, both acceptance and concern have been evident in how projects have been received to date. For example, direct-use applications, such as heated pools and day spas, have received little attention, however, intense community concern was reported in relation to a proposed geothermal power plant near Geelong in Victoria [15].

To understand how the technology will be perceived as its profile increases in Australia and what factors will potentially characterise acceptance or concern, this research aimed to identify:

(1) the extent of agreement with geothermal technology before and after information, including media reports focusing on a range of the technology's attributes; and (2) how the characteristics of individuals with different levels of agreement vary. The provided information, included sources that are readily available on the internet, such as media reports on the technology. Based on existing research of societal acceptance of low emission energy technologies, it was anticipated that acceptance was likely to differ depending on an individual's knowledge of the technology [16], how individuals perceive the benefits and risks of such technology [17], the location of energy projects, local community views of such projects [18], and individuals' demographics [19].

1.1. Knowledge of the Technology

The role of knowledge is widely investigated in technology acceptance studies with previous research indicating that subjective and objective knowledge about an energy technology can have different effects on technology acceptance [16]. While some studies, considered in a review by Huijts *et al.* [16] reported positive interactions between objective knowledge and acceptance in the context of carbon capture and storage (e.g., [20]), hydrogen technology (e.g., [21,22]), Ellis *et al* [23] found little evidence of a relationship between objective knowledge of wind power and its acceptance.

Other research has found technology acceptance to be influenced by subjective and not objective knowledge, for both genetically modified foods [24] and rooftop solar photovoltaic systems [25]. Such findings highlight that acceptance is informed by more than objective facts about the technology. In addition, individuals will not necessarily develop an in-depth understanding of every technology and other factors, such as benefit and risks perceptions, are likely to play a key role in technology acceptance of energy projects.

1.2. Perceived Benefits and Risks of the Technology

Perceptions of benefits and risks of a technology are critical to the public's support of a technology's implementation [17]. Previous research has shown that individuals are more likely to support a technology when they perceive that the benefits of such technology outweigh the risks [16,17] and this has been demonstrated in the context of controversial technology such as nanotechnology, biotechnology and stem cell research [26]. Furthermore, emerging technologies are often inherently uncertain and therefore their associated risks can be largely unknown [27].

Risk perception is a social construct and thus previous research has shown that individuals are likely to react to hazards differently, with each individual's characteristics or specific technical knowledge resulting in quite different judgments about benefits and risks [27,28]. Likewise, the perceptions of the benefits of energy technologies depend on factors such as individuals' level of trust in institutions [16], their subjective knowledge, values and beliefs [25], and therefore individuals might perceive the same benefits differently. For example, a previous study about societal acceptance of carbon capture and storage found that while some people evaluated the benefits as being greater than the risks, others evaluated those same risks as being greater than the benefits [29]. Similarly, Cacciatore *et al.* [30] found that individuals that could make a link between the concept of nanotechnology and examples of how the technology could be applied were more likely to take risks into account.

Energy technologies also present a complex combination of benefits and risks at both the local and global level, for example, there is disparity between local risks and global benefits for nuclear energy [31] and carbon capture and storage [29]. Previous research into public perceptions of geothermal energy in Australia has shown similar results, as concerns about geothermal energy technology are mostly local: water usage and seismic activity instigated by geothermal drilling [14]. Whereas benefits of geothermal energy technology commonly identified by the community, such as low emission energy have an effect at a larger scale, either nationally or globally.

1.3. Project Location and Procedural Justice

Historically, geothermal energy projects have demonstrated that closeness to the end-use application is critical to achieving efficient heat transfer when projects are intended for direct-use or combined direct-use and electricity generation. Research into societal acceptance of energy projects has suggested that the location of energy technology demonstrations can impact acceptance, with concerns that such projects might threaten the locality or its safety [18]. In the context of renewable energy projects, this concept is usually referred as the "Not In My Backyard" (NIMBYism) phenomenon. The assumption underlying NIMBYism is that individuals hold more negative attitudes when the project is proposed for their local residential area than they would normally [18]. However, further research has suggested that opposition towards a particular technology is broader than proximity alone and involves a "range of social and personal factors affecting human interactions with social and political institutions" [32]. As argued by Devine-Wright [18], a multidimensional framework concept is required to understand how contextual, social, economical and personal factors shape public perceptions of energy technologies.

Research has also shown that it is important to engage with the public in the early stages of technology development [29,33]. Early engagement with the community, through a variety of mechanisms has emerged as the best approach to facilitate meaningful participation, to empower the community and to build trust of the institutions deploying the technology. An effective engagement process happens when the decision process is judged to be fair, also known as procedural justice [34]. For example, in the context of carbon capture and storage, Bradbury et al. [35] found that the public was concerned about whether the project implementation would be fair and transparent, including having mechanisms for voicing concerns.

1.4. Demographics

Several studies have reported demographics interacting with the acceptance of energy technologies. For example Carr-Cornish et al. [36] segmented a sample of responses from an Australian population and found that the segment that preferred renewable energy were more likely to consist of individuals that were female and have low to moderate household incomes. Whereas individuals in the segment that supported a range of technologies, were more likely to be male, tertiary educated, middle aged, with moderate to high household incomes. A study that reported explicitly on geothermal, found that gender impacted support for a geothermal facility on the Greek island of Nisyros, where women were less likely to support the technology than men [19]. In addition, previous studies have indicated that women tend to show more concern with the risks associated with technologies than men [37].

2. Method

The exploratory research questions of this study were addressed using online focus groups and a mixed methods approach. The mixed method approach afforded both in-depth explorations of participants' perceptions through typed dialogue, as well as questionnaires which allowed comparison of participants' responses [38]. Online focus groups were utilised because, as with offline focus groups, discussion is immediate, free-flowing and allows for affect—it also allowed participants to attend who may not have been able to travel to a physical focus group setting [39]. The sample could also be accessed in a timely manner and a complete record of the discussion data was immediately available for analysis.

2.1. Sample

A total of 136 individuals participated in the online focus groups, combined these participants had similar age, gender and location characteristics to that of the Australian population. However, the sample reported on in this paper consisted of the 101 participants that completed at least 95% of both pre- and post-questionnaires. The remaining sample had characteristics which were consistent with the Australian population, providing a relatively representative sample, although some demographics were over- or under-represented [40] and the data collection method was biased toward internet users. For example, participants ranged from 20 to 68 years, with a mean age of 43 years ($SD = 12.91$) higher than the national median of 37 years. The Australian population consists of 49% males and 51% females, and the same proportion of male and females were sampled. While participants reported a range of education levels, 35% of the survey sample had a bachelor/honours degree, compared to14% of the Australian population [40]. Similar to the Australian population, participants were from a range of employment situations with 39% of participants employed full-time (40%; [40]). Participants' incomes ranged from less than $20,000 to $150,000 or more, and the median was $60,000 to $79,999, which is slightly higher than the national average of $58,375. At least one participant was from each of Australia's states and territories except Western Australia. The distribution was similar to the population; 36% of participants were from New South Wales, 29% from Victoria, 25% from Queensland, 4% from South Australia, 3% from Tasmania, and 1% respectively from the Australian Capital Territory and the Northern Territory.

2.2. Material

The online focus groups were conducted using an online qualitative research (OQR) platform, called Revelation│Next [41]. At the start and end of the online focus groups participants were asked to complete a questionnaire which included measures such as agreement with the use of geothermal energy technology in Australia; self-rated knowledge of geothermal; perceptions of the benefits and risks of geothermal energy; preferences for project location and procedural justice; and demographic information—specifically age, gender, education, employment and income. Amongst other institutions, trust in the research organisation (CSIRO) conducting the study was also measured to provide insight into participants' perceptions of the research process.

Agreement with the technology was measured by asking "Please select the option that best matches how strongly you agree or disagree with using the following energy sources and technologies in Australia". A range of technologies were listed and responses to geothermal were reported in this study. Participants could respond from 1—strongly disagree, 3—neither disagree nor agree, to 5—strongly agree and "I have no idea". This measure was adapted from a survey by Hobman *et al.* [10] of the Australian public's preference for energy sources and related technologies. To compare participants with different attitudes toward the use of geothermal energy technology, the attitude measure was re-coded into three attitude groups: "Disagree", "Unsure" and "Agree". The "Disagree" group included participants with ratings of 1—strongly disagree and 2—disagree. The "Unsure" group included participants with ratings of 3—neither disagree nor agree or "I have no idea". The "Agree" group included participants with ratings of 4—agree and 5—stongly agree.

Self-rated knowledge was measured by asking participants to "Please rate your knowledge from 1—no knowledge to 5—high knowledge of the following energy sources and technologies in Australia". Again a range of technologies were listed and responses to geothermal were reported in this study. The self-rated measure of knowledge was also adapted from Hobman *et al.* [10].

To measure benefits and risks, participants were asked to rate their agreement with five benefit statements and four risk statements. Specifically participants were asked: "Please select the option that best matches how strongly you agree or disagree that the development of geothermal resources in Australia will": e.g., "Reduce greenhouse gas emissions" and "Induce earthquakes". Responses could range from 1—strongly disagree, 3—neither disagree nor agree, to 5—strongly agree. This question was informed by research of the actual benefits and risks of geothermal energy technology [7] and followed methodology previously used in risk assessment research [17].

Two questions were asked to measure participants' preferences regarding project location. At both the start and finish a question was asked of the distance projects should be from built-up areas. Participants were asked: "Please indicate the distance a geothermal project should be from built-up areas in your community (e.g., houses, businesses)". The response options were: less than 1km, at least 1 km, at least 5 km, at least 10 km, at least 50 km, at least 100 km and greater than 100 km. Only in the end questionnaire participants were asked about how concerned they would be if a project was proposed for their community. Participants were specifically asked: "On a scale of 1 (not at all concerned) to 5 (very much concerned) please select the number that best matches how you would feel if a geothermal project was proposed within 1km of built areas in your local community". These measures were adapted from the Special Eurobarometer 364 [42] on public awareness and acceptance of carbon capture and storage.

In the final questionnaire participants were also asked about procedural justice, the extent they believed they could participate in decisions about a project. They were asked: "Please rate from 1 (not at all) to 5 (very much) the extent to which you believe people in your community would have the opportunity to participate in decisions about geothermal energy projects". Also, to measure trust in a range of institutions, including CSIRO, participants were asked: "Please rate from 1 (not at all) to 5 (very much) the extent to which you trust [institution specified] to provide honest information about geothermal resources and projects in Australia".

Between completing the pre- and post-questionnaires participants were asked to view and discuss their reaction to four pieces of information about geothermal energy technology that are currently available on the internet. A geothermal energy researcher with industry experience, assisted with the identification of relevant materials. Participants were first presented with an overview of the technology and Australia's industry. This was done by providing participants with a CSIRO authored factsheet [43] about the technology and a YouTube video recording of a 7 min TV news segment that aired on the Australian Broadcasting Commission (ABC) in 2011 [44]. Following this, participants were presented with two articles, one that was positively framed and focused on the potential for enhanced geothermal systems and a second that was more negatively framed and focused on the hydraulic fracturing aspect of enhanced geothermal systems. The predominately positive news article was "Enhanced geothermal systems: Have a little faith" and published on the renewable energy news website, Renewable EnergyWorld.com [45]. The more negatively framed article was "France fractured by fracking-like geothermal projects" and published online in The Age [46].

2.3. Procedure

A market research firm was used to recruit participants for the nine online focus groups. Each focus group was moderated by one of two CSIRO researchers and the initial group was moderated by both to ensure consistency of approach. All groups followed the same procedure though the order in which the positive and negative news articles were presented varied to negate the news articles having a primacy or recency effect on responses. From the 101 participants that completed the questionnaires, 60 viewed the positive article first and 41 viewed the negative article first. The groups commenced when participants responded to their email invitation to log into the Revelation|Next platform. To maintain their privacy they used only their first name or an alias. The groups ran for approximately 2 h in which participants completed seven activities. First participants completed the questionnaires as presented in Table 1 followed by Activity 2 which was a written discussion of their awareness of geothermal energy. Activity 3 to 5 all involved the provision of information and discussion. Activity 6 was the final discussion, which was followed by a questionnaire.

Table 1. Online focus group activities, including information provided and discussion prompts.

Activity schedule	Purpose	Questions
1. Pre-questionnaire	Collect quantitative data on existing attitudes and demographics	See Section 2.2. Materials
2. Initial group discussion	Collect qualitative data on existing attitudes towards geothermal energy	Please write down what comes to your mind when you think about geothermal energy
3. What is geothermal energy?	Provide information about technology. Participants were asked to read the CSIRO factsheet and view the ABC news segment	Facilitated discussion. - Was this information new to you? - Was this information clear? - What particular points are most relevant to you?
4/5. Geothermal projects 2012 article 5/4. Geothermal projects 2013 article	Provide information on the risks and benefits of geothermal technology presented in media articles. One article was positively framed and one article was negatively framed	Facilitated discussion. - What did you think about the information presented in the article? - What do you think the project and technology discussion in that article?
6. What do you think?	Explore participants' overall opinions and attitudes towards geothermal technology	Facilitated discussion. Given the information you have been provided on geothermal technology, the industry in Australia and examples from around world, overall: (a) What do you think about Australia's effort to date to develop geothermal resources? (b) What would be important to you if geothermal projects are implemented across Australia?
7. Post-questionnaire	Collect quantitative data on participants attitudes after provision of information	See Section 2.2. Materials

All of the questionnaire responses and discussion interactions were collected securely online. The questionnaire responses were analysed using the Statistical Package for the Social Sciences (SPSS) version 20. Descriptive statistics were used to identify the distribution of responses. To compare responses collected at the start with those at the end, two-tailed paired sample t-tests were used. Analysis of variance (ANOVA) was used to compare the responses of each attitude group ("Disagree", "Unsure", "Agree") on continuous variables and cross-tabulations with Pearson's chi-squared tests were used to make comparisons on categorical variables. Differences were considered significant at $p < 0.05$. Qualitative data from the facilitated discussion were thematically coded using NVivo 10, a form of Computer Assisted Qualitative Data Analysis Software.

3. Results

Overall the questionnaire responses confirmed that there was considerable agreement with the use of geothermal energy technology in Australia. On the recoded scale of 1—disagree to 3—agree, the mean measure of agreement was high at both the start, 2.58 (SD = 0.50) and end, 2.67 (SD = 0.60), to the extent there was no significant difference, $t(99)$ = −1.38, p = 0.171.

Also, the perceptions of those that viewed the positive article first were similar to those that viewed the negative article first. For example, at the start the mean agreement of these groups was only marginally significant, $t(83)$ = 2.034, p = 0.045 and at the end the difference was definitely not significant, $t(64)$ = 1.855, p = 0.068. Additionally there was no statistically significant difference in the mean change of participants that viewed the positive article first, $t(59)$ = −1.230, p = 0.224 or of those that viewed the negative article first, $t(39)$ = −0.684, p = 0.498. Trust in CSIRO increased significantly during the process from 3.94 (SD = 0.952) at the start to 4.18 (SD = 0.833), $t(99)$ = −3.129, p < 0.05. This increase suggests that providing both negative and positive information from media sources did not have an adverse effect on the trust participants had in the organisation conducting the research.

Although there was no significant change in mean ratings of participants' agreement with the technology, the proportion of participants that either disagreed, were unsure or agreed with the technology did significantly change [$\chi^2(2, N = 100) = 10.71, p = 0.005$]. Shown in Table 2 is the percentage of participants that reported disagree, unsure and agree at both the start (last column) and end (last row). At the start, participants were either unsure of the technology or agreed with the technology, however, by the end there was more variation in responses; a small percentage of participants disagreed, fewer participants were unsure, and more agreed.

Table 2. Attitudes toward geothermal energy technology being used in Australia.

Start	End			
	Disagree	Unsure	Agree	Total
Disagree	–	–	–	–
Unsure	5% (5)	13% (13)	24% (24)	42% (42)
Agree	2% (2)	6% (6)	50% (50)	58% (58)
Total	7% (7)	19% (19)	74% (74)	100% (100)

Consistent with participants' attitudes that were measured through the questionnaire, qualitative responses collected in the last facilitated discussion (Activity 6), confirmed that although mean agreement was high, participants expressed a range of agreement levels with use of the technology in Australia. For example, the following quote reflects strong agreement:

I'm pleased that Australia is making some effort to explore and develop its geothermal resources, but I'd like to see more action given our enormous potential for energy derived through geothermal technology. I sense we are somewhat lagging other parts of the world in this respect, which is disappointing.

The following quotes demonstrate expressions of agreement, but with conditions, such as safety, no hydraulic fracturing and measured funding:

I think that Australia can play a significant role in the development of geothermal energy and should continue to do so if it can be proven to be 'safe'. To me, I am very happy to see geothermal projects in Australia, but importantly for me, we must not implement fracking or anything similar. Geological disturbance must be avoided; we simply don't know the potential long term consequences on geological disturbance.

It should be slowly funded as it is until we know we are not throwing money down a hole.

Some participants indicated they did not have enough information to form a judgment and that they perceived the need for more research:

I struggle with this right now, simply because we have incomplete information. If this were to be implemented here in Australia, I would want to know that there was going to be no disastrous or potentially disastrous results. In theory it's a great idea, and one that appears to be a long term supply, but none of that will matter if we end up with earthquakes and poisoned water!

Expressions of disagreement emphasised the need for more research and also alternative technologies:

I think a lot more testing, scientific discussion and research is required before Australia can step into this kind of technology. I feel there are several easier, sustainable and less dangerous alternatives to geothermal energy systems available to us at the moment. Let's utilise these options first.

3.1. Self-Rated Knowledge

Figure 1 shows the mean self-ratings of knowledge for each attitude group, both at the start and end of the processes. Overall the mean rating significantly changed from the start to the end, $t(97) = -16.440, p < 0.001$. The mean at the start was low, 1.89 ($SD = 0.93$) and at the end the mean was moderate, 3.65 ($SD = 0.79$). The mean self-ratings were significantly higher for participants that agreed with using the technology in Australia compared to those that were either unsure or disagreed, at both the start, $F(1, 99) = 25.971, p < 0.001$, and end, $F(2, 95) = 5.012, p < 0.05$.

Similar to the questionnaire responses, the qualitative responses collected in the first facilitated discussion (Activity 2) confirmed that the majority of participants stated limited to moderate awareness of the technology, especially in Australia. The quotes below reflect a participant expressing low awareness and another expressing moderate awareness:

Until the opportunity has opened for this discussion, I wasn't aware that Australia was involved in any projects to develop any business opportunities or were conducting public company business.

This is a subject that I have only heard about in the past two or three years, so I'm still learning the intricacies about it.

Figure 1. Mean self-ratings of knowledge of geothermal by attitude group.

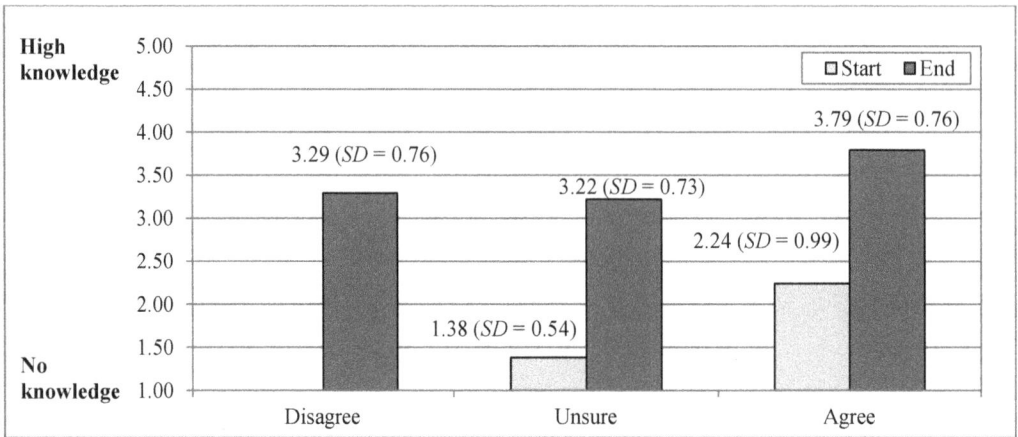

Participants also showed interest in being better informed:

I'm really interested to learn more about it, because it's important to advocate for the development and implementation of safe sustainable energy sources.

No, I don't feel well informed at all. It is a shame that it is not more widely discussed/debated as is coal seam gas and coal extraction for electricity.

Participants reported to recognise the technology from a range of information sources, most often from a media source:

I have heard it discussed in the media, but I am not sure if it's operational or how widespread it is.

I have seen a bit about if overseas on some of the grand design shows on the ABC they have used it as heating for their homes instead of other sources it is interesting.

3.2. Agreement with Benefits and Risks

The questionnaire results showed that different perceptions of the benefits and risks of the technology were linked with different levels of agreement with the use of the technology. At the start of the online focus groups, the mean ratings of four benefit and two risk statements were significantly different between participants that were unsure with the use of geothermal technology in Australia and those that agreed. The significant differences were identified using analysis of variance (ANOVA) and the results for benefit statements were: $F(1, 98) = 16.360$, $p < 0.001$ for "have benefits that outweigh the risks"; $F(1, 99) = 27.558$, $p < 0.001$ for "benefit future generations"; $F(1, 99) = 23.110$, $p < 0.001$ for "reduce greenhouse gas emissions"; and $F(1, 99) = 25.347$, $p < 0.001$ for "improve energy security". The results for the risks statements were: $F(1, 99) = 4.856$, $p < 0.05$ for "induce earthquakes"; and $F(1, 98) = 4.763$, $p < 0.05$ for "negatively impact on groundwater".

Figure 2 illustrates how the mean ratings of benefits by participants in agreement with the technology were consistently higher and risks were significantly lower, when compared to the responses of the participants that were unsure about the technology. Two risk statements that did not show any statistically significant differences were: $F(1, 98) = 0.199$, $p = 0.657$ for "have risks that are unknown" with a mean of 3.41 ($SD = 0.805$); and $F(1, 99) = 0.181$, $p = 0.671$ for "increase the price of electricity" with a mean of 2.86 ($SD = 0.861$). One statement about benefits also did not show any statistically significant differences: $F(1, 98) = 1.289$, $p = 0.259$ "have benefits that are unknown" with a mean of 3.51 ($SD = 0.611$).

Figure 2. Mean ratings of benefits and risk by attitude group at the start.

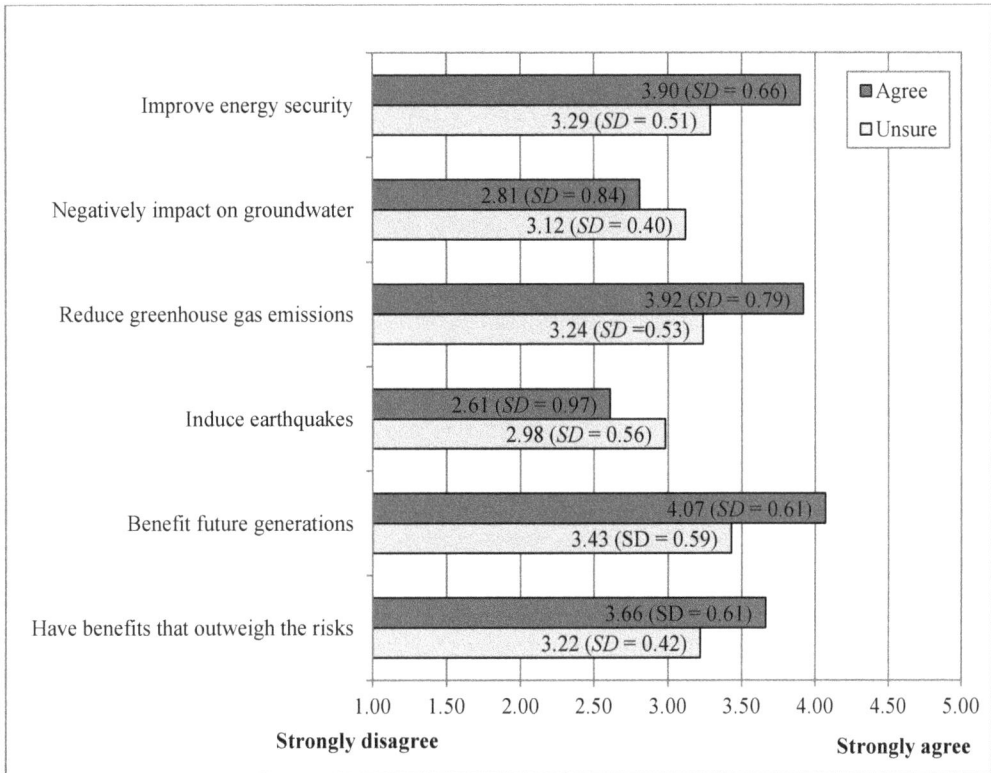

Figure 3 shows the mean ratings of benefits and risks for each attitude group in the post-questionnaire. Similar to the pre-questionnaire, there were significant differences: $F(2, 96) = 24.721$, $p < 0.001$ for "have benefits that outweigh the risks"; $F(2, 96) = 23.779$, $p < 0.001$ for "benefit future generations"; $F(2, 97) = 7.945$, $p < 0.05$ for "induce earthquakes"; $F(2, 97) = 11.997$, $p = 0.001$ for "reduce greenhouse gas emissions"; $F(2, 97) = 112.976$, $p < 0.001$ for "negatively impact on groundwater"; and $F(2, 96) = 12.976$, $p < 0.001$ for "improve energy security". Additionally there was a significant difference in agreement with 'have risks that are unknown', $F(2, 98) = 4.275$, $p < 0.05$. There were no significant differences between the responses of each attitude group to: $F(1, 98) = 0.49$, $p = 0.952$ for "have benefits that are unknown" with an overall mean of 3.53 ($SD = 0.881$); and $F(1, 97) = 2.627$, $p = 0.077$ for "increase the price of electricity" with an overall mean of 2.84 ($SD = 0.987$). Overall, similar to the pre-questionnaire, participants' that agreed with the use of the technology compared to participants that were unsure or disagreed rated the benefits of using the technology higher and the risks lower.

Figure 3. Mean ratings of benefit and risk statements by attitude group at the end.

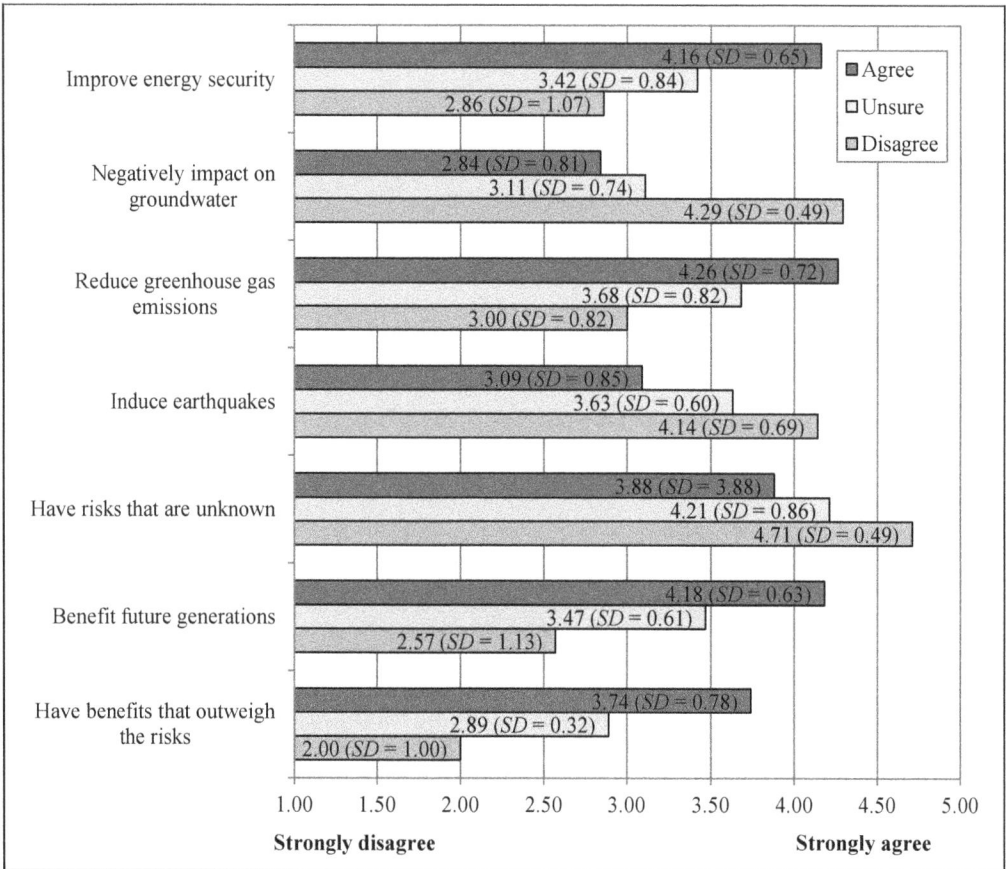

During the final facilitated discussion (Activity 6), when participants were asked to share their perspectives on the technology being used in Australia, a range of benefits and risks were discussed. The potential for seismic activity best illustrated in the quotes below:

Possible seismic activity, as reported from international projects and negative effect on the Artesian Basin.

NO FRACKING. The method to get through the rock has to be safe and sustainable. The projects have to be far enough away from cities and the populous so that it does not cause injury or environmental concerns.

There were mixed responses to the implications for ground water, some participants did not foresee negative effects, others did:

The water gets pumped in, it comes out hot, then gets pumped back in cold. It is great that the water can be reused—there is always a worry for water shortages.

I would treat it in a similar manner to coal seam gas production. I would not want it near residential areas or homes. It would need to be shown to be safe: i.e., no chance of it producing seismic activity or polluting groundwater or the environment.

Participants also perceived benefits such as low emissions, an abundant supply and potential cost savings:

I think Australia needs to do more to develop its geothermal resources. We can't keep going the way we are. Putting aside the greenhouse emissions and environmental factors, we Aussies are struggling with our power bills that keep rising.

A common conclusion amongst participants was that the benefits outweighed the risks:

Clean renewable resource that we could all benefit from. More jobs and a better economy. With the way electricity prices are going up we have to do something. I really think the benefits outweigh the negatives.

Another frequent conclusion was that the technology had potentially hazardous unknown risks:

I think Australia is right to be cautious for now. At least until we know more about the risks and benefits.

3.3. Project Location and Procedural Justice

As shown in Figure 4, participants reported similar preferences at the start and end of the process regarding the distance geothermal projects should be from built-up areas in their community. To meet the minimum distance preference of approximately half of the sample (55% in the start questionnaire; 57% in the end questionnaire) projects needed to be 'at least 50 km' and to meet the minimum distance preference of three-quarters of the sample (78% in the pre-questionnaire; 71% in the post-questionnaire) projects needed to be 'at least 100 km'. Regarding the differences between groups, at the start, the participants from the 'Unsure' and 'Agree' attitude groups rated their distance preferences similarly, $\chi^2(2, N = 101) = 6.30$, $p = 0.327$. At the end the preferences were more distinct for each attitude group, though the differences were not significant, $\chi^2(2, N = 100) = 20.20$, $p = 0.063$.

At the end of the focus group process participants were asked to rate their concerns if a geothermal project was proposed within 1km of built-up areas in their local community. Mean ratings of concern varied significantly between participants that disagreed, were unsure and agreed with the technology's use in Australia, $F(2, 97) = 9.478$, $p < 0.001$. The mean rating of participants in the "Disagree" and "Unsure" attitude groups were higher, 4.71 ($SD = 0.286$) and 4.63 ($SD = 0.684$), compared to the participants of the "Agree" group, 3.66 ($SD = 1.114$).

The facilitated discussion (Activity 6) captured some of the participants concerns about the technology and the possibility of it being used close to communities:

I don't think this technology should be utilised at all let alone in any populated area.

Some participants were more accepting of the technology though maintained the technology should be away from their community:

Definitely would not be happy if it was proposed in my area. I'm not sure how far away would make me feel better.

Others reported to be potentially comfortable with projects being in or near their community, though they had conditions such as safety and being consulted:

Provided I was consulted along the way and I was sure it was completely safe—go ahead!

After reading all the information I think I would be ok if they were to start a project in my area, I am not sure how far they should be, far away enough that there is minimal noise, traffic congestion, and an eyesore to the environment?

Figure 4. Percentage of agreement with distance of geothermal projects from built-up areas in the community.

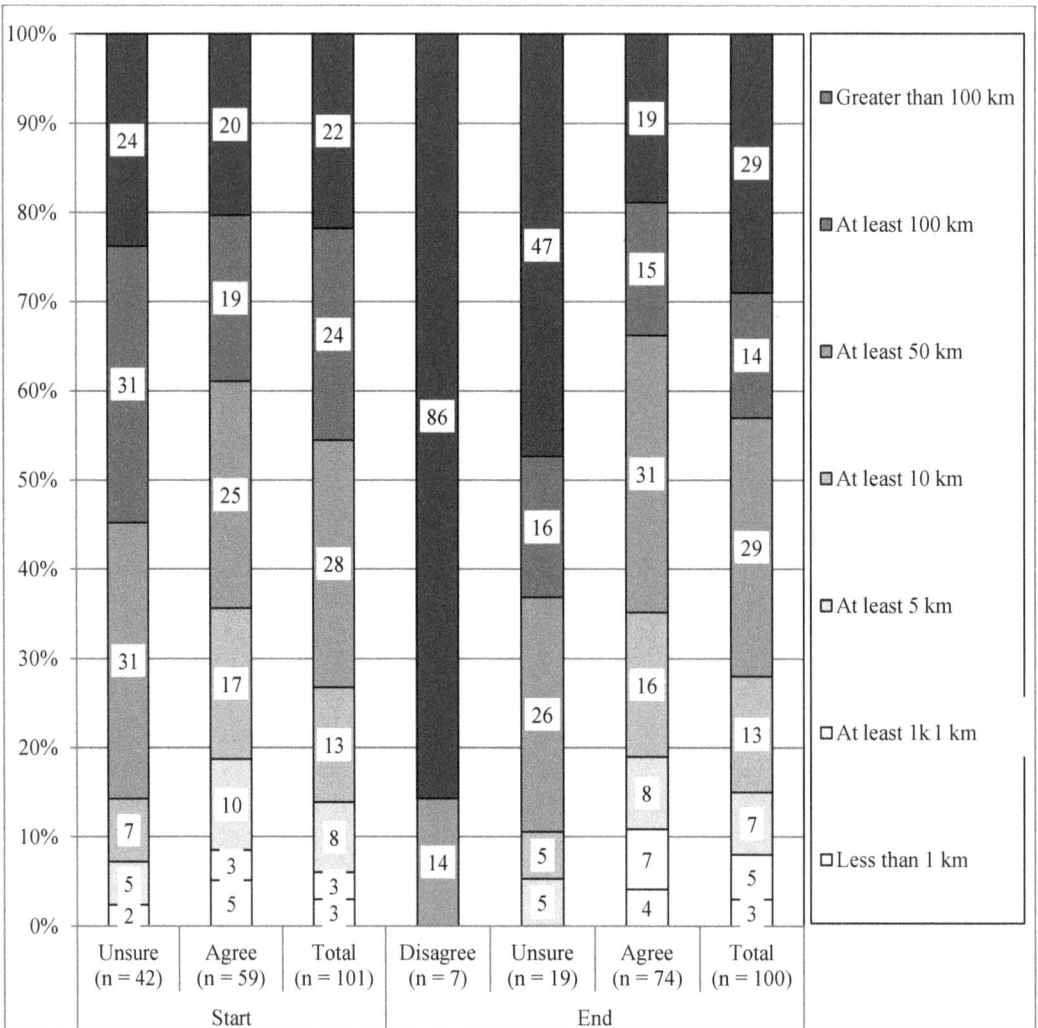

The final questionnaire results also showed that participants have concerns about the procedural justice elements of such projects. Participants were asked to rate from 1 (not at all) to 5 (very much) the extent to which they believed people in their community would have the opportunity to participate in decisions about geothermal energy projects. This question emphasises the participants' belief in having the opportunity to participate, whether participants would actually participate, would be subject to additional factors regarding the particular project and community. The difference in responses between attitude groups was significant, $F(2, 96) = 6.132$, $p < 0.05$. The mean rating of participants in the "Disagree" and "Unsure" groups were low, 2.00 ($SD = 1.000$) and 2.63 ($SD = 1.116$), whereas the mean response of the "Agree" group was moderate, 3.19 ($SD = 1.276$). Similarly, in the last facilitated discussion (Activity 6) participants indicated the need for more information to be available to the public:

Similar to the first handout, concise information of what Geothermal energy is. Also a table to highlight the benefits versus the negatives. With the negatives, it will be useful to describe how they can be prevented, mediated or dealt with. More media coverage will be useful, as it will reach out to more Australian public. Knowledge eases concerns.

Others emphasised addressing the risks, including hydraulic fracturing and involve the community:

I would be concerned on the fracking issue so wouldn't want it in my backyard unless all concerns were addressed.

If implemented, it would be important to involve (and actually involve and listen to, not just pay lip service!) local communities who are affected by having drilling near them. There would need to be strict regulation of the companies who explore and initiate drilling of sites, to make sure the sites are environmentally sound.

However, I do think that the area that is used to obtain the geothermal energy should be benefited the most from it, I don't like the way that big mining companies take all the profits offshore or away from the area's that they get there resources from.

3.5. Demographics

Based on their attitude toward the technology, participants were compared across age, gender, education and income. There was only a significant difference for gender. The difference was significant both at the start, $\chi^2(1, N = 100) = 11.45$, $p < 0.001$, and end, $\chi^2(1, N = 100) = 6.86$, $p < 0.05$. Figure 5 shows that there were a similar proportion of males to females, and that consistently, those that were unsure or disagreed were more likely to be female, whereas participants that agreed were more likely to be male.

Figure 5. Gender by attitude group.

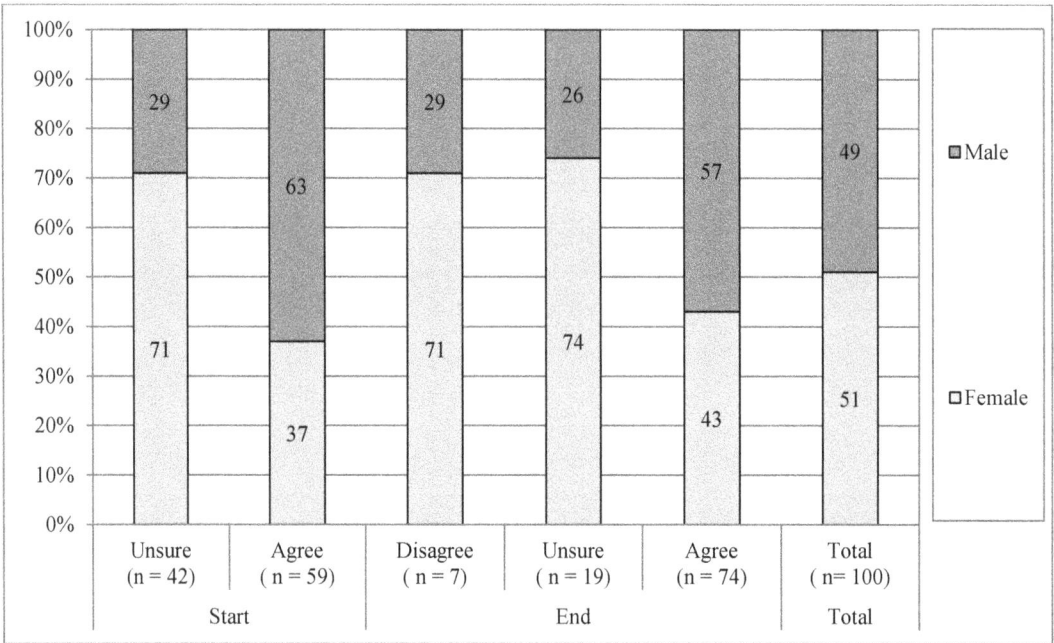

4. Discussion

The findings from this research appear consistent with previous studies about societal acceptance and awareness of geothermal energy technology in Australia [10,14]. This study, specifically found that at the start, prior to additional information, participants were either unsure of the technology or agreed with the technology. However, after the provision of information, a greater number of participants agreed with the use of the technology in Australia, fewer participants were unsure but a small number of participants disagreed with its use.

Additionally the study considered whether individuals that disagreed, were unsure or agreed with the technology's use in Australia had varying characteristics. Overall, participants that agreed with the technology both at the start and end of the online focus groups were more likely to be male, rate their knowledge as high, report stronger agreement with the technology's benefits and less agreement with the risks. They also indicated the technology could be located closer to their community, with only 34% of those respondents requiring geothermal projects to be located over 100 km as opposed to 63% of unsure respondents, or 100% of those who disagreed with the use of geothermal energy. In addition, respondents who agreed with the technology reported less concern about the technology being proposed for their community and believed they would be able to participate in the decision-making process.

The findings were consistent with previous research that suggests men are more likely to agree with the use of geothermal energy technology [19] than women, and that in general women are more inclined to be concerned with the risks of technology [37]. Also consistent with existing research reviewed by Huijts et al. [16], participants who reported higher subjective knowledge were more likely to agree with using the technology in Australia. Furthermore, after information there was an increase in both self-rated knowledge of the technology and agreement with the use of the technology.

Similar to the risk assessment literature, the findings showed that support for geothermal energy technology is dependent on an individual's perception of the technology benefits outweighing the risks [16,17,26]. The questionnaire results showed that after the provision of information, participants in the "Disagree", "Unsure" and "Agree" groups significantly differed in response to the statement 'that risks are unknown', confirming Slovic's [27] work on the implications of risk uncertainty for emerging technology. The risks participants focused on were consistent with the information presented, however, they were also consistent with the risks previously identified by Dowd *et al.* [14] and Reith *et al.* [13] and included seismicity, water usage and pollution. Similar to previous research of the acceptance of carbon capture and storage, benefits identified by participants were mainly global in nature such as geothermal being a low emission energy technology [29,31]. The change in ratings of benefits and risks, before and after information, demonstrated how individuals can change their perceptions due to available information, including knowledge of advances in the technology.

Consistent with previous research regarding the potential location of energy technologies [18] the majority of focus group participants would prefer that geothermal projects be located at least 100 km away from their community. This finding suggests the important role community engagement could have for direct-use or combined direct-use and electricity generation projects, which need to be proximal to the end-use application. Additionally, the "Disagree" and "Unsure" participants also reported less agreement with the notion that people in their community would have an opportunity to participate in decisions about such projects. This finding is consistent with previous research [13,29,33] and indicates the importance of early and transparent engagement as a means for overcoming community preconceptions and addressing concerns.

Overall this research demonstrates how acceptance of emerging energy technology can be further understood by comparing the characteristics of individuals with different levels of acceptance and at least two other directions for future research are evident. The first is to further explore the effect of information provision and framing on technology acceptance, investigating the influence of information source and trust, as well as the effect of messaging framing about benefits, risks and project location. While the second avenue for future research is limited in Australia due to the lack of commercial hot rocks geothermal plants, future research could extend the findings of this study by surveying both individuals that have had exposure to the technology and those who have not. Similar to Reith *et al.* [13] such analysis would explore the effect of exposure to the technology on risks and benefits perceptions as well as on technology support.

5. Conclusions

This study explored the level of agreement with the use of geothermal energy technology in Australia and how perceptions are impacted by media reports of the technology that are readily available on the internet. In conclusion the findings suggest that while the majority of participants agreed with geothermal technology use in Australia and agreement increased after information, concerns about the potential risks of the technology are present and the dominant preference is for the technology to be deployed away from communities. The reluctance to have the technology near communities could present a challenge for direct-use and combined direct-use and electricity projects which need to be located close to the end-use application. Participants concerns were not addressed at the time they were raised, which may have contributed to participants that were initially unsure or in agreement with the technology, disagreeing with the technology at the end. However, individuals do not have the opportunity to have their concerns addressed while reading news media or other content that is readily available on the internet. Thus the results highlight the importance of responding to uncertainty about the technology's risks and suggest a role for policy-makers and industry in engaging with Australians ahead of large-scale demonstration of the technology.

Acknowledgments

This research was conducted with strategic funding from Commonwealth Scientific Industrial Research Organisation (CSIRO)'s Petroleum and Geothermal Resources Portfolio.

Author Contributions

Simone Carr-Cornish and Lygia Romanach developed the study concept and the study design. Lygia Romanach managed and conducted data collection. Simone Carr-Cornish conducted data analysis and lead the writing of the manuscript. Lygia Romanach provided critical revisions of the manuscript. All authors approved the final version of the paper for submission.

Conflicts of Interest

The authors declare no conflict of interest.

References

1. Batel, S.; Devine-Wright, P.; Tangeland, T. Social acceptance of low carbon energy and associated infrastructures: A critical discussion. *Energy Policy* **2013**, *58*, 1–5.
2. Pickett, S.E. Japan's nuclear energy policy: From firm commitment to difficult dilemma addressing growing stocks of plutonium, program delays, domestic opposition and international pressure. *Energy Policy* **2002**, *30*, 1337–1355.
3. Hall, N.; Ashworth, P.; Devine-Wright, P. Societal acceptance of wind farms: Analysis of four common themes across Australian case studies. *Energy Policy* **2013**, *58*, 200–208.
4. Van Noorden, R. Buried trouble. *Nature* **2010**, *463*, 871–873.
5. Popovski, K. Political and public acceptance of geothermal energy. In Proceedings of the Intergovernmental Conference Short Course Geothermal Training Programme, The United Nations University, Reykjavík, Iceland, 14–17 September 2013; pp. 31–41.
6. Wüstenhagen, R.; Wolsink, M.; Baürer, M.J. Social acceptance of renewable energy innovation: An introduction to the concept. *Energy Policy* **2007**, *35*, 2683–2691.

7. Geoscience Australia, Australian Bureau of Agriculture and Resource Economics. *Australian Energy Resource Assessment*; Department of Resources, Energy and Tourism: Canberra, Australia, 2010.

8. Ergon Energy. Renewable Energy Sources. Available online: http://www.ergon.com.au/energy-conservation/what-are-we-doing/renewable-energy-sources (accessed on 22 August 2013).

9. Bureau of Resources and Energy Economics. *Energy in Australia 2013*; Department of Resources Energy and Tourism: Canberra, Australia, 2013.

10. Hobman, V.; Ashworth, P.; Graham, P.; Hayward, J. *The Australian Public's Preference for Energy Sources and Related Technologies*; Commonwealth Scientific Industrial Research Organisation (CSIRO): Brisbane, Australia, 2012.

11. Insightrix Research Inc. *Public Awareness and Acceptance of Carbon Capture and Storage in Canada*; International Performance Assessment Centre for Geological Storage of Carbon Dioxide (IPAC CO$_2$): Saskatchewan, Canada, 2011.

12. Eurobarometer. *Special Eurobarometer 262, Energy Technologies: Knowledge, Perception, Measures*; European Commission: Brussels, Belgium, 2007.

13. Public Acceptance of Geothermal Electric Production. Available online http://www.geoelec.eu/wp-content/uploads/2013/07/Deliverable_4-4_final-public-acceptance-mmi1.pdf (accessed on 16 September 2013).

14. Dowd, A.-M.; Boughen, N.; Ashworth, P.; Carr-Cornish, S. Geothermal technology in Australia: Investigating social acceptance. *Energy Policy* **2011**, *39*, 6301–6307.

15. Greenearth Energy Limited. Greenearth Committed to Addressing Community Concerns, 2010. Available online: http://www.greenearthenergy.com.au/investorcentre (accessed on 2 May 2013).

16. Huijts, N.M.A.; Molin, E.J.E.; Steg, L. Psychological factors influencing sustainable energy technology acceptance: A review-based comprehensive framework. *Renew. Sustain. Energy Rev.* **2012**, *16*, 525–531.

17. McComas, K.A.; Besley, J.C.; Yang, Z. Risky business: Perceived behavior of local scientists and community support for their research. *Risk Anal.* **2008**, *28*, 1539–1552.

18. Devine-Wright, P. Beyond NIMBYism: Towards an integrated framework for understanding public perceptions of wind energy. *Wind Energy* **2005**, *8*, 125–139.

19. Polyzou, O.; Stamataki, S. Geothermal energy and local societies—A NIMBY syndrome contradiction? In Proceedings of World Geothermal Congress, Bali, Indonesia, 25–29 April 2010; pp. 1–10.

20. Duan, H. The public perspective of carbon capture and storage for CO$_2$ emission reductions in China. *Energy Policy* **2010**, *38*, 5281–5289.

21. O'Garra, T.; Mourato, S. Public preferences for hydrogen buses: Comparing interval data, OLS and quantile regression approaches. *Environ. Resour. Econ.* **2011**, *36*, 389–411.

22. O'Garra, T.; Mourato, S.; Pearson, P. Investigating attitudes to hydrogen refuelling facilities and the social cost to local residents. *Energy Policy* **2008**, *36*, 2074–2085.

23. Ellis, G.; Barry, J.; Robinson, C. Many ways to say 'No', different ways to say 'Yes': Applying Q-methodology to understand public acceptance of wind farm proposals. *J. Environ. Plan. Manag.* **2007**, *50*, 517–551.

24. House, L.; Lusk, J.; Jaeger, S.; Traill, W.B.; Moore, M.; Valli, C.; Morrow, B.; Yee, W.M.S. Objective and subjective knowledge: Impacts on consumer demand for genetically modified foods in the United States and the European Union. *AgBio Forum* **2004**, *7*, 113–123.

25. Ashworth, P.; Romanach, L.; Contreras, Z. Understanding Australian householders' willingness to participate in the solar distributed energy market. In Proceedings of Energy Systems in Transition: Inter- and Transdisciplinary Contributions Conference, Karlsruhe, Germany, 9–11 October 2013.

26. Weaver, D.A.; Lively, E.; Bimber, B. Searching for a frame: News media tell the story of technological progress, risk and regulation. *Sci. Commun.* **2009**, *31*, 139–166.

27. Slovic, P. Perception of risk. *Sci. Commun.* **1987**, *236*, 280–285.

28. Savadori, L.; Savio, S.; Nicotra, E.; Rumiati, R.; Finucane, M.; Slovic, P. Expert and public perception of risk from biotechnology. *Risk Anal.* **2004**, *24*, 1289–1299.

29. Ashworth, P.; Bradbury, J.; Wade, S.; Feenstra, C.F.J.Y.; Greenberg, S.; Hund, G.; Mikunda, T. What's in store: Lessons from implementing CCS. *Int. J. Greenh. Gas Control* **2012**, *9*, 402–409.

30. Cacciatore, M.A.; Scheufele, D.A.; Corley, E.A. From enabling technology to applications: The evolution of risk perceptions about nanotechnology. *Public Underst. Sci.* **2011**, *20*, 385–404.

31. Poetz, A. A systems view of decision-making for risky technologies: From global to local and local to global. *Risk Hazards Crisis Public Policy* **2011**, *2*, 1–25.

32. West, J.: Bailey, I.; Winter, M. Renewable energy policy and public perceptions of renewable energy: A cultural theory approach. *Energy Policy* **2010**, *38*, 5739–5748.

33. Ashworth, P.; Cormick, C. Emerging Legal and Regulatory Issues. In *Carbon Capture and Storage*; Havercroft, I., Macrory, R., Stewart, R.B., Eds.; Hart: Oxford, UK, 2011; pp. 251–263.

34. Lind, E.A.; Tyler, T.R. *The Social Psychology of Procedural Justice*; Plenum: London, NY, USA, 1988.

35. Bradbury, J.; Ray, I.; Peterson, T.; Wade, S.; Wong-Parodi, G.; Feldpausch, A. The role of social factors in shaping public perceptions of CCS: Results of multi-state focus group interviews in the US. *Int. J. Greenh. Gas Control Technol.* **2010**, *9*, 4665–4672.

36. Carr-Cornish, S.; Ashworth, P.; Gardner, J.; Fraser, S. Exploring the orientations which characterise the likely public acceptance of low emission energy technologies. *Clim. Chang.* **2011**, *107*, 549–565.

37. Siegrist, M. The influence of trust and perceptions of risks and benefits on the acceptance of gene technology. *Risk Anal.* **2000**, *20*, 195–203.

38. Teddie, C.; Tashakkori, A. Mixed Methods Research: Contemporary Issues in an Emerging Field. In *The Sage Handbook of Qualitative Research*; Denzin, N.K., Lincoln, Y.S.S., Eds.; Sage: Washington, DC, USA, 2011; pp. 285–300.

39. Stewart, K.; Williams, M. Researching online populations: The use of online focus groups for social research. *Qual. Res.* **2005**, *5*, 395–416.
40. Australian Bureau of Statistics. *2011 Census QuickStats*; Australian Bureau of Statistics: Canberra, Australia, 2011.
41. Revelation│Next. Platform and Services. Available online: http://www.revelationglobal.com/applications/ (accessed on 12 April 2012).
42. Eurobarometer. *Special Eurobarometer 364, Public Awareness and Acceptance of CO_2 Capture and Storage*; European Commission: Brussels, Belgium, 2011.
43. CSIRO. What is Geothermal Energy? Available online: http://www.csiro.au/en/Portals/Publications/Brochures--Fact-Sheets/geothermal-energy.aspx (accessed on 12 April 2012).
44. Australian Broadcasting Corporation. Geothermal Industry Pushes for More Power. Available online: https://www.youtube.com/watch?v=Yvl2g1GjYsc\ (accessed on 12 April 2012).
45. Cichon, M. Enhanced Geothermal Systems: Have a Little Faith. Available online: http://www.renewableenergyworld.com/rea/news/article/2012/03/enhanced-geothermal-systems-have-a-little-faith (accessed on 12 April 2012).
46. Patel, T. France Fractured by Fracking-Like Geothermal Projects. Available online: http://www.smh.com.au/business/france-fractured-by-frackinglike-geothermal-projects-20130407-2heul.html (accessed on 12 April 2012).

Geophysical Methods for Monitoring Temperature Changes in Shallow Low Enthalpy Geothermal Systems

Thomas Hermans, Frédéric Nguyen, Tanguy Robert and Andre Revil

Abstract: Low enthalpy geothermal systems exploited with ground source heat pumps or groundwater heat pumps present many advantages within the context of sustainable energy use. Designing, monitoring and controlling such systems requires the measurement of spatially distributed temperature fields and the knowledge of the parameters governing groundwater flow (permeability and specific storage) and heat transport (thermal conductivity and volumetric thermal capacity). Such data are often scarce or not available. In recent years, the ability of electrical resistivity tomography (ERT), self-potential method (SP) and distributed temperature sensing (DTS) to monitor spatially and temporally temperature changes in the subsurface has been investigated. We review the recent advances in using these three methods for this type of shallow applications. A special focus is made regarding the petrophysical relationships and on underlying assumptions generally needed for a quantitative interpretation of these geophysical data. We show that those geophysical methods are mature to be used within the context of temperature monitoring and that a combination of them may be the best choice regarding control and validation issues.

Reprinted from *Energies*. Cite as: Hermans, T.; Nguyen, F.; Robert, T.; Revil, A. Geophysical Methods for Monitoring Temperature Changes in Shallow Low Enthalpy Geothermal Systems. *Energies* **2014**, *7*, 5083-5118.

1. Introduction

Geothermal heat pumps have a lot of advantages over standard heating and/or cooling systems such as gas or oil boilers and cooling machines which render them sustainable [1,2]. They allow large primary energy savings (e.g., [3,4]), strong reductions of CO_2 and other greenhouse gases emissions ([2,5] and references therein), and the use of energy stored in the subsurface (soil or groundwater). With regards to air-source heat pumps, geothermal heat pumps have the advantage of the ground temperature, which is far more constant than air temperatures. This energy is considered to be renewable as long as it is used reasonably and rationally (e.g., [6] and references therein). For instance, among shallow and deep geothermal resources, geothermal heat pumps accounted for 47.2% of thermal energy use and 68.3% of the total installed capacity in the World in 2010 [7]. Geothermal energy resources therefore constitute an essential field of research and development in the diversification of energy resources to hinder global warming (e.g., [8]).

Two main techniques exist to exploit shallow geothermal energy systems. In *closed systems*, heat is collected through a heat exchanger (vertical or horizontal), generally plastic pipes with a circulating fluid. This technology is generally referred as ground source heat pumps (GSHP). Such systems can be used for heating in winter and cooling in summer, suing the heat stored in the ground. Such a closed system is called borehole thermal energy storage (BTES). In *open systems*, groundwater circulates between production and injection wells and/or towards surface water. This is referred as groundwater heat pumps (GWHP) [9]. This kind of system requires a relatively high permeability to allow large pumping volumes. When heated or cooled water is infiltrated directly in the aquifer, one can take advantage of this energy storage as long as hydrogeological requirements are met (a weak hydraulic gradient for example). In this specific case, we rather speak of (seasonal) aquifer thermal energy storage (ATES) systems [4].

Very low-temperature systems (<30 °C) are generally located at depths between 0 and 100 m. They are thus much more easily accessible and involve lower implementation costs than deeper, high temperature, systems. The cost difference is essentially due to the drilling costs. Moreover, very low-temperature reservoirs, such as shallow aquifers, are relatively abundant in alluvial or coastal plains where urban development is concentrated. From 0 to less than 100 m depth, groundwater has an average temperature ranging from 5 to 30 °C and may be used for domestic or industrial cooling and/or heating (e.g., [10,11]).

Designing such systems requires a multidisciplinary approach including geological and hydrogeological studies. The most common approach is to model the subsurface using a coupled groundwater, heat flow and transport simulator. However, such models require the knowledge of the parameters governing groundwater flow (e.g., hydraulic conductivity and specific storage) and heat transport (e.g., thermal conductivity, volumetric heat capacity, flow rate). *In situ* tests, such as thermal response tests [12,13] or laboratory measurements [14] are sometimes possible, but the values obtained may deliver only well-centered information or may not always (if not at all) be representative of *in situ* conditions at a larger scale. Such data are often scarce if not missing and authors often have to rely on standard calculation charts, values found in the literature, or simply default values implemented in standard software (e.g., [15–18]). In addition, the heterogeneity of the material properties and their potential anisotropy, which are difficult to detect with standard integration methods, make the problem more complex. The direct consequence is a lower confidence in the predictions of models leading to inadequate design of the heat pumps.

Besides the characterization of shallow geothermal systems themselves (e.g., for dimensioning purposes), the impact on the evolution of groundwater temperatures is also an important research topic both economically and environmentally. From an environmental point of view, the exploitation of geothermal heat pumps yields cold and hot plumes [19–21] which may influence aquifer properties such as groundwater chemistry (e.g., [22–24]) and microbiology [25]. Moreover, geothermal systems can only be qualified as renewable if there is a global annual thermal equilibrium. These potential environmental effects must be (and will increasingly be) studied because they can have strong economic repercussions such as lowering the global life of the system with side effects such as aquifer freezing and scaling on heat exchangers and wells [6,26,27]. From an economical point of view, a better knowledge of the thermal affected zone (TAZ) [28,29] can prevent shallow geothermal systems competing with each other (thermal feedback or thermal recycling) [30,31] or better, to take benefit of multiple ATES systems with mutual energy storage zones to enhance the global systems efficiency [22].

Haehnlein *et al.* [6,11] pointed out that while policies exist in some countries to limit the temperature difference caused by the use of geothermal systems, the development of anomalies is rarely monitored. With the growth of the demand for renewable energy (more than 200 new ATES systems are currently being installed each year in The Netherlands, for example) [22], we can expect that regulations will become more severe and control of installations more common. New monitoring technologies will therefore be needed to better conceptualize, design, and then control shallow operating geothermal systems.

Thermal tracing experiments have been performed for decades in hydrogeology [32–34]. Such experiments are used to improve the characterization of hydrogeological parameters (e.g., hydraulic conductivity and hydrodynamic dispersivity tensors), but the same methodologies may also be used to study the thermal properties of shallow geothermal systems [35–37]. However, the heterogeneity of geothermal and hydrogeological systems may be too complex to be fully captured by classic thermal or solute tracer experiments alone [38,39], with only punctual measurements in wells.

In this context, new technologies are clearly needed to monitor the spatial and temporal distributions of temperature in the shallow geothermal system (Figure 1) to: (1) better design the geothermal system and the monitoring network, (2) prevent any thermal feedback/recycling, and (3) image and control the thermal affected zone. Among these technologies, we will emphasize in our review three emerging geophysical techniques to monitor geothermal systems (Figure 1). The first one is called electrical resistivity tomography (ERT) or electrical resistivity imaging (ERI). It provides 2D or 3D tomograms of the resistivity distributions of the subsurface. Time-lapse variations of the electrical resistivity can be used to map changes in temperature. The second approach is the self-potential method (SP), which is used to map or monitor the electrical potential at the ground surface or in wells. The self-potential anomalies can be associated with ground water flow and temperature variations. The last method is the use of an *in situ* fiber optic distributed temperature sensor (DTS). This method provides linear measurements of temperature with centimeter resolution in boreholes with a precision of around 0.1 °C.

Figure 1. Emerging geophysical technologies to measure the temperature distribution in the subsurface due to heat injection in a fully non-invasive manner (electrical tomography and self-potential) to spatially continuous measures in boreholes (distributed fiber optics).

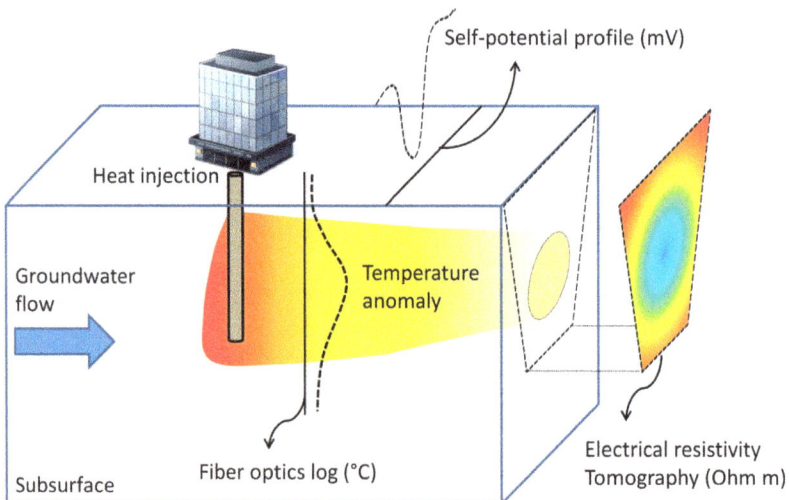

The paper is organized as follows: first, we describe the principles of the three methods with a focus on the information provided in the framework of geothermal systems, hypothesis made and set up. Then, a review of the literature for the three methods with emphasis on practical consideration and limitations is made. The paper ends with conclusions, perspectives of development and applications of these methods to geothermal heat pump systems.

2. Electrical Resistivity Tomography

Electrical resistivity tomography (ERT) is a method that images the bulk electrical resistivity distribution of the subsurface (Figure 2) in two, three, or four dimensions (the three dimensions of space plus time when monitoring is performed) from the meter to the hundreds of meter scale depending on the electrode spacing. When conducted from the surface only, the method is non-invasive. Electrical resistivity is a property depending on several textural properties of the porous material (such as porosity and pore shape), the presence of clay minerals (mineralogy and weight fractions), the properties of the pore water (saturation, salinity), and environmental variables such as temperature [40–42]. As a consequence, ERT is often used to infer these subsurface properties but require being able to separate the different contributions affecting the measurements. Quantitative interpretation remains indeed difficult without additional information, generally in the form of a few additional *in situ* measurements or/and the use of additional geophysical methods (for instance GPR or induced polarization). Electrical resistivity tomography (ERT) has proven its efficiency to image and/or monitor spatial phenomena [43] such as salt water intrusions [44,45], variations in moisture content (e.g., [46,47]), biodegradation of hydrocarbons (e.g., [48,49]) and salt tracer experiments (e.g., [50,51] and references therein). An in-depth review of electrical properties of rocks can be found in Schön [52] or Revil *et al.* [53], whereas a description of electrical methods can be found in Binley and Kemna [54].

The tomography of Figure 2 shows a background snapshot of the resistivity distribution at a site in an alluvial plain [55]. The changes in resistivity reflect changes in the lithology. To obtain such an image, a series of electrical current injections into the soil was performed, generally between two electrodes called current electrodes and the resulting electrical potential difference was measured simultaneously between two other electrodes, called potential electrodes. Given Ohm's law, the ratio between the measured difference of electrical potentials and the known current intensity equals the electrical resistance. A value of electrical resistance is therefore assigned to the used quadrupole and the process is repeated automatically hundreds to thousands of time along a profile (or a panel for cross-borehole application) to acquire a full data set. Electrical resistance is not an intrinsic property of earth because it is linked to the volume of subsurface that is scanned by electrical current lines. The resistance data set must be inverted to find an inverse model of electrical resistivity distribution that explains collected data. This inverse model is the electrical image/tomogram that is then interpreted physically in terms of temperature using a petrophysical transformation (see Section 2.2). A common way to solve such inverse problems [54] is to add a regularization constraint to the least-square problem [56]. The problem is then to minimize, through an iterative process, an objective function of the form:

$$\psi(m) = \psi_d(d,m) + \lambda\psi_m(m,m_0) \tag{1}$$

In the above equation, m represents the (unknown) model vector (*i.e.*, a vector containing the resistivity values for the cells used to discretize the subsurface), $\psi_d(d,m)$ is a measure of the data misfit (difference between the measured and modeled data according to a given norm), the model functional $\psi_m(m,m_0)$ defines desired model constraints and can include a prior resistivity model m_0, and λ is the so-called regularization parameter, which balances the two terms. This regularization parameter can also be optimized in the inversion process. Minimizing the objective function with respect to the model vector m is a non-linear problem in ERT. The solution of this problem can be achieved using an iterative process such as the Gauss-Newton algorithm.

Figure 2. Background tomography obtained using 28 electrodes in two boreholes. The inverse model shows resistivity values varying between 100 and 200 $\Omega\cdot$m. The panel seems slightly heterogeneous and reflects lithological changes. Dots show the position of borehole electrodes. Average groundwater temperature is around 13 °C (modified after [55]).

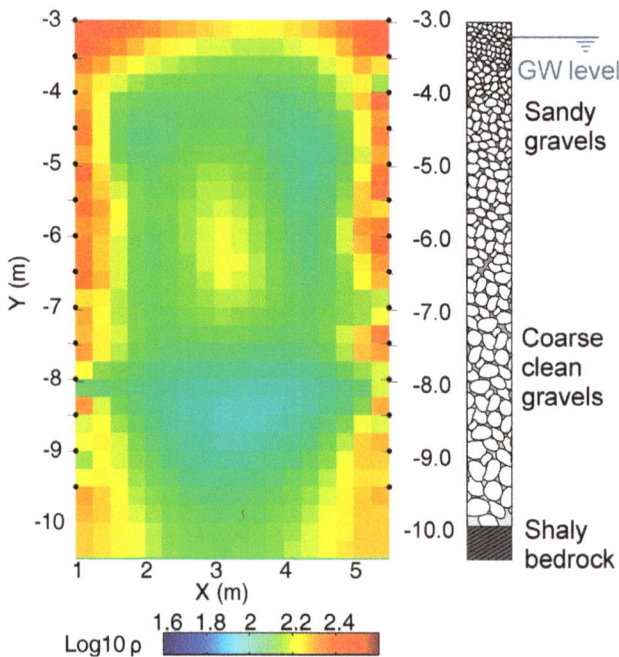

2.1. Time-Lapse ERT

Bulk electrical resistivity of saturated soil/rock samples decreases with temperature [40,41,57]. This correlation reflects the change in conductivity of water contained in the pores but also in the surface conductivity of grains. These effects are linked to the modification of viscosity with temperature which modifies the mobility of charge carriers. By extension, the temperature changes observed on operating GWHP/ATES systems [4] are typically in the range of temperature changes that could be detected by ERT (~2 °C and more, see [55]).

During a monitoring study, we acquire at least one snapshot, which is linked to a background or reference state, such as the one shown in Figure 2. Data acquisition is then repeated over time with the same sequences, parameters and so on. Data sets are finally inverted and compared to the reference image to visualize resistivity changes over time. These changes are in turn connected to the physical process we want to follow (temperature changes in our case, Figure 3). One advantage of such time-lapse ERT over static ERT is that it can be used to monitor processes involving only one or a few of the parameters influencing electrical resistivity. This makes the interpretation of time-lapse ERT more easily quantitative.

Figure 3. Tomographies of changes in bulk electrical resistivity after injecting hot water (3 m^3/h at a mean temperature of 38 °C). The background tomography is the one shown in Figure 2. Hot water injected in a well located at 8m upgradient, on the whole aquifer thickness, can be seen as negative resistivity changes (blue). It is preferentially found in the deeper part of the aquifer because groundwater flows preferentially in this high hydraulic conductivity zone (gravel). This figure compares time-lapse inversions using the standard smoothness constraint regularization (**left**), minimum gradient support (**middle**) and geostatistical inversion (**right**) (modified after [65]).

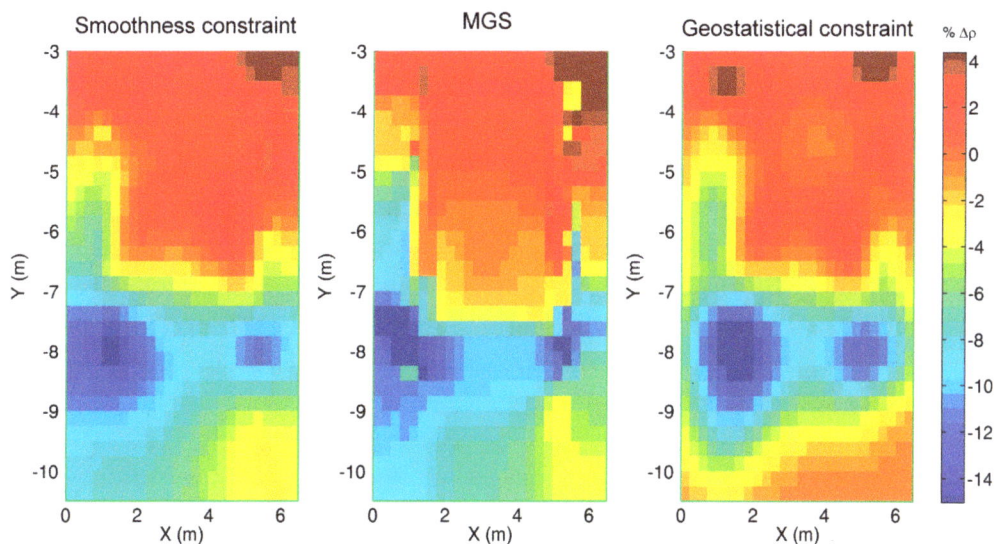

In most monitoring studies, temperature effects are undesirable and therefore, often considered as noise that may create artefacts and misinterpretation of the resulting images. As a result, temperature corrections in time-lapse (monitoring) series may be necessary to correct electrical resistivity tomography results in order to avoid misinterpretation when explaining resistivity changes linked to other physical processes such as changes in contamination or porosity [40,58–60].

With time-lapse data sets, we are interested in imaging the changes in electrical conductivity/resistivity with respect to a reference in time. Generally, the process of inversion is adapted in order to improve inversion results [61]. Three main procedures, with several variants, exist to invert for time-lapse ERT data [62]: namely independent inversion, difference inversion, and time-constrained or reference model inversion.

In independent inversion, inversion results obtained separately are simply subtracted [63], which should eliminate systematic errors but amplify uncertainties in the data. The difference inversion scheme [50,64] formulates the problem in terms of variations for both data and model, *i.e.*, the data differences are inverted to calculate the model changes to apply to the reference state. This process eliminates the systematic errors on the data linked to modeling, measuring device or contact resistances. The background state is used as a starting model for subsequent time-lapse, which generally leads to a fast convergence of the algorithm.

For temporally constrained schemes, a regularization operator is added in the time dimension (under the form of a reference model) in addition to the space dimensions, to minimize changes between successive tomograms [66,67]. This provides a 4D inversion scheme, which has proven to be efficient in tracer tests [68].

According to the location of electrodes on the field, the inversion is made using a true 3D model. The subsurface is divided in cells in the three dimensions and the model is inverted considering current flow in all directions [69]. When the collected data results from electrodes situated on a plane, the inversion generally uses a 2.5D scheme, where the resistivity distribution is considered as constant in the direction perpendicular to the section [70]. The latter inversion scheme may yield inversion artefacts when used in time-lapse mode. Indeed, in the case of a moving plume of heat, it may be imaged before crossing the ERT section because it already influences current lines. This phenomenon is known as the shadow effect [71,72].

To compare the successive models in the monitoring study, it is important that all data sets be inverted with the same level of data misfit corresponding to the expected noise level and the same approach to optimize the regularization parameter of Equation (1). Indeed, over-fitting the data may create artefacts of inversion in the corresponding image, whereas the contrary would result in an over-smoothed inverted section [73]. Similarly, using different approaches to optimize the regularization parameter (L-curve, fixed, iteratively decreasing) may introduce undesired artifacts.

Another issue is the assessment of the propagation of noise in the inversion. Robert *et al.* [51] proposed the use of two background data sets corresponding to a common state of the subsurface to study this phenomenon. They inverted one data set with the inversion of the other as reference. The resulting changes in resistivity are small and the maximum change is considered as the limit to interpret resistivity change. Hermans *et al.* [74] estimated the level of noise with reciprocal measurements and generated 100 new data background data sets. Then, they evaluated the mean and standard deviation of the 100 inverted electrical resistivity models and proposed a conservative cut-off (two times the maximum standard deviation) to interpret time-lapse images. Hermans *et al.* [55] used a simple physical interpretation to assess the error level. They expected only decreases in resistivity in their time-lapse sections. However, increases up to 3% were observed. They thus considered the changes between −3% and +3% has not interpretable and chose −3% as the limit of quantification for ERT in their specific case. This corresponded to a change in temperatures of 1.2 °C.

A major drawback of traditional ERT inversion is the smoothing effect induced by the regularization operator. To avoid this, coupled hydrogeophysical inversions are possible [39,75,76].

Resistance data are directly incorporated in the inversion of (thermo-) hydrogeological data, which avoids the inversion steps. Besides, many efforts have been made in the last decade to improve static ERT inversion by incorporating prior information in the inversion process. New constraints have been developed including blocky inversion [77], minimum gradient support [78], structural inversion [79], geostatistical inversion [44] or guiding images [80]. These constraints have proved to be efficient in many field cases (e.g., [81]).

Except for the reference model constraint used in the spatio-temporal and difference inversion scheme, few specific regularization operators have been developed for time-lapse ERT inversion. Globally, the same constraints as for static image could be used. Nguyen *et al.* [65] proposed to adapt minimum gradient support (MGS) and geostatistical inversions for time-lapse inversion. Figure 3 proposes a comparison of these approaches with standard smoothness constraint within the context of difference inversion. The models were inverted with the same error level to the same inversion error. Data correspond to the heat tracing experiments performed by Hermans *et al.* [55] in an alluvial aquifer with cross-borehole ERT. Hot water was injected in a well located 8 m upgradient on the whole thickness of the aquifer. The results obtained with the standard smoothness constraint (Figure 3, left) showed a good agreement with direct and DTS measurements. Hot water flows preferentially in the deeper part of the aquifer (gravel) due to its higher hydraulic conductivity. In this case, MGS yields focused resistivity changes and avoid smoothing (Figure 3, middle), which is efficient to image transport in fractures, karsts or along faults, but not really appropriate in an alluvial aquifer where flow is supposed to be relatively homogeneous. Geostatistical inversion in time-lapse allows integrating direct measurements in boreholes, such has DTS temperature logs, to build the regularization operator (Figure 3, right). Since, spatio-temporal variograms are possible [82], this could be included in a 4D inversion scheme as well. The geostatistical inversion reduces the smoothing effect of standard inversion to the suggested correlation length and is the most coherent with observed temperatures.

Qualitatively, the three inversions proposed in Figure 3 are similar. They enable to locate the zone of preferential flow path, related with stronger decrease of resistivity [55]. However, quantitatively, the variations in resistivity are slightly different which would in turn modify the estimation of temperature. It is thus important to consider the regularization operator which best fits the flow process. Direct measurements are helpful in this process. Even if those issues (noise and inversion) are not specific to the monitoring of temperature, they may play an important role in the future in the development of time-lapse ERT, and may thus influence its use within the context of geothermal systems.

2.2. Petrophysical Considerations

In the framework of ERT, the aim of the petrophysical relationship is to quantify the link between bulk electrical conductivity and temperature. The bulk electrical conductivity σ_b is generally expressed as a function of porosity ϕ and tortuosity (often joined in a term called the formation factor F), saturation S_w, fluid electrical conductivity σ_f and surface conductivity σ_s due to electrical conduction in the electrical double layer coating the surface of the grains in contact with water [83]:

$$\sigma_b = \frac{1}{F} S_w^n \sigma_f + \frac{1}{F\phi} S_w^{n-1} \sigma_s \tag{2}$$

where n, the saturation exponent, is an empirical exponent close to 2. Surface conductivity is related to the cation exchange capacity of the matrix and is significant for shaly and clayey sediments. Description of experimental models investigating this term can be found in Waxman and Smits [42] or Revil et al. [41,84]. Archie's law (Equation (3)) describes the relationship between F and the connected porosity ϕ with a power law through the porosity-exponent m [85]:

$$\frac{1}{F} = \phi^m \tag{3}$$

The formation factor F may vary spatially (both laterally and with depth), depending on the lithology. Revil [83] showed that it may be reasonable to take $m = n$. In this case Equations (2) and (3) becomes:

$$\sigma_b = \theta^m \sigma_f + \theta^{m-1} \sigma_s \tag{4}$$

where θ is the water content. Revil [83] showed that the surface conductivity can be related to the cation exchange capacity or the specific surface area of the material. Both parameters can be easily measured in the laboratory for core samples. Surface conductivity can also be determined through induced polarization measurements, which can be measured with the same instrument used to do resistivity measurements [86].

In saturated conditions ($S_w = 1$), Equation (2) simplifies to:

$$\sigma_b = \frac{\sigma_f}{F} + \frac{\sigma_s}{F\phi} \tag{5}$$

When surface conductivity is negligible, bulk electrical conductivity is directly proportional to fluid electrical conductivity:

$$\sigma_b = \frac{\sigma_f}{F} \tag{6}$$

When limited temperature intervals are considered (a few tens of degrees), a linear dependence between water electrical conductivity and temperature can be assumed, it is called the ratio model. Equation (7) expresses the linear relation around a temperature of reference T_{ref} [40,87]:

$$\frac{\sigma_{f,T}}{\sigma_{f,T_{ref}}} = m_{f,T_{ref}} (T - T_{ref}) + 1 \tag{7}$$

where $\sigma_{f,T}$ is the fluid electrical conductivity at temperature T (in °C), $\sigma_{f,Tref}$ is the fluid electrical resistivity at the temperature of reference T_{ref} (typically 25 °C), and $m_{f,Tref}$ is the linear temperature dependence of electrical conductivity with temperature (expressed in °C^{-1}). Equation (7) corresponds to a first-order Taylor expansion of the conductivity dependence with temperature around the reference temperature. The value of $m_{f,25}$ can be experimentally determined and varies according to the composition of the fluid. A value between 0.018 °C^{-1} and 0.025 °C^{-1} is often found [40].

Surface conductivity variations with temperature can be expressed by similar equations, with a different fractional change per degree Celsius $m_{s,Tref}$ ($°C^{-1}$):

$$\frac{\sigma_{s,T}}{\sigma_{s,T_{ref}}} = m_{s,T_{ref}}(T - T_{ref}) + 1 \tag{8}$$

Hayley et al. [40] applied this model on the temperature range 0–25 °C and found $m_{s,25}$ around 0.018 $°C^{-1}$ and $m_{f,25}$ equal to 0.0187 $°C^{-1}$. These values are similar and lead globally to a linear temperature dependence for the bulk electrical conductivity. It signifies that Equation (7) is also valid for bulk electrical conductivity which varies linearly with temperature. This is generally acceptable for temperature intervals below 40 °C but may be wrong for higher temperature intervals [41].

In the case of ERT monitoring studies, we measure bulk electrical resistivity at different time steps and compare it to a reference state, called the background. If we take the ratio of Equation (5) between a specific time-step (state 2) and the reference background (state 1), we have:

$$\frac{\sigma_{b1}}{\sigma_{b2}} = \frac{\sigma_{f1} + \dfrac{\sigma_{s1}}{\phi}}{\sigma_{f2} + \dfrac{\sigma_{s1}}{\phi}} \tag{9}$$

and the relation is not dependent on the formation factor but on porosity only. This can be done only if the formation factor is supposed to be independent from electrical conductivity and constant in time which is generally true. If surface conductivity can be neglected, Equation (9) simplifies further to:

$$\frac{\sigma_{b1}}{\sigma_{b2}} = \frac{\sigma_{f1}}{\sigma_{f2}} \tag{10}$$

Through Equation (10), we see that the variation in bulk electrical conductivity in the saturated zone is related only to a variation of the fluid electrical conductivity. When we consider temperature variations, Equation (10) is also valid when surface conductivity is non-negligible if the temperature linear dependence of water and surface conductivity are the same [41]. Indeed, in this case the variations of water, surface and bulk electrical conductivity have the same slope, which allows writing Equation (10) as well.

Equation (10) may be interesting to use to derive subsurface temperature from water conductivity. Indeed, in many cases, temperature measurements will be only accessible through groundwater inside boreholes and the fractional change per degree will be determined from a water sample. In Equation (10), σ_{b1} and σ_{b2} are determined using ERT after inversion of resistance data and σ_{f1} can be measured on a sample from formation water of the aquifer before the experiment (calibration process). The only unknown in Equation (10) is thus the fluid electrical conductivity at state 2 σ_{f2}, which can be expressed as:

$$\sigma_{f2} = \frac{\sigma_{b2}}{\sigma_{b1}}\sigma_{f1} \tag{11}$$

If we assume that the salinity of the fluid remains constant during the experiment, the water electrical conductivity depends only on temperature.

Introducing Equation (7) into Equation (11), we can express the temperature T (in °C) according to bulk electrical conductivity of the background and of the considered time-lapse section, to water electrical conductivity at the temperature of reference and at the temperature of the background and to the fractional change per degree Celsius:

$$T = \frac{1}{m_{f,25}} \left[\frac{\sigma_{b2,T}}{\sigma_{b1}} \frac{\sigma_{f1}}{\sigma_{f,25}} - 1 \right] + T_0 \tag{12}$$

where $\sigma_{b2,T}$ represents the bulk electrical conductivity at the time-step for which we try to determine the temperature and $T_0 = 25$ °C.

In Equation (12), the ratio $\dfrac{\sigma_{f1}}{\sigma_{f,25}}$ is only dependent on the temperature and the fractional change per degree Celsius. We can thus deduce the absolute temperature in the aquifer according to the initial temperature T_1:

$$T = \frac{\sigma_{b2,T}}{\sigma_{b1}}(T_1 - T_0) + \frac{\dfrac{\sigma_{b2,T}}{\sigma_{b1}} - 1}{m_{f,25}} + T_0 \tag{13}$$

Equation (13) is similar to Equation (7) expressed for the bulk electrical resistivity at two different temperatures. The result of applying these petrophysical relationships to changes in ERT can be seen in Figure 4.

The developments above consider that changes in bulk electrical conductivity are only related to direct temperature changes. It does not take into account modifications in equilibrium and kinetics that may arise with an increase in temperature.

Figure 4. Time-lapse tomography of changes in temperature derived from ERT using Equation (13) on Figure 3, left. The image of temperature is limited to the saturated zone, because Equation (5) is not valid in the unsaturated zone (modified after [55]).

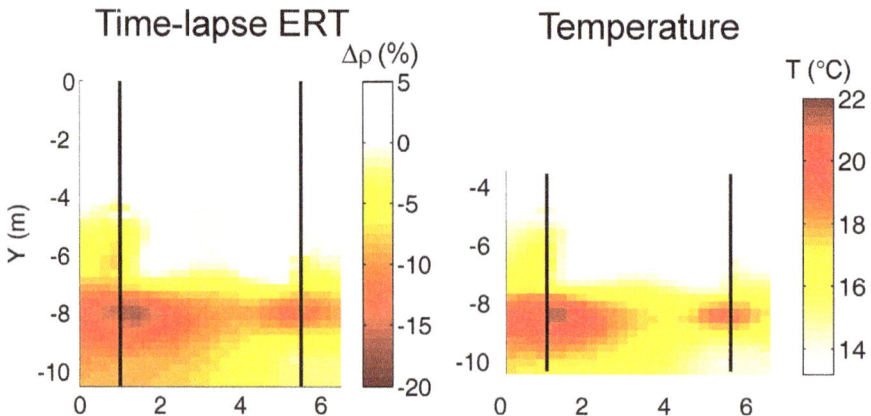

3. Self-Potential Method

The self-potential method is a passive geophysical method that is sensitive to the generation of electrical current densities in the ground. The method consists in mapping or monitoring passively electrical potentials on the ground surface or in wells using a set of non-polarizing electrodes (e.g., Cu/CuSO$_4$ or Pb/PbCl$_2$ electrodes). One of these electrodes is used as a reference, preferably located far from the area that is being monitored. The instrument used for the measurement is a voltmeter with a sensitivity of at least 0.1 mV and an input impedance of 10 to 100 MΩ or higher.

These currents include, within the context of geothermal systems, the streaming current associated with the flow of the ground water and the thermoelectric current associated with a temperature gradient. The sum of these two terms corresponds to the total source current density $\mathbf{j}_S$:

$$\mathbf{j}_S = \hat{Q}_V \mathbf{u} + C_T \sigma \nabla T \tag{14}$$

where $\mathbf{u}$ (in m·s^{-1}) denotes the Darcy velocity, $\hat{Q}_V$ (in C·m^{-3}) denotes the excess of electrical charge at saturation that is carried along with the flow of the pore water, σ denotes the (saturation-dependent) electrical conductivity of the porous material, and C_T is the thermoelectric coupling coefficient defined below. For pH values between 5 and 8, Jardani et al. [88] found that the $\hat{Q}_V$ is controlled by the permeability at saturation k (in m^2) and they developed the following empirical relationship [89]:

$$\log_{10} \hat{Q}_V = -9.2 - 0.82 \log_{10} k \tag{15}$$

In conductive materials, the source current density $\mathbf{j}_S$ is responsible for an electrical field and the tangential component of this electrical field is measured at the ground surface. With respect to the macroscopic electrical field, the generalized Ohm's law for the total current density $\mathbf{j}$ (A·m^{-2}) is written as:

$$\mathbf{j} = \sigma \mathbf{E} + \mathbf{j}_S \tag{16}$$

where $\mathbf{E} = -\nabla \varphi$ denotes the electrical field and φ the self-potential field, and σ denotes the bulk electrical conductivity of the porous material. From Equations (14) and (16), the streaming potential and thermoelectric coupling coefficients C_S and C_T are defined by:

$$C_S = \left(\frac{\partial \varphi}{\partial h} \right)_{j=0, \nabla T=0} \tag{17}$$

$$C_T = \left(\frac{\partial \varphi}{\partial T} \right)_{j=0, \mathbf{u}=0} \tag{18}$$

and are expressed in V·m^{-1} and V·K^{-1} (or V·°C^{-1}) respectively, where h is the piezometric level. In Equation (18), the thermoelectric coupling coefficient is only properly defined in absence of flow and when the total current density is zero. An experimental procedure is described in Leinov et al. [90]. In a recent work, Revil et al. [91] obtained a value of the thermoelectric coupling coefficient of -0.5 mV·°C^{-1}. The negative polarity implies that positive temperature anomalies (increase in temperature) should be associated with negative self-potential anomalies.

Equation (16) is combined with a conservation equation for the electrical charge that is written as $\nabla \cdot \mathbf{j} = 0$ in the quasi-static limit of the Maxwell equations [92]. The combination of these equations yields the following elliptic partial differential equation for the self-potential φ (in V):

$$\nabla \cdot (\sigma \nabla \varphi) = \nabla \cdot \mathbf{j}_S \qquad (19)$$

The right-hand side of Equation (19) corresponds to the self-potential source term associated with the Darcy velocity, with the temperature distributions, and with the heterogeneity in the distribution of the volumetric charge density, thermoelectric coupling coefficient, and electrical conductivity.

Figures 5 and 6 show the simulation of the self-potential signals associated with the passage of an electrical potential anomaly in a preferential flow channel. We can see a negative self-potential anomaly at the ground surface indicating the presence of the thermal anomaly (Figure 5). This anomaly is highly correlated to the temperature changes (Figure 6).

Figure 5. Time-lapse simulation of electric potential in a tank, relative to the background, following the hot injection. 3D illustration of growth of the negative self-potential anomaly in the channel at the surface, due to the temperature distribution in the channel in the subsurface (modified after [93]).

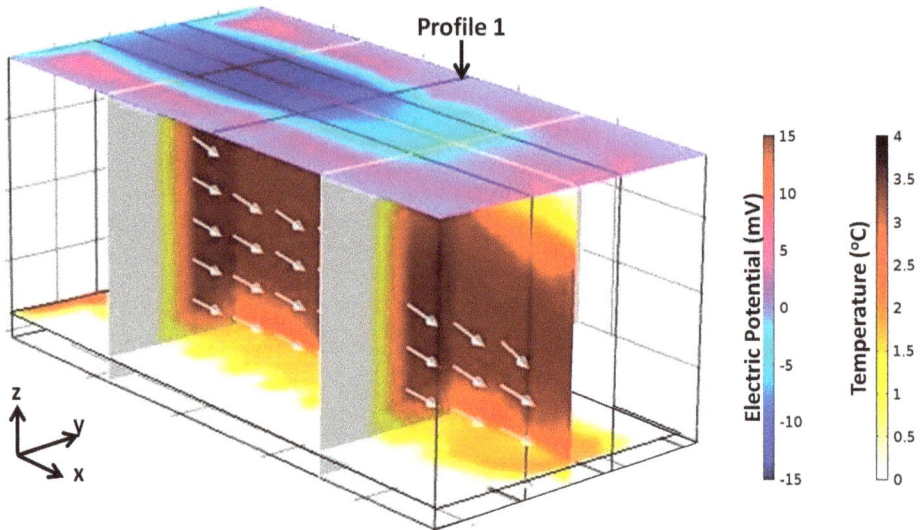

3D simulation of temperature and electric potential anomalies

Figure 6. Simulated temperature and electric potential changes relative to background along Profile 1 (see position in Figure 5) following the injection of hot water in an upstream reservoir. (**a**) Simulated temperature change relative to background. The temperature anomaly is confined primarily to the permeable channel; (**b**) Simulated electric potential change relative to background following the hot injection. The electric potential anomaly is negative, achieves a peak amplitude of approximately −13 mV, and is confined primarily to the permeable channel; (**c**) Simulated relationship between temperature change and electric potential change along profile 1. The relationship is linear and has a slope of −4.9 mV·K^{-1}, which is approximately equivalent to the thermo-electric (intrinsic) coupling coefficient of $C_T = -5$ mV·K^{-1} incorporated into the model, indicating the potential anomaly is due to the temperature change in the tank for this simulation (modified after [93]).

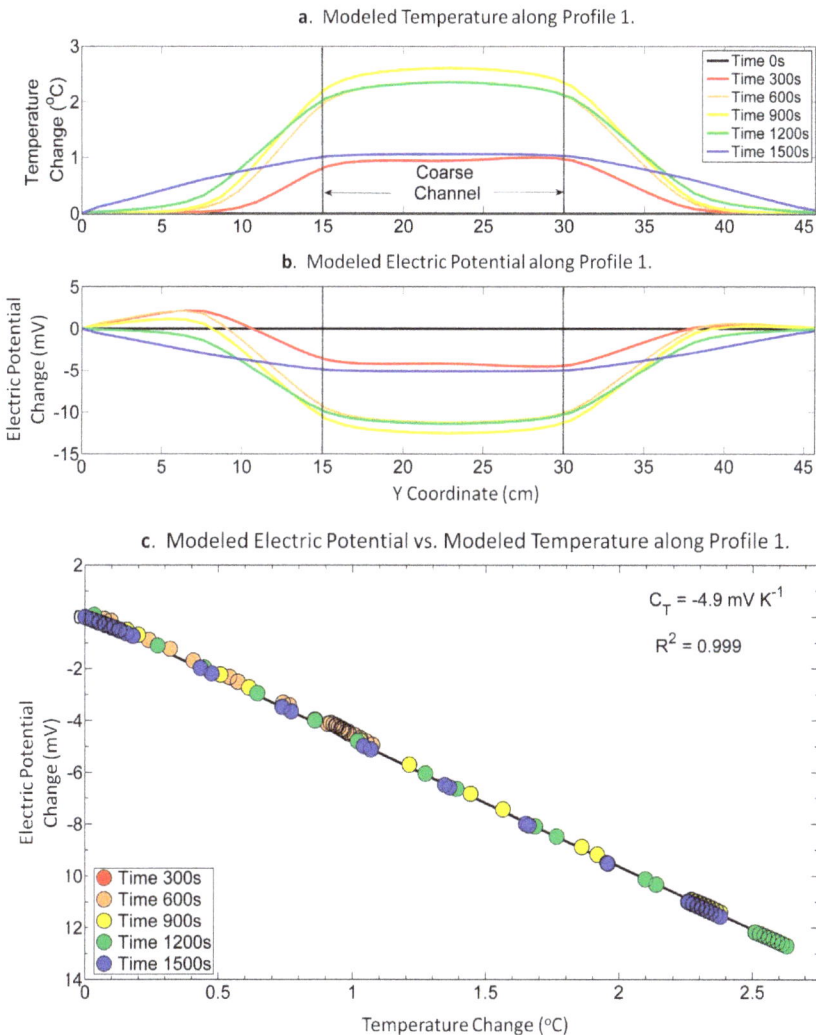

a. Modeled Temperature along Profile 1.

b. Modeled Electric Potential along Profile 1.

c. Modeled Electric Potential vs. Modeled Temperature along Profile 1.

4. Distributed Temperature Sensing

Distributed temperature sensing is based on Raman scattering effect in fiber optic cables. Raman scattering is an inelastic scattering resulting from the interaction of an incident ray of light with the electrons in the molecular bond. Inelastic interaction means that the frequencies of incident and scattered photons are different (Figure 7). Frequency shifts correspond to the vibration frequencies of the electrons. Two states are possible: photons scattered to lower frequencies, corresponding to energy absorbed by the molecules, are called Stokes lines, whereas photons with higher frequencies, corresponding to a loss of energy from the molecules are called anti-Stokes lines. A detailed description of the effects can be found in Selker *et al.* [94].

Figure 7. Sketch of the Raman scattering effect in fiber-optic. Incident light is scattered due to the interaction of photons with electrons. The Anti-Stokes amplitude varies with temperature. Note that Brillouin scattering [94] has been ignored on the sketch.

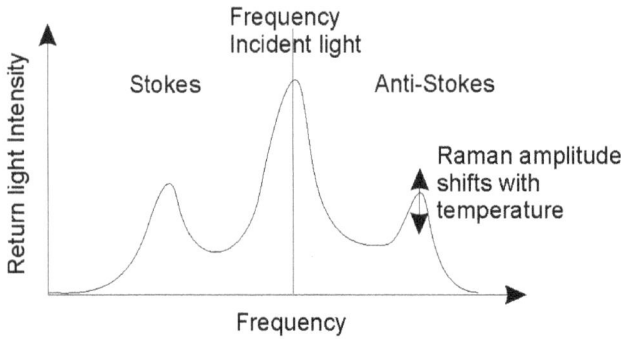

The temperature dependence of Raman scattering is linked to frequency relations between incident and scattered photons. When temperature increases, the number of electrons in high energy states increases, too. Consequently, the ratio of anti-Stokes photons relative to Stokes photons will be higher. The ratio R_r of anti-Stokes to Stokes intensity in the back-scattered light is [95]:

$$R_r = \left(\frac{\lambda_S}{\lambda_A}\right)^4 \exp\left(-\frac{hc\Delta\upsilon}{kT}\right) \qquad (20)$$

where λ_S and λ_A are Stokes and anti-Stokes wavelengths, c is the speed of light in vacuum, $\Delta\upsilon$ is the frequency shift with incident light, h is the Planck's constant, T is the absolute temperature (in Kelvin) and k is the Boltzman constant. This ratio is dependent on the temperature at the position corresponding to the two-way travel time only (it means that it is independent of light intensity). Practically, the diffusion of light and the measuring time result in an integrative rather than punctual measure of temperature. To achieve a high spatial resolution, the speed of light must be precisely known. Commercial fibers generally have a speed around 0.2 m/ns. It means that to 1 m of fiber corresponds 5 ns of signal to analyze with a delay of 10 ns/m for the backscattered signal to arrive (two-way travel time) at the measuring device. To avoid dispersion of light effects, the first and last parts of the signal are trimmed, leading to signal loss. This is a bigger issue for small spatial measuring interval, since the ratio trimmed signal over total signal is higher. The signal strength of Stokes and anti-Stokes photons is also a limit on the precision of the method. Longer integration times, averaging several pulses, will increase the number of photons and thus the resolution of the method (from around 0.1 °C down to 0.01 °C). The use of DTS to determine temperature will thus always be a compromise between, temperature precision, spatial and time resolutions.

Tyler *et al.* [96] provide guidelines for the use of DTS to measure temperature in hydrological studies. DTS allows achieving meter to centimeter resolution over long distances in 1D but requires the installation of boreholes. The measured temperature is that of the groundwater in open systems (Figure 8).

Figure 8. Comparaison of DTS and ERT-derived temperatures in borehole during heat tracing experiment (Figure 4). The agreement is very good which validates the ability of ERT. The difference in the upper part of the aquifer likely results from convection in the borehole. This effect is less important in ERT which provide a temperature on an integrated volume (modified after [55]).

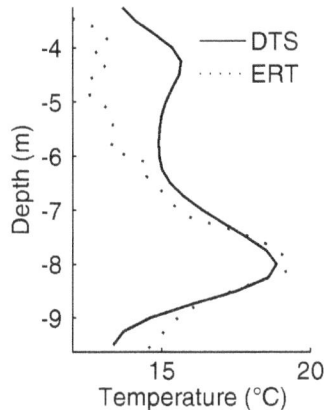

Tyler *et al.* [96] also point out the importance of the calibration procedure to obtain accurate temperature data, taking into account signal attenuation (cables and connections dependent) and temperature offset (laser and sensor dependent). This is generally done using an ice bath at constant temperature (0 °C). Compared to thermocouples and traditional temperature loggers, DTS systems offer the advantage of providing distributed measurements of temperature using a single cable, making its implementation easier. If the measuring device is considerably more expensive in the case of the DTS relative to the thermocouple where a simple voltmeter is sufficient, the cost and installation of the DTS sensors is significantly lower per meter of installation.

5. Previous Works

5.1. Using ERT to Monitor Temperature Changes

ERT has already been applied to study heat reservoirs where hydrothermal fluids generate high resistivity contrasts due to their temperature often exceeding 150 °C. In those situations, ERT can detect the reservoir itself, map preferential flow paths, and be useful to characterize rock properties [97–100]. Recently, several studies were carried out to image volcano hydrothermal systems with very long resistivity cables, showing that ERT is a reliable tool to detect hydrothermal features [101,102].

However, to our knowledge, few studies have used time-lapse ERT to estimate the temperature distribution during heat injection and storage experiment. The first geothermal test using geoelectrical methods was conducted by Benderitter and Tabbagh [103]. They carried out an experiment where the injection of heated water (40 °C) in a 4 to 7 m deep confined aquifer was monitored with DC resistivity measurements. The first potential electrode was fixed and used as a reference; the second one was covering the research area. At the time, the authors produced qualitative anomaly maps using percentage changes in electrical potential. These maps were interpreted using electrical forward modelling calculated for simple geometric subsurface models determined according to the injected volume and the expected temperature. They explained the existence of an electrical anomaly in bulk electrical resistivity (−33%) as the result from the injection of heated water (40 °C).

During the nineties, the development of automated acquisition systems resulted in a strong increase in the use of geoelectrical tomography methods in many contexts. Ramirez et al. [104] used cross-borehole time-lapse ERT to monitor a steam injection during a restoration process. Electrical resistivity being influenced by temperature and saturation effect, it was not possible to derive directly temperature from their measurements. Resistivity was initially expected to increase due to water displacement related to steam injection (decrease in saturation). However, a global decrease in resistivity was observed on the field. This change is explained by an increase in the conductance of exchange cations of clay minerals and increase in the ionic content of water. An US patent was even delivered for the application of the method for relatively high temperature variations in clayey soils [105].

LaBrecque et al. [106] monitored temperature changes within the context of Joule heating combined to vapour extraction during a remediation process with cross-borehole time-lapse ERT. They compared their results with temperature measurements but did not proceed to a conversion of ERT results into temperature. They analysed the variation in conductivity between background and time-lapse series in a partially saturated clay layer. During the first part of the monitoring, the mean temperature increased by 17 °C, and the corresponding conductivity change was coherent with expectations. Then, the temperature reached 100 °C and the change in conductivity was slightly smaller than expected by temperature effects only. It was explained by a decrease in saturation. At the end of heating, conductivity values were much below the background values showing an important loss of water produced by desaturation.

The two examples above show attempts to estimate the temperature effects on electrical resistivity in deposits dominated by clay, where the cation exchange capacity is responsible for the major part of electrical conductivity during strong heating procedure. At the opposite, Hermans et al. [74] went back to the idea developed by Benderitter and Tabbagh [103] to monitor changes in temperature related to the injection of heated water in aquifers. They successfully monitored with a surface profile the 3 days of injection of heated water (48 °C) at a relatively low rate (87 L/h) in a homogeneous sandy aquifer (10.5 °C, 2.5 m thick). Surface time-lapse ERT proved to be able to qualitatively follow such experiments by highlighting heat flow and diffusion around the well. Quantitatively, ERT-derived temperatures were similar to temperatures predicted by a thermo-hydrogeological model developed by Vandenbohede *et al.* [36]. However, a correction term accounting for the discrepancy between formation and injected (tap) water conductivity had to be calculated. This experiment proved that ERT also has a potential for temperature monitoring of clay-free sediments with low temperature variations. These conditions are typical of shallow open-loop geothermal systems.

Hermans *et al.* [55] extended this work for deeper and more complex reservoirs by implementing a new experiment in a heterogeneous sandy gravel alluvial aquifer (7 m thick). A heat injection (3 m^3/h of formation water heated at 38 °C during 24 h) and pumping experiment (30 m^3/h, 13 °C) was designed with a cross-borehole ERT panel crossing the main direction of flow. The ERT-derived temperatures in the panel (maximum change about 8 °C) were coherent with DTS temperatures and groundwater temperature loggers, allowing a spatially and temporally distributed quantitative estimation of temperature within the aquifer. ERT results, consolidated by an important amount of direct measurements, confirmed the heterogeneous nature of the aquifer (e.g., preferential flow paths). ERT proved its ability to detect temperature changes below 1.5 °C and then to follow lower incremental changes. The design of cross-borehole experiment allows more flexibility regarding the depth of the reservoir and the resolution since a major drawback of surface ERT is the loss of resolution/sensitivity with increasing depth [51,107].

Firmbach *et al.* [108] studied heat transport with ERT in an experimental box (1 m × 0.4 m × 0.4 m) with various levels of saturation. This experimental design enables to control temperature and heat fluxes within the box. The qualitative correlation between electrical resistivity and temperature was confirmed for two different media, but they did not provide a quantitative use of ERT to derive temperatures.

Qualitative and quantitative temperature monitoring have presently only been successfully attempted in favourable conditions (shallow aquifers) regarding the resolution/sensitivity of ERT and its depth of investigation. However, plenty of studies showed the ability of ERT in general to follow dynamic processes in less favourable conditions. As an example, Kemna *et al.* [50] and then Müller *et al.* [107] successfully used ERT in the Krauthausen test site in Germany to follow different types of tracer migration within a 10m thick heterogeneous and layered sandy aquifer. Supper *et al.* [109] demonstrated the ability of the method to study seasonal variations in permafrost. Auken *et al.* [110] showed the ability of 3D surface ERT to qualitatively detect changes in groundwater chemistry linked to the injection of CO_2 (gas) in an aquifer to simulate CO_2 leakage in the framework of CO_2 geological storage. Firstly, the resistivity of groundwater decrease linked to the increasing water mineralisation caused by CO_2 dissolution. Then, the resistivity started to increase when some scaling occurred (e.g., calcite). The authors confronted their 3D images with lots of ground truth data and showed that 3D time-lapse images reproduced well the affected areas.

Robert *et al.* [51] were able to qualitatively follow the injection of a salt solution in a fractured area in carboniferous limestone. With the help of two parallel surface ERT profiles, they managed to find the groundwater flow direction and the preferential flow paths at a depth of 20 m by taking great care during the experiment dimensioning and the data acquisition.

These studies present all the ability of non-invasive techniques such as ERT to follow dynamic processes occurring in aquifers with different geological conditions but also different depths that could be met during geothermal site prospection/characterization.

5.2. Petrophysical Considerations Regarding Electrical Conductivity

Equation (13) is the key to provide a quantitative estimation of temperatures using electrical resistivity temperatures. In addition to ERT measurements, it requires one to determine the fractional change per degree Celsius. A common practice is to collect a sample of formation water and to verify the linear relationship in the laboratory. The value of $m_{f,25}$ is generally around 0.02. The representativeness of such a lab experiment may be questionable since the conditions are not representative of the subsurface systems. The test is sufficient to estimate the increase of electrical resistivity due to the viscosity effect. However, in presence of chemical and kinetics effects, the results may depend on the duration of the test (how quickly the temperature increases) and may neglect reactions occurring between the pore water and the solid matrix. These effects are not expected to be important for low temperature systems (<30 °C), at least for short experiments.

Another parameter to determine in Equation (13) is the initial temperature in the aquifer T_1, in geothermal studies, such values will generally be available through direct measurements. Applying Equation (13) to a whole ERT section requires a few assumptions. First, it requires determining an initial temperature everywhere in the section. The aim of ERT being to provide a spatial distribution of temperature, such value is generally not available. It is thus necessary to consider a constant temperature in the aquifer or to propose an interpolation or geostatistical estimation of the temperatures based on available measurements in boreholes.

Another assumption is that $m_{f,25}$ is constant everywhere in the section. As stated previously, $m_{f,25}$ is mainly related to the variation of viscosity of water and should not largely vary. However, small changes are possible, related to the chemistry of the pore water, and thus to its specific electrical conductivity $\sigma_{f,25}$. However, considering a unique reservoir with relatively constant properties and the range of variations of $m_{f,25}$, taking a constant value should not lead to strong discrepancies. This should be more deeply investigated for contrasted reservoirs such as polluted sites or coastal aquifers.

Static ERT provides an estimate of electrical resistivity, but it is rarely able to provide quantitative estimates of indirect parameters because it depends on many factors (Equation (2)). When considering temperature estimates in saturated soil/rock, the hidden assumption is often that the specific electrical conductivity of water does not vary with time. If it is not the case, a correction term must be applied before deriving temperatures. The process is similar to temperature corrections applied to time-lapse ERT results when monitoring other phenomena [40,58–60].

As an example, Hermans et al. [74] had to correct their ERT-derived temperatures because a difference in specific electrical conductivity existed between formation and injection waters. They calculated the correction term using a simulation of the injection process, injecting less conductive water into the aquifer. A side effect of the correction term was to partly counterbalance the smoothing effect of ERT inversion. Such methodology has to be applied when it is not possible to use formation water for injection. In practice, even if formation water is reinjected, scaling may occur in the heat exchange process and such corrections might have to be made.

It must also be kept in mind that changes in temperature may have side effects that are not taken into account in Equation (12). Indeed, an increase in temperature does not influence water electrical conductivity only, it also influences chemical and physical processes such as reaction constants or kinetics of reaction (e.g., [22,23]).

Hermans et al. [111] observed that for a longer term experiment, Equation (13) was not able to reproduce correctly temperatures even if electrical resistivity were correctly retrieved as shown by electromagnetic logs. This behavior was subsequently investigated in the laboratory by Robert et al. [112]. Sand and water samples were collected on the site investigated by Hermans et al. [74] to reproduce the heat and storage experiment in a saturated soil column.

The heating experiment consisted of increasing the temperature of the column from 20 °C (ambient temperature) to 60 °C, whereas the temperature of injection on the field was about 50 °C. The experiment was performed on a column filled with formation water in equilibrium with the soil.

The monitoring of the column shows an increase in conductivity with the increase of temperature. Figure 9 compares the observed behavior during the test and the expected behavior according to Equation (13). Up to 30 °C, the increase in conductivity is coherent with the proposed law. Then, the error becomes more and more important; the measured resistivity is too low compared to the expected one. The reason for this behavior lies in the decrease in solubility of calcium carbonates, corroborated with chemical analysis of water samples taken before and after the experiment.

Figure 9. The calculated bulk electrical conductivity (Equation (7)) is not coherent with the measured conductivity due to chemical reactions in the sample (modified after [112]). The missing data correspond to a bad electrical contact on one electrode of the column.

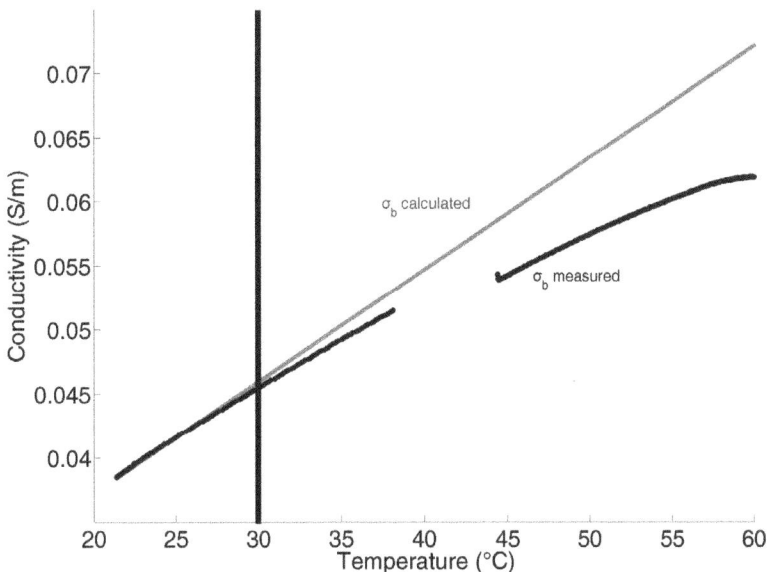

An effect of temperature can also appear on the linear relationship between water electrical conductivity and temperature (Figure 10). A test was performed on a sample with a tendency to precipitate calcium carbonate. Up to 40 °C, a linear relationship is coherent as the samples show a constant slope. Above 45 °C, the increase in electrical conductivity is smaller than expected, this yields a decrease in $m_{f,25}$ with temperature. In such case, $m_{f,25}$ is dependent on the temperature and Equation (18) becomes non-linear which makes more difficult the prediction of temperature from ERT:

$$T = \frac{\sigma_{b2,T}}{\sigma_{b1}}(T_1 - 25) + \frac{\dfrac{\sigma_{b2,T}}{\sigma_{b1}} - 1}{m_{f,25}(T)} + 25 \tag{21}$$

Using electrical resistivity to estimate temperatures is thus possible. However, it is necessary to verify if the assumptions made to model the petrophysical relationship are rational. Otherwise, one needs to further model chemical reactions to correct for these effects. This is not a simple task since the problem of retrieving the temperature from resistivity changes may become non-linear (Equation (21)).

Figure 10. Water electrical conductivity increases with temperature but with a non-linear behaviour. Consequently, the fractional change per degree Celsius m_f is not constant with temperature.

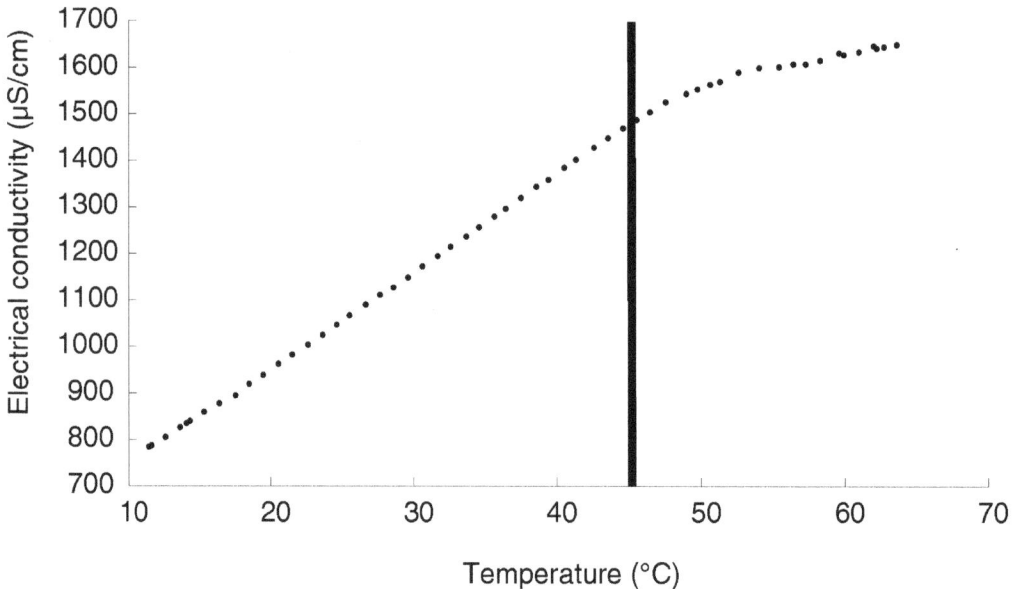

5.3. ERT Survey Design

A distinction has to be done between surface and cross-borehole ERT. Surface ERT is particularly suited for monitoring application. It is non-invasive since it only requires electrodes at the ground surface. However, surface measurements suffer from poor resolution at depth [45,113,114] even if Robert *et al.* [51] successfully managed to follow a salt tracer in fractures at a depth of 20 to 30 m. Hermans *et al.* [74] propose guidelines deduced from their study case to design surface arrays for monitoring studies. Using 62 electrodes with an electrode spacing a, they successfully imaged a heat plume $3.33a$ thick, $4a$ wide and at a depth of $4a$. The minimum temperature variations detected is dependent on the propagation of noise which was evaluated to about 10%.

Cross-borehole ERT enables to maintain sufficient resolution at depth. Electrodes can be fixed at the outer edge of the borehole or mounted on cables with the borehole fluid ensuring electrical contact (require a screen all along the borehole). In the latter case, the fluid contained in the borehole influences the measurement (borehole fluid effect, see [115]). The sensitivity is maximum near the borehole and decrease in the middle of the section. A minimum aspect ratio (equipped length over distance between boreholes) has to be maintained to ensure enough resolution in this part of the model. Using 13 electrodes in each borehole with an electrode spacing a, Hermans *et al.* [55] imaged a heterogeneous heat plume $5a$ thick and $4a$ wide, with an aspect ratio (ratio of the distance between boreholes over length equipped with electrode) of 0.75.

As an imaging technology resulting from a deterministic regularized inversion process, ERT is subject to limitations due to resolution patterns [116]. For surface ERT, the resolution pattern is strongly depth-dependent and the ability of the method to image temperature variations will rapidly decrease with depth. For cross-hole ERT, difference may appear in zones close to boreholes compare to the centered part of the channel. Those aspects have to be taken into account in the interpretation of tomograms transformed with petrophysical models.

5.4. Sensitivity of Self-Potential Signals to Temperature

We first describe a sandbox experiment to estimate the amplitude of the self-potential anomalies associated with a heat source in absence of ground water flow. The sandbox was filled with a silica sand and demineralized water. We used two non-polarizing $Pb/PbCl_2$ (Petiau) electrodes and an MX20 voltmeter (sensitivity 0.1 mV, internal impedance 100 MΩ) for the self-potential measurements. The reference electrode (ref) was located on the corner of the sandbox and its temperature was monitored over time (19.2 ± 0.1 °C). The other electrode was used to scan the electrical potential at the surface of the sand. The measurements were done in such a way that the temperature of the scanning electrode was kept constant in order to avoid artefacts in the experiment due to the difference of temperature between the scanning and the reference electrodes. Thermal probes and a digital thermometer were used to measure the temperature distribution. The thermal probes were inserted at a depth of 20 ± 1 cm prior the beginning of the experiment with a spacing of 5 cm (see Figure 11). The temperature was measured with an accuracy of 0.2 °C.

Self-potential and temperature data were gathered 5 min prior the introduction of the heat source to get reference profiles. The reference temperatures were 20.8 ± 0.1 °C. An amount of 9 g of a chemical heater was introduced in the tank at $t = 0$. The chemical heater was put dry at the bottom of a tube (2 cm in diameter) with dry sand above and a cap at the bottom. The tube was poured in the sandbox at a depth of 20 cm (see positions in Figure 11). These experiments will be referred as Experiments #1 and #2, respectively. The cap was pushed and the chemical heater poured in the sandbox and the tube removed. The chemical heater is a Flameless Ration Heater (FRH), which is a water-activated exothermic heater. The exothermic chemical reaction is:

$$Mg + 2H_2O \rightarrow Mg(OH)_2 + H_2 + Heat \qquad (22)$$

with about 350 kilojoules of heat produced per kg. Then the temperature was monitored (at a depth of 20 cm) and the self-potential profiles were repeated for 80 min. We also monitored potential changes in the electrical conductivity of the pore water close to the chemical heater. Once corrected for temperature, we found no notable changes in the pore water conductivity.

A negative self-potential anomaly was observed above the heat source. Modeling indicates that the thermoelectric coupling coefficient was −0.5 mV per °C. This indicates that heat pulses can be measured non-intrusively and the result in self-potential signals can be inverted to retrieve some characteristic of the heat source or material properties such as the thermal diffusivity of the material.

We describe now a field example showing how self-potential and resistivity can be used in concert to locate a heat source in the ground. This heat source corresponds to a shallow coal seam fire located near Denver (CO, USA). The burning front is located in the Gorham subbituminous coal formation, located at a depth of about 10 m (Figure 12). Revil *et al.* [91] obtained new self-potential (with an anomaly of −50 mV) and resistivity data along the profile shown in Figure 12a,b. These data were analyzed jointly to localize along a single profile the position of the burning front (Figure 12c). This case study shows how self-potential and resistivity can be used to localize and eventually monitor a shallow heat source.

Figure 11. Distribution of temperature (at a depth of 20 cm), self-potential (at the top of the tank) at a given time after the introduction of the heat source Q_H at a depth of 20 cm in a sandbox. We have removed the temperature and self-potential distributions recorded prior the introduction of the heat source (modified after [91]).

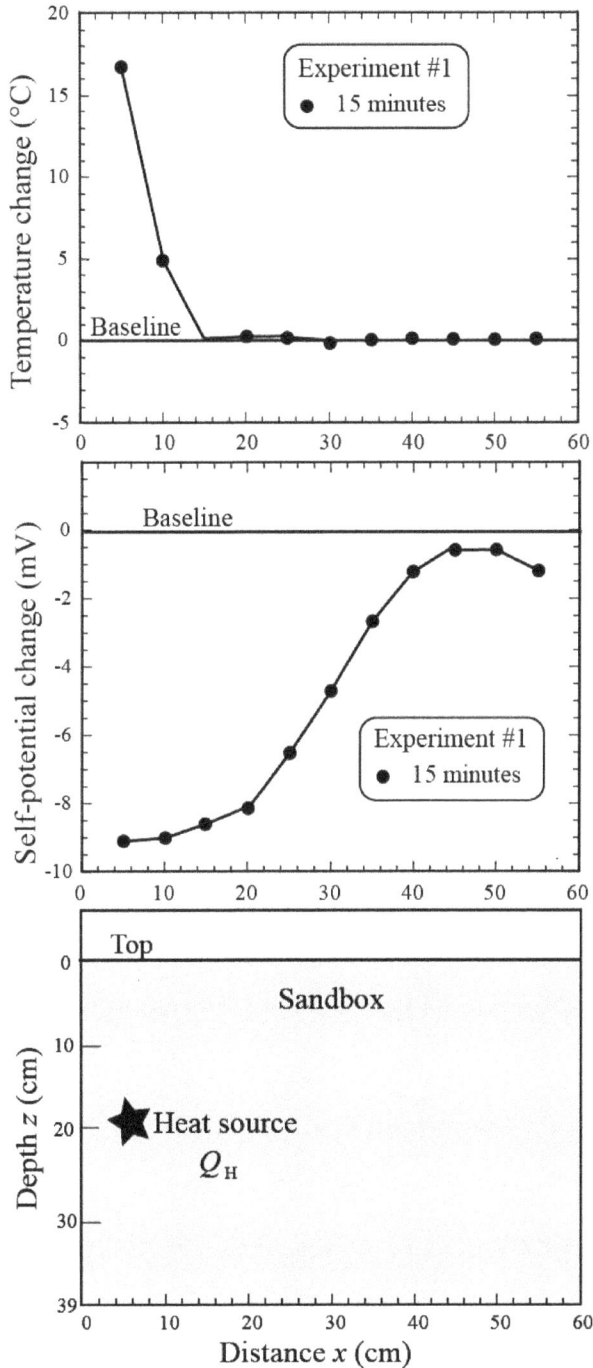

Figure 12. Self-potential (SP) and resistivity data. (**a**) Self-potential data (79 stations performed at the ground surface) showing a self-potential anomaly at about 175 mV over the burning area; (**b**) Electrical resistivity tomogram (714 apparent resistivity data, Wenner-α, 79 electrodes, inverted with a Gauss-Newton algorithm with isotropic smoothing). Note the low resistivity anomaly (2 Ω·m) at about 175 m below the negative self-potential anomaly (−50 mV); (**c**) Burning front index based on the self-potential and resistivity values. High values correspond to a high probability zone in terms of recovering the position of the burning front. The depth of the maximum of the NBI coincides with the depth of the coal bed (approximately 10 m) (modified after [91]).

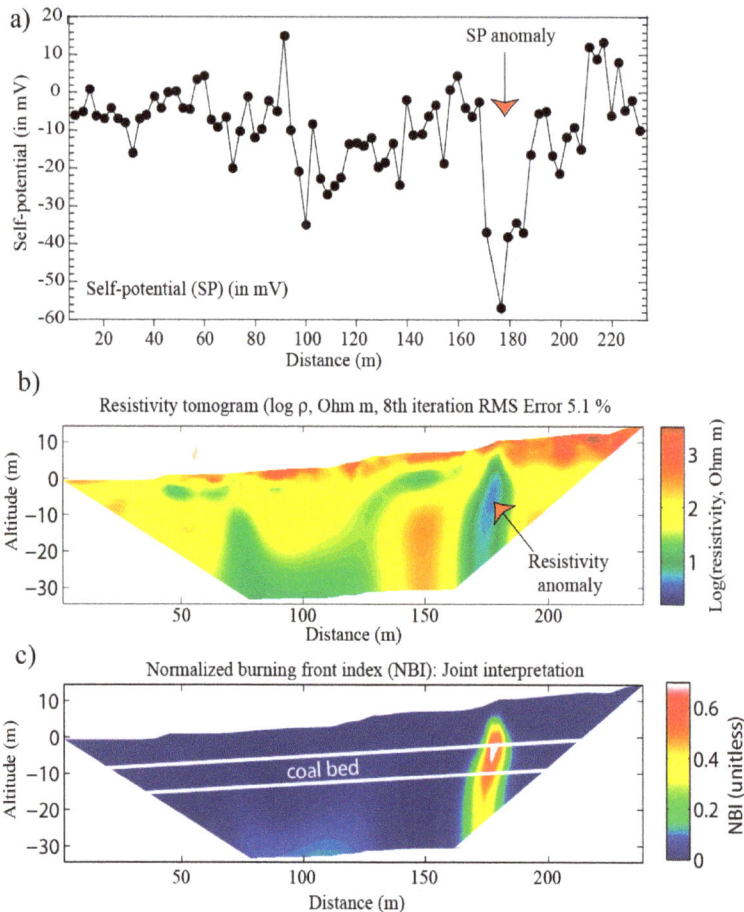

5.5. Using DTS to Measure Temperature

DTS systems were first used for fire or pipeline monitoring and used existing communication utilities rather than dedicated cables. However, in the last decade, the development of designed measuring systems in hydrogeology and geothermal systems has grown.

In the last decade, the method gained popularity among scientists concerned by hydrology. One of the main utilization of DTS in environmental hydrological studies concerns the surface water/groundwater interface. Groundwater temperature being almost constant with time, groundwater inflows may be detected with decrease in stream temperature in summer and increase in winter. Selker *et al.* [117] used the Raman-backscatter DTS along multimode fiber-optic cable with a temperature precision of 0.01 °C every meter to estimate stream temperature dynamics and groundwater inflows. Such a resolution is obtained by integrating signals over one hour time period. They used their results to derive groundwater temperatures and relative groundwater inflows. Lowry *et al.* [118] investigated groundwater discharge in a wetland stream using DTS. They repeated measurements every 15 min to propose a high resolution spatio-temporal monitoring of temperatures with the fiber buried in streambed sediments. They observe that temperature anomalies correlated with groundwater discharges with constant position in time. DTS temperatures correlated well with temperature data loggers.

DTS was also used to derive seepage rates in streams by determining the vertical temperature profiles. Vogt *et al.* [119] used DTS to get high resolution vertical temperature profiles in surface water sediments in order to derive seepage rates over depth and time. They wrapped the fiber around a PVC tube installed in the streambed sediments, as recommended by Selker *et al.* [94]. This configuration enables to refine the spatial resolution to about 5 mm instead of 1m which is the standard order of magnitude. Similarly, Mamer and Lowry [120] studied groundwater discharge to streams using paired fiber-optic cables. They propose a set up where two cables are parallel, one on the top of the other with a small vertical separation and tested this configuration in 10 m long sandbox. The overlapping time series measurements are then used to estimate fluxes along the stream longitudinally with the amplitude-shift method.

Applications in lakes, although more complex due to surface discharge area, exist too. Sebok *et al.* [121] used DTS to map spatial and temporal changes in temperature on a lakebed area to confirm the presence of a relatively high discharge of groundwater. The cable was spread out an area of 25 m × 6 m. The measurement through the seasons showed that the extent of the discharge zone was changing as well as its position relative to the shore. They also used a looped layout for multilevel lake temperature measurements.

The use of DTS in such applications has become standard. Consequently, many efforts are now done for the improvement of processing and interpretation of the spatio-temporal data sets. Lauer *et al.* [122] tested the fiber-optic DTS method to detect groundwater discharge to streams to assess uncertainty and limit of detections of the method. They implemented artificial upstream discharge in a stream with controlled inflow rates and temperatures. The sensitivity of the method appeared to be relatively high since DTS was able to detect discharge of 2% of natural flow. Krause and Blume [123] determined the impacts of seasonal variability in signal strengths and monitoring modes on the accuracy of fiber-optic DTS systems used to analyze thermal patterns in aquifer-river interfaces. They found that the stability in signal strength was better in winter and that two-way single ended averaging surveys were the best suited for monitoring. Mwakanyamale *et al.* [124] used fiber-optic DTS temperature measurements and propose a new approach combining spectral analysis and discriminant analysis to process DTS data and detect zones of exchange between surface water and groundwater more objectively. Blume *et al.* [125] compare upscaling approaches utilizing lacustrine groundwater discharge rates derived from fiber-optic DTS measurements. The major issue for lacustrine groundwater discharges lies in the identification of 2D patterns. Two transfer functions integrating DTS transects were developed and compared to a simple exponential decline of discharge rates perpendicular to the shoreline.

Ciocca *et al.* [126] used a fiber optic system to estimate soil moisture in alysimeter experiment. They used the metal around the fiber optic cable as an electrical resistance heater to generate heat pulse. The temperature of the soil was then monitored with DTS to derive the thermal conductivities of the soil. The latter were used to estimate water content through a calibrated petrophysical relationship that was compared to standard measurements of soil moisture with a relatively good agreement in wet conditions.

However, application of DTS in hydrogeology is not limited to surface or near-surface applications. Borehole applications, that are logistically similar to open loop systems, are also common. Hurtig *et al.* [127] reported one of the first utilization of DTS in hydrogeology. Cables were placed into a 40 m deep inclined borehole during injection of hot and cold water. With temperature profiling, a fracture was detected at the position of a sharp decrease in temperature. Macfarlane *et al.* [128] evaluated aquifer properties by heating fluid and stimulating flow (forced gradient) between wells. Temperatures were recorded using DTS in transient thermal conditions in injection and production wells, with short screen intervals. This thermal tracer test highlighted a zone of higher hydraulic conductivity between injection and pumping wells. Yamano and Goto [129] used DTS for long-term monitoring (6 years) of a deep borehole to investigate an active fault in Japan. They used a spatial resolution of 1 m with an expected resolution of 0.1 K. They coupled temperature measurement with cold water injection to detect leakage zones. Leaf *et al.* [130] monitored advective heat movement in borehole dilution tests with DTS. They tested several thermal tracer dilution experimental designs (continuous, pulse injection, single and multiple locations). Large screened intervals enabled them to use vertical temperature profiles as an indicator of vertical heterogeneity in the aquifer and of inflows from fractures and porous media. Read *et al.* [131] performed heat transport tests in fractured media using DTS for temperature monitoring in injection and pumping wells. They monitored thermal dilution tests to detect cross-flowing fractures and assess the cross flow rate. A cross well thermal tracer test was performed to identify the connection between boreholes through the fracture network. Banks *et al.* [132] used fiber-optic DTS to measure temperature profiles in open groundwater wells in fractured rock. They used electrical heating cables to heat the water in the well and create a temperature difference with the surrounding aquifer. Temperature profiles were used to identify active fracture zones within the aquifer. In open systems, the use of DTS to measure the temperature variations allows avoiding or at least minimizing forced convection effects in boreholes, which can occur when performing multilevel well sampling.

It is only very recently that such measures began to be made to characterize and eventually design geothermal systems, in particular for closed loop systems. Fujii *et al.* [133] first proposed to study thermal response test (TRT) with distributed temperature measured by optic fiber. However, they located the fiber outside of the U-pipe used for TRT. This configuration is not optimal because it is difficult to control the closeness of the fiber with the U-pipe in the ground material. Fujii *et al.* [134] included DTS measurements as a part of a thermal response test to determine ground thermal conductivities from 1 to 2 m thick layers. This time the cables are located inside the U-pipe. They show the reliability of the technique for the interpretation of thermal response test. Acuna *et al.* [135] similarly performed a distributed thermal response test with the fiber optic cables enclosed in the U-pipe. Such measures have helped to demonstrate and quantify a significant difference between the average thermal resistance of drilling obtained by thermal response tests and estimated thermal resistances along the drilling using distributed measurements.

With some adaptations, the use of DTS for deep geothermal systems is also possible. Reinsch *et al.* [136] installed a fiber-optic cable behind the cemented annulus of the casing of a high temperature geothermal well. Repeated measurements were used to detect mechanical, thermal and chemical degradation of the fiber. At high temperature, geothermal fluids may contain hydrogen which degrades the fiber and distort the optical signal through absorption. Mechanical stress may generate bending loss in the signal. Those conditions required to use specific cables with polyamide and hermetic carbon coatings. The measurements showed that the cable was damaged at several positions during installation but without preventing temperature measurements.

There is no limitation in the possible utilizations of the DTS technology. Selker *et al.* [94] reported several applications, including temperature measurements in a decommissioned mine shaft in Czech Republic and air-snow and air-water interfaces studies. Yilmaz and Karlik [95] incorporated DTS measurements for the monitoring of power cables.

6. Conclusions

Groundwater contributes to a major part in the production of geothermal energy, rather directly (GWHP or ATES systems) or indirectly (GSHP or BTES systems), by taking advantage of its inherent temperature stability for the operation of heat pumps. Very low temperature systems (<30 °C) are generally located at depth between 0 and 100 m, thus they are much more easily accessible and involve lower implementation costs than deeper high temperature systems (*i.e.*, drilling costs). Moreover, very low temperature reservoirs, such as shallow aquifers, are relatively abundant in alluvial or coastal plains where urban development concentrates. At these depths, the subsurface is very heterogeneous and complex to characterize. The relatively wide range of natural variations both in terms of thermal and hydraulic properties makes designing efficient geothermal systems and predicting its impact on the environment challenging tasks.

In this context, new technologies are clearly needed to monitor the spatial and temporal distribution of temperature in the reservoir to: (1) better design the geothermal system and the monitoring network; (2) prevent any thermal feedback/recycling; and (3) image and control the thermal affected zone. Three emerging geophysical techniques in this field have been reviewed in this paper: electrical resistivity tomography (ERT) also known as electrical resistivity imaging (ERI) which allows to obtain 2D or 3D images of the temperature variations in the subsurface non-invasively; self-potential method (SP) which is a very fast method to map anomalies of the ambient electrical potential at the surface corresponding to temperature anomalies in the subsurface; and fiber-optic distributed temperature sensor (DTS) which provides linear measurements of temperature with centimetric resolution in boreholes. Whereas the latter provides a direct estimation of the temperature with little uncertainties, it requires the installation of boreholes to set up the fiber optics. The former two methods are mature enough to estimate temperature variations with respect to a given background in an almost non-invasive manner in the short term.

Challenges remain in terms of improving ERT imaging through advanced inversion algorithm and further strengthening the petrophysical relationships needed to obtain temperature changes for long term experiment. In this regard, research perspectives should focus on the incorporation of chemical and kinetic effects in the interpretation of the geophysical data.

Acknowledgments

We thank the F.R.S.-FNRS (Fonds de la Recherche Scientifique) and the Fondation Roi Baudouin—Prix Ernest Dubois (grant No. 2013-8126501-F002) for their financial support of Thomas Hermans. We also thank three anonymous reviewers for their comments and suggestions.

Author Contributions

Thomas Hermans contributed to the redaction of the sections related to ERT and time-lapse ERT, petrophysical considerations, DTS and previous works in ERT and DTS. He also ensured the coordination of the paper. Frédéric Nguyen contributed to the redaction of sections dedicated to ERT and time-lapse ERT as well as DTS and previous works in DTS. He also reviewed other sections of the paper. Tanguy Robert mainly contributed to the sections concerning ERT, time-lapse ERT and previous works in ERT. He also reviewed other parts of the paper. Finally, the major contributions of Andre Revil were the sections dedicated to SP, previous works in SP, and petrophysical considerations. He also reviewed other sections of the paper.

Conflicts of Interest

The authors declare no conflict of interest.

References

1.   Bayer, P.; Rybach, L.; Blum, P.; Brauchler, R. Review on life cycle environmental effects of geothermal power generation. *Renew. Sustain. Energy Rev.* **2013**, *26*, 446–463.
2.   Saner, D.; Juraske, R.; Kübert, M.; Blum, P.; Hellweg, S.; Bayer, P. Is it only CO_2 that matters? A life cycle perspective on shallow geothermal systems. *Renew. Sustain. Energy Rev.* **2010**, *14*, 1798–1813.
3.   Andersson, O. ATES utilization in Sweden—An overview. In Proceedings of the MEGASTOCK'97 7th International Conference on Thermal Energy Storage, Sapporo, Japan, 18–20 June 1997; pp. 925–930.
4.   Vanhoudt, D.; Desmedt, J.; van Bael, J.; Robeyn, N.; Hoes, H. An aquifer thermal storage system in a Belgian hospital: Long-term experimental evaluation of energy and cost savings. *Energy Build.* **2011**, *43*, 3657–3665.
5.   Blum, P.; Campillo, G.; Kölbel, T. Techno-economic and spatial analysis of vertical ground source heat pump systems in Germany. *Energy* **2011**, *36*, 3002–3011.

6. Haehnlein, S.; Bayer, P.; Ferguson, G.; Blum, P. Sustainability and policy for the thermal use of shallow geothermal energy. *Energy Policy* **2013**, *59*, 914–925.

7. Lund, J.W. Direct utilization of geothermal energy. *Energies* **2010**, *3*, 1443–1471.

8. European Commission. Horizon 2020: The EU Framework Programme for Research and Innovation. Secure, Clean and Efficient Energy. Available online: http://ec.europa.eu/programmes/horizon2020/en/h2020-section/secure-clean-and-efficient-energy (accessed on 11 August 2014).

9. Lund, J.W.; Freeston, D.H.; Boyd, T.L. Direct application of geothermal energy: 2005 worldwide review. *Geothermics* **2005**, *34*, 691–727.

10. Allen, A.; Milenic, D. Low-enthalpy geothermal energy resources from groundwater in fluvioglacial gravels of buried valleys. *Appl. Energy* **2003**, *74*, 9–19.

11. Haehnlein, S.; Bayer, P.; Blum, P. International legal status of the use of shallow geothermal energy. *Renew. Sustain. Energy Rev.* **2010**, *14*, 2611–2625.

12. Mattsson, N.; Steinmann, G.; Laloui, L. Advanced compact device for the *in situ* determination of geothermal characteristics of soils. *Energy Build.* **2008**, *40*, 1344–1352.

13. Raymond, J.; Therrien, R.; Gosselin, L.; Lefebvre, R. A review of thermal response test analysis using pumping test concepts. *Groundwater* **2011**, *49*, 932–945.

14. Haffen, S.; Geraud, Y.; Diraison, M.; Dezayes, C. Determination of fluid-flow zones in a geothermal sandstone reservoir using thermal conductivity and temperature logs. *Geothermics* **2013**, *46*, 32–41.

15. Busby, J.; Lewis, M.; Reeves, H.; Lawley, R. Initial geological considerations before installing ground source heat pump systems. *Q. J. Eng. Geol. Hydrogeol.* **2009**, *42*, 295–306.

16. De Paly, M.; Hecht-Méndez, J.; Beck, M.; Blum, P.; Zell, A.; Bayer, P. Optimization of energy extraction for closed shallow geothermal systems using linear programming. *Geothermics* **2012**, *43*, 57–65.

17. Liang, J.; Yang, Q.; Liu, L.; Li, X. Modeling and performance evaluation of shallow ground water heat pumps in Beijing plain, China. *Energy Build.* **2011**, *43*, 3131–3138.

18. Lo Russo, S.; Civita, M.V. Open-loop groundwater heat pumps development for large buildings: A case study. *Geothermics* **2009**, *38*, 335–345.

19. Molson, J.W.; Frind, E.O.; Palmer, C.D. Thermal energy storage in an unconfined aquifer: 2. Model development, validation and application. *Water Resour. Res.* **1992**, *28*, 2857–2867.

20. Palmer, C.D.; Blowes, D.W.; Frind, E.O.; Molson, J.W. Thermal energy storage in an unconfined aquifer: 1. Field injection experiment. *Water Resour. Res.* **1992**, *28*, 2845–2866.

21. Warner, D.L.; Algan, U. Thermal impact of residential ground-water heat pumps. *Groundwater* **1984**, *22*, 6–12.

22. Bonte, M. Impacts of Shallow Geothermal Energy Storage on Groundwater Quality—A Hydrochemical and Hydromicrobial Study of the Effects of Ground Source Heat Pumps and Aquifer Thermal Storage. Ph.D. Thesis, VU University Amsterdam, Amsterdam, The Netherlands, 2013.

23. Garnier, F. Contribution à l'évaluation Biogéochimique des Impacts liés à l'exploitation Géothermique des Aquifères Superficiels: Expérimentations et Simulations à l'échelle d'un Pilote et d'installations Réelles. Ph.D. Thesis, Bureau de Recherche Géologique et Minière—Institut des Sciences de la Terre d'Orléans, Université d'Orléans, Orléans, France, 2012. (In French)

24. Jesußek, A.; Grandel, S.; Dahmke, A. Impacts of subsurface heat storage on aquifer hydrogeochemistry. *Environ. Earth Sci.* **2013**, *69*, 1999–2012.

25. Brielmann, H.; Griebler, C.; Schmidt, S.I.; Michel, R.; Lueders, T. Effects of thermal energy discharge on shallow groundwater ecosystems. *FEMS Microbiol. Ecol.* **2009**, *68*, 273–286.

26. Lerm, S.; Westphal, A.; Miethling-Graff, R.; Alawi, M.; Seibt, A.; Wolfgramm, M.; Würdemann, H. Thermal effects on microbial composition and microbiologically induced corrosion and mineral precipitation affecting operation of a geothermal plant in a deep saline aquifer. *Extremophiles* **2013**, *17*, 311–327.

27. Vetter, A.; Mangelsdorf, K.; Wolfgramm, M.; Rauppach, K.; Schettler, G.; Vieth-Hillebrand, A. Variations in fluid chemistry and membrane phospholipid fatty acid composition of the bacterial community in a cold storage groundwater system during clogging events. *Appl. Geochem.* **2012**, *27*, 1278–1290.

28. Lo Russo, S.; Gnavia, L.; Roccia, E.; Taddia, G.; Verda, V. Groundwater Heat Pump (GWHP) system modeling and Thermal Affected Zone (TAZ) prediction reliability: Influence of temporal variations in flow discharge and injection temperature. *Geothermics* **2014**, *51*, 103–112.

29. Lo Russo, S.; Taddia, G.; Verda, V. Development of the thermally affected zone (TAZ) around a groundwater heat pump (GWHP) system: A sensitivity analysis. *Geothermics* **2012**, *43*, 66–74.

30. Gao, Q.; Zhou, X.-Z.; Jiang, Y.; Chen, X.-L.; Yan, Y.-Y. Numerical simulation of the thermal interaction between pumping and injecting well groups. *Appl. Therm. Eng.* **2013**, *51*, 10–19.

31. Milnes, E.; Perrochet, P. Assessing the impact of thermal feedback and recycling in open-loop groundwater heat pump (GWHP) systems: A complementary design tool. *Hydrogeol. J.* **2013**, *21*, 505–514.

32. Anderson, M.P. Heat as a ground water tracer. *Groundwater* **2005**, *43*, 951–968.

33. Wikdemeersch, S.; Jamin, P.; Orban, P.; Hermans, T.; Nguyen, F.; Brouyère, S.; Dassargues, A. Coupling heat and chemical tracer experiments for estimating heat transfer parameters in shallow alluvial aquifers. *J. Contam. Hydrol.* **2014**, submitted.

34. Saar, M.O. Review: Geothermal heat as a tracer of large-scale groundwater flow and as a means to determine permeability fields. *Hydrogeol. J.* **2011**, *19*, 31–52.

35. Giambastiani, B.M.S.; Colombanii, N.; Mastrocicco, M. Limitation of using heat as a groundwater tracer to define aquifer properties: Experiment in a large tank model. *Environ. Earth Sci.* **2013**, *70*, 719–728.

36. Vandenbohede, A.; Hermans, T.; Nguyen, F.; Lebbe, L. Shallow heat injection and storage experiment: Heat transport simulation and sensitivity analysis. *J. Hydrol.* **2011**, *409*, 262–272.

37. Vandenbohede, A.; Louwyck, A.; Lebbe, L. Conservative solute *versus* heat transport in porous media during push-pull tests. *Transp. Porous Med.* **2009**, *76*, 265–287.

38. Brouyère, S. Etude et Modélisation du Transport et du Piégeage des Solutés en Milieu Souterrain Variablement Saturé. Ph.D. Thesis, University of Liege, Liege, Belgium, 2001. (In French)

39. Jardani, A.; Revil, A.; Dupont, J.P. Stochastic joint inversion of hydrogeophysical data for salt tracer test monitoring and hydraulic conductivity imaging. *Adv. Water Resour.* **2013**, *52*, 62–77.

40. Hayley, K.; Bentley, L.R.; Gharibi, M.; Nightingale, M. Low temperature dependence of electrical resistivity: Implications for near surface geophysical monitoring. *Geophys. Res. Lett.* **2007**, *34*, doi:10.1029/2007GL031124.

41. Revil, A.; Cathles, L.M.; Losh, S.; Nunn, J.A. Electrical conductivity in shaly sands with geophysical applications. *J. Geophys. Res.* **1998**, *103*, 23925–23936.

42. Waxman, M.H.; Smits, L.J.M. Electrical conductivities in oil-bearing shaly sands. *Soc. Pet. Eng. J.* **1968**, *8*, 107–122.

43. Vereecken, H.; Binley, A.; Cassiani, G.; Revil, A.; Titov, K. *Applied Hydrogeophysics. NATO Science Series: IV Earth and Environmental Sciences*; Springer: Berlin, Germany, 2006.

44. Hermans, T.; Vandenbohede, A.; Lebbe, L.; Martin, R.; Kemna, A.; Beaujean, J.; Nguyen, F. Imaging artificial salt water infiltration using electrical resistivity tomography constrained by geostatistical data. *J. Hydrol.* **2012**, *438–439*, 168–180.

45. Nguyen, F.; Kemna, A.; Antonsson, A.; Engesgaard, P.; Kuras, O.; Ogilvy, R.; Gisbert, J.; Jorreto, S.; Pulido-Bosch, A. Characterization of seawater intrusion using 2D electrical imaging. *Near Surf. Geophys.* **2009**, *7*, 377–390.

46. Binley, A.; Winship, P.; West, L.J.; Pokar, M.; Middleton, R. Seasonal variation of moisture content in unsaturated sandstone inferred from borehole radar and resistivity profiles. *J. Hydrol.* **2002**, *267*, 160–172.

47. Chambers, J.E.; Gunn, D.A.; Wilkinson, P.B.; Meldrum, P.I.; Haslam, E.; Holyoake, S.; Kirkham, M.; Kuras, O.; Merrit, A.; Wragg, J. 4D electrical resistivity tomography monitoring of soil moisture dynamics in an operational railway embankment. *Near Surf. Geophys.* **2014**, *12*, 61–72.

48. Atekwana, E.A.; Sauck, W.A.; Werkema, D.D., Jr. Investigations of geoelectrical signatures at a hydrocarbon contaminated site. *J. Appl. Geophys.* **2000**, *44*, 167–180.

49. Chambers, J.E.; Wilkinson, P.B.; Wealthall, G.P.; Loke, M.H.; Dearden, R.; Wilson, R.; Allen, D.; Ogilvy, R.D. Hydrogeophysical imaging of deposit heterogeneity and groundwater chemistry changes during DNAPL source zone bioremediation. *J. Contam. Hydrol.* **2010**, *118*, 43–61.

50. Kemna, A.; Kulessa, B.; Vereecken, H. Imaging and characterization of subsurface solute transport using electrical resistivity tomography (ERT) and equivalent transport models. *J. Hydrol.* **2002**, *267*, 125–146.

51. Robert, T.; Caterina, D.; Deceuster, J.; Kaufmann, O.; Nguyen, F. A salt tracer test monitored with surface ERT to detect preferential flow and transport paths in fractured/karstified limestones. *Geophysics* **2012**, *77*, B55–B67.

52. Schön, J.H. Physical Properties of Rocks, Fundamentals and Principles of Petrophysics. In *Handbook of Geophysical Exploration, Seismic Exploration*; Helbig, K., Treitel, S., Eds.; Elsevier: Amsterdam, The Netherlands, 2004; Volume 18, pp. 1–600.

53. Revil, A.; Karaoulis, M.; Johnson, T.; Kemna, A. Review: Some low-frequency electrical methods for subsurface characterization and monitoring in hydrogeology. *Hydrogeol. J.* **2012**, *20*, 617–658.

54. Binley, A.; Kemna, A. DC resistivity and induced polarization methods. In *Hydrogeophysics*; Springer: Berlin, Germany, 2005; pp. 129–156.

55. Hermans, T.; Wildemeersch, S.; Jamin, P.; Orban, P.; Brouyère, S.; Dassargues, A.; Nguyen, F. Quantitative temperature monitoring of a heat tracing experiment using cross-borehole ERT. *Geothermics* **2015**, *53*, 14–26.

56. Tikhonov, A.N.; Arsenin, V.A. *Solution of Ill-Posed Problems*; Winston & Sons: Washington, DC, USA, 1977.

57. Sen, P.N.; Goode, P.A. Influence of temperature on electrical conductivity on shaly sands. *Geophysics* **1992**, *46*, 781–795.

58. Hayley, K.; Bentley, L.R.; Pidlisecky, A. Compensating for temperature variations in time-lapse electrical resistivity difference imaging. *Geophysics* **2010**, *75*, WA51–WA59.

59. Ma, R.; McBratney, A.; Whelan, B.; Minasny, B.; Short, M. Comparing temperature correction models for soil electrical conductivity measurement. *Precis. Agric.* **2011**, *12*, 55–66.

60. Sherrod, L.; Sauck, W.; Werkema, D.D., Jr. A low-cost, in-situ resistivity and temperature monitoring system. *Ground Water Monit. Remediat.* **2012**, *32*, 31–39.

61. Loke, M.H.; Dahlin, T.; Rucker, D.F. Smoothness-constrained time-lapse inversion of data from 3D resistivity survey. *Near Surf. Geophys.* **2014**, *12*, 5–24.

62. Miller, C.; Routh, P.; Brosten, T.; McNamara, J. Application of time-lapse ERT imaging to watershed characterization. *Geophysics* **2008**, *73*, G7–G17.

63. Daily, W.; Ramirez, A.; LaBrecque, D.; Nitao, J. Electrical resistivity tomography of vadose water movement. *Water Resour. Res.* **1992**, *28*, 1429–1442.

64. LaBrecque, D.J.; Yang, X. Difference inversion of ERT data: A fast inversion method for 3-D *in-situ* monitoring. *J. Environ. Eng. Geophys.* **2001**, *6*, 83–90.

65. Nguyen, F.; Hermans, T.; Robert, T. Minimum gradient support and geostatistics regularization approaches for inverting time-lapse data. In Proceedings of 2nd International Workshop on Geoelectrical Monitoring, Vienna, Austria, 4–6 December 2013.

66. Karaoulis, M.; Tsourlos, P.; Kim, J.-H.; Revil, A. 4D time-lapse ERT inversion: Introducing combined time and space constraints. *Near Surf. Geophys.* **2014**, *12*, 25–34.

67. Karaoulis, M.C.; Kim, J.H.; Tsourlos, P.I. 4D active time constrained resistivity inversion. *J. Appl. Geophys.* **2011**, *73*, 25–34.

68. Revil, A.; Skold, M.; Karaoulis, M.; Schmutz, M.; Hubbard, S.S.; Mehlhorn, T.L.; Watson, D.B. Hydrogeophysical investigations of the former S-3 ponds contaminant plumes, Oak Ridge Integrated Field Research Challenge site, Tennessee. *Geophysics* **2013**, *78*, EN29–EN41.

69. Pidlisecky, A.; Haber, E.; Knight, R. RESINVM3D: A 3D resistivity inversion package. *Geophysics* **2007**, *72*, H1–H10.

70. Kemna, A. Tomographic Inversion of Complex Resistivity: Theory and Application. Ph.D. Thesis, University of Bochum, Bochum, Germany, 2000.

71. Nimmer, R.E.; Osiensky, J.L.; Binley, A.M.; Williams, B.C. Three-dimensional effects causing artifacts in two-dimensional, cross-borehole, electrical imaging. *J. Hydrol.* **2008**, *359*, 59–70.

72. Vandenborght, J.; Kemna, A.; Hardelauf, H.; Vereecken, H. Potential of electrical resistivity tomography to infer aquifer transport characteristics from tracer studies: A synthetic case study. *Water Resour. Res.* **2005**, *41*, doi:10.1029/2004WR003774.

73. LaBrecque, D.J.; Miletto, M.; Daily, W.; Ramirez, A.; Owen, E. The effects of noise on Occam's inversion of resistivity tomography data. *Geophysics* **1996**, *61*, 538–548.

74. Hermans, T.; Vandenbohede, A.; Lebbe, L.; Nguyen, F. A shallow geothermal experiment in a sandy aquifer monitored using electric resistivity tomography. *Geophysics* **2012**, *77*, B11–B21.

75. Irving, J.; Singha, K. Stochastic inversion of tracer test and electrical geophysical data to estimate hydraulic conductivities. *Water Resour. Res.* **2010**, *46*, doi:10.1029/2009WR008340.

76. Kowalsky, M.B.; Finsterle, S.; Peterson, J.; Hubbard, S.; Rubin, Y.; Majer, E.; Ward, A.; Gee, G. Estimation of field-scale soil hydraulic and dielectric parameters through joint inversion of GPR and hydrological data. *Water Resour. Res.* **2005**, *41*, doi:10.1029/2005WR004237.

77. Loke, M.H.; Acworth, I.; Dahlin, T. A comparison of smooth and blocky inversion methods in 2D electrical imaging surveys. *Explor. Geophys.* **2003**, *34*, 182–187.

78. Blaschek, R.; Hördt, A.; Kemna, A. A new sensitivity-controlled focusing regularization scheme for the inversion of induced polarization based on the minimum gradient support. *Geophysics* **2008**, *73*, F45–F54.

79. Doetsch, J.; Linde, N.; Pessognelli, M.; Green, A.G.; Günther, T. Constraining 3-D electrical resistance tomography with GPR reflection data for improved aquifer characterization. *J. Appl. Geophys.* **2012**, *78*, 68–76.

80. Zhou, J.; Revil, A.; Karaoulis, M.; Hale, D.; Doetsch, J.; Cuttler, S. Image-guided inversion of electrical resistivity data. *Geophys. J. Int.* **2014**, *197*, 292–309.

81. Caterina, D.; Hermans, T.; Nguyen, F. Case studies of prior information in electrical resistivity tomography: Comparison of different approaches. *Near Surf. Geophys.* **2014**, *12*, 451–465.

82. Kyriakidis, P.C.; Journel, A.G. Geostatistical space-time models: A review. *Math. Geol.* **1999**, *31*, 651–684.

83. Revil, A. Effective conductivity and permittivity of unsaturated porous materials in the frequency range 1 mHz–1 GHz. *Water Resour. Res.* **2013**, *49*, 306–327.

84. Revil, A.; Schwaeger, H.; Cathles, L.M.; Manhardt, P.D. Streaming potential in porous media: 2. Theory and application to geothermal systems. *J. Geophys. Res.* **1999**, *104*, 20033–20048.

85. Archie, G.E. The electrical resistivity log as an aid in determining some reservoir characteristics. *Trans. AIME* **1942**, *146*, 54–62.

86. Revil, A. On charge accumulations in heterogeneous porous materials under the influence of an electrical field. *Geophysics* **2013**, *78*, D271–D291.

87. Sorensen, J.A.; Glass, G.E. Ion and temperature dependence of electrical conductance for natural waters. *Anal. Chem.* **1987**, *59*, 1594–1597.

88. Jardani, A.; Revil, A.; Bolève, A.; Dupont, J.P.; Barrash, W.; Malama, B. Tomography of groundwater flow from self-potential (SP) data. *Geophys. Res. Lett.* **2006**, *33*, doi:10.1029/2006GL027458.

89. Revil, A.; Mahardika, H. Coupled hydromechanical and electromagnetic disturbances in unsaturated clayey materials. *Water Resour. Res.* **2013**, *49*, 744–766.

90. Leinov, E.; Vinogradov, J.; Jackson, M.D. Salinity dependence of the thermoelectric coupling coefficient in brine-saturated sandstones. *Geophys. Res. Lett.* **2010**, *37*, doi:10.1029/2010GL045379.

91. Revil, A.; Karaoulis, M.; Srivastava, S.; Byrdina, S. Thermoelectric self-potential and resistivity data localize the burning front of underground coal fires. *Geophysics* **2013**, *78*, B259–B273.

92. Sill, W.R. Self-potential modeling from primary flows. *Geophysics* **1983**, *48*, 76–86.

93. Ikard, S.J.; Revil, A. Self-potential monitoring of a thermal pulse advecting through a preferential flow path. *J. Hydrol.* **2014**, *2014*, doi:10.1016/j.jhydrol.2014.07.001.

94. Selker, J.S.; Thévenaz, L.; Huwald, H.; Mallet, A.; Luxemburg, W.; van de Giesen, N.; Stejskal, M.; Zeman, J.; Westhoff, M.; Parlange, M.B. Distributed fiber-optic temperature sensing for hydrologic systems. *Water Resour. Res.* **2006**, *42*, doi:10.1029/2006WR005326.

95. Yilmaz, G.; Karlik, S.E. A distributed optical fiber sensor for temperature detection in power cables. *Sens. Actuators A* **2006**, *125*, 148–155.

96. Tyler, S.W.; Selker, J.S.; Hausner, M.B.; Hatch, C.E.; Torgersen, T.; Thodal, C.E.; Schladow, S.G. Environmental temperature sensing using Raman spectra DTS fiber-optic methods. *Water Resour. Res.* **2009**, *45*, doi:10.1029/2008WR007052.

97. Arango-Galván, C.; Prol-Ledesma, R.M.; Flores-Márquez, E.L.; Canet, C.; Villanueva Estrada, R.E. Shallow submarine and subaerial, low-enthalpy hydrothermal manifestations on Punta Banda, Baja California, Mexico: Geophysical and geochemical characterization. *Geothermics* **2011**, *40*, 102–111.

98. Bruno, P.P.G.; Paoletti, V.; Grimaldi, M.; Rapolla, A. Geophysical exploration for geothermal low enthalpy resources in Lipari Island, Italy. *J. Volcanol. Geotherm. Res.* **2000**, *98*, 173–188.

99. Garg, S.K.; Pritchett, J.W.; Wannamaker, P.E.; Combs, J. Characterization of geothermal reservoirs with electrical surveys: Beowave geothermal field. *Geothermics* **2007**, *36*, 487–517.

100. Pérez Flores, M.A.; Gomez Trevino, E. Dipole-dipole resistivity imaging of the Ahuachapan-Chipilapa geothermal field, El Salvador. *Geothermics* **1997**, *26*, 657–680.

101. Revil, A.; Finizola, A.; Ricci, T.; Delcher, E.; Peltier, A.; Barde-Cabusson, S.; Avard, G.; Bailly, T.; Bennati, L.; Byrdina, S.; *et al.* Hydrogeology of Stromboli volcano, Aeolian Islands (Italy) from the interpretation of resistivity tomograms, self-potential, soil temperature and soil CO_2 concentration measurements. *Geophys. J. Int.* **2011**, *186*, 1078–1094.

102. Revil, A.; Johnson, T.C.; Finizola, A. Three-dimensional resistivity tomography of Vulcan's forge, Vulcano Island, southern Italy. *Geophys. Res. Lett.* **2010**, *37*, doi:10.1029/2010GL043983.

103. Benderitter, Y.; Tabbagh, J. Heat storage in a shallow confined aquifer: Geophysical tests to detect the resulting anomaly and its evolution with time. *J. Hydrol.* **1982**, *56*, 85–98.

104. Ramirez, A.; Daily, W.; LaBrecque, D.; Owen, E.; Chesnut, D. Monitoring an underground steam injection process using electrical resistance tomography. *Water Resour. Res.* **1993**, *29*, 73–87.

105. Ramirez, A.; Chesnut, D.A.; Daily, W.D. Using Electrical Resistance Tomography to Map Subsurface Temperatures. U.S. Patent US5346307 A, 3 June 1993.

106. LaBrecque, D.J.; Ramirez, A.L.; Daily, W.D.; Binley, A.M.; Schima, S.A. ERT monitoring of environmental remediation processes. *Meas. Sci. Technol.* **1996**, *7*, 375–383.

107. Müller, K.; Vanderborght, J.; Englert, A.; Kemna, A.; Huisman, J.A.; Rings, J.; Vereecken, H. Imaging and characterization of solute transport during two tracer tests in a shallow aquifer using electrical resistivity tomography and multilevel groundwater samplers. *Water Resour. Res.* **2010**, *46*, doi:10.1029/2008WR007595.

108. Firmbach, L.; Giordano, N.; Comina, C.; Mandrone, G.; Kolditz, O.; Vienken, T.; Dietrich, P. Experimental heat flow propagation within porous media using electrical resistivity tomography x(ERT). In Proceedings of the European Geothermal Congress 2013, Pisa, Italy, 3–7 June 2013.

109. Supper, R.; Ottowitz, D.; Jochum, B.; Römer, A.; Pfeiler, S.; Kauer, S.; Keushnig, M.; Ita, A. Geoelctrical monitoring of frozen ground and permafrost in alpine areas: Field studies and considerations towards an improved measuring technology. *Near Surf. Geophys.* **2014**, *12*, 93–115.

110. Auken, E.; Doetsch, J.; Fiandaca, G.; Christiansen, A.V.; Gazoty, A.; Cahill, A.G.; Jakobsen, R. Imaging subsurface migration of dissolved CO_2 in a shallow aquifer using 3-D time-lapse electrical resistivity tomography. *J. Appl. Geophys.* **2013**, *101*, 31–41.

111. Hermans, T.; Daoudi, M.; Vandenbohede, A.; Robert, T.; Caterina, D.; Nguyen, F. Comparison of temperature estimates from heat transport model and electrical resistivity tomography during a shallow heat injection and storage experiment. *Ber. Geol. Bundesanst.* **2012**, *93*, 43–48.

112. Robert, T.; Hermans, T.; Dumont, G.; Nguyen, F.; Rwabuhungu, D.E. Reliability of ERT-derived temperature: Insights from laboratory measurements. *Near Surf. Geosci.* **2013**, *2013*, doi:10.3997/2214-4609.20131373.

113. Caterina, D.; Beaujean, J.; Robert, T.; Nguyen, F. A comparison study of different image appraisal tools for electrical resistivity tomography. *Near Surf. Geophys.* **2013**, *11*, 639–657.

114. Perri, M.T.; Cassiani, G.; Gervasio, I.; Deiana, R.; Binley, A. A saline tracer test monitored via both surface and cross-borehole electrical resistivity tomography: Comparison of time-lapse results. *J. Appl. Geophys.* **2012**, *79*, 6–16.

115. Doetsch, J.A.; Coscia, I.; Greenhalgh, S.; Linde, N.; Green, A.; Günther, T. The borehole-fluid effect in electrical resistivity imaging. *Geophysics* **2010**, *75*, F107–F114.

116. Day-Lewis, F.D.; Singha, K.; Binley, A.M. Applying petrophysical models to radar travel time and electrical resistivity tomograms: Resolution-dependent limitations. *J. Geophys. Res. Solid Earth* **2005**, *110*, doi:10.1029/2004JB003569.

117. Selker, J.; van de Giesen, N.; Westhoff, M.; Luxemburg, W.; Parlange, M.B. Fiber optics opens window on stream dynamics. *Geophys. Res. Lett.* **2006**, *33*, doi:10.1029/2006GL027979.

118. Lowry, C.S.; Walker, J.F.; Hunt, R.J.; Anderson, M.P. Identifying spatial variability of groundwater discharge in a wetland stream using a distributed temperature sensor. *Water Resour. Res.* **2007**, *43*, doi:10.1029/2007WR006145.

119. Vogt, T.; Schneider, P.; Hahn-Woernle, L.; Cirpka, O.A. Estimation of seepage rates in a losing stream by means of fiber-optic high-resolution vertical temperature profiling. *J. Hydrol.* **2010**, *380*, 154–164.

120. Mamer, E.A.; Lowry, C.S. Locating and quantifying spatially distributed groundwater/surface water interactions using temperature signals with paired fiber-optic cables. *Water Resour. Res.* **2013**, *49*, 7670–7680.

121. Sebok, E.; Duque, C.; Kazmierczak, J.; Engesgaard, P.; Nilsson, B.; Karan, S.; Frandsen, M. High-resolution distributed temperature sensing to detect seasonal groundwater discharge into Lake Vaeng, Denmark. *Water Resour. Res.* **2013**, *49*, 5355–5368.

122. Lauer, F.; Frede, H.-G.; Breuer, L. Uncertainty assessment of quantifying spatially concentrated groundwater discharge to small streams by distributed temperature sensing. *Water Resour. Res.* **2013**, *49*, 400–407.

123. Krause, S.; Blume, T. Impact of seasonal variability and monitoring mode on the adequacy of fiber-optic distributed temperature sensing at aquifer-river interfaces. *Water Resour. Res.* **2013**, *49*, 2408–2423.

124. Mwakanyamale, K.; Day-Lewis, F.D.; Slater, L.D. Statistical mapping of zones of focused groundwater/surface-water exchange using fiber-optic distributed temperature sensing. *Water Resour. Res.* **2013**, *49*, 6979–6984.

125. Blume, T.; Krause, S.; Meinikmann, K.; Lewandowski, J. Upscaling lacustrine groundwater discharge rates by fiber-optic distributed temperature sensing. *Water Resour. Res.* **2013**, *49*, 7929–7944.

126. Ciocca, F.; Lunati, I.; van de Giesen, N.; Parlange, M.B. Heated optical fiber for distributed soil-moisture measurements: A lysimeter experiment. *Vadose Zone J.* **2012**, *11*, doi:10.2136/vzj2011.0199.

127. Hurtig, E.; Groswig, S.; Jobmann, M.; Kuhn, K.; Marschall, P. Fibre-optic temperature measurements in shallow boreholes: Experimental application for fluid logging. *Geothermics* **1994**, *23*, 355–364.

128. Macfarlane, A.; Förster, A.; Merriam, D.; Schrötter, J.; Healey, J. Monitoring artificially stimulated fluid movement in the Cretaceous Dakota Aquifer, Western Kansas. *Hydrogeol. J.* **2002**, *10*, 662–673.

129. Yamano, M.; Goto, S. Long-term monitoring of the temperature profile in a deep borehole: Temperature variations associated with water injection experiments and natural groundwater discharge. *Phys. Earth Planet. Inter.* **2005**, *152*, 326–334.

130. Leaf, A.T.; Hart, D.J.; Bahr, J.M. Active thermal tracer tests for improved hydrostratigraphic characterization. *Groundwater* **2012**, *50*, 726–735.

131. Read, T.; Bour, O.; Bense, V.; le Borgne, T.; Goderniaux, P.; Klepikova, M.V.; Hochreutener, R.; Lavenant, N.; Boschero, V. Characterizing groundwater flow and heat transport in fractured rock using fiber-optic distributed temperature sensing. *Geophys. Res. Lett.* **2013**, *40*, 1–5.

132. Banks, E.W.; Shanafield, M.A.; Cook, P.G. Induced temperature gradients to examine groundwater flowpaths in open boreholes. *Groundwater* **2014**, *2014*, doi:10.1111/gwat.12157.

133. Fujii, H.; Okubo, H.; Itoi, R. Thermal response tests using optical fiber thermometers. *Geotherm. Resour. Counc. Trans.* **2006**, *30*, 545–551.

134. Fujii, H.; Okubo, H.; Nishi, K.; Itoi, R.; Ohyama, K.; Shibata, K. An improved thermal response test for U-tube ground heat exchanger based on optical fiber thermometers. *Geothermics* **2009**, *38*, 399–406.

135. Acuña, J.; Mogensen, P.; Palm, B. Distributed thermal response test on a U-pipe borehole heat exchanger. In Proceedings of the EFFSTOCK—The 11th International Conference on Energy Storage, Stockholm, Sweden, 14–17 June 2009.

136. Reinsch, T.; Henninges, J.; Ásmundsson, R. Thermal, mechanical and chemical influences on the performance of optical fibres for distributed temperature sensing in a hot geothermal well. *Environ. Earth Sci.* **2013**, *70*, 3465–3480.

Chapter 3:
Operational Performance of Geothermal Energy Systems

On the Design and Response of Domestic Ground-Source Heat Pumps in the UK

Chris Underwood

Abstract: The design and response of ground source heat pumps coupled to vertical closed loop arrays in UK domestic applications are investigated in this article. Two typical UK house types are selected as the vehicle for the study and a detailed dynamic thermal modelling method is used to arrive at time-series heating demands for the two houses. A new empirical heat pump model is derived using experimental data taking into account the deteriorating performance of the heat pump during periods of light load. The heat pump model is incorporated into an existing numerical ground model and completed with a classical effectiveness type heat exchange model of the closed loop array. The model is used to analyse array sizing and performance over an extended time period, as well as sensitivity of the design to soil conductivity and borehole heat exchanger resistance and sensitivity to over-sizing and part-load behavior of the heat pump. Results show that the UK's standard for ground source design (the Microgeneration Certification Scheme) may lead to under-estimated array sizes and that heating system over-sizing and deleterious part-load heat pump performance can add up to 20% to the electrical consumption of these systems.

Reprinted from *Energies*. Cite as: Underwood, C. On the Design and Response of Domestic Ground-Source Heat Pumps in the UK. *Energies* **2014**, *7*, 4532-4553.

1. Introduction

During the first phase of recent UK field trials on 81 domestic air- and ground-source heat pumps the median seasonal performance factor (*SPF*) of the sample of 54 ground source heat pumps was found to be 2.2 [1]. (In this context, the seasonal performance factor is taken to mean the heat delivered to the space heating and domestic hot water service over a complete annual operating cycle divided by the corresponding electricity required by the heat pump and its associated source and sink circulating pumps.) The performance was well below expectations and was attributed to a multitude of factors including system sizing, type of source, building efficiency, user behavior and installation practices. Following the first phase of these trials, several site interventions were planned and improvements to the Microgeneration Certification Scheme (MCS) [2] were made to inform future users of heat pumps on procedures for design and installation. The interventions ranged from minor measures (e.g., adjustments to controls) to major interventions (in some cases, involving a re-installation of the heat pump or radiators where they were considered inappropriately sized). Following the interventions, a sample of 32 improved installations were monitored for one further year of which 21 were ground source heat pumps. The median *SPF* for this sample was found to be 3.1 [3]. Whilst the improvement in seasonal coefficient of performance to 3.1 is welcome, it is still short of the potential for these systems. Most of the domestic heat pumps available at present use a remarkably similar kit of parts. These include brazed plate heat exchangers for the evaporator and condenser (which tend to give better heat exchange "pinch" than the first-generation shell and tube heat exchangers), scroll fixed-speed compressors and mechanical ("thermostatic") expansion devices. A buffer water store on the heating side is usually included either as part of the heat pump package or plumbed-in separately and the control of the heat pump is usually by means of a heating thermostat mounted in the heating flow from the heat pump (though often in the heating return especially when a buffer tank is not used). Therefore, it is to be expected that variations of the kind identified in [1,2] are most likely to be due to design, installation and operational issues rather than issues concerned with the heat pump itself.

The UK experiences a quite different climate than that experienced in continental regions in that its weather is governed by maritime conditions with mild winters and frequent episodes of abrupt variations in weather from day to day. Partly because of this, lower standards of thermal insulation tend to be used in UK building construction than is the case in colder climatic regions. It is also generally accepted that the UK lags behind continental Europe and America in its exploitation of ground source heat (though the recent introduction of tariff incentives is likely to see that situation change quite abruptly in the coming years). The UK's geological formations (many of which are at least moderately water-bearing) are also quite different to many of the well-drained sites in America where closed loop ground source heat pumps have been widely used and reported. To account for this, reference will be made in this work to a new and large data set of ground formation conductivities from a large number of sites across the UK. Thus, UK conditions for the use of ground source heating offer a number of unique features which make detailed research and evaluation particularly timely.

In this work, the impact of heat pump capacity and ground array design for domestic ground source heat pumps operating in UK conditions are explored in detail with a view to establishing design criteria that might lead to improvements in operating performance. This will be achieved through the following objectives:

- Detailed dynamic modelling of two typical UK house types of differing sizes and energy demands.
- Development of a new empirical heat pump model which accounts for the degradation in part-load operating performance evident in these systems.
- Design of vertical ground loop arrays using the simulated house energy demands and the new heat pump model.
- Investigations into the sensitivity of the array designs to variations in soil and borehole heat exchanger properties and an analysis of the robustness of the designs over an extended operating time horizon (5 years).
- Investigations into the sensitivity of the seasonal heat pump electrical energy use and house comfort conditions to the under-sizing and over-sizing of the heating system.

Previous Work

Bagdanavicius and Jenkins [4] modelled electrical energy demands for a community of 96 two, three and four-bedroom houses with the assumption that all of the houses used a ground source heat pump sized according to the MCS heat pump standard [2]. They found domestic hot water energy use to have a major bearing on results but their results were restricted to just one winter week. In an experimental study, Blanco *et al.* report on the performance of a variable speed compressor-driven heat pump at domestic scale [5]. Variable speed heat pumps are yet to be widely adopted at domestic scale and are likely to bring part-load performance improvements as well as dispensing with the need for sink-side buffer heat storage. They found that, on average, the electrical consumption when operating with heating temperatures of 35 °C was 29% lower than when operating at 45 °C. Wood *et al.* carried out an experimental investigation into a vertical ground array consisting of single "U-tubes" cast into 10 m-deep piles [6]. They used 21 such piles making an overall array size of 210 m and demonstrated a typical heating delivery rate of 6 kW with a seasonal performance factor of 3.62. So called "energy piles" can provide strong economic advantages compared with conventional vertical ground source arrays though this method of foundation construction is seldom used in house construction. Boait *et al.* investigated the performance of a group of bungalows equipped with ground source heat pumps [7]. They obtained results that were consistent with the first phase of the Energy Saving Trust's trials mentioned earlier [1] and, again, revealed performances that were below observations from other field trials carried out elsewhere in continental Europe. This conclusion is also comprehensively arrived at by a detailed analysis of a number of European field trials on both air- and ground-source heat pumps carried out by Gleeson and Lowe [8]. One of a number of reasons for the performance shortfall mentioned in the Boait *et al.* study was the limited availability of very low capacity heat pumps for small well-insulated UK dwellings (such as bungalows) meaning that the larger capacity heat pumps that are used tend to operate for long periods at light load. However the overarching finding from all of these sources is that the reasons for the disappointing UK performances are complex and multi-faceted and further work is needed.

In conclusion, evidence is beginning to emerge which suggests that domestic ground source heat pumps in the UK are performing below expectations and below comparable installations in other parts of Europe. It appears that there are many reasons for this but a key consideration would appear to be the capacity of the heat pumps used in the UK in relation to the pattern of domestic energy demand. Matters are complicated by the requirement to generate domestic hot water through the summer months when space heating is usually not required. This is often at temperatures that are higher than would be required for space heating due to the need to ensure safe hygiene standards in hot water storage and plumbing systems (though this can be conveniently addressed with minimal loss in performance through the use of a two-zone de-superheater/condenser [5,9]).

2. Energy Demand Modelling in a Sample of UK Houses

2.1. House Type Selection and Seasonal Modelling

Nearly 55% of the 2.8 million UK houses surveyed in 2010 and reported in the National Energy Efficiency Data Framework (NEED) [10] had gross floor areas of between 51 m^2 and 100 m^2 and the next largest group (29%) had gross floor areas of between 101 m^2 and 150 m^2. The mean gas consumptions of these two groups in 2010 were 13,200 kWh and 18,000 kWh, respectively, which, with a well-maintained gas boiler efficiency of 0.85, can be considered to translate to thermal energy demands of 11,220 kWh and 15,300 kWh, respectively. Two house types with gross floor areas of 75 m^2 and 125 m^2 were therefore chosen to fall within this range. It was assumed that the first of these would be a two-story mid-terrace house and the second would be a two-story detached house. Simplified seasonal energy demand modelling was carried out using BREDEM 2012 [11].

Briefly, BREDEM 2012 [11] "Building Research Establishment Domestic Energy Model" is a calculation procedure to estimate all forms of energy consumption in houses. It is a monthly calculation method based on monthly-averaged weather data. The house to be modelled is split into two zones—a main living zone (usually the living room or main reception room) and the rest of the house forms a second balance zone. A simple time-constant-based model is used to estimate heating energy use whereas regression-fitted models based on observations of a large number of UK house types are used to calculate hot water and electrical equipment energy demands. Though the electrical demands are of no direct interest in the present work, they are indirectly used to provide information on internal heat gains which are used to adjust space heating demands. Mainly because of the regression models for energy uses that would otherwise be very difficult to calculate theoretically, BREDEM 2012 [12] tends to give highly representative and accurate results over longer-term averaging periods in strictly UK conditions but is not suitable for energy studies over short time periods. The methods described in BREDEM 2012 [11] form the primary methods used in all UK domestic energy planning and design evaluation work.

Adjustments were made to house layout, glazing and construction details such that results from the BREDEM modelling provided values that were similar to the two mean gas consumptions from the NEED data [10] with the intention of arriving at two "very typical" UK house types as judged by energy demand performance. The details arrived at in this way are consistent with houses that were either constructed or fully refurbished to standards prevailing at about the beginning of this century.

Thermal properties and other details of the two house types can be found in Appendix A (Tables A1 and A2).

Results of the BREDEM modelling using a London site selection for both houses are summarized in Table 1. The totals given in Table 1 can be seen to be of the order of the NEED results mentioned above and can thus be considered to be representative of commonly occurring UK house types of this scale.

Table 1. Simulated annual energy demand for two typical UK house types.

House Type	Space Heating (kWh)	Domestic Hot Water (kWh)	Total (kWh)
Mid-terrace	7,764	2,983	10,747
Detached	13,556	3,377	16,933

2.2. Dynamic Thermal Modelling

The BREDEM results give a good representation of annual energy demand in UK housing which is why the methods used are linked to the Standard Assessment Procedure (SAP) for the purposes of energy efficiency policy development, design and planning applications in the UK. The method is, however, limited in that for the detailed analysis of renewable, low carbon and other complex microgenerators, more granular time-series results of energy demand data are required (e.g., half-hourly or hourly data) whereas BREDEM is only able to provide results at monthly (minimum) intervals. For loads that can be considered to follow a daily average pattern such as domestic hot water this is of no consequence but for the dominant energy demand due to space heating which varies due to both climate and user activity this is a serious limitation. Indeed the need to achieve a better understanding of the response of the heat pump and heating system with the dynamics of the building were emphasised by Boait *et al.* in their field trial study [7]. The two house types were thus remodeled using a dynamic thermal modelling procedure and the results compared for accuracy with the BREDEM reference results.

Re-modelling of the space heating energy use for both houses was carried out using a superset of bespoke building energy modeling components developed for Simulink in the Matlab environment—the Simulink HVAC (heating, ventilating and air conditioning) Blockset. Details of the blockset library can be found in Appendix B (Figures B1 and B2) and details of the mathematical derivations of the most relevant blocks used in the present work can be found in the literature [12,13]. All dimensional and other details as were used in the BREDEM modelling were also used in the dynamic thermal modelling (Appendix A). A current test reference year weather file for London was used from the Chartered Institution of Building Services Engineers (CIBSE) Future Weather Years set [14]. (Note that monthly-average values from this weather file were also used in the earlier BREDEM modelling.)

For domestic hot water energy, the original BREDEM monthly predictions were broken down into daily average values. The operating daily schedules were then applied which consisted of one 2 h heating period each weekday morning following by one 7 h heating period during each weekday evening. On weekend days, one single 16 h heating period was applied. The domestic hot water loads were allocated to the morning and evening periods in the ratio of 1:2, respectively (this pattern was assumed to apply on both weekdays and weekend days).

Annual total energy demands predicted using the dynamic thermal model were found to be 10,529 kWh for the mid-terrace house and 16,994 kWh for the detached house which agree very favourably with the totals from the reference model (BREDEM) given in Table 1. The annual distributions of these demands generated using the Simulink dynamic thermal model are plotted (including hot water demands) in Figure 1.

Figure 1. Simulated distributions of energy demand due to heating and hot water.

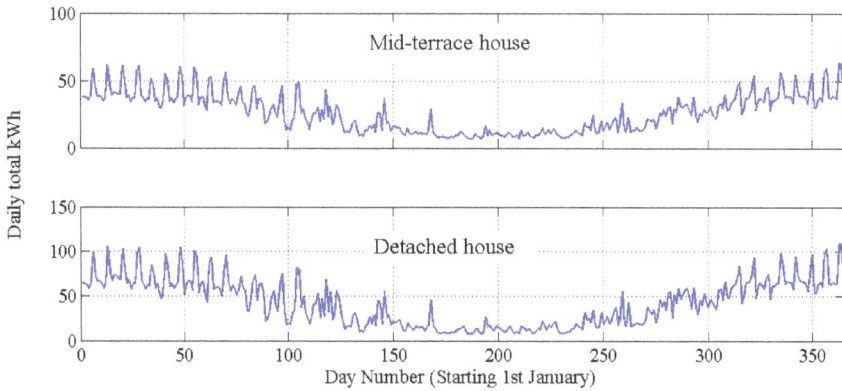

3. Development of a Ground-Source Heat Pump Model

One of the stated objectives of this work is to take account of the decline in heat pump performance when operating at part load. Conventionally, a heat pump model capable of describing part load performance would require being fully dynamic and, consequently, very computationally demanding. A simpler model is needed particularly when the heat pump merely forms part of a more extensive modelling problem (*i.e.*, treatment of the ground array as discussed in the next section). Historically, a simple (and conditionally accurate) modelling approach often used is the so-called "catalogue fit" model. In this, one or more dependent variables (such as electricity consumption and coefficient of performance) are fitted to a manufacturer's performance data set using multiple-regression, e.g., [15–17]. These models are accurate (at least as far as that particular manufacturer's product is concerned) but are limited in that the various standards from which these data are prepared assume static operation at the declared boundary conditions specified in the catalogue. In other words, they assume that the heat pump operates at full continuous capacity at the stated conditions. In practice, heat pumps like other energy generating plants will operate according to some control conditions and will spend large parts of the operating cycle at part load. When operating intermittently to meet a varying load under a thermostat, the first few seconds or minutes of operation as the heat pump starts is merely recovering losses prior to raising the heating temperature to a point where it can contribute to the prevailing load. Thus, in this initial phase of operation, the energy delivered by the heat pump constitutes a loss. This loss accumulates according to the number of thermostat starts the heat pump makes over a full operating cycle. Since most domestic heat pumps in use today for space heating are controlled by thermostats, losses can be significant and clearly become amplified when the heat pump capacity is larger than it needs to be (*i.e.*, when it is over-sized). To capture this behaviour whilst also retaining a simple model structure, an empirical model is developed based on a domestic-scale laboratory pilot.

3.1. Description of the Heat Pump Test Rig

The heat pump test rig consists of a conventional domestic scale water-to-water heat pump using refrigerant R410A with a rated (nominal) heating capacity of 9.5 kW. The heat pump was controlled using a thermostat (with an adjustable set-point) mounted in the heating system return connection at the inlet to the heat pump. Detailed physical information about the heat pump can be found in Appendix C (Table C1).

The heat pump is sourced from three 100 m (deep) plus one 65 m (deep) vertical closed loop borehole heat exchangers giving a total array capacity of 365 m. All four heat exchangers are connected in parallel and each can be individually isolated to enable a variable capacity array. Each borehole heat exchanger comprises a 32 mm high density polyethylene U-tube inside a 130 mm diameter borehole with the inner spaces grouted using thermally-enhanced bentonite. The array lies in Coal Measures and a thermal response test carried out shortly after installation gave a mean soil conductivity of 2.45 $W \cdot m^{-1} \cdot K^{-1}$ and a borehole thermal resistance of 0.162 $m \cdot K \cdot W^{-1}$. The undisturbed soil temperature was 12.7 °C. The source fluid is water-ethylene glycol mixture (10% ethylene glycol by volume).

The heat pump outputs to four equally-sized double panel convector-radiators which provide heating to the local laboratory environment. Each has a rated emission of 1.07 kW with reference to a mean water temperature of 40 °C and a local air temperature of 20 °C. Though the heat pump does not have a buffer tank, the heating system has a significant amount of "natural" buffering due to a low-loss header and significant runs of larger diameter (32 mm) steel piping upstream of the heating system. It is estimated that, collectively, these features provide approximately 50 L of system-side buffering that would not normally be present in a domestic installation.

Instrumentation consisted of a current clamp and voltage transducer on the incoming electrical connections to the heat pump. The power factor was measured separately using an electric circuit analyser and found to have an average value of 0.924 with very minimal variance. Source and sink heats were measured using a pair of resistance-wire temperature detectors and time-of-flight ultrasonic flow meters on both sides of the heat pump. Compound measurement uncertainties were assessed to be on average ±0.8 kW on heating loads (typically <10% of measured heat) and ±0.12 kW on active electricity use (typically <5% of measured electricity use). The laboratory air temperature was measured using several K-type thermocouples and subsequently averaged at each reporting time row.

3.2. Experimental Procedure

Forty-eight heat pump capacity tests were carried out using the following variations in plant configuration:

- 100 m, 200 m, 300 m and 365 m of source array capacity.
- 1, 2, 3 and 4 convector-radiators turned on.
- Nominal heating thermostat set point temperatures of 38 °C, 40 °C and 50 °C.

All variables were monitored at 5 s intervals and averaged in sets of six for reporting at 30 s intervals. The first and final rows of data during each on-phase were discarded in order to remove data spikes arising from incompleteness in the data during a reporting interval in which the thermostat's status changed. Individual borehole tests were carried out on separate days to allow for soil stabilization between tests and the first thermostat cycle of results between convector-radiator adjustment events were discarded to allow for heating system stabilization. Note that the test heat pump in this case is monovalent and the logged electrical consumption includes the source pump but not the sink (heating system) pump. Thus, when used for seasonal performance evaluations, the results based on the present work would lead to a seasonal performance factor that has become referred to in some of the literature as "SPF_{H2}" [3,8].

3.3. Results and Model-Fitting

It is well established that the performance of a heat pump depends *inter alia* on the source and sink temperatures. In addition, the part-load ratio (P) of the heat pump is defined here as the ratio of actual heating delivered to the maximum heating delivered. This can be determined from the results in one of two ways. Either by calculating the total heating delivered over time (in kWh) and dividing by the continuous average heating that would have been delivered over the same time period if the heat pump had not been operating intermittently; or by calculating the radiator emission from the measured heating water temperatures and laboratory air temperature. (Both methods were used and differences between them were found to be minor.)

To identify the influence of the array capacity and P on the heat pump performance, each set of test results was averaged and the heat pump coefficient of performance plotted against P for each discrete array capacity. Results, given in Figure 2, show a strong influence of both of these variables on performance. It is particularly noted that the heat pump performance falls sharply at P values of ≤ 0.5. Therefore, the fitted model was selected to account for variations in source and sink temperature, array capacity and P. For convenience, the source and sink temperature were reduced to a temperature differential, ΔT_{sosi}, between heat pump outlet sink water temperature and heat pump outflow source fluid temperature. Thus, for individual fitting over each array capacity, there will be two dependent variables; P and ΔT_{sosi}.

Two alternative model forms were tested; bi-linear (Equation (1)) and bi-quadratic (Equation (2)):

$$CoP = \left(a_1 + b_1 \times P\right) \times \left(c_1 + d_1 \times \Delta T_{sosi}\right) \tag{1}$$

$$CoP = \left(a_2 + b_2 \times P + c_2 \times P^2\right) \times \left(d_2 + e_2 \times \Delta T_{sosi} + f_2 \times \Delta T_{sosi}^2\right) \tag{2}$$

(in which $a_1...d_1$ and $a_2...f_2$ are regression constants).

Figure 2. Part load performance results from the heat pump test rig.

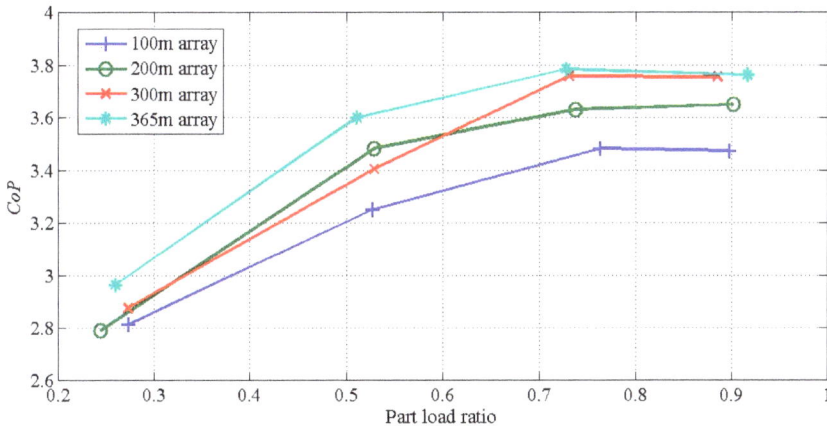

Results of multiple-regression fitting over all viable data using both models are summarized in Figures 3 and 4. For the bi-linear model, most of the target/model error values across all data are within ±10%. There is no improvement in goodness-of-fit when adopting the slightly more complex bi-quadratic model and the target/model error values here are mostly contained within the higher error range of ±20%. Hence, the bi-linear model is adopted in this work. Multiplying out Equation (1) gives the following for which values of the fitted constants, A, B, C, and D, for all array capacities investigated can be found in Table 2:

$$CoP = A + B \times P + C \times \Delta T_{sosi} + D \times P \times \Delta T_{sosi} \tag{3}$$

Figure 3. Bi-linear model fitting results. (**a**): Regression plot; (**b**): Target-model errors (%).

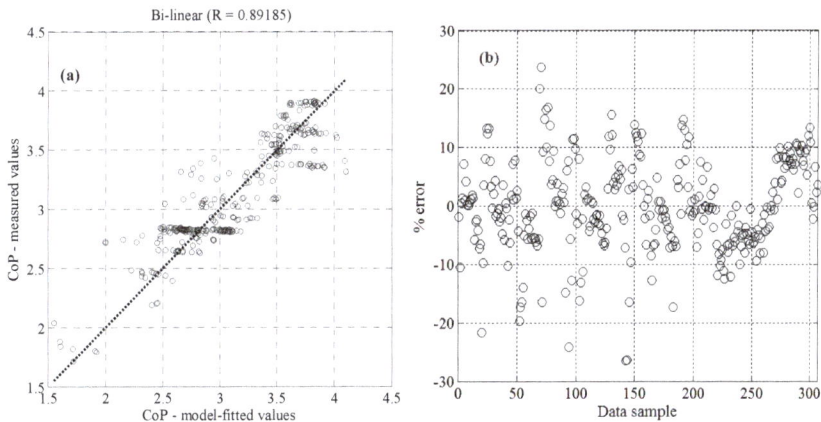

Figure 4. Bi-quadratic model fitting results. (**a**): Regression plot; (**b**): Target-model errors (%).

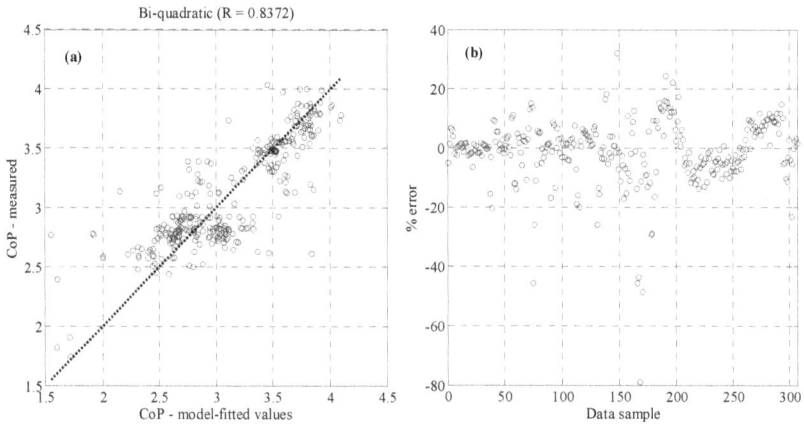

Table 2. Fitted coefficients for the heat pump model.

Array Size (m)	A	B	C	D
100	2.852525	2.868282	−0.017015	−0.037951
200	7.936727	−1.206375	−0.154520	0.067780
300	3.315421	2.816345	−0.021494	−0.042534
365	3.188746	3.414232	−0.017961	−0.062144

4. Ground Array Modelling

The closed loop vertical array was modelled as a conventional heat exchanger problem using Kays and London's classical "number of transfer units" (*NTU*) method [18]. This involved calculating the number of heat exchange transfer units from which the array effectiveness could be determined (Equations (4) and (5)):

$$NTU_{array} = \frac{L_{array}}{R_{bhx}\, m_{f,array}\, c_{pf,array}} \tag{4}$$

$$E_{array} = 1 - \exp(-NTU) \tag{5}$$

Note that the specific heat capacity of the array fluid, $c_{pf,array}$, is calculated using the properties of ethylene glycol solution at the user-defined ethylene glycol concentration. In the present work, the concentration was set at 20% (by volume). The heat exchange (*i.e.*, array) effectiveness is defined as the heat transfer achieved divided by the maximum theoretically possible heat transfer. Thus, the array outlet temperature can be calculated at each time row in the simulation from Equation (6):

$$T_{f,array,out} = T_{array,soil} + \Delta T_{f,max}\, P \times \left(1 - 1/E_{array}\right) \tag{6}$$

(where $T_{array,soil}$ is the current time row average of the soil temperature along the paths of the array heat exchangers).

For the soil domain, a three-dimensional (in space) dynamic numerical solution of the Energy Equation was used through a uniformly discretized grid cast throughout a defined soil domain.

Full details of this model can be found in [19]. The house time-series energy demands described in Section 2.2 are read into this model and used to calculate the energy balance on the new heat pump model described in Section 3.3 and the array model described above. This leads to a heat pump source flux at each time row which is then uniformly imposed on the defined array path in the soil domain model. In effect, this amounts to imposing a variable line source of heat on the numerical soil domain model.

The soil domain model used a 50 m × 50 m × 150 m (deep) domain size with a uniform 1 m grid mesh size. The grid mesh size was found to be acceptable for long time horizon simulations (such as in the present work) but a smaller size is needed for short-time simulations (a further discussion about this can be found in [19]). The domain size was found to give good results for up to 5 years of simulation duration. A comparison with a 100 m × 100 m × 150 m domain size was made and differences in results in the key variable of heat pump electricity usage were found to be negligible (the larger domain requiring almost five-times the computation effort as the smaller domain). A uniform undisturbed earth temperature distribution (see Section 5) was imposed throughout the domain as an initial condition and the boundaries of the domain were held at these values. A uniform time step of 1 h was used. The computation time on a quad-core workstation was found to be 4.7 h of simulation time per 1 s of computer elapsed time. The entire ground, heat pump and array model was implemented as a bespoke Matlab function.

5. Results and Discussion

Reviews of soil properties based on thermal response tests carried out in the UK have received attention recently. Underwood [19] reported on 13 such tests from a variety of sites giving lower quartile, median and upper quartile soil thermal conductivities of 2.40 $W \cdot m^{-1} \cdot K^{-1}$, 2.47 $W \cdot m^{-1} \cdot K^{-1}$ and 3.08 $W \cdot m^{-1} \cdot K^{-1}$, respectively and lower quartile, median and upper quartile borehole resistances of 0.152 $m \cdot K \cdot W^{-1}$, 0.162 $m \cdot K \cdot W^{-1}$ and 0.216 $m \cdot K \cdot W^{-1}$, respectively. Banks et $al.$ [20] reported on a much larger sample of 61 UK sites. Correspondingly, conductivities were found to be 1.86 $W \cdot m^{-1} \cdot K^{-1}$, 2.25 $W \cdot m^{-1} \cdot K^{-1}$ and 3.00 $W \cdot m^{-1} \cdot K^{-1}$, respectively, and borehole thermal resistances of 0.09 $m \cdot K \cdot W^{-1}$, 0.11 $m \cdot K \cdot W^{-1}$ and 0.14 $m \cdot K \cdot W^{-1}$, respectively [20]. There are some similarities in the conductivity results between the two sources though there are differences between borehole resistance values most likely due to different methods being used to arrive at the results from basic measurements (extraction using a bespoke optimisation algorithm in [19] and extraction by conventional line source theory in [20]). Since [20] represents the larger sample of data, these results will be used in the evaluative modelling that follows. In addition, the undisturbed ground temperatures at the lower quartile, median and upper quartile of 11.7 °C, 12.3 °C and 13.2 °C [20] are used.

5.1. Array Design

The median soil property and borehole resistance data reported in [20] were used for initial array design:

Soil thermal conductivity: $2.25 \text{ W·m}^{-1}\text{·K}^{-1}$

Borehole thermal resistance: 0.11 m·K·W^{-1}

Undisturbed soil temperature: 12.3 °C

In addition, the volume heat capacity for the soil was assumed to be 2.4 $\text{MJ·m}^{-3}\text{·K}^{-1}$ which is appropriate for a wide range of soil and rock types. Reference is made to the house design heat losses detailed in Appendix A, Table A2; a notional design coefficient of performance of 2.82 as suggested by the Energy Saving Trust's most recent field trials [3]; and the array design look-up tables contained in the Microgeneration Certification Scheme (MCS) [2]. Using this information, recommended array designs of 83 m for the mid-terrace house and 138 m for the detached house were arrived at. These array sizes would therefore be used in normal practice.

For design evaluation, simulations using the foregoing data together with the time-series energy demands detailed in Section 2.2 were carried out for both house types based on a range of array sizes starting with the MCS values of 83 m and 138 m for the respective house types. Results in the form of alternative heat pump seasonal performance factors (*SPF*) are summarized in Table 3. Note that seasonal performance factors used in the present work are what have become referred to in some of the literature as "SPF_{H2}" [3,8]. They are applicable to a monovalent heat pump (including source fluid pump) electricity use but exclude the heating system (sink) pump. For all simulations, the heat pump heating water outlet temperature set point was fixed at 42 °C which is within the range of the test results described in Section 3.2 upon which the heat pump model is based. It was assumed that all space heating and domestic hot water loads are delivered at this temperature from the heat pump buffer store of capacity 50 L. No allowance has been made in the present work for additional direct electric heating due to periodic hot water pasteurization (if used).

Table 3. Design array sizes and corresponding heat pump *SPF*s.

Array Size (m)	Mid-Terrace House		Detached House	
	SPF	Improvement	*SPF*	Improvement
83 (MCS, [2])	3.05	-	-	-
138 (MCS, [2])	-	-	3.02	-
200	3.32	8.9%	3.14	4.0%
300	3.38	1.8%	3.33	6.1%
400	3.35	-	3.31	-

Comments: The performances based on array sizes recommended by the MCS are broadly consistent with the median performance of the actual installations monitored by the Energy Saving Trust [3] after improvements had been made (*i.e.*, 3.1). For both house types, there are improvements of up to around 10% in the heat pump *SPF* by increasing the array size significantly. For the mid-terrace house, increasing the array size by a little over 100 m beyond the value recommended by the MCS gives a performance improvement of 8.9% whereas a further increase by 100 m results in a much lower improvement of 1.8% and there is no improvement at higher array sizes. Therefore, a design array size of 200 m would seem appropriate for this case. For the detached house, there is a performance improvement of 4% when increasing from the MCS recommended array size to 200 m and the improvement increases to 6.1% when increasing from

200–300 m after which there is no further improvement with increasing array size. Thus, a 300 m array would seem to be the appropriate choice here.

In the subsequent analyses, the array size for the mid-terrace house will be set at 200 m and, for the detached house, 300 m.

5.2. Sensitivity to Time Horizon

To test the designs over an extended operating period, the array sizes and other design conditions set out in Section 5.1 were applied to five-year simulations for each house. It was assumed that the annual demand patterns (Figure 1) remained the same in each year. Results of the daily minimum array inlet temperatures over the five-year duration are provided in Figures 5 and 6 for both houses. Results of the daily mean heat pump coefficient of performance over the same duration are given in Figures 7 and 8 also for both houses.

Comments: As is to be expected, there is a decline in minimum array fluid temperature over the extended operating time horizon though the rate of decline reduces over time particularly after the first year. What is important with these array designs is that the array fluid temperature for both houses barely falls below 4 °C which suggests that fresh water (rather than ethylene glycol solution) might safely be used with corresponding performance, cost and environmental advantages. The trends in heat pump coefficient of performance (Figures 7 and 8) show a small decline over the five-year period as the ground temperature reduces and a more pronounced decline in the middle period of each year. This is because light loads are being met at these times (*i.e.*, domestic hot water only) with the consequence that the heat pump is operating highly intermittently. (A key assumption in the modelling is that the thermostatically-controlled heat pump operates at the same set point temperature (42 °C) at all times for both heating and hot water delivery and so the heat pump performance is governed by intermittent operation and seasonal variations in the source temperature.) Equally, though the performance is inferior at this time of the year, the heating delivered (and therefore electrical energy consumed by the heat pump) will be lower than in winter and this mitigates the inferior heat pump performance to some extent. The decline in heat pump performance over the five-year horizon is more pronounced with the mid-terrace house than with the detached house and the detached house exhibits a more pronounced dip in summer heat pump performance. The reason for this is that the domestic hot water loads are similar for both houses whereas the detached house has a significantly higher heat load due to space heating. The correspondingly larger array size for the detached house recovers better in summer when demands are light than the mid-terrace house which has the smaller array size. However, a lower pattern of P for the detached house due to higher heat load but only marginal increases in summer hot water load results in a lower summer CoP than experienced by the mid-terrace house. Again, because relatively low amounts of energy are generated in summer, the inferior heat pump performance is mitigated to some extent.

Figure 5. Minimum daily array inlet temperatures (mid-terrace house).

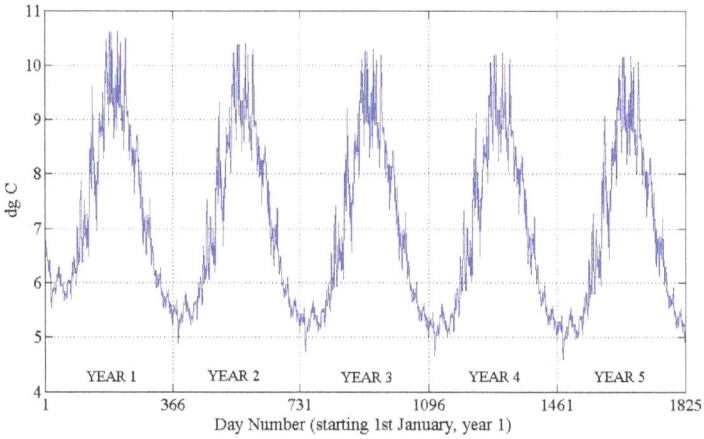

Figure 6. Minimum daily array inlet temperatures (detached house).

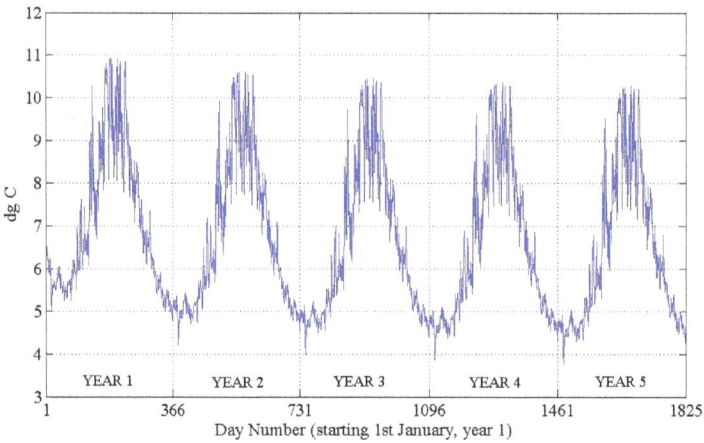

Figure 7. Mean daily heat pump CoP (mid-terrace house).

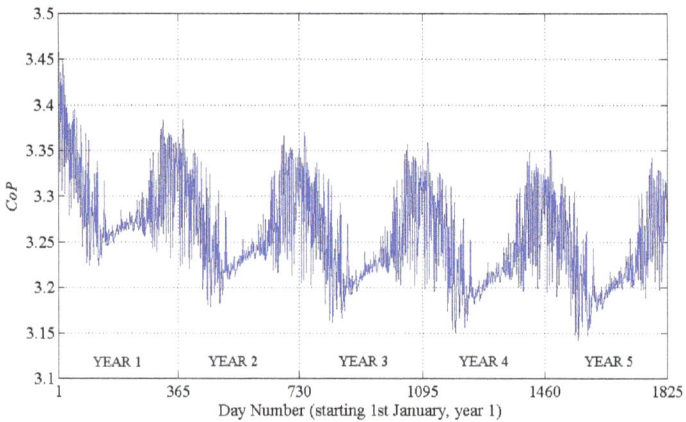

Figure 8. Mean daily heat pump CoP (detached house).

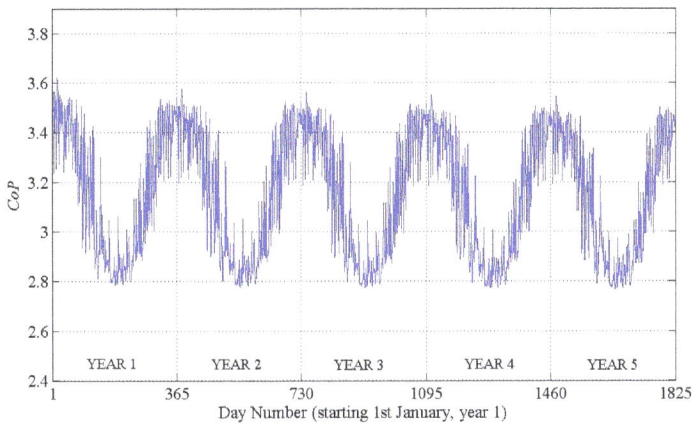

5.3. Sensitivity to Soil Conductivity and Array Resistance

To evaluate the sensitivity of the designs to variations in soil conductivity and borehole heat exchanger resistance values, a simulation was conducted using alternative soil conductivities of 1.86 $W \cdot m^{-1} \cdot K^{-1}$ (*i.e.*, the lower quartile value from the Banks *et al.* review [20]) and a further simulation was conducted using this lower conductivity and a higher borehole resistance value of 0.216 $m \cdot K \cdot W^{-1}$. In the choice of the latter value, the upper quartile resistance from [19] was used instead of from [20] since the range of values in [19] was wider. Results of the original seasonal performance factors compared with the new values arising from these changes in soil and borehole properties are given in Table 4.

Comments: The results show that the performance is not significantly affected by a "typical" range of soil and resistance properties found in UK conditions. Here, the results appear to be more sensitive to borehole resistance than soil conductivity though it should be stressed that this illustration involved a relatively minor reduction in conductivity and a substantial (near doubling) increase in borehole resistance. However, it is becoming clear that the range of soil conductivities across many UK applications is relatively low and this prompts the need to consider carefully whether expensive thermal response testing is needed in every case when ground geology is known with a reasonable degree of confidence. It should also be pointed out that, in most cases, the existence of groundwater flows (not considered in this work) will actually improve performance.

Table 4. Sensitivity to variations in soil conductivity and borehole resistance.

Property Choices	SPF	
	Mid-Terrace House	Detached House
Median (design) values	3.32	3.33
Reduced k	3.31	3.33
Reduced k & increased R_{bhx}	3.17	3.26

5.4. Sensitivity to Heating System Sizing

A final analysis was carried out into the impact of heating system sizing. First, the simulation was re-run with P (in Equation (3)) fixed at a constant value of 1. This will show how much additional energy is required due to the deleterious part-load performance of the heat pump thus revealing the potential for improved control over the heat pump at light loads. Second, the simulation was re-run for both houses with the heating system capacity (Table A2, Appendix A) reduced by 20%, and then increased by 20% and 40%.

Results, compared with the original design results for reference, are given in Table 5. Also included are the averages of the comfort (operative) temperature of the whole house averaged over all periods when the space heating system is active only. Note that the operative temperature as used here is the average of the internal air and mean radiant temperatures.

Table 5. Sensitivity to heating system sizing and comfort.

Capacity Options	Mid-Terrace House			Detached House		
	SPF	kWh·m^{-2}	T_{OP} (°C)	*SPF*	kWh·m^{-2}	T_{OP} (°C)
Perfect tracking	3.56	36.1	19.97	3.75	33.0	19.74
Design	3.32	38.6	19.97	3.33	37.1	19.74
20% under-sizing	3.34	35.9	18.68	3.35	34.5	18.46
20% over-sizing	3.31	40.8	20.61	3.31	38.9	20.54
40% over-sizing	3.30	42.4	21.19	3.29	40.3	21.04

(kWh·m^{-2}: annual electricity usage due to heat pump and source fluid pump divided by house gross floor area).

Comments: A 40% heating system over-sizing margin will not translate to a 40% increase in energy because the heating system controls will act to regulate the system to the required comfort conditions. However, oversizing will lead to an increase in energy because an over-sized control system will not be able to track the required control condition perfectly. In particular, the use of simple proportional controls used in domestic "thermostatic" radiator valves will exhibit offset (a sustained difference between set point and actual value). Thus, an over-sized system will lead to an increase in both energy use and comfort temperature. Consequently, the seasonal performance factor is not significantly affected by over-sizing (or under-sizing), however the electricity consumed has increased in all over-sizing cases. If the heat pump was able to operate over all load patterns without loss in performance it would operate with a *SPF* in excess of 3.5 for both houses. This falls to around 3.3 for both houses when its part load behavior is accounted for and the electrical consumption increases by 7% (mid-terrace house) and 12% (detached). If the heating is oversized by up to 40% the energy consumption over a heat pump well matched to the load at all times will increase by almost 18% (mid-terrace house) and 22% (detached house). These should be considered as viable targets for further improvements in domestic heat pump performances in the UK where evidence of equipment over-sizing is plentiful. A particularly notable result in Table 5 is revealed that when the heating system is under-sized by 20% the energy use falls whilst comfort remains very close to acceptable limits. Note here that the area-weighted target comfort temperature for both houses based on 21 °C in the main living space and 18 °C in all other spaces is 18.5 °C—almost precisely met on average by a heating system that is under-sized by 20%. The reason for this is that conventional heating sizing is carried out using steady-state calculations at boundary conditions that, in many winters, will never be realized or if they are, will be of short duration. There is a compelling case for the use of dynamic thermal modelling for the sizing of complex systems such as heat pumps and other microgenerators.

6. Conclusions

The aim of this work was to investigate the impact of heat pump capacity and ground array design for domestic ground source heat pumps operating in UK conditions with a view to establishing design criteria that might lead to improvements in operating performance.

The work has been largely based on numerical modelling supported with the introduction of a new empirical heat pump model which takes into account the decline in heat pump performance during periods of light load. Results have been drawn from two exemplar houses which have been configured to be typical of common terraced and detached houses in the UK and use has been made of an existing numerical model of a closed loop vertical ground array.

The results of this work suggest vertical ground loop array sizes for the two typical house types investigated of around 2.5 m of array length per m^2 of house gross floor area. The recommended allowance using the Microgeneration Certification Scheme recommendations would be around 1.1 m/m^2. Furthermore, the array sizes proposed in this work show that it will be possible to operate the array safely using fresh water rather than ethylene glycol solution (or some other form of antifreeze) which is beneficial for performance, cost and the environment.

As data on ground thermal conductivities in the UK start to become more abundant, it is becoming clear that many sites have mean conductivities of around 2–2.5 $W \cdot m^{-1} \cdot K^{-1}$ and modest variations about this figure have little effect on heat pump performance. However, the more uncertain values of borehole heat exchanger resistance do have an influence on performance.

The impact of both deteriorating part-load performance of thermostatically-controlled ground source heat pumps and heating system over-sizing (by up to 40%) has been shown to increase energy use by up to 18%–22% for the two typical house types considered.

Evidence in this work points to a strong potential for a better matching of the capacity of ground source heat pumps to the required building load pattern through the use of dynamic thermal modelling instead of conventional steady-state design methods.

Further work is needed in the following areas:

- Development of verifiably accurate and easy-to-use tools for the design and seasonal performance evaluation of ground source heat pumps suitable for use by practitioners.
- Consideration of the impact of groundwater flow over closed loop arrays in UK conditions.
- Investigations into variable speed drives, electronic expansion devices and improved controls for domestic scale heat pumps.
- Development of alternative dynamically-based methods for system design and capacity-sizing as an alternative to conventional steady-state sizing methods.

Glossary

$a_1...d_1$	Regression constants
$a_2...f_2$	Regression constants
$A...D$	Regression constants
CoP	Coefficient of performance (dimensionless)—instantaneous or short-time average heat output divided by instantaneous or short-time average heat pump plus source fluid pump electrical consumption.
$c_{pf,array}$	Specific heat capacity of array fluid ($J \cdot kg^{-1} \cdot K^{-1}$)
E_{array}	Array (heat transfer) effectiveness
k	Soil mean thermal conductivity ($W \cdot m^{-1} \cdot K^{-1}$)
L_{array}	Total array length (m)
$m_{f,array}$	Array fluid mass flow rate ($kg \cdot s^{-1}$)
NTU_{array}	Array number of (heat) transfer units
P	Part load ratio (current heating demand divided by seasonal maximum heating demand)
R	Correlation coefficient
SPF	Seasonal performance factor (dimensionless)—seasonal heat pump heating energy divided by seasonal heat pump plus source fluid pump electrical consumption
R_{bhx}	Borehole heat exchanger thermal resistance ($m \cdot K \cdot W^{-1}$)
$T_{f,array,out}$	Array fluid outlet temperature (°C)
$T_{array,soil}$	Average soil temperature along the entire array path (°C)
T_{OP}	Room space operative temperature (°C)
$\Delta T_{f,max}$	Maximum (design) array inlet/outlet fluid temperature difference (K)
ΔT_{sosi}	Heat pump nominal temperature lift (difference between the heating outlet temperature (K or °C) and the source fluid outlet temperature (K or °C)

Symbols used in Tables A1 and A2

MT	mid-terrace
DET	detached
GFA	gross floor area
LS	main living space
B	balance of ground floor space
FF	first floor

Appendix A. Seasonal Energy Demand Modelling

Table A1. Seasonal house energy modeling—main parameters.

Type	Areas (m²)		Footprint		Orientation	Windows	Glazing (m² *)	Roof
	GFA	LS	Width (m)	Depth (m)				
MT	75	12	6.1	6.1	N-S	Clear double	10.4	Pitched
DET	125	20	7.9	7.9	N-S	Clear double	17.2	Pitched

* Based on the UK Standard Assessment Procedure: $0.1382 \times GFA - 0.027$. An additional allowance of one single door (rear) and one single door (front) of 2 m² (each) is added to the window areas.

Table A2. Seasonal house energy modeling—thermal properties.

U-values (W·m²·K⁻¹)				Design Temperatures (°C)			Ventilation (h⁻¹)	Design Loss (kW *)
Wall	Roof	Gr-floor	Window	LS	B	FF		
0.45	0.35	0.45	2.2	21	18	18	0.5	3.09
0.45	0.35	0.45	2.2	21	18	18	0.5	5.95

* Design heat loss with an external design temperature of −3 °C. Calculations include a pre-heat margin of 25% and were carried out in accordance with the methods set out in the CIBSE Guide. The standards are those that would be expected of UK houses either constructed or refurbished at around 2002.

Appendix B. Dynamic Thermal Modelling—Simulink HVAC Blockset

The Simulink HVAC Blockset is a generic Simulink library that can be used to construct detailed dynamic models of buildings including HVAC plant and controls and certain embedded renewable energy systems. Several such libraries exist for modelling energy in buildings such as SIMBAD [21] and CARNOT [22] but the advantage of the HVAC Blockset used here is that all component models are fully dynamic, enabling more accurate control and system response modelling to be carried out. (The HVAC Blockset used here is made freely available by the author for other users.) Note that only certain component model selections were used in the present work as detailed below.

Components used to remodel the two house types:

Building envelope:	Plant and controls:	Utility:
Zone heat balance	Generic emitter [2]	Schedule [4]
Exposed element	Detector [3]	Solar simulator [5]
Internal element	Control valve [3]	
Window	PID controller [3]	
Ventilation [1]		

1. Infiltration due to wind and stack effect during winter with closed windows. The block was set to open windows by 50% of their opening capacity when internal temperatures reached 26 °C and by a further 50% to fully-open when temperatures reached 28 °C (*i.e.*, in summer when the heating is off).
2. The "generic emitter" was used to represent panel convector-radiator heating.
3. The "detector", "control valve" and "PID controller" blocks were combined to represent "thermostatic" radiators valves attached to each zone radiator. Only the proportional term of the PID controller block was enacted. The detector was set with a long time constant (3 min) to help represent the long time delay in these valves. (Note that the widely-used term "thermostatic" is a misnomer in this context since these control valves actually modulate the hot water flow rate in practice.)
4. Separate "schedule" blocks were used to represent switching of occupant activity and plant activity.
5. The "solar simulator" was used to generate in-plane irradiances on each exposed window and opaque surface.

Figure B1. Simulink HVAC Blockset.

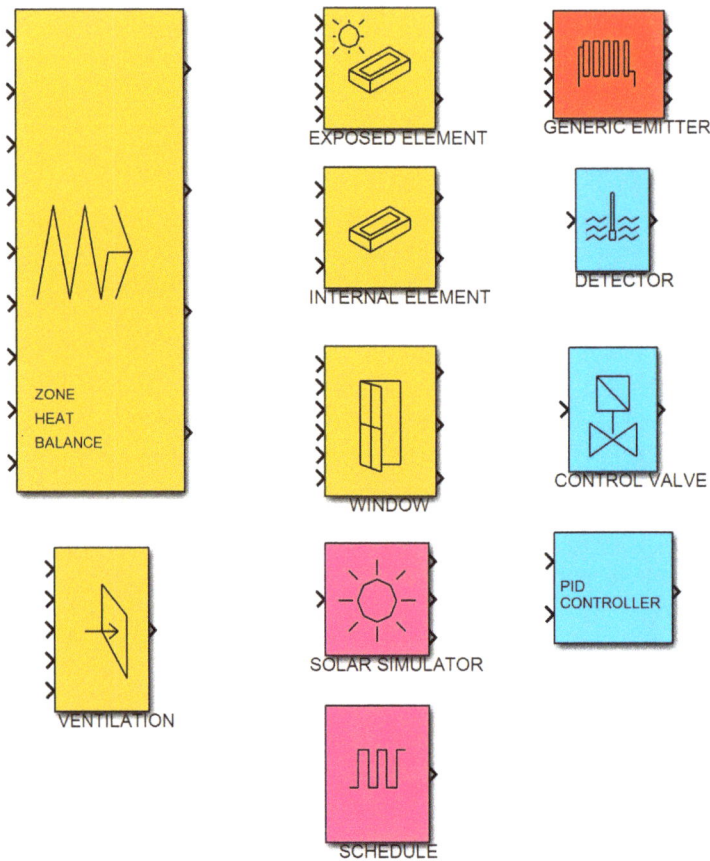

The highest level of the Simulink model created for the mid-terrace house is shown as an example in Figure B2, showing all parent blocks and information flows between them. The three main blocks at the centre represent the parent blocks of the living room (top), balance of ground floor spaces (middle) and first floor (bottom). Beneath these parent blocks are all the component blocks (and their information flow connections) for the room elements, zone energy balance, ventilation, heating system and heating controls. The large block to the left in Figure B2 is the parent block which contains all microclimate modelling (*i.e.*, a file-read utility which reads in weather data and solar simulating blocks for all orientations forming the mid-terrace house).

For further details of the individual block descriptions and derivations, see [12,13].

Note that the other blocks shown in the generic library and not specifically referred to above were not used in the present work. Furthermore, the heat pump model in the generic library (which is a simple manufacturer's catalogue-fit type model) was not used in the present work either.

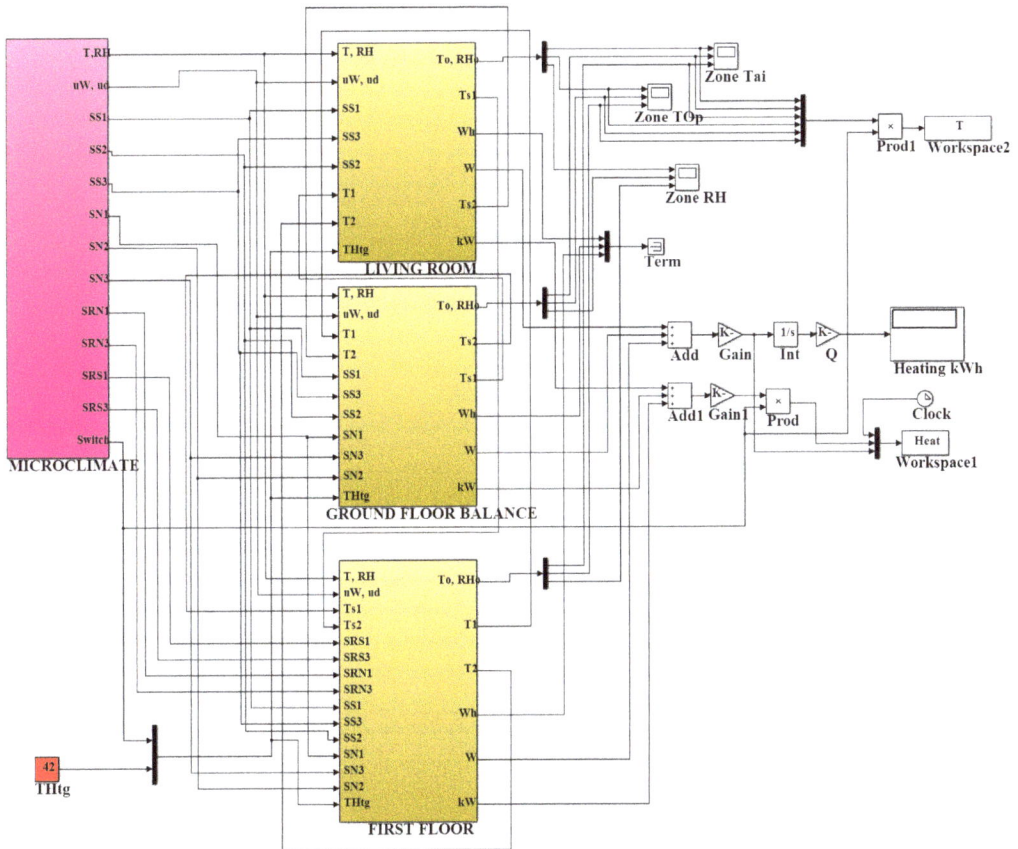

Figure B2. Top-level block diagram model for the mid-terrace house.

Appendix C. Details of the Test Ground Source Heat Pump

Table C1. Details of the test heat pump.

System Component	Parameter	Value
Compressor	Type	Scroll
	Refrigerant	R410A
	Electrical supply	1-phase, 220 V
	Displacement	5.34 m$^3\cdot$h^{-1}
Evaporator and condenser	Type	Brazed plate
	Plate material	Corrugated stainless steel
	Number of plates	13
	Plate width	118.4 mm
	Plate height	440 mm
	Plate spacing	2.24 mm
	Plate thickness	0.4 mm
	Volume, refrigerant side	0.57 L
	Volume, water side	0.66 L
Expansion device	Type	Mechanical, thermostatic

Conflicts of Interest

The author declares no conflict of interest.

References

1. Getting Warmer: A Field Trial of Heat Pumps. Available online: http://www.heatpumps.org.uk/ PdfFiles/TheEnergySavingTrust-GettingWarmerAFieldTrialOfHeatPumps.pdf (accessed on 11 December 2013).
2. Microgeneration Installation Standard MIS 3005—Requirements for Contractors Undertaking the Supply, Design, Installation, Set to Work, Commissioning and Handover of Microgeneration Heat Pump Systems, Issue 4.0. Available online: http://www.microgenerationcertification.org/images/ MIS%203005%20Issue%204.0%20Heat%20Pump%20Systems%20%202013.12.16%20FINAL3 .pdf (accessed on 25 April 2014).
3. The Heat is on. Available online: http://www.energysavingtrust.org.uk/Organisations/ Technology/Field-trials-and-monitoring/Fieldmonitoring/Field-trial-reports/Heat-pump-field-trials (accessed on 25 April 2014).
4. Bagdanavicius, A.; Jenkins, N. Power requirements of ground source heat pumps in a residential area. *Appl. Energy* **2013**, *101*, 591–600.
5. Blanco, D.L.; Nagano, K.; Morimoto, M. Experimental study on a monovalent inverter-driven water-to-water heat pump with a desuperheater for low energy houses. *Appl. Therm. Eng.* **2013**, *50*, 826–836.
6. Wood, C.J.; Liu, H.; Riffat, S.B. An investigation of the heat pump performance and ground temperature of a pile foundation heat exchanger system for a residential building. *Energy* **2010**, *35*, 4932–4940.
7. Boait, P.J.; Fan, D.; Stafford, A. Performance and control of domestic ground-source heat pumps in retrofit installations. *Energy Build.* **2011**, *43*, 1968–1976.
8. Gleeson, C.P.; Lowe, R. Meta-analysis of European heat pump field trial efficiencies. *Energy Build.* **2013**, *66*, 637–647.
9. Fernández-Seara, J.; Pereiro, A.; Bastos, S.; Dopazo, J.A. Experimental evaluation of a geothermal heat pump for space heating and domestic water simultaneous production. *Renew. Energy* **2012**, *48*, 482–488.
10. National Energy Efficiency Data-Framework (NEED), Table 1. Available online: https://www.gov.uk/government/collections/national-energy-efficiency-data-need-framework (accessed on 28 April 2014).
11. BREDEM 2010—A Technical Description of the BRE Domestic Energy Model, Version 1.0. Available online: http://www.bre.co.uk/filelibrary/bredem/BREDEM-2012-specification.pdf (accessed on 28 April 2014).
12. Gouda, M.M.; Danaher, S.; Underwood, C.P. Building thermal model reduction using nonlinear constrained optimization. *Build. Environ.* **2002**, *37*, 1255–1265.

13. Gouda, M.M.; Underwood, C.P.; Danaher, S. Modelling the robustness properties of HVAC plant under feedback control. *Build. Serv. Eng. Res. Technol.* **2003**, *24*, 271–280.

14. CIBSE TM48. *Use of Climate Change Scenarios for Building Simulation: The CIBSE Future Weather Years*; Chartered Institution of Building Services Engineers: London, UK, 2009.

15. Underwood, C.P.; Yik, F.W.H. *Modelling Methods for Energy in Buildings*; Blackwell: Oxford, UK, 2004; Chapter 4, pp. 135–147.

16. Fisher, D.E.; Rees, S.J.; Padhmanabhan, S.K.; Murugappan, A. Integration and validation of ground source heat pump system models in an integrated building and system simulation environment. *HVAC&R Res.* **2006**, *12*, 693–710.

17. Kinab, E.; Marchio, D.; Rivière, P.; Zonghaib, A. Reversible heat pump model for seasonal performance optimization. *Energy Build.* **2010**, *42*, 2269–2280.

18. Kays, W.M.; London, A.L. *Compact Heat Exchangers*, 2nd ed.; McGraw-Hill: New York, NY, USA, 1964.

19. Underwood, C.P. Ground-source heat pumps: Observations from United Kingdom ground thermal response tests. *Build. Serv. Eng. Res. Technol.* **2013**, *34*, 123–144.

20. Banks, D.; Withers, J.G.; Cashmore, G.; Dimelow, C. An overview of the results of 61 *in situ* thermal response tests in the UK. *Q. J. Eng. Geol. Hydrogeol.* **2013**, *46*, 281–291.

21. SIMBAD—Simulation of Buildings and Devices. Available online: http://www.simbad-cstb.fr/ (accessed on 20 June 2014).

22. CARNOT—Conventional and Renewable Energy Siystems Optimisation Blockset. Available online: http://mv.fh-duesseldorf.de/d_pers/Adam_Mario/a_lehre/gm_allg_down/Carnot-Hilfe.pdf (accessed on 20 June 2014).

Temperatures and Heat Flows in a Soil Enclosing a Slinky Horizontal Heat Exchanger

Pavel Neuberger, Radomír Adamovský and Michaela Šeďová

Abstract: Temperature changes and heat flows in soils that host "slinky"-type horizontal heat exchangers are complex, but need to be understood if robust quantification of the thermal energy available to a ground-source heat pump is to be achieved. Of particular interest is the capacity of the thermal energy content of the soil to regenerate when the heat exchangers are not operating. Analysis of specific heat flows and the specific thermal energy regime within the soil, including that captured by the heat-exchangers, has been characterised by meticulous measurements. These reveal that high concentrations of antifreeze mix in the heat-transfer fluid of the heat exchanger have an adverse impact on heat flows discharged into the soil.

Reprinted from *Energies*. Cite as: Neuberger, P.; Adamovský, R.; Šeďová, M. Temperatures and Heat Flows in a Soil Enclosing a Slinky Horizontal Heat Exchanger. *Energies* **2014**, *7*, 972-987.

1. Introduction

The basic low-potential sources of energy for heat pump evaporators used for heating and cooling are air, water, and ground, *i.e.*, soil. When air is used, the installation of the energy system is easier and cheaper. A disadvantage here is the lower amount of energy savings resulting from the lower seasonal performance factor (SPF). The use of surface or ground water as the source for heat pumps is very advantageous in view of the energy effect of the system. Installations of such systems, however, are limited especially by environmental requirements for the protection of such water sources. Soil or rock mass appears efficient, especially in relation to the ambient temperature. In winter, the mass temperature is higher than the ambient temperature and, conversely, lower in summer. This mass and ambient temperature ratio is useful in winter when the heat pump is employed for heating and in summer for cooling. The performance and economic comparison of air, soil and rock mass acting as low-potential sources of energy for heat pumps has been addressed by Petit and Meyer [1]. They state that the highest performance is achieved with vertical rock mass exchangers. Horizontal ground heat exchangers deliver a better heating factor and the best economic parameters of all three sources of energy. Air has received the worst rating as the source of energy for heat pumps.

Lund *et al.* [2] prepared an overview of the use of geothermal energy for direct consumption in 78 countries. The total installed thermal power capacity of geothermal sources amounted to 48,493 MW in 2009. Those sources delivered 423,830 TJ of heat energy per year. Approximately 47.2% of that energy was acquired from ground-to-water heat pumps. The total installed output of ground-to-water heat pumps was 33,134 MW in 2009. The number of installed energy systems with ground-to-water heat pumps in 2009 was double and quadruple compared to 2005 and 2000, respectively.

Heat from the soil or rock mass is removed using horizontal or vertical heat exchangers. Vertical ground heat exchangers deliver a high efficiency in performance and have the minimum requirements for the ground area. However, they present an investment that is considerably higher than that of horizontal heat exchangers, due to design and installation of vertical borehole heat exchangers. Horizontal ground heat exchangers represent a compromise between high efficiency and investment costs of the heat exchanger. They are available in three basic configurations: linear, spiral and coil type [3]. The 30–50 mm pipes of these heat exchangers are buried at a depth of 1.5–2.0 m under the surface, depending on the thermal characteristics of the ground mass. The results of measurement of ground mass temperature [4] have indicated that the area up to 1 m deep is highly sensitive even to short-term variations of weather. In the summer months (July–September), the density of transferred heat flow from the ground mass surface to deeper strata amounts to 3.6 $W \cdot m^{-2}$. At the end of September, the temperature gradient drops to zero and the heat flow reverses. Heat is transferred towards the surface of the mass.

A number of publications have explored the experimental as well as numerical analyses over the last several years. De Swardt and Meyer [5] compared two low-potential sources of energy for heat pumps: air and water from the community water mains. Water mains may be regarded as a horizontal ground heat exchanger. The results demonstrated lower consumption of electricity and higher seasonal performance factor when mains water was used. The Research Center for Energy and Environment in Lecce [6] verified the basic configuration of horizontal ground source heat exchangers. Throughout the year the researchers there measured the temperatures of the soil and the thermal output discharged by the heat exchangers. The results of the test showed that the primary parameters fundamentally affecting heat transfer in the soil are the thermal conductivity coefficient of the soil and the speed at which the heat-transfer fluid flows through the pipes of the heat exchanger. The spacing of the heat exchanger pipes and the depth at which they were laid did not play a significant role.

Song et al. [7] analysed the most important parameters affecting the thermal conductivity coefficient of the soil. Their experiments have shown that within the temperature range of 10–40 °C in a dry state the thermal conductivity coefficient of the soil is 0.55–0.6 $W \cdot m^{-1} \cdot K^{-1}$; with normal moisture content the average is 2.3 $W \cdot m^{-1} \cdot K^{-1}$ and with wet soil the figure is 2.7 $W \cdot m^{-1} \cdot K^{-1}$. When the moisture content of the soil increases, the coefficient also increases to a certain specific figure. When the moisture content rises above this specific level, the thermal conductivity coefficient of the mass is almost constant. The differing thermal conductivity coefficients of water in a liquid state (0.58 $W \cdot m^{-1} \cdot K^{-1}$) and of ice (2.25 $W \cdot m^{-1} \cdot K^{-1}$) show that the properties of the frozen soil are different. Experimental measurements [7], for example, have shown that the thermal conductivity coefficient of clay at plus-zero temperatures is 1.616 $W \cdot m^{-1} \cdot K^{-1}$, but while the clay is frozen it is 2.454 $W \cdot m^{-1} \cdot K^{-1}$. Leong et al. [8] have demonstrated a strong correlation of heat pump performance to moisture content and mineralogical composition of the ground mass. The findings of their experiments have proven that any reduction of moisture in the ground mass below 12.5% has a devastating impact on the performance of the heat pump. Ground mass moisture above 25% significantly improves the heat pump performance, albeit moisture levels in excess of 50% have an insignificant impact on the pump performance. The effect of the flow of the heat-transfer fluid in the pipes of the heat exchangers on the heat-transfer process in the soil is described by Tarnawski [9]. He has conducted computer simulations of horizontal linear heat exchangers installed at depths of 0.5 m and 1.0 m in single and two overlapping layers. He states that with the laminar flow of the heat-transfer fluid the heat-transfer process depends on the flow rate and the pipe diameter, as well as on the density, thermal conductivity and specific heat capacity of the heat-transfer fluid. He also states that heat transfer between the fluid and the pipe wall is more intense at lower temperatures and concentrations of the water and anti-freeze mixture, with shorter heat exchanger pipes, wider pipe diameters and faster fluid flow rates. He also confirms that horizontal ground heat exchangers have little impact on thermal degradation of the ground mass.

Researchers at the Department of Earth Resources Engineering at Kyushu University in Japan [10] checked various configurations of slinky-type horizontal heat exchangers. They checked heat exchangers with a loop diameter of 0.8 m, pipe diameters of 0.034 m and 0.024 m and with a spacing of 0.4 m, 0.6 m and 0.8 m; in order to assess the heat-transferring process between the soil and the heat exchanger pipes they used the ratio $\Delta t/q_\tau$, where $\Delta t = /t_k - t_0/$ (K), t_k (°C) is the average temperature of the heat-transfer fluid, t_0 (°C) is the temperature of the reference soil measured at a distance of at least 5 m from the heat exchanger and at the same depth as the heat exchanger, and q_τ (W·m^{-1}) is the specific thermal output of the ground source heat exchanger. They also used the finite element method and FEFLOW simulator to simulate the thermal output of a horizontal slinky-type ground source heat exchanger. During the simulation, attention was focused on the energy balance on the surface of the soil, the temperatures of the heat-transfer medium and the surrounding soil. The accuracy of the simulation model was confirmed by calculations and the results of the test.

Rezaei *et al.* [11] investigated the effect of surface cover on soil containing a horizontal heat exchanger on temperature distribution and heat flows discharged to the soil. When the surface of the soil is covered by an insulating layer of recycled tyres, the heat flux discharged from the soil during winter increased by 17%.

Researchers at the Department of Civil, Geological and Mining Engineering, École Polytechnique de Montréal [12] presented a new analytical model of a ground source heat exchanger based on the line source of heat method applicable for all types of horizontal ground source heat exchangers, including spiral and slinky heat exchangers. The model also takes account of possible changes in the water phase in the soil around the pipe of the horizontal heat exchanger. This model monitors the impact that the length of the pipe, the depth it is placed at and the spacing of the heat exchanger pipes have on the discharged thermal output and the risk of the soil thawing around the heat exchanger pipes.

Slinky-type horizontal heat exchangers require much smaller area of land. So it is assumed a greater interest in their implementation. Another motive for pursuing this type of horizontal exchangers, compared to linear horizontal exchangers is the relative lack of published knowledge.

The aim of our work was to monitor temperatures and analyse temperature changes in soil enclosing a slinky horizontal heat exchanger. Also, to assess the potential regeneration of the energy potential of the soil when the heat exchangers are not operational, and to determine the specific heat flows and specific thermal energy discharged from the soil.

2. Material and Methods

2.1. Theoretical Analysis

The basic equation for heat transfer by a cylindrical linear source in a solid mass was published by Carslaw and Jaeger [13]. Their aim was to determine the temperature t at any point at a distance of r from the linear source at the time τ after the source was started up. The equation at the coordinates x, y, z has the following basic form:

$$\frac{\partial^2 t}{\partial x^2} + \frac{\partial^2 t}{\partial y^2} + \frac{\partial^2 t}{\partial z^2} = \frac{C}{\lambda}\frac{\partial t}{\partial \tau} \tag{1}$$

In Equation (1):

C: volumetric specific heat capacity of mass $(J \cdot m^{-3} \cdot K^{-1})$;
λ: thermal conductivity coefficient of mass $(W \cdot m^{-1} \cdot K^{-1})$.

For the following conditions the authors give Equation (1) in the form:

$$t - t_0 = \frac{q}{4\pi\lambda} E(u) \tag{2}$$

Equation (2) applies for the following conditions:

* at time $\tau = 0$ $t = t_0$ for all r values (at the beginning of the measurement the temperature of the soil is constant at all distances from the linear source of heat);
* for $r \to \infty$ $t = t_0$ for all τ values (the distance from a linear source of heat approaching infinity, the temperature of the soil is constant independently of the operating line source).

In Equation (2):

t_0: constant temperature of thermally unaffected mass (°C);
q: constant discharged (supplied) specific heat flux $(W \cdot m^{-1})$;
$E(u)$: exponential integral function u.

If the specific thermal resistance of the heat exchanger pipes R_p $(K \cdot m \cdot W^{-1})$ is taken into account, Equation (2) becomes:

$$t - t_0 = q R_p + \frac{q}{4\pi\lambda} E(u) \tag{3}$$

After the heat exchanger has been started up, the pipes start to discharge heat from the soil in a radial direction into the heat-transfer medium in a pipe. After a certain amount of time, the heat exchanger begins to react to the temperature of the soil surface. The temperature of the soil surface thus becomes the dominant boundary condition. It may take several months for this to take effect with horizontal ground source heat exchangers.

To determine heat conduction through a semi-defined mass with a number of pipes the resources methods and the principles of superposition of the temperature field as described by Šorin [14] and later applied in the fundamental publication by Banks [15] should be used. This method is based on the above assumption that after a certain amount of time a pipe at a depth of z discharges heat flux q_τ corresponding to heat flux q_τ on the surface of the soil. When the superposition method is used, the heat flux q_τ on the surface of the soil is expressed with an imaginary pipe at a distance of z above the surface of the soil. According to Banks [15], Equation (3) then becomes:

$$t_0 - t_k = q_\tau R_p + \frac{q_\tau}{4\pi\lambda} E(u) - \frac{q_\tau}{4\pi\lambda} E(u') \tag{4}$$

where:

$$u = \frac{r_p^2 . C}{4\lambda.\tau} \quad \text{(for a real heat exchanger)} \tag{5}$$

$$u' = \frac{(2z)^2 . C}{4\lambda.\tau} \quad \text{(for an imaginary heat exchanger)} \tag{6}$$

In Equations (4) to (6):

  r_p: is the outer radius of the heat exchanger pipe (m);
  τ: time after starting the heat exchanger (s);
  t_k: average temperature of the heat-transfer fluid (°C).

When using a logarithmic approximation, according to Banks, Equation (6) then becomes:

$$t_0 - t_k = q_\tau R_p + \frac{q_\tau}{4\pi\lambda}\left[\ln\left(\frac{4\lambda\tau}{r_p^2 C}\right) - 0.5772\right] - \frac{q_\tau}{4\pi\lambda}\left[\ln\left(\frac{4\lambda\tau}{(2z)^2 C}\right) - 0.5772\right] = q_\tau R_p + \frac{q_\tau}{2\pi\lambda}\ln\left(\frac{2z}{r_p}\right) \tag{7}$$

where 0.5772–Euler constant.

The thermal resistance of the heat exchanger pipe R_p may be expressed using the equation:

$$R_p = R_t + R_\alpha \tag{8}$$

In Equation (8):

  R_t: thermal resistance of conduction through pipe wall (K·m·W^{-1});
  R_α: thermal resistance of convection between pipe wall and heat-transfer fluid (K·m·W^{-1}).

The calculation of the thermal resistances R_t and R_α is shown in the article by Šedová et al. [16].

2.2. Measurement Methods

The ground source heat exchangers tested are sources of energy for three heat pumps. Prior to 10 July 2012 these were 2× IVT Greenline HT PLUS E17 (Industriell Värme Teknik, Tnanas, Sweden) with a heat output of 16.2 kW and 1× IVT Premiumline X15 with a heat output of 11.7 kW. The IVT Premiumline X15 heat pump was replaced on 10 July 2012 with an IVT Premiumline EQ13 with a heat output of 13.3 kW. Heat output is determined at temperatures of 0/35 °C. These and another three heat pumps are used to heat the administrative building and operational halls of VESKOM s.r.o., based in Dolní Měcholupy.

A plan of a horizontal slinky-type ground source heat exchanger is shown in Figure 1. The heat exchanger was made from PE 100RC 32 × 2.9 mm polyethylene tubing (Luna Plast a.s., Hořín,

Czech Republic) resistant to point loads and cracking. It is not laid in a sand bed. The heat exchanger pipes, 200 m in total length, are installed at a depth of 1.5 m in 53 loops twisted into a circle with a loop spacing of 0.38 m. The soil to a depth of approximately 2 m consists of dark-brown sandy loam soil, coarse gravel, rubble and brick fragments. The sensors for measuring the temperature of the soil were installed at a distance of 4 m from the start of the heat exchanger. The heat transfer fluid flowing through the heat exchanger is a mixture of 33% (volumetric) ethanol and 67% water.

Figure 1. Plan of slinky-type heat exchanger and location of temperature sensors (exchanger dimensions in millimeters).

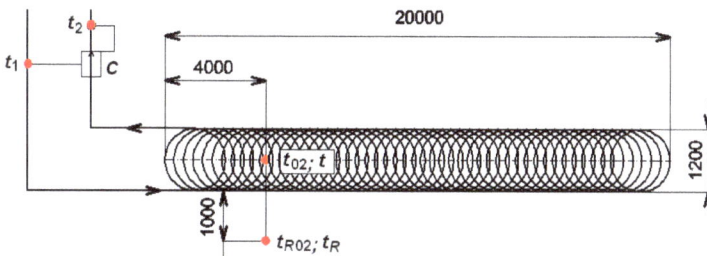

t: temperature sensor located at a depth of 1.5 m in the vicinity of the heat exchanger; t_R: reference temperature sensor located 1.0 m from the heat exchanger at a depth of 1.5 m; t_{02}: temperature sensor located at a depth of 0.2 m above the heat exchanger; t_{R02}: reference temperature sensor located 1.0 m from the heat exchanger at a depth of 0.2 m; t_1: temperature sensor for measuring the temperature of heat transfer fluid exiting the evaporator of the heat pump; t_2: temperature sensor for measuring the temperature of heat transfer fluid entering the evaporator of the heat pump; C: measure the volumetric flow of heat transfer fluid.

The temperatures of the soil were measured using PT 1000A RTD temperature sensors (manufactured by Greisinger Electronic GmbH, Regenstauf, Germany) and recorded in 15-min intervals. The ambient temperatures t_e were measured at a height of 2 m above the ground and at a distance of 20 m from the horizontal ground source heat exchangers. MTW 3 electronic heat consumption meters (manufactured by Itron Inc., Liberty Lake, WA, USA) were used to measure the total heat flow discharged by the horizontal heat exchangers. Electronic meter heat consumption works on the principle of integration of the heat transfer fluid flow and heat transfer fluid temperature difference between the inlet and the outlet of the evaporator of the heat pump.

The thermal characteristics of the soil, thermal conductivity coefficient λ (W·m^{-1}·K^{-1}), volumetric specific heat capacity C (J·m^{-3}·K^{-1}) and temperature conductivity coefficient a (m^2·s^{-1}) were determined using an Isomet 2104 (manufactured by Applied Precision, Bratislava, Slovakia) at temperature t (°C) and volumetric moisture w (%). Isomet 2104 is a portable instrument designed for the direct measurement of the thermal conductivity and specific volumetric heat capacity. For the measurement uses replaceable needle or a flat probe with integrated memory and the known calibration constants. Soil moisture in the ground heat exchanger was measured PR2—Profile Probe (manufacturer Delta-T Devices, Cambridge, UK).

The primary aim was to monitor the temperature of the soil and the environment during the heating period and the period of stagnation of ground heat exchanger. Furthermore, to determine in the following heating season (with valid recorded data) temperature of the heat transfer medium and the specific heat dissipated from soil.

3. Results and Discussion

The results of the measurement of the thermal characteristics of the soil are given in Table 1. The measurements were taken directly in the soil, when the heat exchanger was idle during the summer, on 6 June 2012. The thermal characteristics are within a range corresponding to Cambisol, the most widespread type of soil [17] in the Czech Republic.

Table 1. Thermal characteristics of the soil.

Depth (m)	t (°C)	w (%)	λ (W·m⁻¹·K⁻¹)	$10^6 \cdot C$ (J·m⁻³·K⁻¹)	$10^{-6} \cdot a$ (m²·s⁻¹)
0.06	13.36	36.65	1.21	2.08	0.583
0.22	12.76	26.25	1.29	2.15	0.602
0.30	12.42	30.70	1.35	2.11	0.640
0.60	12.66	31.55	1.24	1.82	0.678
0.90	12.73	29.30	1.48	2.15	0.688
1.20	12.65	31.60	1.39	2.08	0.672
1.50	13.64	39.00	1.58	2.24	0.704
1.60	13.83	-	1.57	2.16	0.727

t: temperature of the soil; w: volumetric moisture; λ: thermal conductivity coefficient; C: volumetric specific heat capacity; a: temperature conductivity coefficient.

This article presents results of the tests of a horizontal slinky-type ground source heat exchanger performed between 7 September 2011 and 22 April 2013. The graph in Figure 2 shows the average daily temperature trends of the soil t and the temperature of the surrounding air t_e when the heat exchanger is operational in the period τ 7 September 2011–16 September 2012. The temperatures t_{02} (°C) and reference temperatures t_R (°C) of the soil are not shown for reasons of clarity. These dependences may be expressed using an equation based on the equation for the free undamped oscillation of a mass point [18]:

$$t = \bar{t} + \Delta t_A . \sin(\Omega \cdot \tau + \varphi) \tag{9}$$

where:

t: temperature (°C);

$\bar{t}$:   average temperature (°C);

Δt_A :   oscillation amplitude around temperature $\bar{t}$ (°C);

τ :   number of days from start of measurement (days);

φ: initial phase of oscillation (rad);

Ω :   angular velocity $= 2 \cdot \pi \cdot 365^{-1}$ (rad·day⁻¹).

Figure 2. Temperatures of a soil containing a slinky heat exchanger from 7 September 2011 to 16 September 2012.

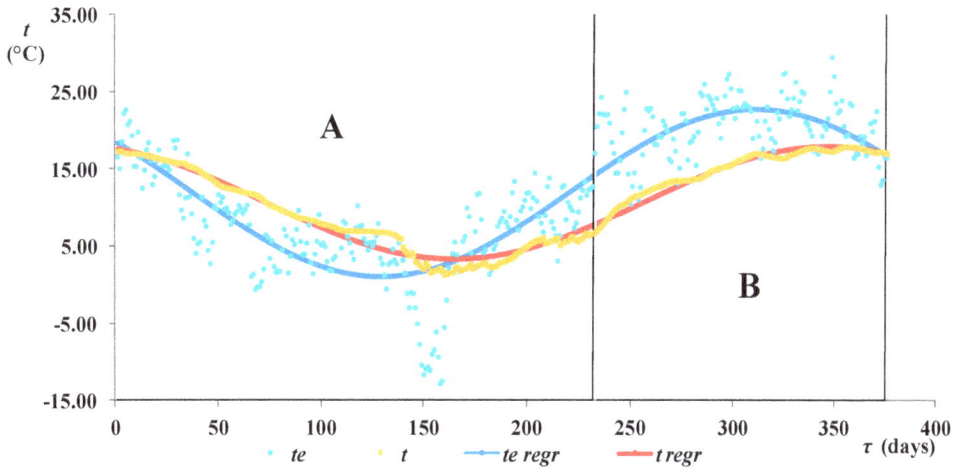

This is the non-linear regression of y to x, and is therefore used to determine the degree of dependence between two random variables of the determination index I_{yx}^2 (-), [19].

The trend in the temperature of the soil mass and that of the surrounding air when the heat exchanger is operational may be expressed using following equations:

$$t = 10.646 + 7.303 \cdot \sin(\Omega \cdot \tau + 1.88) \qquad (I_{t,\tau}^2 = 0.975) \qquad (10)$$

$$t_{0.2} = 10.897 + 9.762 \cdot \sin(\Omega \cdot \tau + 2.24) \qquad (I_{0.2}^2 = 0.958) \qquad (11)$$

$$t_R = 10.646 + 7.287 \cdot \sin(\Omega \cdot \tau + 1.919) \qquad (I_{t_R \tau}^2 = 0.981) \qquad (12)$$

$$t_e = 11.928 + 10.839 \cdot \sin(\Omega \cdot \tau + 2.493) \qquad (I_{t_e \tau}^2 = 0.778) \qquad (13)$$

The period in question (376 days) may be divided up into the heating part of period A (7 September 2011–26 April 2012, 233 days) and that part of the period in which the system was idle B (27 April 2012–16 September 2012, 143 days). In part of period B the energy potential of the soil regenerates for the next heating season. The temperatures measured and Equations (10) and (12) show that the temperature difference $\Delta t = t_R - t$ was insignificant in both parts of the period, $\Delta t_{p,A} = -0.1 \pm 0.7$ K, $\Delta t_{p,B} = 0.1 \pm 0.3$ K. The negligible temperature difference Δt is, however, affected by the position of the reference temperatures sensor in the soil a mere 1 m from the heat exchanger. One important finding is that during part of period A the temperature of the soil around the heat exchanger was above zero. Lower temperatures of the soil result in lower-temperature evaporation in the evaporator, which may have an adverse effect on the heat pump performance factor.

The minimal temperature of the soil mass in vicinity of the heat exchanger $t_{min} = 0.83$ °C was determined as the average of four temperatures measured between 10:00 to 11:00 am 13 February 2012 ($\tau = 160$ days). The minimal average daily temperatures of the soil in the vicinity of the heat exchanger $t_{min.\ day} = 1.27$ °C also occurred on 13 February 2012.

The temperature of the soil t_{02} at a depth of 0.2 m above the heat exchanger is influenced particularly by the temperature and speed of the surrounding air, the intensity of incident solar radiation, and falls of rain and snow. The heat flux discharged by the heat exchanger also has a massive effect on the temperature, however. This is confirmed by Equations (10), (11) and (13). $\bar{t} < \bar{t}_{02} < \bar{t}_e$ is applicable for the average temperature of the soil at a depth of 0.2 m and $\Delta t_A < \Delta t_{A02} < \Delta t_{Ae}$ for the oscillation amplitude. The initial phase of temperature oscillation at a depth of 0.2 m ($\varphi_{02} = 2.24$ rad) is influenced by the surrounding environment ($\varphi_e = 2.493$ rad). In the vicinity of the heat exchanger and at the reference point the difference in the initial phases ($\Delta\varphi = 0.039$ rad) may be considered insignificant.

The minimal average daily temperatures calculated from Equations (10) to (13) are shown in Table 2.

Table 2. Calculated minimal average daily temperatures.

Temperature	Minimal temperature (°C)	Time of heat exchanger operation τ (days)
t_e	1.09	129
t_{02}	1.14	144
t	3.34	164
t_R	3.36	164

t_e: average daily temperature of the surrounding air; t_{02}: temperature of the soil at a depth of 0.2 m; t: average daily temperature of the soil at a depth of 1.5 m; t_R: average daily reference temperature of the soil at a depth of 1.5 m.

From the summary in Table 2 it is apparent that the minimal temperatures occur in a logical sequence. The graph in Figure 3 displays the results of measuring the temperature of the soil in the vicinity of the heat exchanger t, the ambient temperature t_e and specific heat q_d discharged from the soil in the heating season 17 September 2012–22 April 2013. The temperatures of the soil in the vicinity of the heat exchanger and the ambient temperatures are given by the equations:

$$t = 8.325 + 7.666 \cdot \sin(\Omega \cdot \tau + 2.073) \qquad (I^2_{t\tau} = 0.961) \qquad (14)$$

$$t_e = 10.049 + 9.348 \cdot \sin(\Omega \cdot \tau + 2.559) \qquad (I^2_{t_e\tau} = 0.558) \qquad (15)$$

$$t_R = 9.064 + 6.971 \cdot \sin(\Omega \cdot \tau + 1.993) \qquad (I^2_{t_R\tau} = 0.974) \qquad (16)$$

During the heating period 17 September 2012–22 April 2013, $\Delta t = t_R - t = 1.2 \pm 0.6$ K. The temperature difference is thus slightly higher than in the previous heating season 7 September 2011–26 April 2012, when it was $\Delta t_{p,A} = t_R - t = -0.1 \pm 0.7$. It is assumed that the cause of higher temperature difference is lower average temperature t_e (about 1.53 K) in the heating period 17 September 2012–22 April 2013 than in the period 7 September 2011–26 April 2012. There is thus a higher heat transfer of the earth mass, resulting in a lower average temperature of the soil mass t (about 2.79 K).

As Equation (14) only portrays the temperatures of the soil in the vicinity of the heat exchanger during the heating season, the average daily temperature $\bar{t}$ is lower than in Equation (10), which also includes the idle period. The oscillation amplitudes Δt_A around the temperature $\bar{t}$ are almost the same in both equations. The same is true of Equations (13) and (15) describing the ambient temperature t_e. However, the oscillation amplitudes show greater differences.

In Equation (15) the determination index is <0.8, which is below the normal level of dependence. The lower determination index is caused by the considerable dispersion of the ambient temperature.

As in the 2011–2012 heating season, during this period the temperature of the soil in the vicinity of the heat exchanger was above zero. The minimum temperature of the soil mass in vicinity of the heat exchanger $t_{min.} = 0.44$ °C was determined as the average of four temperatures measured from 8:00 to 9:00 am 2 April 2013 ($\tau = 197$ days). The minimal average daily temperature of the soil $t_{min.day} = 0.83$ °C occurred on 28 March 2013. During the 2012–2013 heating season the minimal temperature of the soil in the vicinity of the heat exchanger occurred 37 days later than in the 2011–2012 heating season. The graphs in Figure 2 and Figure 3 show that the ambient temperature dropped in a single wave in the 2011–2012 heating season, after the heat exchanger had been operational for approximately 153 days.

Figure 3. Temperatures of the soil and heat discharged to a soil containing a slinky heat exchanger in the heating season 17 September 2012–22 April 2013.

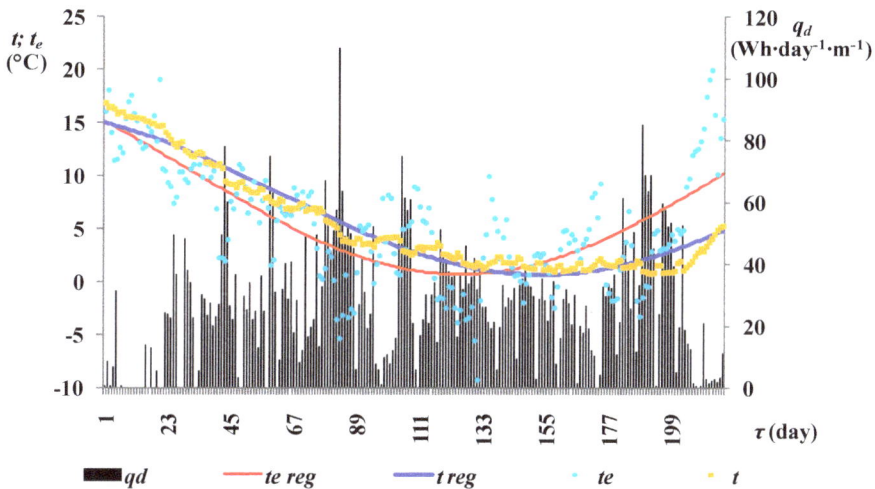

The temperature of the soil in the vicinity of the heat exchanger subsequently also fell to the minimum. In the 2012–2013 heating season the ambient temperature dropped to higher than in the previous heating season in three waves. It was not until after the third wave that the temperature of the soil fell to the minimum with a relatively large discharge of heat.

The energy potential regeneration capacity when the heat exchanger is idle may be gauged on the basis of the starting and final temperatures of the soil in the vicinity of the heat exchanger over several heating seasons. In particular, a reduction in the temperature of the soil at the start of the heating season would be a clear sign of the gradual reduction of the energy potential of the mass and also therefore of the amount of time the mass could be used to power a heat pump. The energy potential of the soil could be increased during the summer by modulating the operation of the heat pump, *i.e.*, by the site cooling and the generated heat accumulating in the soil. The measurements are given in Table 3.

Table 3. Average daily temperatures of the soil at the start and end of the heating seasons.

Phase of heating season	Heating season	Date	t (°C)	Δt (K)
	2010–2011	30 August 2010	18.40	
				−0.04
Start of heating season	2011–2012	7 September 2011	18.44	
				1.54
	2012–2013	17 September 2012	16.90	
	2010–2011	22 March 2011	4.45	
				−1.58
End of heating season	2011–2012	22 March 2012	6.03	
				0.73
	2012–2013	22 April 2013	5.30	

t: temperature of the soil; Δt: temperature difference.

The differences in the temperature of the soil at the start and end of the heating seasons are within the range of measurement accuracy. The results of more than three years of validation indicate that the slinky-type horizontal ground heat exchanger can be considered as stable energy source for heat pumps.

The heat q_d (Wh·m^{-1}·day^{-1}) discharged by the ground source heat exchanger is displayed in the columns in Figure 3. Average value of heat discharged during the period in question was 26.57 Wh·m^{-1}·day^{-1}; the maximum was 109.93 Wh·m^{-1}·day^{-1}. During the entire heating season of 217 days, 1 m of heat exchanger pipe discharged 5831.54 Wh·m^{-1} of heat to the soil. A water storage tank was connected to the condenser of the heat pump, meaning that it is impossible to assess the direct link between the heat discharged and the temperature of the surrounding air.

During the heating period 17 September 2012–22 April 2013 the average temperature of the heat transfer fluid at the outlet of the ground heat exchanger was 4.17 °C, minimal temperature was 0.39 °C. The average temperature of the heat-transfer fluid at the heat pump outlet was 2.52 °C; the minimal temperature was −2.02 °C.

An example of average thermal outputs q_τ (W·m^{-1}) discharged to the soil by the heat exchanger, the temperature t (°C) of the soil in the vicinity of the heat exchanger and the ambient temperature t_e (°C) is shown in the graph in Figure 4. During the 24 h of that day 1 m of heat exchanger pipe discharged 109.87 Wh·m^{-1} to the soil. The average specific thermal output of the heat exchanger was $q_\tau = 4.58$ W·m^{-1}; the maximum was 8.21 W·m^{-1}. The circulation pump pumping the ground source heat exchanger's heat-transfer fluid worked at a higher rate ($1.72 \cdot 10^{-4}$ m^3·s^{-1}) at the intervals 3–7 a.m., 5–10 p.m. and at 12 a.m. (a total of 12 h). The pump was switched off or worked at a lower rate at the intervals 1–2 a.m., 8 a.m.–5 p.m. and at 11 p.m. (also a total of 12 h). Although the operation of the circulation pumps was influenced by the accumulation of heat from the heat pump condenser, the graph in Figure 4 clearly shows a link between the specific heat flux q_τ (W·m^{-1}) discharged by the ground source heat exchanger and the ambient temperature t_e (°C). The temperature of the soil in the vicinity of the heat exchanger reacts to changes in specific thermal output only within the range $t = 4.48 \pm 0.4\ °C$.

Figure 4. Temperatures of the soil and specific thermal output of the heat exchanger on a typical winter day 8 December 2012.

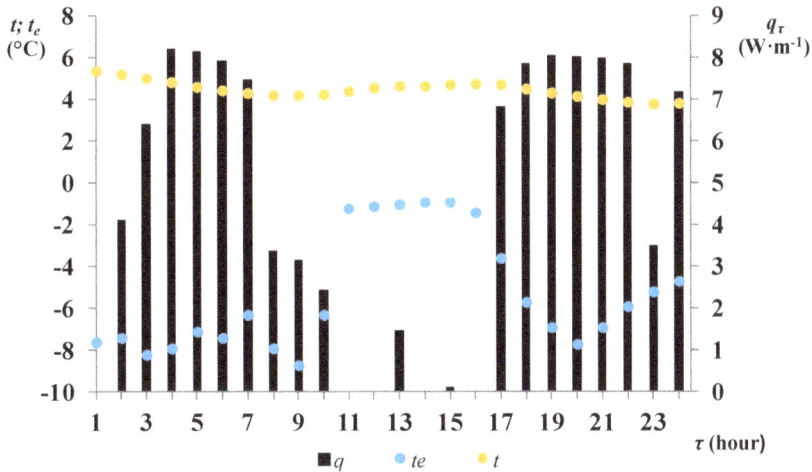

Validity of the Equation (7) designed for the linear type of heat exchanger, was in case of example depicted in Figure 4 verified for the Slinky-type heat exchanger. Towards determination of flow type of the heat transfer fluid, form of the criteria equation and calculation of the heat transfer coefficient α between the tube wall of heat exchanger and heat transfer fluid was used Reynolds and Nusselt criteria. The Reynolds criterion Re (-) of the heat-transfer fluid was in the range (586.6–1544.4). Computation of the Nusselt criterion Nu (-) is based on the equation given by Prof. Schramek [20], which is applicable for the laminar fluid flow in the pipe:

$$Nu = \frac{\alpha \cdot d_h}{\lambda_k} = \left(49.028 + 4.173 \cdot Re \cdot Pr \cdot \frac{d_h}{L} \right)^{0.333} \tag{17}$$

In Equation (17):

Nu: Nusselt criterion (-);

α: convective heat-transfer coefficient between the wall of the heat exchanger pipe and the heat-transfer fluid $(W \cdot m^{-2} \cdot K^{-1})$;

d_h: hydraulic diameter of pipe (m);

λ_k: thermal conductivity coefficient of heat-transfer fluid $(W \cdot m^{-1} \cdot K^{-1})$;

Pr: Prandtl criterion (-);

L: length of heat exchanger pipe (m).

Equation (17) shows Nu within the range (4.507–4.508). Spitler [21], using the GHELPRO software for laminar calculation of the thermal resistance between heat transfer fluid and the inner wall of the exchanger tube in laminar flow conditions ($Re \le 2,100$), achieves $Nu = 4.364$. Determining the thermal conductivity of the soil for the Equation (7) we used the established relationship between change of the thermal conductivity λ (W·m^{-1}·K^{-1}) and volumetric moisture of soil mass w (%), the soil type Cambisol in the area. For the estimated volumetric moisture soil mass, in vicinity of the heat exchanger, w = 35%–45% (control measurement), the thermal conductivity of the soil, for a given soil type Cambisol, is in the range of λ = 1.45–1.61 W·m^{-1}·K^{-1}. In the calculation of the Equation (7) is considered λ = 1.54 W·m^{-1}·K^{-1}.

The temperature difference (t_0 − t_k) (K) computed from Equation (7) and the measured temperatures are within the range $\Delta(t_0 - t_k)$ = 0.85 ± 0.90 K. If the secondary circulation of the heat-transfer fluid caused by the centrifugal force generated by the curvature of the pipe is taken to the account, the heat-transfer coefficient α calculated from Equation (17) is multiplied by the pipe curvature coefficient [22]:

$$\varepsilon_R = 1 + 1.77 \cdot \frac{d}{r} \tag{18}$$

In Equation (18):

    d: diameter of heat exchanger pipe (m);
    r: loop radius of heat exchanger (m).

The curvature coefficient ε_R calculated from Equation (18) is 1.0944. Curvature thus increases the heat-transfer coefficient α by approximately 10%. The temperature difference is then $\Delta(t_0 - t_k)$ = 0.92 ± 0.90 K. Both these differences $\Delta(t_0 - t_k)$ (K) show that Equation (7) may be applied with sufficient accuracy to slinky-type heat exchangers.

The thermal resistance of convection R_a (m·K·W^{-1}) between the wall of the pipe and the heat-transfer fluid is, together with the thermal conductivity coefficient of the soil λ (W·m^{-1}·K^{-1}), a crucial factor affecting heat flux q_τ (W·m^{-1}) discharged to the soil. If R_a (m·K·W^{-1}) increases excessively, q_τ (W·m^{-1}) is reduced and the temperature of the heat-transfer fluid t_k (°C), may drop to unfavourable minus temperatures. In the horizontal heat exchanger (measurement 8 December 2012) in question the relatively high R_a value (0.165–0.166) m·K·W^{-1} is due to the low heat-transfer coefficient α (73.28–73.81) W·m^{-2}·K^{-1} caused by the laminar flow of the heat-transfer fluid. A better turbulent flow will be achieved by increasing the flow rate, although this consumes more energy to drive the circulation pump, which reduces the SPF (seasonal performance factor). It seems more effective to reduce the concentration of anti-freeze in the heat-transfer fluid. Reducing this concentration would reduce kinematic viscosity, which, especially at low temperatures, would have a positive impact on Re (-) and thus also on Nu (-) and R_a (-). According to The Engineering ToolBox [23], the freezing point of the heat-transfer fluid (33% C_2H_6O + 67% H_2O) is −17.4 °C. The results of our measurements showed that the minimal temperature of the heat-transfer fluid at the outlet of the heat pump was −2.02 °C. Therefore, for the heat exchanger in question a lower concentration would be more suitable, e.g., 20% ethanol, when the freezing point of the mixture is −9.0°C. At the same volumetric flow of the heat-transfer fluid (8 December 2012 measurement) the thermal resistance R_a (-) would be (0.145–0.146) m·K·W^{-1}. Convective heat transfer coefficient α may increase for 13.62%. Decrease of concentration of ethyl alcohol in heat transferring liquid from 33% to 20% results in rise of specific heat capacity for 5.41% as well as rise of density for 1.22%. This will positively influence thermal output of ground heat exchanger. Xu and Spitler [24] state that with a 20% concentration of propylene glycol in the heat-transfer fluid and with a fluid temperature of -5 °C, the Re (-) figure would be a mere 39% of the figure achieved at the same flow rate and at a temperature of 20 °C.

4. Conclusions

The temperature trends in soil containing a slinky heat exchanger, the temperature of the reference mass and the ambient temperature as shown in the graphs in Figures 2 and 3 may be expressed using Equations (10)–(16). During the entire heating season (217 days) the amount of heat discharged to the soil through 1 m of heat exchanger pipe totalled a relatively significant 5831.54 Wh·m^{-1}. Therefore, if the total length of the heat exchanger pipes is 200 m, 1166.30 kWh was discharged to the soil in the heating season. The maximum specific thermal output of the heat exchanger was 12.33 W·m^{-1}. The meaning of work for design practice and implementation of Slinky-type horizontal heat exchangers can be summarized in the following points:

- The results of the test showed that in both the heating seasons in question the temperatures of the soil in the vicinity of the heat exchanger were above zero. The temperatures of the heat-transfer fluid at the heat pump outlet also dropped to below zero in the second half of the heating season (minimum -2.02 °C);
- The results of more than three years of validation indicate that the slinky-type horizontal ground heat exchanger can be considered as stable energy source for heat pumps. The temperature difference at the beginning and end of the heating seasons (Table 3) did not exceed 2 K in the four years that measurements were taken;
- Equation (7), designed for a linear heat exchanger, may also be used for slinky-type heat exchangers with a sufficient degree of accuracy.
- The thermal performance of horizontal ground heat exchanger has a dominant influence to thermal conductivity of the soil λ and thermal resistance $R\alpha$ between the pipe wall and the heat transfer fluid;
- The maximum specific thermal output of the heat exchanger was 12.33 W·m^{-1};
- The temperatures of the heat-transfer fluid at the heat pump outlet proved that the concentration of ethanol (33%) in the heat-transfer fluid is unnecessarily high. High concentration of ethanol is due to the higher value of kinematic viscosity and a lower value of thermal conductivity of heat transfer fluid the predominant cause of laminar flow of heat transfer fluid, and high values of thermal resistance of convection $R\alpha$ (m·K·W^{-1}) between the tube wall and the heat transfer fluid;
- Further research phase of horizontal ground slinky-type heat exchanger will focus on:
  * Creating a mathematical model of the temperature field in the ground mass exchanger;
  * Verification of lower concentrations of ethanol in the heat transfer fluid to the heat exchanger performance;
  * Verification of use of ground heat exchanger for cooling the building in summer on the energy potential of the soil mass and heat transfer rate in the heating season.

Acknowledgments

Authors acknowledge the financial support of the Technology Agency of the Czech Republic (Project No. TA02020991 "Optimization of the energetic parameters of the horizontal ground heat exchangers with respect to soil and hydrological conditions at the specific location").

Conflicts of Interest

The authors declare no conflict of interest.

References

1. Petit, P.J.; Meyer, J.P. Techno-economic analysis between the performances of heat source air conditioners in South Africa. *Energy Convers. Manag.* **1998**, *39*, 661–669.
2. Lund, J.W.; Freeston, D.H.; Boyd, T.L. Direct utilization of geothermal energy 2010 worldwide review. *Geothermics* **2011**, *40*, 159–180.
3. Brandl, H. Energy foundations and other therma-active ground structures. *Géotechnique* **2006**, *2*, 81–122.
4. Popiel, C.; Wojtkowiak, J.; Biernacka, B. Measurements of temperature distribution in ground. *Exp. Therm. Fluid Sci.* **2001**, *25*, 301–309.
5. De Swardt, C.A.; Meyer, J.P. A performance comparison between an air-coupled and a ground-coupled reversible heat pump. *Int. J. Energy Res.* **2001**, *25*, 810–899.
6. Congedo, P.M.; Colangelo, G.; Starace, G. CFD simulations of horizontal ground heat exchangers: A comparison among different configurations. *Appl. Therm. Eng.* **2012**, *33–34*, 24–32.
7. Song, Y.; Yao, Y.; Na, W. Impacts of Soil Pipe Thermal Conductivity on Performance of Horizontal Pipe in a Ground-Source Heat Pump. Energy Systems Laboratory. Available online: http://hdl.handle.net/1969.1/5465 (accessed on 12 November 2006).
8. Leong, W.H.; Tarnawski, V.R.; Aittomaki, A. Effect of soil type and moisture content on ground heat pump performance. *Int. J. Refrig.* **1998**, *21*, 595–606.
9. Tarnawski, V.R.; Leong, W.H.; Momose, T.; Hamada, Y. Analysis of ground source heat pumps with horizontal ground heat exchangers for northern Japan. *Renew. Energy* **2009**, *34*, 127–134.
10. Fujii, H.; Nishia, K.; Komaniwaa, Y.; Choub, N. Numerical modeling of slinky-coil horizontal ground heat exchangers. *Geothermics* **2012**, *41*, 55–62.
11. Rezaei, B.A.; Kolahdouz, E.M.; Dargush, G.F.; Weber, A.S. Ground source heat pump pipe performance with tire derived aggregate. *Int. J. Heat Mass Transfer.* **2012**, *55*, 2844–2853.
12. Fontainea, P.O.; Marcottea, D.; Pasquiera, P.; Thibodeaub, D. Modeling of horizontal geoexchange systems for building heating and permafrost stabilization. *Geothermics* **2011**, *40*, 211–220.
13. Carslaw, H.S.; Jaeger, J.C. *Conduction of Heat in Solids*, 2nd ed.; Oxford University Press: London, UK, 1948; pp. 239–272.

14. Šorin, S.N. *Transmission of Heat*, 1st ed.; State Publishing of Technical Literature: Prague, Czech Republic, 1968; pp. 142–148. (in Czech)

15. Banks, D. *An Introduction to Thermogeology: Ground Source Heating and Cooling*, 2nd ed.; John Wiley & Sons: Chichester, West Sussex, UK, 2012; pp. 332–344.

16. Šeďová, M.; Adamovský, R.; Neuberger, P. Analysis of ground massif temperatures with horizontal heat exchanger. *Res. Agric. Eng.* **2013**, *59*, 91–97.

17. Němeček, J.; Vokoun, J.; Smejkal, J.; Macků, J.; Kozák, J.; Němeček, K.; Borůvka, L. Taxonomic Classification System of Soils in the Czech Republic. Available online: http://klasifikace.pedologie.czu.cz/index.php?ac (accessed on 10 October 2013). (in Czech)

18. Beer, F.P.; Johnston, E.R., Jr. *Vector Mechanics for Engineers: Statics and Dynamics*, 5th ed.; McGraw-Hill: New York, NY, USA, 1988; pp. 943–946.

19. Bowerman, B.L.; O'Connell, R.T. *Applied Statistics: Improving Business Processes*, 1st ed.; Richard D. Irwin Inc.: Boston, MA, USA, 1997; pp. 712–723.

20. Recknadel, H.; Sprenger, E.; Schramek, E.R. *Handbook for Heating and Air Conditioning*, 73rd ed.; Oldenbourg Industrieverlag: München, Germany, 2007; pp. 150–156. (in Deutsch)

21. Spitler, J.D. GLHEPRO 4.0 for Windows User'S Guide. International Ground Source Heat Pump Association, Oklahoma State University, Stillwater. Available online: http://www.hvac.okstate.edu (accessed on 17 July 2007).

22. Kalčík, J.; Sýkora, K. *Technical Thermodynamics*, 1st ed.; Academia: Prague, Czech Republic, 1973; pp. 448–453. (in Czech)

23. The Engineering ToolBox: Ethanol Freeze Protected Water Solutions. Available online: http://www.engineeringtoolbox.com/ethanol-water-d_989.html (accessed on 2 July 2013).

24. Xu, X.; Spitler, J. Modeling of Vertical Ground Loop Heat Exchangers with Variable Convective Resistance and Thermal Mass of the Fluid. In Proceedings of 10th International Conference on Thermal Energy Storage Ecostock 2006, Stockton, NJ, USA, 31 May–2 June 2006.

Thermoeconomic Analysis of Hybrid Power Plant Concepts for Geothermal Combined Heat and Power Generation

Florian Heberle and Dieter Brüggemann

Abstract: We present a thermo-economic analysis for a low-temperature Organic Rankine Cycle (ORC) in a combined heat and power generation (CHP) case. For the hybrid power plant, thermal energy input is provided by a geothermal resource coupled with the exhaust gases of a biogas engine. A comparison to alternative geothermal CHP concepts is performed by considering variable parameters like ORC working fluid, supply temperature of the heating network or geothermal water temperature. Second law efficiency as well as economic parameters show that hybrid power plants are more efficient compared to conventional CHP concepts or separate use of the energy sources.

Reprinted from *Energies*. Cite as: Heberle, F.; Brüggemann, D. Thermoeconomic Analysis of Hybrid Power Plant Concepts for Geothermal Combined Heat and Power Generation. *Energies* **2014**, *7*, 4482-4497.

1. Introduction

For low-enthalpy geothermal resources binary power plants like the Organic Rankine Cycle (ORC) or the Kalina Cycle (KC) are suitable [1,2]. Combined heat and power generation (CHP) is a promising approach to improve the economic conditions for geothermal energy generation. An additional heat supply could be realized in various types of power plant configurations. In general, serial or parallel circuit of power and heat generation are considered [3]. Furthermore, innovative concepts like hybrid power plants are a promising approach to increase the thermodynamic and economic efficiency. For this purpose, geothermal power plants are typically coupled with an alternative energy source like a biogas cogeneration unit, solar thermal panels, solid biomass or fossil fuels [4–11]. In climatic zones where solar thermal systems are not practical, but renewable CHP is still favoured, a hybrid power plant consisting of a geothermal heat source and a biogas engine seems to be a suitable concept. In this paper different configurations for hybrid power plants based on ORC-technology are compared to conventional geothermal CHP and separate use of the energy sources. For geothermal water temperatures of 120 °C, the electricity produced annually, second law efficiency and economic parameters are calculated. Sensitivity analyses are performed concerning ORC working fluid, supply temperature of the heating network and geothermal conditions.

2. Methodology

The annual power output for the considered CHP concepts is calculated using quasi-steady-state considerations, consisting of ORC process simulations and approximation of the annual duration curve of the heat demand. The most efficient power plant configurations are identified based on second law efficiency, internal rate of return and cumulative cashflow. Therefore, thermodynamic and economic boundary conditions are defined.

2.1. Process Simulations

Geothermal CHP for low-enthalpy resources is investigated in parallel or serial configuration of power unit and heat generation. A scheme of both power plant concepts is shown in Figure 1. For serial circuit, a bypass pipe provides sufficient geothermal water temperatures in case of high supply temperatures of the heating network and low ambient temperatures, respectively.

Figure 1. (a) Scheme of geothermal CHP in parallel circuit; **(b)** Scheme of geothermal CHP in serial circuit with bypass pipe.

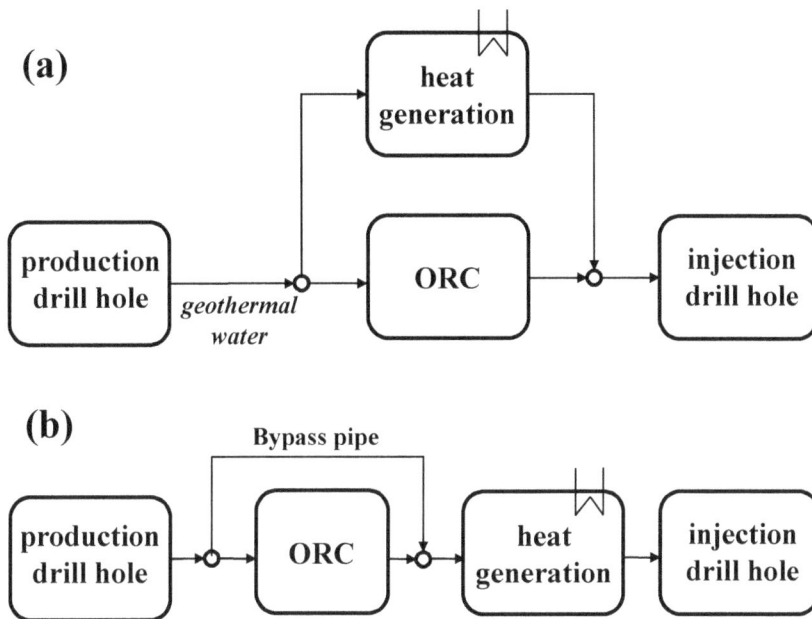

A hybrid power plant for CHP is also feasible in parallel and serial configuration. Figure 2 shows the parallel power unit and heat generation circuit. According to heat demand the geothermal water mass flow is split and the ORC operates under partial load. A higher geothermal water temperature at the inlet of the ORC-unit is obtained by utilizing the exhaust gases of the gas engine. The engine coolant provides heat for the heating network in a first step. If necessary, a higher amount of heat or higher supply temperatures are obtained in a second heat exchanger. The serial configuration of the hybrid power plant is analogue to the serial geothermal CHP in Figure 1b. Finally, a separate use of geothermal heat source and biogas cogeneration unit is examined. In this case, the exhaust gases of the gas engine are simply used for heat generation instead of coupling with the geothermal water.

Figure 2. Scheme of geothermal hybrid power plant in parallel circuit.

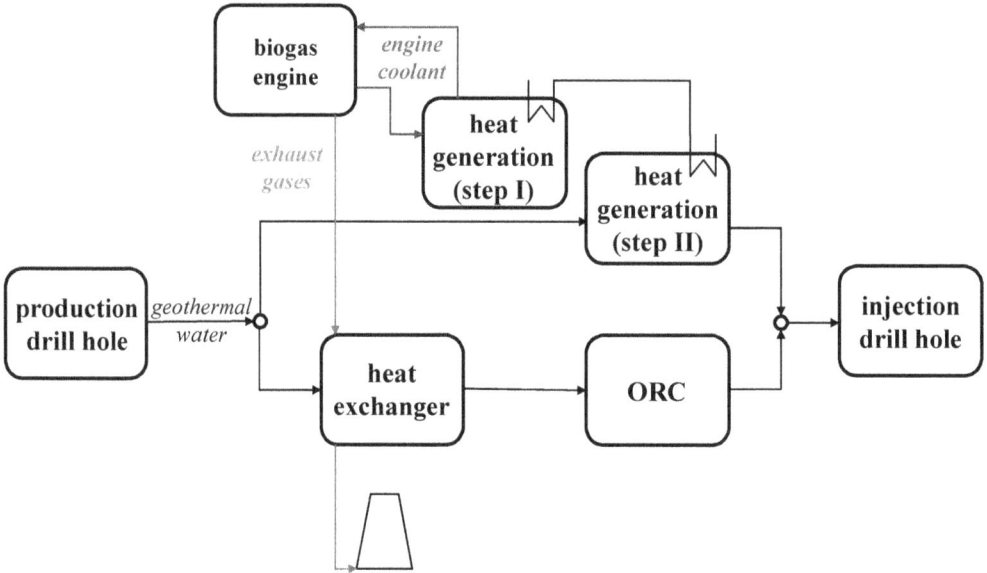

The ORC is calculated using the software Cycle Tempo [12] and fluid properties are based on REFPROP 9.1 [13]. According to Figure 3a the ORC working fluid is forced by the pump to a higher pressure level (1→2) followed by the coupling with the geothermal heat source, in the preheater (2→3) first, and then in the evaporator (3→4). A saturated cycle is assumed, so in state point 4 no superheating arises. For the considered working fluids R245fa (1,1,1,3,3-pentafluoropropane), isopentane and isobutane, all so-called dry fluids, there is no danger of turbine erosion due to the positive slope of the dew line in the T,s-diagram. In the next step the working fluid is expanded in the turbine (4→5). The condensation (5→1) closes the cycle. Figure 3b shows exemplarily the changes of states in a T,s-diagram for an ORC using the working fluid isopentane. In Table 1 the boundary conditions of the ORC like isentropic efficiency of the rotating equipment η_i, temperature difference at the pinch point ΔT_{PP} in the condenser and evaporator, cooling temperature at the inlet $T_{CW,in}$ and temperature difference of the cooling water ΔT_{CW} are outlined. Due to a high content of dissolved salts in the geothermal fluid, mineral deposits could occur for low temperatures. To avoid such scalings in the heat exchangers, in particular the preheater, the reinjection temperature of geothermal water is set to 60 °C. Regarding the hybrid power plant, the biogas cogeneration unit (a GE Jenbacher JMS 620 GS-B.L.) is coupled with the geothermal heat source. All relevant parameters of the gas engine like electric power P_{el}, thermal power $\dot{Q}$, outlet temperature of cooling water $T_{CW,out}$, mass flow of cooling water $\dot{m}_{CW}$ or outlet temperature of the exhaust gases $T_{EG,out}$ are shown in Table 2.

Figure 3. Scheme of ORC-unit (**a**) and corresponding T,s-diagram for ORC with the working fluid isopentane (**b**).

(a)

(b)

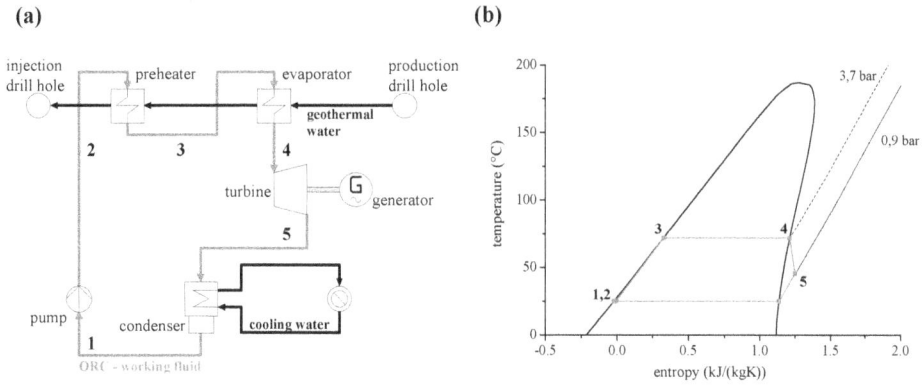

Table 1. Boundary conditions for the ORC power plant.

Parameter	Unit	
Isentropic efficiency of the ORC-turbine $\eta_{i,T}$	%	80
Generator efficiency η_G	%	95
Isentropic efficiency of the ORC-pump $\eta_{i,P}$	%	75
$\Delta T_{PP,EVP}$	K	5
$T_{CW,in}$	°C	15
ΔT_{CW}	K	5

Table 2. Operational parameters of the biogas cogeneration unit (JMS 620 GS-B.L.).

Parameter	Unit	
Electrical power output P_{el}	kW	2717
Thermal power output $\dot{Q}$	kW	1315
Engine coolant outlet temperature $T_{CW,out}$	°C	87.8
Engine coolant inlet temperature $T_{CW,in}$	°C	65.5
Engine coolant mass flow rate $\dot{m}_{CW}$	kg/s	19.9
Exhaust gas outlet temperature $T_{EG,out}$	°C	463.9
Exhaust gas mass flow rate $\dot{m}_{EG}$	kg/s	4.35

In the case of heat generation, a heating network which supplies a settlement of 8000 inhabitants is considered. A distribution of 30% single-family houses and 70% multi-family houses is assumed. The heat demand for each housing unit is calculated based on load profiles for typical climatic patterns (zone 13) according to VDI 4655 [14]. For a thermal power higher than 6000 kW a peak load boiler is considered. In total a thermal energy of 23.9 GWh is coupled to the heating network. For a quasi-steady-state calculation of power and heat generation, the annual duration curve is approximated by 10 load steps, which correspond to the averaged ambient temperature of the typical climate patterns (see Figure 4). In addition, a linear dependence of supply and return temperature of the heating network on ambient temperature between −14 °C and 16 °C is taken into account. The maximum supply temperature is 90 °C and the minimal value is 60 °C. The temperature difference between supply and return temperature is set constant to 20 K.

Figure 4. Annual duration curve of the heating network and approximation by load steps corresponding to the averaged ambient temperature of the considered climatic patterns.

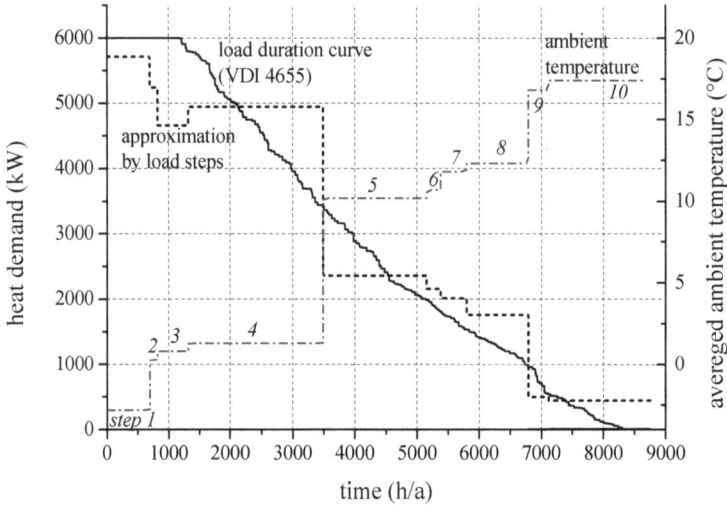

As a fixed criterion for the process simulations, the heat demand is fully covered by all CHP concepts. Hence the annual amount of produced electricity is suitable to compare the considered concepts under thermodynamic aspects.

2.2. Second Law Analyses

Next to the annual amount of produced electricity, the second law efficiency η_{II} is calculated. In case of single power generation or consideration of the ORC-unit in a CHP-system, the net power output P_{Net} is divided by the exergy flow rate of the geothermal water $\dot{E}_{GW}$:

$$\eta_{II} = \frac{P_{Net}}{\dot{E}_{GW}} \tag{1}$$

The exergy flow rate of the geothermal source is obtained by multiplying the specific exergy e with the mass flow rate of geothermal water $\dot{m}_{GW}$. For the analysis, the specific exergy e is based on:

$$e = h - h_0 - T_0(s - s_0) \tag{2}$$

$$\dot{E}_{GW} = \dot{m}_{GW}\, e \tag{3}$$

The state variables T_0, p_0 and s_0 are related to ambient conditions. In case of CHP the numerator of Equation (1) is extended by the exergy flow rate of the heating network $\dot{E}_{HN}$ and in case of a hybrid power plant the exergy flow rate of the biogas $\dot{E}_{BG}$ has to be considered in the denominator according to Equation (4):

$$\eta_{II} = \frac{P_{Net} + \dot{E}_{HN}}{\dot{E}_{GW} + \dot{E}_{BG}} \tag{4}$$

To calculate the exergy flow rate of the biogas $\dot{E}_{BG}$, the molar exergy of the biogas $E_{m,BG}$ is defined as:

$$E_{m,BG} = \sum_{i=1}^{N} \tilde{x}_i E_{m,i} + R_m T_0 \sum_{i=1}^{N} \tilde{x}_i \ln(\tilde{x}_i) \tag{5}$$

Here R_m is the universal molar gas constant, $\tilde{x}_i$ describes the molar fraction for each component and $E_{m,i}$ is the molar exergy of each component according to Baehr and Kabelac [15]. A gas mixture of 65% methane and 35% carbon dioxide is assumed. In the following, second law efficiency for a certain power plant concept is calculated by evaluating each load step and finally rating according to the annual contribution.

2.3. Economic Analyses

For a comprehensive analysis of different plant concepts or potential ORC working fluids an additional economic evaluation is of steadily growing importance [7,16–18]. In this study cumulated cashflow and internal rate of return are calculated as economic parameters. According to Equation (6) the cashflow Cf for a period is calculated by the difference between revenues R and total costs C. Therefore Cf describes the inflow of available funds within a certain time period t:

$$Cf_t = R_t + T_t \tag{6}$$

Equation (7) shows the cumulated cashflow Cf_{cum} at a certain point in time T, which is obtained by summarizing Cf of previous time periods:

$$Cf_{cum} = \sum_{t=0}^{T} Cf_t \tag{7}$$

In addition, the internal rate of return IRR is calculated for the considered power plant concepts. This parameter is the interest rate r, at which the net value of the investment is equal zero:

$$0 = -C_0 + \sum_{t=0}^{T} (R_t - C_t) \cdot (1+r)^{-t} \tag{8}$$

For the economic evaluation of the power plant concepts the specific costs listed in Table 3 are estimated. Drilling costs of 18 million € and insurance of 2 million € are assumed [19]. Costs for operation and maintenance, including personnel costs, are set to 4% of the total investment costs for a separate use of geothermal heat source and biogas engine [7]. In case of a hybrid concept, this value is reduced to 2% due to the cost savings in personnel and administrative costs. The lifetime of the power plant is 30 years and the interest rate is 6.5% [20]. The credit period is 12 years and the rate of borrowed capital is 80%. For the biogas cogeneration, maize silage (30 €/t) is assumed as energy source [21]. The length of the heating network is 8 km. The heating price is 0.05 €/kWh [22]. German feed-in tariffs for geothermal power generation (0.25 €/kWh) and biomass power generation (0.11 €/kWh) are considered [23]. Furthermore, an electricity price of 0.12 €/kWh for auxiliary power requirements, like working fluid pump, downhole pump, condensation system or table-top coolers for the engine coolant in the summer period, is assumed [22]. The annual price increase for electricity and heat supply as well as the considered inflation rate is 2%.

Table 3. Specific costs for power plant units and components.

Parameter	Unit	
ORC power plant [2]	€/kW$_{el}$	3500
Table-top cooler [24]	€/kW$_{th}$	14.8
Heating network [25]	€/km	500,000
Peak load boiler [26]	€/kW$_{th}$	200
Biogas engine [27]	€/kW$_{el}$	225
Heat exchanger hybrid power plant [26]	€/m^2	125

3. Results and Discussion

In the standard case, a mass flow rate of 100 kg/s and a temperature of 120 °C are assumed for geothermal fluid. This corresponds to the characteristic conditions of the Southern German Molasse Basin near Munich. R245fa is chosen as ORC working fluid. In the thermodynamic results, thermal and electric power of the units are present depending on different load steps. In addition, the annual amount of generated electricity and the second law efficiency is shown. The economic results compare the cashflow and *IRR* for the considered power plant concepts. Finally, the economic effects of varying selected boundary conditions are discussed.

3.1. Thermodynamic Results

Regarding a geothermal CHP in parallel circuit, the heating network has to be fully supplied by the geothermal water. For a hybrid power plant the heat demand could partly be covered by the engine coolant. Furthermore, the power generation of the ORC-unit is more efficient due to the temperature increase of the geothermal water by coupling with the exhaust gases of the gas engine. In this context, for a geothermal CHP in parallel circuit, the electric power of the ORC-unit $P_{el,ORC}$ as well as the total thermal power of the heating network $P_{th,HN}$ depending on the assumed load steps are shown in Figure 5a. In addition, the part of thermal power supplied by geothermal water $\dot{Q}_{Geo}$ pointed out. For geothermal CHP the heat demand is supplied completely by the geothermal fluid. Therefore in Figure 5a the values for $P_{th,HN}$ and $\dot{Q}_{Geo}$ are equal. For higher load steps the thermal power of the heating network decreases and a higher amount of thermal energy is coupled to the ORC. As a result the power output of the ORC increases. In Figure 5b these parameters are shown for a hybrid power plant in parallel circuit, extended by electric power of the gas engine $P_{el,GE}$ and part of thermal power supplied by engine coolant $\dot{Q}_{EC}$. In case of the hybrid power plant, the biogas engine operates 8000 h/a with a maximum electrical power of 2717 kW. The electrical power of the ORC-unit increases for higher load steps which correspond to higher ambient temperatures and less heat demand. The engine coolant supplies the heating network partly for all load steps. Finally, for load steps 8 to 10, corresponding to 2952 h/a, the heating network is fully supplied by engine coolant. In this period, the geothermal water is not required for heat generation. Therefore, the complete geothermal mass flow rate can be coupled to the ORC-unit for power generation. In addition, in case of a hybrid power plant, geothermal water temperature is increased. As a result, higher process pressures of the ORC can be reached and the efficiency of the ORC-unit is about 3% higher. In this context, the ORC pressure at condensation and evaporation for the geothermal CHP and the hybrid power plant are listed in Table 4.

Figure 5. Electrical and thermal power of the power plant units (**a**) Geothermal CHP in parallel circuit; (**b**) Hybrid power plant in parallel circuit.

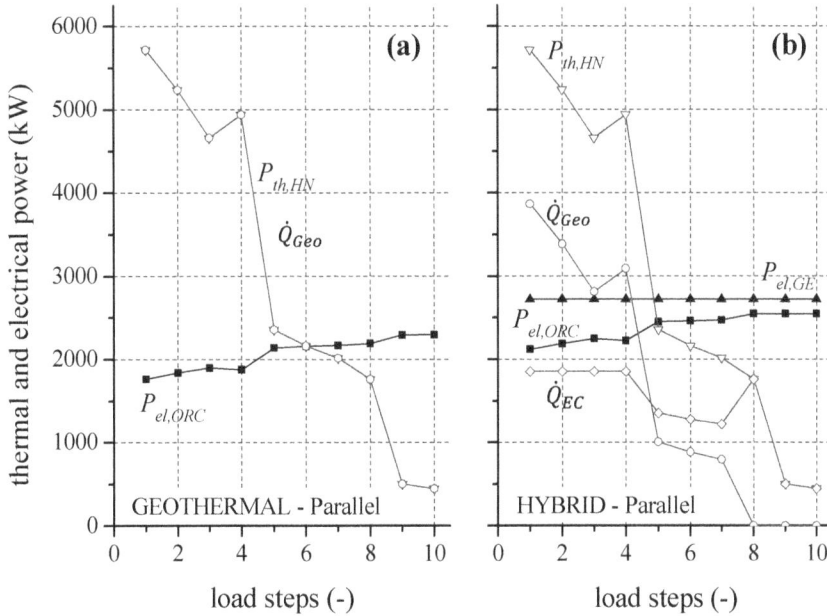

Table 4. Condensation and evaporation pressure.

Parameter	R245fa-GeoCHP	R245fa-Hybrid	Isopentane-GeoCHP	Isopentane-Hybrid
p_1 (bar)	1.47	1.47	0.90	0.90
p_2 (bar)	6.53	6.94	3.67	3.85

The annual amount of generated gross electricity for all considered power plant concepts is shown in Figure 6. In case of the hybrid power plant, a distinction is made between ORC-unit and gas engine. The generated electricity of the gas engine is equal for the hybrid concepts and separate use. In case of the hybrid power plant in parallel circuit, the highest amount of generated electricity per year is obtained. In comparison, a separate use of geothermal water and biogas engine leads to a 4.7% lower amount of generated electricity. This difference is due to the efficiency increase of the ORC-unit by increasing the geothermal water temperature within the hybrid concept. The hybrid power plant in serial circuit leads to an 11% lower amount of electricity compared to the parallel circuit. In case of the serial circuit, a higher geothermal mass flow is needed to obtain the required supply temperature and heat load. For the first load step, 39.6% of the geothermal water mass flow are required to supply the heating network, while in parallel circuit only 18.5% are sufficient. This difference occurs up to load step 7 and leads to a significantly lower electricity generation for the serial circuit. In case of geothermal CHP, the electricity generation is up to 23% lower compared to the hybrid power plant in parallel circuit. Due to the heat supply which has to be fully covered by the geothermal heat source, a considerable reduction occurs. CHP in parallel circuit is 11.3% more efficient compared to CHP in serial circuit.

Figure 6. Annual amount of generated electricity for the investigated power plant concepts.

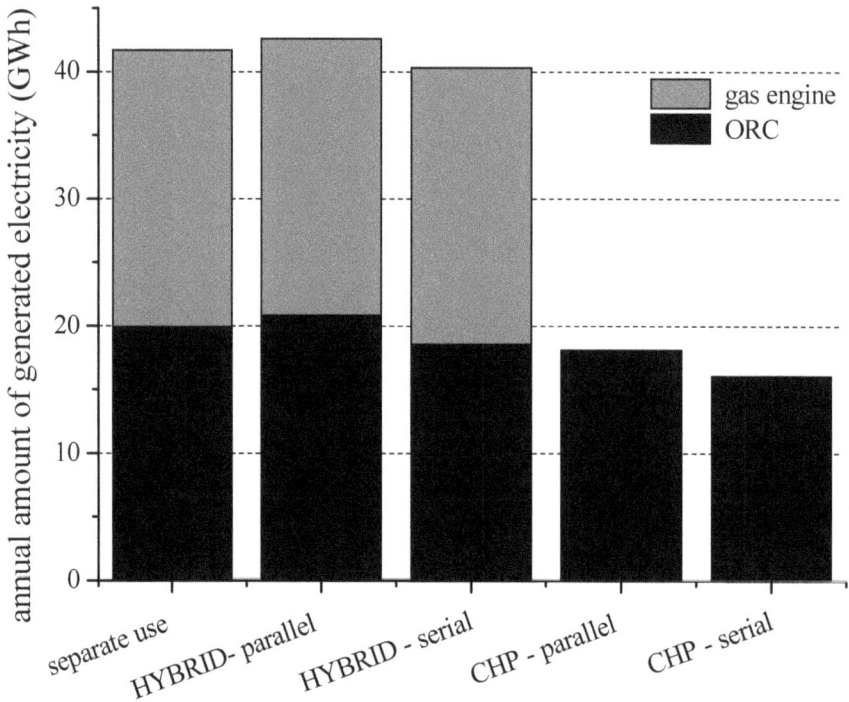

Figure 7 presents the second law efficiency for the analyzed concepts. In general, the results are consistent with the annual amount of generated electricity. The most efficient concept is the hybrid power plant in parallel circuit. A separate use of geothermal heat source and biogas cogeneration unit is 2.1% less efficient. In case of hybrid power plants as well as for geothermal CHP concepts, parallel circuit is more efficient compared to serial circuit. The efficiency increase is between 5.2% and 10.4%. A comparison under thermodynamic aspects based on second law efficiency seems to be more appropriate, since the additional use of biogas as energy resource is considered.

Figure 7. Second law efficiency for the considered power plant concepts.

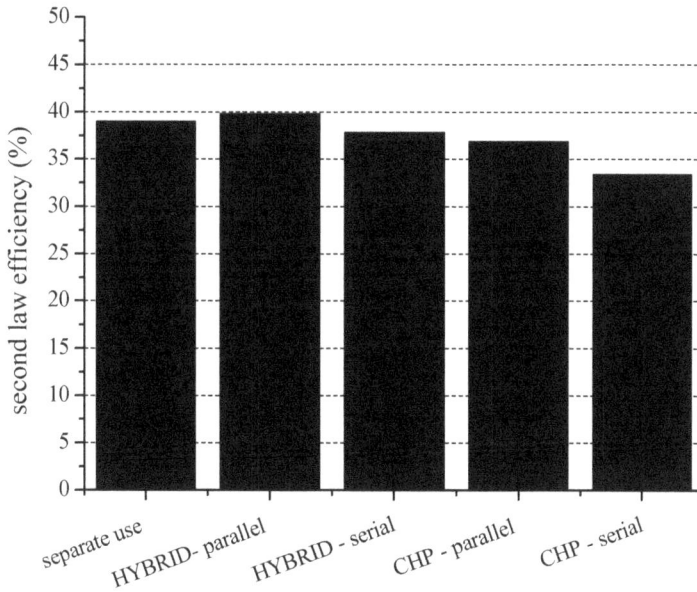

3.1.1. ORC Working Fluid

Regarding second law efficiency, the choice of working fluid has a minor role in these systems. Exemplarily in Figure 8 second law efficiency for R245fa, isobutane and isopentane are shown for separate use, hybrid power plant and geothermal CHP in parallel circuit.

Figure 8. Second law efficiency for different ORC working fluids and selected power plant concepts.

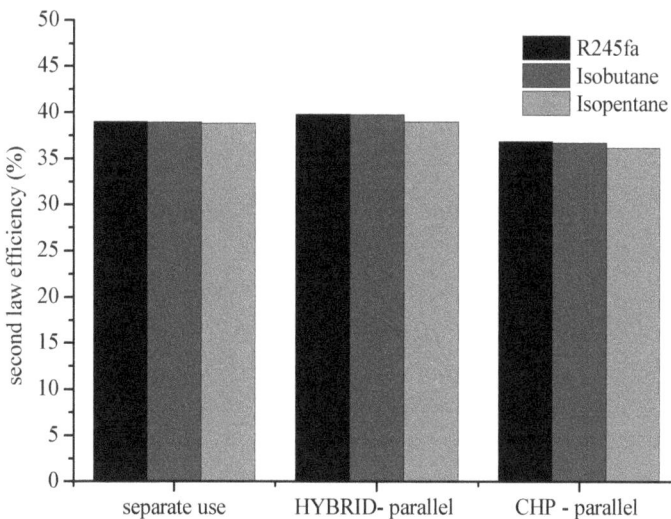

R245fa as ORC working fluid leads to the most efficient power plant concepts. The ORC-unit with isopentane is up to 1.5% less efficient. In case of isobutane the differences are below 0.5%. Therefore the choice of working fluid is more dependent on fluid properties, component design, Global Warming Potential and safety issues than system efficiency.

3.1.2. Geothermal Conditions

In respect to the geothermal resource, the mass flow rate and the temperature are the most important parameters. In case of typical geothermal conditions of the Upper Rhine Rift Valley with a geothermal water temperature of 160 °C and a mass flow rate of 65 kg/s, second law efficiency for the investigated power plant concept is shown in Figure 9. With increasing geothermal temperature, the second law efficiency of the ORC-unit is rising. In the context of a hybrid power plant in parallel circuit, this increase is 10.7% due to the raise of geothermal water temperature from 120 °C to 160 °C. For geothermal CHP the second law efficiency of the ORC-unit increases from 34.4% to 42.3%. A comparison between the different power plant concepts at higher geothermal water temperature shows qualitatively the same results. The hybrid power plant in parallel circuit is the most efficient concept and in general hybrid power plants are favorable compared to geothermal CHP. However, the differences in efficiency of parallel and serial circuit are less pronounced. Due to higher geothermal water temperature a lower partial flow rate is needed to obtain the required supply temperature for the heating network. Comparing again the first load step between the hybrid power plant and the geothermal CHP in parallel circuit, here 24.2% of the geothermal water mass flow are required to supply the heating network in serial circuit, while in parallel circuit 15.6% are sufficient. Therefore, compared to the serial circuit, the parallel configuration is only 2.1% more efficient in case of geothermal CHP and 2.6% for the hybrid concept. Compared to the low-temperature case the efficiency increase for a hybrid power plant in relation to separate use is similar with 2.9%. In both scenarios, an increase of geothermal water temperature due to the coupling with the exhaust gases of the gas engine affects the efficiency of the ORC-unit in a positive manner. In this context, the second law efficiency increases in the range of 2.6% and 3.1%.

3.1.3. Supply Temperature of the Heating Network

The supply temperature of the heating network plays an important role in the energy conversion system. Exemplarily, a raise of the maximum supply temperature to 130 °C (at ambient temperature below −14 °C) is examined. The minimum supply temperature for ambient temperatures higher than 16 °C is 80 °C. Again a linear function for supply temperature depending on ambient temperature is assumed. Figure 10 shows the electrical and thermal power for a hybrid power plant in parallel circuit. In comparison to a maximum supply temperature of 90 °C (see Figure 5b) for load steps 1 to 4, the engine coolant cannot be used for heat generation. In addition, a full supply of the heat demand by the engine coolant is only possible for 2064 h/a, in load steps 9 to 10, respectively. As a result, the amount of generated electricity is reduced by 5.3 MWh/a and the second law efficiency decreases by 1.5%.

Figure 9. Second law efficiency for the investigated power plant concepts considering geothermal mass flow rate of 65 kg/s and geothermal water temperature of 160 °C.

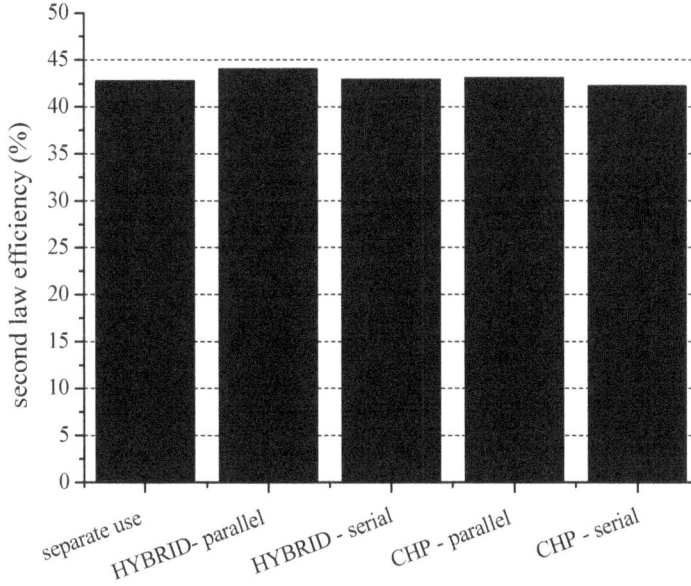

Figure 10. Electrical and thermal power of the power plant units for a maximum supply temperature of the heating network of 130 °C.

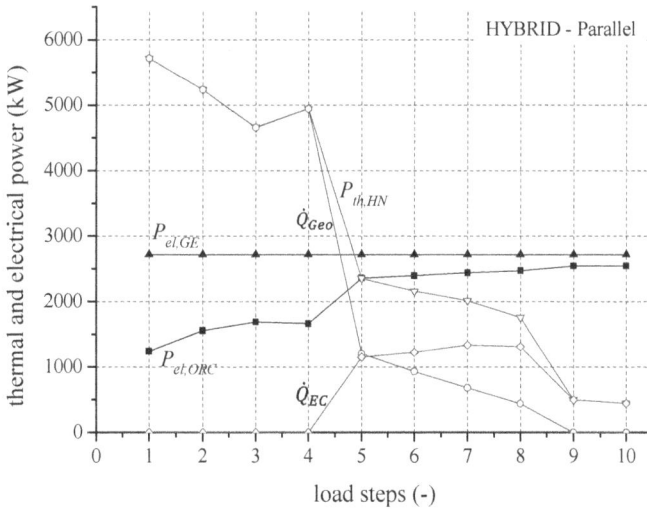

3.2. Economic Results

For an economic evaluation of the examined CHP concepts, investment, operation and maintenance as well as fuel costs have to be considered. On the other hand, the revenues from feeding electricity into the grid and heat sales have an effect on the energy cost balance and economic parameters like the cumulated cashflow and *IRR*. The cumulative cashflow for the selected power plant designs is shown in Figure 11. A construction time of 2 years is assumed, the related investment costs are evenly distributed. In general, during operation the curve of the cumulated cashflow shows unsteadiness. The first change occurs 10 years after initial operation of the power plant. This is due to the assumed payback period. A second one is observed for 20 years of operation and is related to the end of the guaranteed electricity feed-in tariffs. Regarding the investment cost, hybrid power plants are the most expensive concept, at 35.5 million €. A separate use leads to cost savings of 0.6 million·€ and a geothermal CHP to cost savings of 1.6 million €. The hybrid power plant in parallel circuit leads with 46.5 million € to the highest accumulated cashflow at the end of the complete lifetime. A significantly lower cumulated cashflow is obtained for separate use, mainly caused by the higher costs for operation and maintenance and lower efficiency. For the last 10 years the total cost balance even shows negative cashflows. At the end of the life time a cumulative cashflow of 12.5 million € is reached. Also in the economic analysis, a serial circuit for hybrid power plants as well as for geothermal CHP leads to lower results compared to parallel circuit. In case of the hybrid circuit a 28.3% lower cashflow is observed and for geothermal CHP the reduction is 33.5%. Geothermal CHP in parallel circuit is almost competitive compared to a hybrid power plant in serial circuit. The accumulated cashflow after 30 years of operation is only 7.5% lower. The described economic relationships are confirmed by the *IRR*. The highest value with 6.3% is obtained for hybrid power plant in parallel circuit, followed by the serial concept with 4.7% and the geothermal CHP in parallel circuit. Lowest *IRR* are calculated for geothermal CHP in serial circuit (2.7%) and separate use (2.4%).

Figure 11. Cumulated cashflow for the considered power plant concepts.

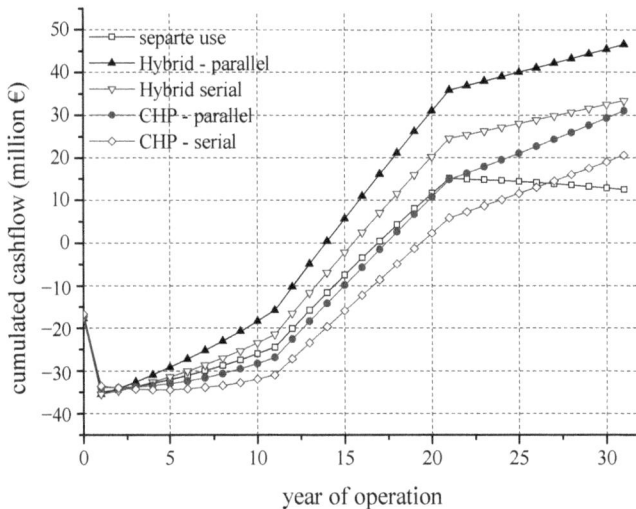

In Figure 12 the cumulated cashflow for an alternative working fluid (isopentane instead of R245fa), a higher maximum supply temperature (130 °C instead of 90 °C), higher operation and maintenance costs for the hybrid power plant (4% of the total investment costs instead of 2%) and different geothermal conditions (T_{GW} = 160 °C; $\dot{m}_{GW}$ = 65 kg/s instead of T_{GW} = 120 °C; $\dot{m}_{GW}$ = 100 kg/s) are shown in addition to the hybrid power plant in parallel circuit.

Figure 12. Cumulated cashflow for the variable boundary conditions.

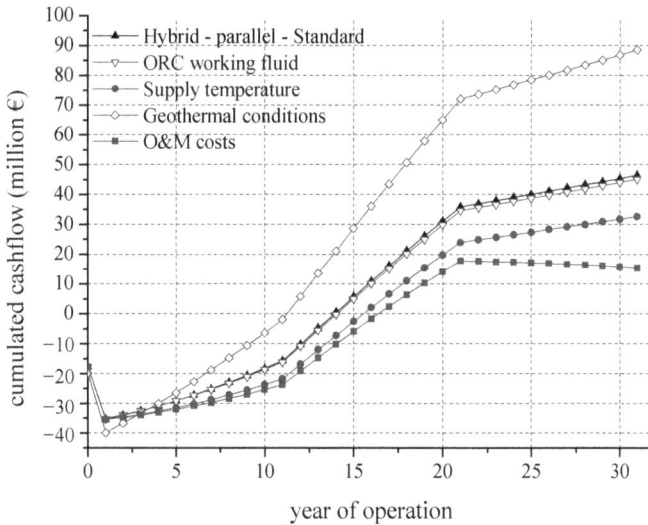

According to the second law efficiency economic parameters are not affected significantly by the choice of working fluid. Isopentane as ORC working fluid leads to a 3% lower accumulated cashflow at the end of the lifetime compared to the use of R245fa. The *IRR* is 6.2% instead of 6.3%. The supply temperature of the heating network has a more obvious effect on economics. Since a higher rate of heat demand has to be supplied by the geothermal heat source, the electricity generation is decreased by 2.4 GWh/a in case of an increase of the maximum supply temperature range from 90 °C to 130 °C. This leads to a reduction of the cumulated cashflow of 27.6% after 30 years of operation. The *IRR* is 4.6%. For a geothermal water temperature of 160 °C and a mass flow rate of 65 kg/s the cumulated cashflow is almost doubled at the end of the lifetime. Compared to the low-temperature case, a more efficient ORC-unit with higher capacity can be realized. In case of 120 °C and 100 kg/s an ORC-unit of 2.5 MW electrical power output results, while 3.8 MW are obtained for a heat source with 160 °C. Therefore, the reduction of geothermal water mass flow can be overcompensated by the increase in temperature. In case of the hybrid power plant in parallel circuit, 10 GWh/a more electricity are generated and the IRR is increased to 9.97%. An increase of operation and maintenance costs lead to a considerable reduction of the economic parameters for the hybrid power plant. However, for an equal cost rate of 4% regarding operation and maintenance the cumulated cashflow at the end of the lifetime is still 23% higher compared to a separate use of the geothermal resource and the biogas CHP-unit.

4. Conclusions

Hybrid power plants are promising concepts for geothermal CHP. Comparisons to the separate use prove the advantages of coupling a geothermal resource and biogas engine. A higher efficiency of the ORC-unit is obtained due to the increase of geothermal water temperature by the exhaust gases. A parallel circuit of power and heat generation is favourable. Compared to conventional geothermal CHP, the second law efficiency is increased by up to 8.0% and the accumulated cashflow at the end of the lifetime is 50% higher. In relation to separate use, the hybrid power plant is 2.1% more efficient and a higher amount of electricity by 943.3 MWh/a could be generated. In addition, advantages regarding costs for operation and maintenance lead to significant economic differences. The cumulative cashflow at the end of the lifetime is more than tripled. Sensitivity analyses show a small influence on efficiency and economic parameters for the choice of the ORC working fluid. In contrast, a higher supply temperature of the heating network leads to a reduced implementation of the biogas-cogeneration unit in the hybrid power plant and a 27.6% lower cumulated cashflow after 30 years of operation is observed. In case of an increase of the geothermal water temperature from 120 °C to 160 °C, second law efficiency is increased by 22.9% and cumulated cashflow is almost doubled. For further work, dynamic simulations are performed, under consideration of part load behavior of pump and turbine as well as variable pinch points in the heat exchanger.

Acknowledgments

This publication was funded by the German Research Foundation (DFG) and the University of Bayreuth in the funding programme Open Access Publishing.

Author Contributions

Florian Heberle is the principle investigator of this work. Final review was done by Dieter Brüggemann.

Conflicts of Interest

The authors declare no conflict of interest.

References

1. Tchanche, B.F.; Lambrinos, G.; Frangoudakis, A.; Papadakis, G. Low-grade heat conversion into power using Organic Rankine Cycles—A review of various applications. *Renew. Sustain. Energy Rev.* **2011**, *15*, 3963–3979.
2. Vélez, F.; Segovia, J.J.; Martín, M.C.; Antolín, G.; Chejne, F.; Quijano, A. A technical, economical and market review of Organic Rankine Cycles for the conversion of low-grade heat for power generation. *Renew. Sustain. Energy Rev.* **2012**, *16*, 4175–4189.
3. Heberle, F.; Brüggemann, D. Exergy based fluid selection for a geothermal Organic Rankine Cycle for combined heat and power generation. *Appl. Therm. Eng.* **2010**, *30*, 1326–1332.
4. Heberle, F.; Preißinger, M.; Brüggemann, D. Thermoeconomic evaluation of combined heat and power generation for geothermal applications. In Proceedings of the World Renewable Energy Congress, Linköping, Sweden, 8–13 May 2011; pp. 1305–1313.

5.   Heberle, F.; Brüggemann, D. Thermoeconomic comparison of designs for geothermal combined heat and power generation. In Proceedings of the European Geothermal Congress, Pisa, Italy, 3–7 June 2013.

6.   Tempesti, D.; Manfrida, G.; Fiaschi, D. Thermodynamic analysis of two micro CHP systems operating with geothermal and solar energy. *Appl. Energy* **2012**, *97*, 609–617.

7.   Astolfi, M.; Xodo, L.; Romano, M.C.; Macchi, E. Technical and economical analysis of a solar-geothermal hybrid plant based on an Organic Rankine Cycle. *Geothermics* **2011**, *40*, 58–68.

8.   Borsukiewicz-Gozdur, A. Dual-fluid-hybrid power plant co-powered by low-temperature geothermal water. *Geothermics* **2010**, *39*, 170–176.

9.   Astina, I.M.; Pastalozi, M.; Sato, H. An improved hybrid and cogeneration cycle for enhanced geothermal systems. In Proceedings of the World Geothermal Congress, Bali, Indonesia, 25–29 April 2010.

10. Kohl, T.; Speck, R. Electricity production by geothermal hybrid-plants in low-enthalpy areas. In Proceedings of the 29th Workshop on Geothermal Reservoir Engineering, Stanford, CA, USA, 26–28 January 2004.

11. Karellas, S.; Terzis, K.; Manolakos, D. Investigation of an autonomous hybrid solar thermal ORC-PV RO desalination system. The Chalki Island case. *Renew. Energy* **2011**, *36*, 583–590.

12. Woudstra, N.; van der Stelt, T.P. *Cycle-Tempo: A Program for the Thermodynamic Analysis and Optimization of Systems for the Production of Electricity, Heat and Refrigeration*; Energy Technology Section, Delft University of Technology: Delft, The Netherlands, 2002.

13. Lemmon, E.W.; Huber, M.L.; McLinden, M.O. *NIST Standard Reference Database 23*, Version 9.1; National Institute of Standards and Technology: Boulder, CO, USA, 2013.

14. Verein Deutscher Ingenieure e.V. *VDI Richtlinie 4655—Referenzlastprofile von Ein- und Mehrfamilienhäusern für den Einsatz von KWK-Anlagen*; Springer-Verlag: Düsseldorf/Beuth, Germany, 2008. (In German)

15. Baehr, H.D.; Kabelac, S. *Thermodynamik: Grundlagen und technische Anwendungen*, 15th ed.; Springer Vieweg Verlag: Auflage/Berlin, Germany, 2012.

16. Quoilin, S.; Declaye, S.; Tchanche, B.F.; Lemort, V. Thermo-economic optimization of waste heat recovery Organic Rankine Cycles. *Appl. Therm. Eng.* **2011**, *31*, 2885–2893.

17. Preißinger, M.; Heberle, F.; Brüggemann, D. Advanced Organic Rankine Cycle for geothermal application. *Int. J. Low-Carbon Technol.* **2012**, *8*, 62–68.

18. Heberle, F.; Bassermann, P.; Preissinger, M.; Brüggemann, D. Exergoeconomic optimization of an Organic Rankine Cycle for low-temperature geothermal heat sources. *Int. J. Thermodyn.* **2012**, *15*, 119–126.

19. Görke, B.; Sievers, A. Gewinnbetrachtung von strom- und wärmegeführten Geothermie-Projekten unter Berücksichtigung der aktuellen EEG Novelle. In Kongressband Geothermiekongress, Karlsruhe, Germany, 11–13 November 2008, Geothermische Vereinigung—Bundesverband Geothermie e.V.: Geeste, Germany, 2008; pp. 147–156.

20. Janczik, S.; Kaltschmitt, M. Kombinierte Nutzung von Geothermie und Klärschlamm. *VGB PowerTech* **2010**, *7*, 84–91.

21. Fachagentur Nachwachsende Rohstoffe e.V. (FNR). Biogas—Pflanzen, Rohstoffe, Produkte. Rostock, 2011. Available online: http://www.fnr-server.de/ftp/pdf/literatur/pdf_175-biogas_broschuere_dina5_nr_175.pdf (accessed on 10 February 2014).

22. Statistisches Bundesamt. Preise—Daten Zur Energiepreisentwicklung, 2014. Available online: https://www.destatis.de/DE/Publikationen/Thematisch/Preise/Energiepreise/Energiepreisentwicklung.html (accessed on 10 February 2014).

23. Bundesregierung. Gesetz für den Vorrang Erneuerbarer Energien (Erneuerbare-Energien-Gesetz—EEG). 22 Dezember 2011. Available online: https://www.clearingstelle-eeg.de/eeg2012 (accessed on 12 February 2014).

24. *EDR Aspen Exchanger Design & Rating*, version 7.3; Aspen Technology, Inc.: Burlington, MA, USA, 2011.

25. Ehrig, R.; Kristöfel, C.; Pointner, C. Operating Figures and Investment Costs for District Heating Systems, 2011. Available online: http://www.afo.eu.com/default.asp?SivuID=28291 (accessed on 10 January 2014).

26. Turton, R.; Bailie, R.C.; Whiting, W.B. *Analysis, Synthesis and Design of Chemical Processes*, 2nd ed.; Prentice Hall: Old Tappan, NJ, USA, 2003.

27. Arbeitsgemeinschaft für sparsamen und umweltfreundlichen Energieverbrauch e.V. (ASUE). BHKW Kenndaten 2011—Module, Anbieter, Kosten. Frankfurt am Main; 2011. Available online: http://asue.de/themen/blockheizkraftwerke/broschueren/bhkw_kenndaten_2011.html (accessed on 4 February 2014).

Thermal-Economic Modularization of Small, Organic Rankine Cycle Power Plants for Mid-Enthalpy Geothermal Fields

Yodha Y. Nusiaputra, Hans-Joachim Wiemer and Dietmar Kuhn

Abstract: The costs of the surface infrastructure in mid-enthalpy geothermal power systems, especially in remote areas, could be reduced by using small, modular Organic Rankine Cycle (ORC) power plants. Thermal-economic criteria have been devised to standardize ORC plant dimensions for such applications. We designed a modular ORC to utilize various wellhead temperatures (120–170 °C), mass flow rates and ambient temperatures (−10–40 °C). A control strategy was developed using steady-state optimization, in order to maximize net power production at off-design conditions. Optimum component sizes were determined using specific investment cost (SIC) minimization and mean cashflow (MCF) maximization for three different climate scenarios. Minimizing SIC did not yield significant benefits, but MCF proved to be a much better optimization function.

Reprinted from *Energies*. Cite as: Nusiaputra, Y.Y.; Wiemer, H.-J.; Kuhn, D. Thermal-Economic Modularization of Small, Organic Rankine Cycle Power Plants for Mid-Enthalpy Geothermal Fields. *Energies* **2014**, *7*, 4221-4240.

1. Introduction

Rural areas worldwide, particularly in developing countries, often lie outside the reach of grid power supplies. In these regions, electricity tends to be supplied via diesel engines which require expensive fuel and are sources of atmospheric pollution. Some rural areas have mid-enthalpy geothermal resources under various geological conditions, whether these are shallow/deep, magmatic/amagmatic or identified/hidden. These kinds of resources comprise 70% of the world's total geothermal resources that are suitable for electricity generation [1]. A geothermally driven, decentralized power plant may, therefore, offer a viable and ecologically sound option for producing electricity in suitable rural and remote regions, such as the Chena Hot Springs in Alaska [2]. Nonetheless, certain requirements must be met. The plant must be capable of meeting small, modulating electricity loads with continuous annual growth and as such it has to be flexible in terms of incremental capacity expansion and frequency control, and have a short construction period to advance energy production, and cash flow starts [3]. It should be able to function efficiently at various different resource and ambient temperatures, and adapt to wellhead temperature changes during power production.

The subject presented in this paper is the development of a modular standardized power plant. Modularity and standardization are expected to lead to cost savings, due to reductions in plant engineering, assembly and installation time and maintenance. These are also expected to improve quality and reliability of the cycles. For example, Volkswagen has managed to save $1.7 billion annually through effective product architecture and component commonality [4].

In this study, the subcritical Organic Rankine Cycle (ORC) system is used as a technology to convert mid-enthalpy geothermal energy into electricity. It is a well-proven technology that has been in commercial use since the beginning of the 1980s [5]. Cycle simplicity and component availability are the main advantages with this technology, particularly in remote area applications. Supercritical ORCs were developed recently in order to achieve higher cycle efficiencies; however, they are not yet sufficiently reliable for widespread use in remote areas. The theoretical advantages of mixture working fluid ORCs have been demonstrated. Nonetheless, pure working fluid ORC power plants remain the most economical and proven technology [5], though they still offer potential for technical improvement.

An example of potential improvements could include advances in component technology, such as turbines with variable nozzle-vanes [6], speed pumps and fans; these would allow the cycle to adapt to a wide range of operating conditions. A control strategy to operate a geothermal ORC system at various wellhead and ambient temperatures has been proposed in [7]. However, the size of the ORC components was not optimized regarding the operation in a wide range of operating conditions. Another study dealt with power plant sizing, taking into consideration wellhead temperature decline during operation, but the control was not optimized [8]. The system is thereby a supercritical ORC with variable speed pump, constant turbine-nozzle and constant fan-speed. The authors of [5] concluded that plant design should be based on the lowest temperature of the geothermal wellhead.

In this study, we propose a thermal-economic modularization technique for a subcritical geothermal ORC, which operate under variable wellhead and ambient temperatures, considering both sizing and control aspects. Off-design steady-state optimization was developed using Covariance Matrix Algorithm-Evolutionary Strategy (CMA-ES) to achieve the maximum net power output. Modularization was tested in three different climate types temperate, tropical, and dry, using two main functions: specific investment-cost (SIC) minimization and mean cash-flow (MCF) maximization.

2. System Description and Methodology

Figure 1 shows the layout of the power system under investigation. The aim of this paper is to propose a methodology for sizing a standardized, modular geothermal ORC power-plant. Consequently, Figure 1 does not describe the system in detail, but rather offers a generic layout. The system consists of six main components, namely evaporator, recuperator, condenser, fan, pump, and turbine. The recuperator helps maintain a high injection temperature; it increases thermal efficiency, and reduces the thermal condenser load. The heat exchangers are represented by a counter-flow shell/tube configuration, with working fluid flowing in the shell of the evaporators and in the tube of the air-cooled condensers. The pump is centrifugal with a variable speed drive. The turbine is equipped with nozzle-vanes, which are also controlled with an electric drive.

Isobutane was used as a working fluid in the system. Working fluid selection is an essential and initial step of the ORC design process, but it is not the main concern of this work. Isobutane was chosen because it has the highest energetic efficiency in medium well-head temperature range [9], low global warming potential, low ozone depleting potential, and good market availability.

Figure 1. Diagram of a recuperative small, modular geothermal Organic Rankine Cycle (ORC) with adaptive control.

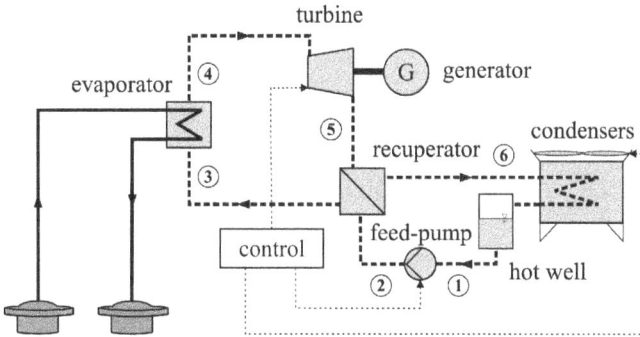

The ORC system considered in this paper is subcritical vapor-cycle, in which heat from a geothermal geofluid is used to heat and evaporate Isobutane. The working fluid vapor then drives the turbine for power generation, and then gets condensed in the air-cooled condenser. The liquid Isobutane is collected in a hot-well and then pumped back to the evaporator to repeat the cycle.

The modular power-plant is designed to work at geothermal wellhead temperatures of 120–170 °C, which is considered a suitable temperature range for Isobutane and also represents typical mid-enthalpy geothermal fields. Currently, more than 150 geothermal binary units with an average capacity between 1 MW and 3 MW are installed world-wide [10]. The design capacity of the modular plant is defined as 1000 kWe. The thermodynamic cycles of the system are shown in Figure 2.

In order to maximize the amount of energy recovered from the geothermal heat and simultaneously consider the installation cost, component size must be optimized. Operation parameters of the ORC should consider the daily and annual course of ambient temperatures, and should be regularly adjusted in the event of changes in wellhead temperature and geofluid flow rate.

Figure 2. (**a**) *P-H* diagram and (**b**) *T-S* diagram of the system for isobutane.

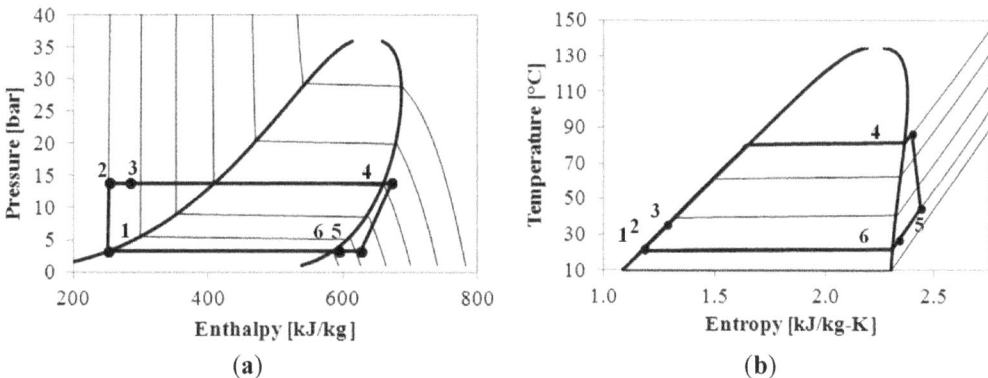

Consequently, a good design will include the following steps:

1. Thermodynamic optimization for a given design-point: normal (design) wellhead and ambient temperature. The components will then be sized using optimum thermodynamic parameters.
2. Mapping the power plant net-power at operation points from the design conditions. This results from an optimal control strategy that maximizes the net power output.
3. Simulation of annual electricity production. Performance is then evaluated using constant exergy input for each off-design condition. The variation of the ambient temperature was examined for three different climate types.
4. Steps 1–3 are repeated for each design-point, and finally the optimal design-point is selected using thermo-economic criteria. Cost correlations of each component are implemented to evaluate the component sizes.

Each step is described in Section 4. The component modeling, which is the basis for all subsequent evaluation steps, will be addressed in the following section.

3. Component Modeling

3.1. Heat Exchangers

The models were implemented in Matlab (The MathWorks, Natick, MA, USA) and the fluid properties computed using Refprop 9.0 (NIST, Gaithersburg, MD, USA). The heat exchangers models were used in two modes: sizing and simulation. These are represented as counter-flow heat exchanger, as shown in Figure 3. In order to consider the property variations of the working fluid and the secondary fluid (geofluid), the entire length of the heat exchangers was divided into three zones with variable lengths of each zone: liquid zone (Liq), two-phase zone (TP), and vapor zone (Vap), with respect to the working fluid.

The heat exchangers were modeled using two energy balance equations. One is the the geofluid heat flow rate. The example following is used for the overall evaporator:

$$\dot{m}_{wf} \cdot (h_4 - h_3) = \dot{m}_g \cdot c_{p,g} \cdot (T_{g,in} - T_{g,out}) \tag{1}$$

The other equation is the heat transfer equation, which uses the weighted temperature difference and an overall heat transfer coefficient. The example following is also shown for the evaporator:

$$\dot{m}_{wf} \cdot (h_4 - h_3) = U_{WTD} A_{tot} \cdot \Delta T_{WTD} \tag{2}$$

Figure 3. Three-zone heat exchanger model (evaporator).

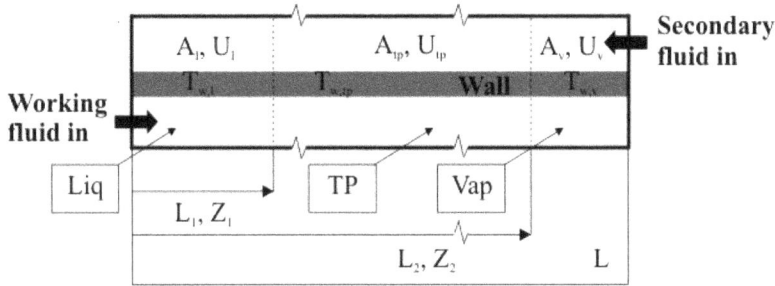

The weighted temperature difference, ΔT_{WTD} is calculated based on heat transfer coefficients and areas of each zone of the exchanger. It is represented by the following equation:

$$\Delta T_{WTD} = \frac{\dot{Q}}{U_1 A_1 + U_{tp} A_{tp} + U_v A_v} \tag{3}$$

Being the partial heat transfer coefficient, for example at the liquid zone, give by:

$$U_1 = \left[\left(\frac{1}{\alpha_{i,1}} + R_{fouling} \right) \frac{AR \cdot D_o}{D_i} + \frac{\ln\left(D_o/D_i\right)}{2\pi \cdot L \cdot k_{wall}} + \frac{1}{\alpha_{o,1}} \right]^{-1} \tag{4}$$

where AR is the area ratio of outer to inner heat transfer area, which is unity for shell/tube heat exchangers. $R_{fouling}$ is the thermal resistance associated with fouling in the heat exchanger tubes ($R_{fouling} = 1.3 \times 10^{-4}$ m$^2 \cdot$K$\cdot$W^{-1}, experiment data for geothermal brine [11]). For the evaporator, the heat transfer area dedicated to liquid zone, A_1 is computed as the similar equation form is applied for the two-phase and vapor zone. The inner tube was assumed to be a standard stainless-steel with the geometry described in Table 1. Simplified layouts of the heat exchangers are illustrated in Figure 4.

$$A_1 = \dot{m}_{wf} \frac{h_{1,sat} - h_3}{U_1 \dfrac{\left(\left(T_{g,1} - T_{wf,sat}\right) - \left(T_{g,out} - T_3\right)\right)}{\ln\left(T_{g,1} - T_{wf,sat}\right)/\left(T_{g,out} - T_3\right)}} = A_{tot} - \left(A_{tp} + A_v\right) \tag{5}$$

Table 1. Geometrical dimensions of heat exchangers.

Component	Type	D_0 [mm]	t [mm]	P_T [mm]	N_{tube}	N_{pass}	L [m]	Width [m]
Evaporator	shell/tube	15.875	1.651	20.64	variable	1	variable	-
Recuperator	shell/tube	31.75	2.11	39.69	variable	1	variable	-
Condenser (1 cell)	fin/tube	25.4	3.3	63.5	192	3	9.14	3.05

Figure 4. (**a**) Layout of shell/tube exchanger (evaporator and recuperator); (**b**) Layout of fin/tube exchanger cell (air-cooled condenser).

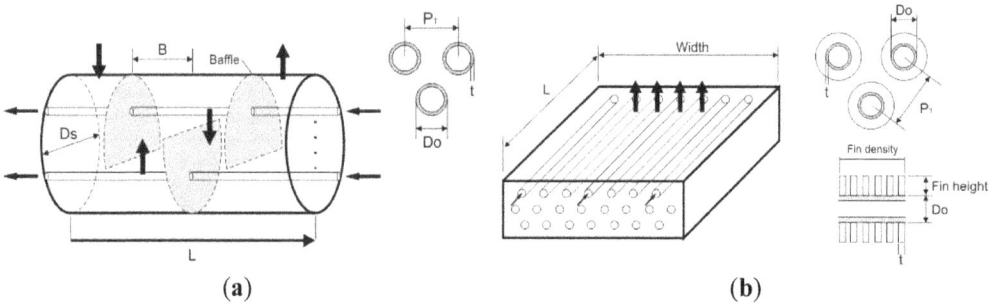

(**a**)   (**b**)

The geometrical dimensions are listed in Table 1. The cooling-system consisted of parallel air-cooled condenser cells that were modeled as a three-zone fin/tube heat exchanger. The fin density was assumed with 393 fins per meter, the fin height was 15.9 mm, and each cell contained 3 induced-draft fans. The dimensions of the heat exchanger model are summarized in Table 1. The variables for the condenser were the cell numbers and fan capacity.

3.1.1. Evaporator and Recuperator Heat Transfer Coefficients and Pressure Drops

Forced convection heat transfer coefficients for single-phase fluid (liquid/vapor) are evaluated by means of the generic correlation:

$$\alpha_{l/v} = C \cdot \mathrm{Re}^i \cdot \mathrm{Pr}^j \times \left(k_1/D_i\right) \tag{6}$$

where the influence of temperature-dependent viscosity-effects was neglected. The constant, C, and exponents i and j were identified according to the Sieder-Tate correlation [12].

The overall boiling heat transfer coefficient was estimated by the Mostinski, and Palen correlations for enhanced heat transfer, due to convection around the bundles, and established for boiling in horizontal tubes without dependency on surface roughness. This heat exchange coefficient is considered to be constant during the whole evaporation process and is calculated by:

$$\alpha_{ev} = 1.167 \times 10^{-8} \cdot p_{critical}^{2.3} \Delta T_{sat}^{2.333} F_p^{3.333} \times F_{bundle} \times \left(k_1/D_{eff}\right) + 250 \left[W/m^2 K\right] \tag{7}$$

Parameters F_p and F_{bundle} were calculated using equations found in the literature [12]. The pressure drops are calculated using the Prandtl-Karman equation as follows:

$$\Delta p = f \frac{G^2 \cdot L}{2 \cdot \rho \cdot D} \times \Phi^2 \tag{8}$$

where f is dependent on flow velocity and tube/shell roughness. For flow inside the tube, f is calculated using the explicit Swamee-Jain correlation [12]. The two-phase multiplier Φ^2 is approximated with the Grant correlation, for two-phase flow crossing tube-bundles [13].

3.1.2. Air Condenser

The single phase working-fluid heat transfer coefficient was calculated in the same manner as in Equation (5). The condensation heat transfer coefficient is estimated using the Dobson-Chato correlation [14], developed for the case of smooth of horizontal tubes:

$$\alpha_{cd} = 0.023 \cdot Re_l^{0.8} \, Pr_l^{0.4} \left[1 + \frac{2.22}{X_{tt}^{0.89}} \right] \times (k_l / D_i) \tag{9}$$

The partial heat transfer coefficient was computed using Equation (4) with $AR = 21.4$, the Gas Processors and Suppliers Association (GPSA) standard. The fouling thermal resistance was assumed with $R_{fouling} = 1.7 \times 10^{-4}$ (GPSA assumption). Heat transfer and pressure drop on the air-side are also approximated based on a GPSA correlation [15]:

$$\alpha_a = 0.019 \cdot G_a^{0.54} \left[W / m^2 K \right] \tag{10}$$

$$\Delta p_a = \frac{1.175 \cdot 10^{-10} \cdot G_a^{1.8} \cdot N_{row}}{\left(\rho_{a,av} / \rho_{21°C} \right)} + \left(\frac{\dot{V}_{a,out}}{10 \cdot D_F^2} \right)^2 \left(\rho_{a,out} / \rho_{21°C} \right) \tag{11}$$

For the calculation of the consumed fan power, a fan efficiency of 70%, and an electrical motor efficiency of 92% were assumed.

3.2. Feed-Pump

The feed-pump and its characteristics are approximated by using the affinity law and the second-order pump characteristics, which can be expressed in the following equations:

$$\Delta p = \Delta p_{V=0} \left[\left(\frac{n}{n_0} \right)^2 - \left(\frac{\dot{V}}{\dot{V}_{\Delta p=0}} \right)^2 \right] \tag{12}$$

The efficiency is calculated from the volumetric flow and rotational speed at the design-point and operating-point, assuming $\eta_0 = 0.8$:

$$\eta_P = \eta_{P0} \left\{ 1 - \exp \left[- \frac{1 - \left(\left| \frac{\dot{V} \cdot n_0}{\dot{V}_{P0} \cdot n} - 1 \right| \right)^a}{c \left(\left| \frac{\dot{V} \cdot n_0}{\dot{V}_{P0} \cdot n} - 1 \right| \right)^b} \right] \right\} \tag{13}$$

where constants a, b, and c are defined as 1.8, 0.58, and 0.68, respectively [16].

3.3. Turbine

A radial turbine was used because it has a better efficiency for small ORC applications compared to axial turbines due to the smaller tip-clearance [17]. In order to calculate the mass flow rate in the ORC during operation, the empirical Stodola steam cone rule is applied in the form of:

$$\dot{m} = \mu_T \cdot C_T \sqrt{p_{in} \cdot p_{in}} \sqrt{1 - \frac{1}{\pi}} \qquad (14)$$

where $\pi = p_{in}/p_{out}$ is the pressure ratio and μ_T is the turbine nozzle position. The turbine constant C_T can be thought of as an equivalent area and has the unit square meters. In off-design operation, the equivalent area was adapted by varying μ_T using variable inlet nozzle guide-vane. The guide vanes are moved in such a way that the flow area between the vanes changes. Thus, the inlet flow area is changed.

In high-pressure ratio operation, where the turbine is choking, the pressure ratio factor $\sqrt{1 - 1/\pi}$ is near unity and, therefore, can be neglected. These equations have been widely used to describe the relation between flow and pressure. Efficiency of the turbine under off-design condition is calculated as:

$$\eta_T = \eta_{T0} \cdot F_{u/c_o} \cdot F_{V_s} \qquad (15)$$

The designed turbine efficiency was 0.75. The first correction factor was related to the variation of u/c_o, ratio of radial velocity to spouting velocity. Spouting velocity, $c_o = \sqrt{2 \cdot \Delta h_{is}}$, is defined as that velocity has an associated kinetic energy equal to isentropic enthalpy drop. At the best efficiency point the value of u/c_o is found at 0.7 [17]. The second correction factor was associated with the variation of the volumetric flow rate from the design value. The two correction factors were then observed in Figure 5b, which is typical for radial turbine characteristics. The design point was pointed at a velocity ratio of 0.7 and volumetric flow rate of 100%.

Figure 5. (a) Stodola's cone rule as a function of nozzle position; (b) Typical turbine efficiency characteristics [18].

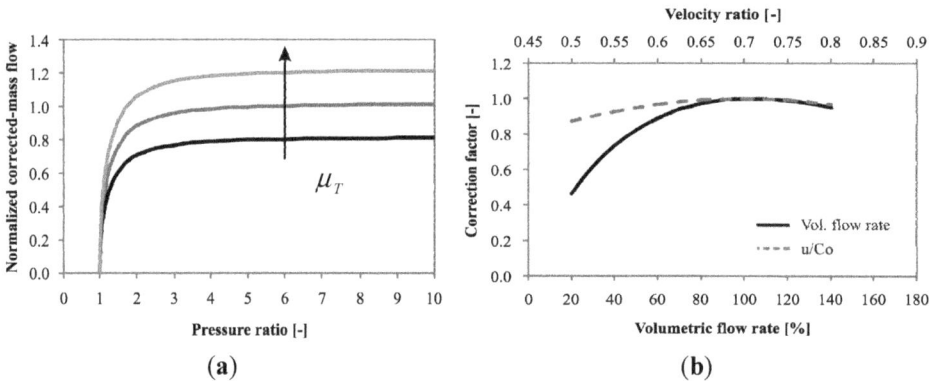

4. Results

Small, modular, subcritical ORCs should deliver good performance under a wide range of operating conditions. Consequently, the optimal component size of the plant needs to be determined. The optimum design-point was found using numerical simulation for operating conditions as follows:

$$120 \leq T_{g,in} \left({}^\circ C \right) \leq 170$$

$$-10 \leq T_{a,in} \left({}^\circ C \right) \leq 40 \tag{16}$$

$$70 \leq T_{g,out} \left({}^\circ C \right)$$

Considering the main component characteristics described in the previous section, there were 11 design variables to be optimized:

1. $D_{shell,ev}$, $D_{shell,re}$, L_{ev}, L_{re}: diameter and length of the shell-and-tube heat exchangers (*i.e.*, evaporator and recuperator).
2. N_{cell}, P_F: cell numbers and fan capacity for air-cooled condensers. These parameters are a function of the condenser load and the air-outlet temperatures.
3. C_T, Δh_{is0}, $\dot{V}_{5,design}$: inlet area constant, isentropic enthalpy drop, and outlet-volumetric flow rate at the design-point. The two latter parameters were used to define the pitch diameter.
4. $\Delta p_{\dot{V}=0}$, $\dot{V}_{P0}$: the shut-off pressure head when the flow is zero which is typically 1.25 times of the design-head, design volumetric flow rate.

The objective of the modularization was to find the optimal configuration of these design variables. Conventional power-plants, such as gas turbines and diesel engines are designed to deliver a specific power output at specific heat source and heat sink temperatures, such as flame and ambient temperatures. Inspired by this approach, two-dimensional optimization was introduced; these are normal (design) wellhead temperature (T_{g0}) and ambient temperature (T_{a0}). The 11 design variables then were a product of the sizing for the design-point (T_{g0}-T_{a0}).

4.1. Component Sizing for Normal-Design: Thermodynamic Optimization

In order to size the components, the thermodynamic cycle must be determined first. Thus, a thermodynamic optimization was carried out to maximize the net power output. The normal (design) condensation temperature is defined as:

$$T_{1,design} = T_{a0} + ITD \tag{17}$$

In low temperature power-plants, lowering condensation temperature benefits power output [19]. An initial temperature difference (ITD) of 14 K was selected as the lower bounding value for practical application [20].

Isobutane can be categorized as a dry fluid (*i.e.*, negative slope of saturated vapor line); hence, at design condition, saturated vapor is the best turbine inlet parameter [21]. The optimal evaporation temperature (OET) as normal (design) evaporation temperature is obtained by solving:

$$\frac{\partial}{\partial T_{4,design}} \left\{ \frac{T_{g0} - T_{4,design} - \Delta T_{pp,ev}}{T_{4,design}} \left(T_{4,design} - T_{1,design} \right) \left[1 + \frac{c_{p,wf} T_{4,design}}{2\gamma} \ln \left(\frac{T_{4,design}}{T_{1,design}} \right) \right] \right\} = 0 \tag{18}$$

The analytical OET results had an accuracy of 2.3%, compared to the numerical OET [21]. The design pinch-point was 5 K for both the evaporator and recuperator. The normal wellhead temperature varied from 120 °C to 170 °C, and the normal ambient temperature varied from

−10–40 °C, with a step of 10 °C, and 12 random points (6 × 6 grid + 12). The sizing results for each normal (design) wellhead temperature are listed in Table 2. The net efficiency is defined as ratio of net power (gross power deducted by feed-pump and fan power) to the heat input.

Table 2. Thermodynamic design of ORC cycles, showing range of optimal sizing results.

T_{g0}	120	130	140	150	160	170
Evaporation temperature (sat.) (°C)	80–87	85–93	91–101	99–111	111–122	115–121 *
Condensation temperature (°C)	4–54	4–54	4–54	4–54	4–54	4–54
Geofluid mass flow rate (kg·s⁻¹)	31.3–95.7	24.8–68.8	20.2–51.1	16.7–38.7	12.9–28.1	10.3–21.8
Isobutane mass flow rate (kg·s⁻¹)	15.6–50.2	14.8–43	13.9–37.4	13–32.8	12.5–27.1	12.8–27.1
Gross power (kW)	1000	1000	1000	1000	1000	1000
Net efficiency (%)	5.1–13.9	5.9–14.5	6.7–15.2	7.6–16	8.8–18.1	8.8–20.6

* The pinch-point was adjusted to set the geofluid after the evaporator to 70 °C.

After determining the optimum thermodynamic cycle conditions, the components were sized. The size of the rotating components (*i.e.*, feed-pump, turbine) was derived using the thermodynamic parameters. The heat exchangers were sized as follows.

1. Evaporator: Evaporation was realized using two parallel evaporators, with one shell/one tube pass configuration. During very low load (<50%) operation, one of the evaporators was fully closed. Both evaporators were sized by determining the shell diameter, and the number of tubes was calculated using "tube counts" based on standardized design parameters described in Table 1. The baffle-spacing was constrained below the shell diameter and maximum-spacing in order to avoid instability caused by vibration. After calculating overall heat transfer coefficients and the total heat transfer area, tube length was computed. By setting the allowable pressure drop on the shell side, the optimum design (or equivalently, shell diameter) with smallest area was selected. This design procedure was also applied to the recuperator.

2. Condenser: An important preliminary step in the condenser design process is outlet air temperature. This parameter has a major effect on exchanger economics [12]. Increasing the outlet air temperature reduces the amount of air required, which reduces the fan power and, therefore, operating cost. However, it also reduces the air-side heat-transfer coefficient and the mean temperature difference in the exchanger, which increases the size of the unit and, therefore, the capital cost. Consequently, optimization with respect to outlet air temperature (or equivalently, air flow rate) was considered an important aspect of air-cooled condenser design.

The optimum condenser air-outlet temperature (or equivalently, pinch-point) was calculated by minimizing the annual cost function. First derivative of this function with respect to air-outlet temperature determines the minimum annual cost. It can be written as follows:

$$\frac{\partial}{\partial T_{a0,out}}\left[\mathrm{CRF}\cdot\left(C_{cd}+C_{F}\right)+\left(0.01\cdot C_{cd}+0.03\cdot C_{F}\right)+CF\cdot P_{F}\cdot C_{el}\right]=0 \tag{19}$$

Air-cooled heat exchanger investment cost C_{cd} and fan investment cost C_F are described in Table 3. The annualization factor, CRF (Capital Recovery Factor) is defined as:

$$CRF = \frac{i(1+i)^{y}}{(1+i)^{y}-1} \tag{20}$$

Heat transfer coefficient and pressure drop were computed from the ratio of design mass flow rate to the reference, which was mass flow rate at air velocity of 3.5 m·s^{-1}, as recommended in the literature [12]. The maintenance cost was assumed to be 1% of the fin/tube heat exchangers cost and 3% of fan-motor cost [22]. CF (capacity Factor) of 0.7, y of 30 years, i of 12%, and electricity price C_{el} of 0.15 \$·kWh^{-1} were assumed.

Increasing the outlet air temperature increases heat transfer area required and conversely, reduces fan power consumption, as shown in Figure 6a. This trade-off resulted in an optimum annual cost of 130-20 (T_{g0}-T_{a0}) at air-outlet temperature of 30.4 °C, approximately 10 K above the inlet air temperature (Figure 6b). Once the optimum air-outlet temperature was established, the heat transfer area (or equivalently, number of cells) and fan capacity were determined.

Table 3. Component cost as function of size.

Component	Cost correlation	Reference
Evaporator	$13,668+658\cdot A^{0.85}$ (Carbon-shell/Stainless-tube)	[22]
Recuperator	$11,256+579\cdot A^{0.8}$ (Carbon-shell/Carbon-tube)	[22]
Air-cooled condensers	$5.6\cdot A$	[23]
Fans	$\left(1887.5+159.95\cdot D_{F}^{2}+3.53\cdot D_{F}+281.25\cdot P_{F}\right)\cdot N_{F}$	[23]
Feed-pump	$4900\cdot\left(P_{P}/30\right)^{0.7}$	[24]
Turbine + generator	$\left(91,200\cdot D_{pitch}^{2.1}+50,800\cdot D_{pitch}^{3}+62,700\cdot D_{pitch}^{2}\right)+680,900\cdot\left(P_{T}/10^{4}\right)^{0.7}$	[25]
Labor	$0.3\times$Total component cost	-

Figure 6. Size optimization based on annual cost of condensers at 130-20 (T_{g0}-T_{a0}) design-point.

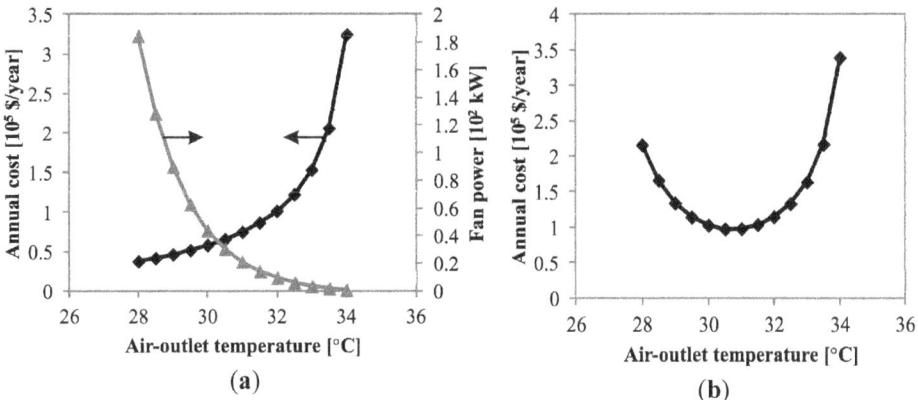

(a)

(b)

4.2. Off-Design Mapping

The off-design performance of the plant may be assessed using the Second Law of thermodynamics by comparing the actual net-power output to the maximum theoretical power that could be produced (energy) from the given geothermal fluid. This involves determining the energy-rate carried into the plant with the incoming geofluid [10]. In order to proportionally evaluate the off-design performance of each design-point, the geofluid mass flow rate at off-design conditions is computed using constant energy rate of 2000 kW at ISO standard ambient temperature of 15 °C:

$$\dot{m}_{g,in\text{-}off} = \frac{2000\,[\text{kW}]}{\left(h_{g,in} - h_{15°C}\right) - 288.15\,[\text{K}]\left(s_{g,in} - s_{15°C}\right)} \quad (\text{off-design}) \tag{21}$$

In order to obtain optimal operating point under off-design conditions, a three-variable control strategy was used. First, evaporation pressure was controlled by the turbine nozzle-opening μ_T. Second, superheating/turbine inlet temperature was controlled by pump-speed n_P (isobutane mass flow rate), and third, condensation temperature by the fan-speed n_F (air volumetric flow rate). Constant sub-cooling was imposed by making use of the static pressure head between the pump and the liquid hot-well (Figure 1). Using this control strategy for a modular ORC system, the net power output was maximized while keeping the injection temperature above scaling temperature to avoid scaling, which is described as:

$$\max_{Tg,out \geq 70°C} \left(\dot{W}_T - \dot{W}_P - \dot{W}_F\right) \tag{22}$$

Scaling temperature is a site-specific problem. It depends on the chemical composition of the geothermal fluid most commonly silica and calcite, and temperature and pressure of the fluid. If the injection temperature of the geofluid falls below this temperature, there is the risk that scales might form in the heat exchanger or the piping system. A minimum bound of 70 °C was selected for this study, based on several works for mid-enthalpy geothermal resources [2,3,26].

The off-design simulation procedure was realized using a set of three heat balance equations, which were solved by using the Trust-Dogleg Region solver. The heat balance equations are:

$$f_1 = \dot{Q}_{re} - \dot{Q}_{re,new} \text{ (function of } \dot{m}_{wf}, p_2, h_2, T_3, T_5, p_5)$$
$$f_2 = \dot{Q}_{ev} - \dot{Q}_{ev,new} \text{ (function of } \dot{m}_{g,in}, T_{g,in}, \dot{m}_{wf}, T_3, p_3, h_4) \tag{23}$$
$$f_3 = \dot{Q}_{cd} - \dot{Q}_{cd,new} \text{ (function of } n_F, T_{a,in}, \dot{m}_{wf}, T_1, T_6, p_6)$$

Where f_1 was determined using the three-zone recuperator model, f_2 the evaporator model, and f_3 the condenser model. Pressure drop in the evaporator was minimized to maintain evaporation temperature drop below 5 K. The equations were solved for given operation parameters to simulate the power-cycle. In order to find the optimum operation parameters for each operating condition, CMA-ES was implemented [27].

After sizing the components for a design-point, the control variables turbine nozzle, pump and fan rotational speed are optimized to achieve maximum net power output during off-design operating conditions (Figure 7a). The system is assumed to be steady-state for the cycle simulation. The net power output of the plant at 36 off-design wellhead and ambient temperatures (Figure 7c) was evaluated. Gridfit algorithm [28] was then used to interpolate the profiles to produce a 2-D net power output surface contour, as shown in Figure 8.

Figure 7. (**a**) Off-design optimization procedure for a design-point (T_{g0}-T_{a0}) and operating condition ($\dot{m}_g$, $T_{g,in}$, $T_{a,in}$); (**b**) Design-point grid; and (**c**) Off-design grid.

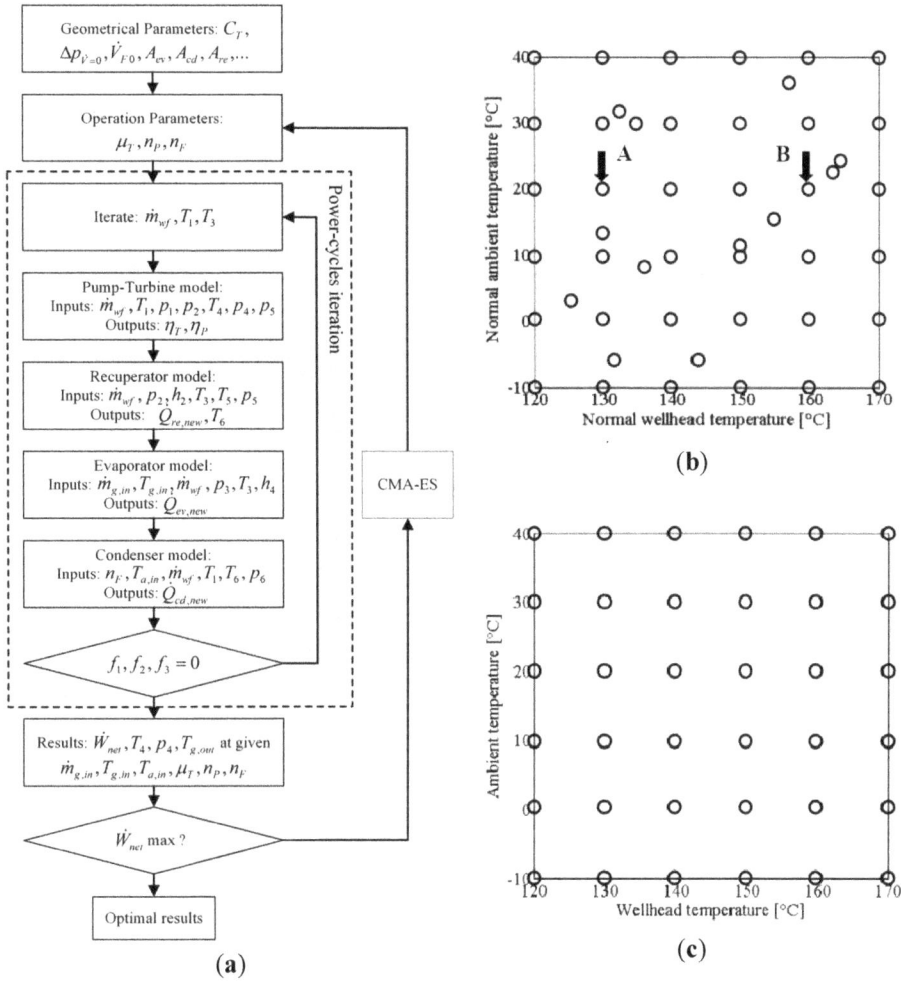

(a)

(b)

(c)

Both design points had constant exergy input, which translated to higher geofluid mass flow rate at lower wellhead temperatures, as previously described in Equation (20). The maximum net power output (978 kW) occurred at $T_{g,in}$ = 120 °C, $T_{a,in}$ = −10 °C for 130-20 (Point A, Figure 7b). While maximum net power output (1025 kW) occurred at $T_{g,in}$ = 160 °C, $T_{a,in}$ = −10 °C for 160-20 (Point B, Figure 7b). It can be observed contradictory net power-output trend between the two design points. For 130-20, by increase of geofluid temperature, the net power output decreases, especially at lower ambient temperature. In contrary, for 160-20, the net power output showed an opposite trend. This was affected mainly on the turbine isentropic efficiency characteristic at off-design. The nominal (design) isentropic enthalpy drop was lower and the nominal volumetric

376

flow rate was higher for 130-20. Hence, if the plant was operated at higher wellhead temperature which has higher enthalpy drop and lower flow rate, the turbine isentropic efficiency would steeply deteriorated (see Figure 5b).

It is also important to note the different net-power dependencies on ambient temperature. When investigating at a constant $T_{g,in}$ at the optimum point, the net power output decreased by 65.1% for 130-20 and 44.5% for 160-20 between $-10\,°C$ and $40\,°C$.

Figure 8. Off-design maps of net power output for (**a**) 130-20; (**b**) 160-20.

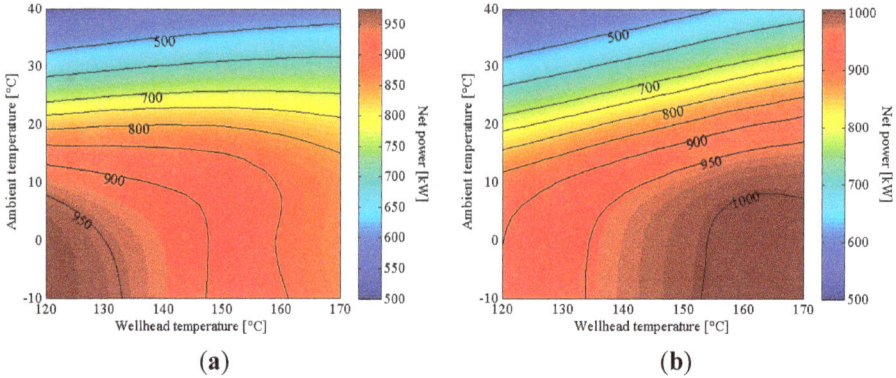

(**a**)   (**b**)

4.3. Annual Simulation and Thermo-Economic Selection

The system is assumed to be at steady-state for the annual simulations, and the heat loss in each component is neglected. The cycle performance is calculated in each time step of 1 h. The steady-state approximation is considered to be reasonably accurate since ambient temperature change is slower than the heat exchanger dynamics in the system. Thermal-economic optimization then was conducted to measure the trade-off between annual energy utilization and cost. Specific component costs are described in Table 3; however, the cost correlations listed are not the exact economic values, since cost can vary strongly depending on market. Nonetheless, the values presented here used as a means to convert geometric design parameters into economic value, and correlations are taken from actual literatures [22–24].

The turbine cost was taken from a model developed by Barber-Nichols [25]. The correlations are corrected to current cost by using the Chemical Engineering Plant Cost Index (CEPCI) [29]. The parameters, A, D_F, N_F, P_F, P_T, and P_P in Table 3 were determined directly from the sizing results. The turbine pitch (average wheel) diameter, D_{pitch}, was derived from a universal functional relationship, for optimum stage efficiency [30] as:

$$\dot{V}_{5,design} = 0.177 \cdot D_{pitch} \cdot \sqrt{\Delta h_{is0}}$$ (24)

Two economic criteria were computed: specific investment cost (SIC) and mean cash flow (MCF). SIC is a typical parameter used in thermal-economic optimization, and is defined as:

$$SIC\left[\$ \cdot kW^{-1}\right] = \frac{\text{Component cost+Labor cost}}{\text{Averaged annual capacity}\left(\bar{P}_{net}\right)}$$ (25)

where $\bar{P}_{net}$ is mean annual net power output calculated as the averaged sum of annual energy production for each wellhead temperature (in kWh) divided by 7008 h. MCF measures the productivity of the power-plant, and is computed as:

$$\text{MCF}\left[\$\cdot\text{year}^{-1}\right]=\text{Revenue}-\text{CRF}\cdot(\text{Component cost}+\text{Labor cost})-C_{O\&M}-\text{Well cost} \qquad (26)$$

where Revenue $= \bar{P}_{net} \times C_{el}$ and the three later terms are particularly annualized cost of electricity, *i.e.*, investment cost, annual operation and maintenance costs of the overall plant which are assumed to be 4% of the investment cost [31], and well cost. Well cost accounted for the geofluid-pumping and drilling costs, which are arbitrary values dependent on site-specific characteristics. It was assumed a well cost equal to zero since it will only shift the MCF to a lower value, and result in an unchanged optimum design-point. The three climates temperate, tropical and dry—chosen for annual simulation were sampled from existing geothermal sites: Upper-Rhine Graben, Germany (temperate climate), Kamojang, Indonesia (tropical climate), and Birdsville, Australia (dry climate). The temperature distributions of each climate are shown in Table 4.

Table 4. Ambient temperature distribution of three climates during generic year.

Temperature [°C]	Temperate climate (T_{av} = 11.6 °C)		Tropical climate (T_{av} = 19.9 °C)		Dry climate (T_{av} = 25.1 °C)	
	Number of hours	% hours	Number of hours	% hours	Number of hours	% hours
−10	266	3.0	0	0	0	0
0	2438	27.8	0	0	17	0.2
10	2926	33.4	351	4.0	1195	13.6
20	2159	24.6	7934	90.6	3139	35.8
30	726	8.3	475	5.4	3154	36.0
40	245	2.8	0	0	1254	14.3

The annual energy production was calculated using the hourly variation of $T_{a.in}$ at each site. This calculation only includes cost, which varies significantly according to the component size. The remaining costs, such as piping, instrumentation and working fluid, were excluded.

Under the conditions assumed for the temperate climate, the optimum points of SIC and MCF optimization was different (Figure 9). SIC minimization yielded 160-6, with a cost value of 1133 $\cdot$kW^{-1}, while MCF maximization yielded 153-10, with a cost value of 761,350 $\cdot$year^{-1}. The SIC and MCF showed large variation, ranging from 1133 $\cdot$kW^{-1} to 5296 $\cdot$kW^{-1}, and 92,224 $\cdot$kW^{-1} to 761,350 $\cdot$kW^{-1}, respectively.

Comparing the two objective functions, SIC minimization resulted in values 5.1%–7.1% lower, relative to plants based on maximizing MCF. By maximizing MCF, values were 2.1%–10.8% higher compared to when SIC was minimized. The temperate climate had the lowest SIC minimum and highest MCF maximum, followed by the tropical and then the dry climate. The optimization results are reported in Table 5. Using SIC minimization, the optimum normal wellhead temperature was constant at T_{g0} of 160 °C across the three climates, and optimum T_{a0} followed lower temperatures of 6 °C, 10 °C and 10 °C. While in MCF maximization, optimum T_{g0} was 153 °C, 163 °C and 163 °C, and T_{a0} followed average temperatures of 10 °C, 22 °C and 23 °C, respectively. Figure 10 shows relative component costs among the three climates.

Figure 9. Design-point based on minimizing SIC (**a**) and maximizing MCF (**b**) in temperate climate.

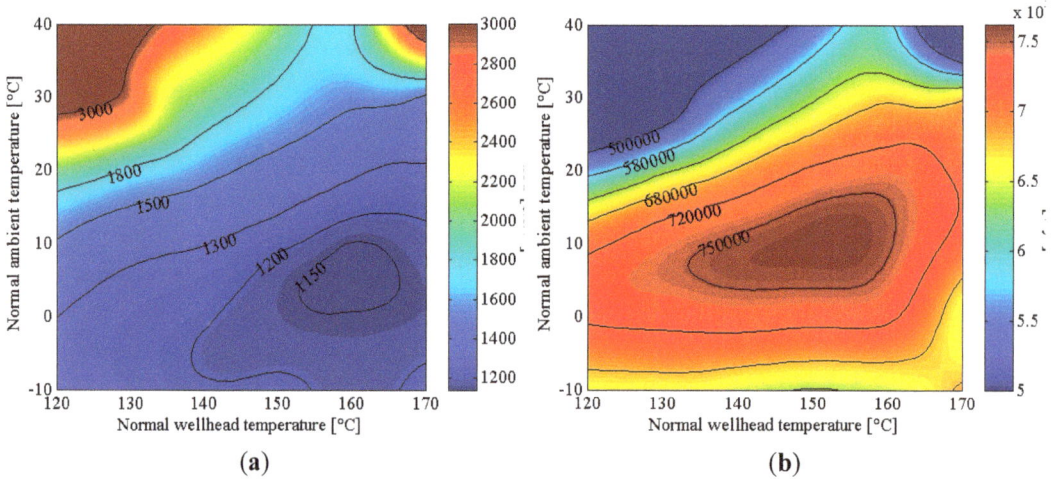

(**a**)   (**b**)

Table 5. SIC and MCF for each optimal design-point and climate type.

Sizing	Design-point [°C]		SIC [$·kW⁻¹]	MCF [$·year⁻¹]
	T_{g0}	T_{a0}		
Temperate climate				
SIC minimization	160	6	1,133	745,770
MCF maximization	153	10	1,198	761,350
Tropical climate				
SIC minimization	160	10	1,303	642,070
MCF maximization	163	22	1,403	683,120
Dry climate				
SIC minimization	161	10	1,520	524,230
MCF maximization	163	23	1,601	580,800

SIC minimization resulted in a investment cost that was 8.2%–13% lower than plants designed using maximized MCF. The cooling-system cost (condenser heat exchangers, fans) dominated the total investment cost. For plants with minimized SIC, the cooling-system cost was 12.5%–16.7% lower than those designed using maximized MCF. In contrast, MCF maximization resulted in a higher mean annual net-power 3.4%–12.7%. This improvement was based on the optimal number of cells and fan capacity, which maintain low condensation pressure and, in turn, result in higher power.

Figure 10. Relative component cost comparison between SIC and MCF optimization under three different climate types.

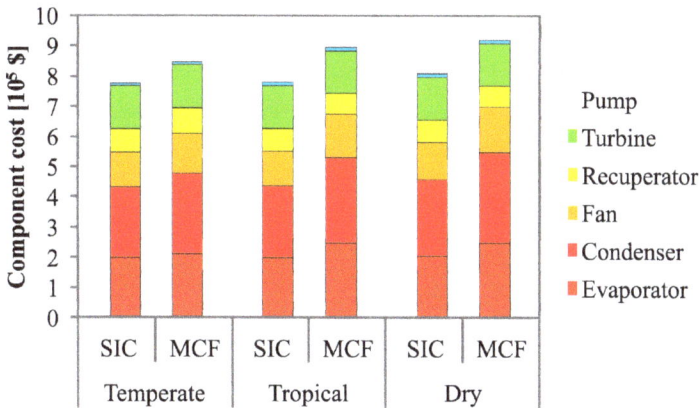

5. Conclusions

The main conclusions of this work are as follows:

1. Using the modularization technique described in this paper, design optimization under three different climates (temperate, tropical, and dry) was derived. Using SIC minimization, the normal ambient temperatures were driven by the lower temperature. Using MCF maximization, the normal ambient temperatures were driven by average temperature in each climate region.

2. When SIC minimization and MCF maximization were compared, average net-power based on MCF maximization was higher. Although investment cost was slightly higher, the revenue or equivalently, the energy utilization was considerably improved. Consequently, MCF maximization is proposed as an optimization function.

3. Concerning the various components analyzed here, the condenser and fan size had the greatest influence on average net power output. The main feature in MCF maximization design was increased size of the cooling-system, which helped maintain low condensation pressure. Using isobutane, the condenser cost amounted to 35%–38% of the investment cost. Enhancing the heat transfer of cooling system technology will reduce the condenser size and, most importantly, the ORC investment cost.

Acknowledgments

This research was financed by the German Ministry of Education and Research (BMBF), under the German-Indonesian cooperation project "Sustainability concepts for exploitation of geothermal reservoirs in Indonesia: Capacity building and methodologies for site deployment" (Grant Number 03G0753A). We thank to Stephanie Frick of German Research Center for Geosciences, Potsdam (GFZ) for the helpful comments in this study.

Nomenclature

$\dot{W}$	Output power [kW]		D	Diameter [mm]
P	Power capacity [kW]		t	Thickness [m]
T	Temperature [°C]		L	Length [m]
p	Pressure [kPa]		P_T	Pitch [mm]
Δp	Pressure drop, head [Pa]		A	Area [m²]
$\dot{m}$	Mass flowrate [kg·s⁻¹]		n	Rotational speed [Hz], indices
$\dot{Q}$	Heat [kW]		i	Interest rate [%]
h	Spec. enthalpy [kJ·kg⁻¹]		y	Depreciation time [yr]
s	Spec. entropy [kJ·kg⁻¹]		N	Number
$\dot{V}$	Volume flowrate [m³·s⁻¹]		η	Efficiency
G	Mass flux [kg·m⁻²·s⁻¹]		C	Constant, cost [$]
c_p	Spec. heat capacity [kJ·kg⁻¹·K⁻¹]		Re	Reynolds number
k	thermal conductivity [W·m⁻¹·K⁻¹]		Pr	Prandtl number
α	Heat transfer coef. [W·m⁻¹·K⁻¹]		f	Fanning friction factor
U	Overall heat transfer coef. [W·m⁻²·K⁻¹]		F	Multiplier factor
Φ^2	Two-phase multiplier		μ_T	Nozzle position [%]
X_{tt}	Turbulent Lockhart-Martinelli parameter		SIC	Specific investment cost [$·kW⁻¹]
u	Wheel tip speed [m·s⁻¹]		MCF	Mean cash flow [$·year⁻¹]
c_o	Spouting-velocity [m·s⁻¹]		CRF	Capital Recovery Factor
γ	Latent heat [kJ·kg⁻¹]		ITD	Initial temperature difference [K]
ρ	Density [kg·m⁻³]			

Subscripts

0	Normal (design)		l	Liquid
1	Pump inlet		tp	Two-phase
3	Evaporator inlet		v	Vapor
4	Turbine inlet		ev	Evaporator
g	Geothermal geofluid		cd	Condenser
a	Air		re	Recuperator
wf	Working-fluid		P	Pump
o	Outer		P	Turbine
i	Inner		F	Fan
s	Shell		sat	Saturated
el	Electrical		WTD	Weighted temperature difference
pp	Pinch-point		O&M	Operation & Maintenance

Author Contributions

Yodha Y. Nusiaputra devised idea and developed the simulation codes used in this paper. Hans-Joachim Wiemer verified the theoretical and practical soundness of the proposed methodology. Final review, including final manuscript rectifications, was done by Dietmar Kuhn.

Conflict of Interest

The authors declare no conflict of interest.

References

1. Stefansson, V. World Geothermal assessment. In Proceedings of World Geothermal Congress 2005, Antalya, Turkey, 24–29 April 2005.
2. Aneke, M.; Agnew, B.; Underwood, C. Performance analysis of the Chena binary geothermal power plant. *Appl. Therm. Eng.* **2011**, *31*, 1825–1832.
3. Bäumer, R.; Kalinowski, I.; Röhler, E.; Schöning, J.; Wachholz, W. Construction and operating experience with the 300-MW THTR nuclear power plant. *Nucl. Eng. Des.* **1990**, *121*, 155–166.
4. Dahmus, J.B. Modular Product Architecture. *Des. Stud.* **2001**, *22*, 409–424.
5. Quoilin, S.; Broek, M.V.D.; Declaye, S.; Dewallef, P.; Lemort, V. Techno-economic survey of Organic Rankine Cycle (ORC) systems. *Renew. Sustain. Energy Rev.* **2013**, *22*, 168–186.
6. Valdimarsson, P. New Development in the ORC Technology. In Proceedings of Short Course VI on Utilization of Low- and Medium-Enthalpy Geothermal Resources and Financial Aspects of Utilization, Santa Tecla, El Salvador, 23–29 March 2014.
7. Manente, G.; Toffolo, A.; Lazzaretto, A.; Paci, M. An Organic Rankine Cycle off-design model for the search of the optimal control strategy. *Energy* **2013**, *58*, 97–106.
8. Gabbrielli, R. A novel design approach for small scale low enthalpy binary geothermal power plants. *Energy Convers. Manag.* **2012**, *64*, 263–272.
9. Augustine, C. Hydrothermal Spallation Drilling and Advanced Energy Conversion Technologies for Engineered Geothermal Systems. Ph.D. Thesis, Massachusetts Institute of Technology, Cambridge, MA, USA, 2009.
10. DiPippo, R. *Geothermal Power Plants: Principles, Applications, Case Studies, and Environmental Impact*; Butterworth-Heinemann: Waltham, MA, USA, 2008.
11. Hernandez-Galan, J.L.; Alberto Plauchu, L. Determination of fouling factors for shell-and-tube type heat exchangers exposed to los azufres geothermal fluids. *Geothermics* **1989**, *18*, 121–128.
12. Serth, R.W. *Process Heat Transfer: Principles and Applications*; Elsevier: Amsterdam, The Netherlands, 2007.
13. Doo, G.H. A Modeling and Experimental Study of Evaporating Two-Phase Flow on the Shellside of Shell-and-Tube Heat Exchangers. Ph.D. Thesis, University of Strathcylde, Scotland, UK, 2005.

14. Dobson, M.K.; Chato, J.C. Condensation in Smooth Horizontal Tubes. *J. Heat Trans.* **1998**, *120*, 193–213.

15. *Engineering Data Book*; Gas Processor Suppliers Association (GPSA): Tulsa, OK, USA, 2004.

16. Shekun, G.D. Approximating the efficiency characteristics of blade pumps. *Therm. Eng.* **2007**, *54*, 886–891.

17. Dixon, S.L. *Fluid Mechanins and Thermodynamics Turbomachinery*; Butterworth Heinemann: Waltham, MA, USA, 2002.

18. Ghasemi, H.; Paci, M.; Tizzanini, A.; Mitsos, A. Modeling and optimization of a binary geothermal power plant. *Energy* **2013**, *50*, 412–428.

19. Frick, S.; Saadat, A.; Kranz, S. Cooling of Low-Temperature Power Plants—Challenges for the Example of Geothermal Binary Power Plants. In Proceedings of 6th International Symposium on Cooling Towers, Cologne, Germany, 20–23 June 2012.

20. *Air-Cooled Condenser Design, Specification, and Operation Guidelines*; The Electric Power Research Institute: Palo Alto, CA, USA, 2005.

21. He, C.; Liu, C.; Gao, H.; Xie, H.; Li, Y.; Wu, S.; Xu, J. The optimal evaporation temperature and working fluids for subcritical Organic Rankine Cycle. *Energy* **2012**, *38*, 136–143.

22. Taal, M.; Bulatov, I.; Klemeš, J.; Stehlík, P. Cost estimation and energy price forecasts for economic evaluation of retrofit projects. *Appl. Therm. Eng.* **2003**, *23*, 1819–1835.

23. Alinia Kashani, A.H.; Maddahi, A.; Hajabdollahi, H. Thermal-economic optimization of an air-cooled heat exchanger unit. *Appl. Therm. Eng.* **2013**, *54*, 43–55.

24. Quoilin, S.; Declaye, S.; Tchanche, B.F.; Lemort, V. Thermo-economic optimization of waste heat recovery Organic Rankine Cycles. *Appl. Therm. Eng.* **2011**, *31*, 2885–2893.

25. Milora, S.L.; Tester, J.W. *Geothermal Energy as a Source of Electric Power*; The MIT Press: Cambridge, MA, USA, 1976.

26. Astolfi, M.; Romano, M.C.; Bombarda, P.; Macchi, E. Binary ORC (Organic Rankine Cycles) power plants for the exploitation of medium–low temperature geothermal sources Part B: Techno-economic optimization. *Energy* **2014**, *66*, 435–446.

27. Hansen, N. Towards a New Evolutionary Computation. In *Studies in Fuzziness and Soft Computing*; Springer: Berlin, Germany, 2006; pp. 75–102.

28. Keim, D.A.; Herrmann, A. The Gridfit Algorithm: An Efficient and Effective Approach to Visualizing Large Amounts of Spatial Data. In Proceedings of Visualization's 98 Research, Triangle Park, NC, USA, 18–23 October 1998; pp. 181–188.

29. Spang, B.; Roetzel, W. C6 Costs and Economy of Heat Exchangers. In *VDI Heat Atlas*; Springer: Berlin, Germany, 2010.

30. Balje, O.E. A Study on Design Criteria and Matching of Turbomachines. *J. Eng. Gas Turbines Power* **1962**, *84*, 83–102.

31. Entingh, D.J.; Eyob, E.; McLarty, L. *Small Geothermal Electric Systems for Remote Powering*; Geothermal Resource Council: Davis, CA, USA, 1994.

Geothermal Power Plant Maintenance: Evaluating Maintenance System Needs Using Quantitative Kano Analysis

Reynir S. Atlason, Gudmundur V. Oddsson and Runar Unnthorsson

Abstract: A quantitative Kano model is used in this study to identify which features are preferred by top-level maintenance engineers within Icelandic geothermal power plants to be implemented in a maintenance tool or software. Visits were conducted to the largest Icelandic energy companies operating geothermal power plants. Thorough interviews with chiefs of operations and maintenance were used as a basis for a quantitative Kano analysis. Thirty seven percent of all maintenance engineers at Reykjavik Energy and Landsvirkjun, responsible for 71.5% of the total energy production from geothermal resources in Iceland, answered the Kano questionnaire. Findings show that solutions focusing on (1) planning maintenance according to condition; (2) shortening documentation times; and (3) risk analysis are sought after by the energy companies but not provided for the geothermal sector specifically.

Reprinted from *EnergiesEnergies*. Cite as: Atlason, R.S.; Oddsson, G.V.; Unnthorsson, R. Geothermal Power Plant Maintenance: Evaluating Maintenance System Needs Using Quantitative Kano Analysis. *EnergiesEnergies* **2014**, *7*, 4169–4184.

1. Introduction

The diminishing access to easily retrieved energy sources will ultimately have a great effect on the quality of life of a large portion of the Earth's population, even larger than it does at present time [1]. It is therefore of the utmost importance that the sources that are now utilized, whether fossil fuels or not, are harvested in the most efficient manner. To do so, power plants providing electricity or other sources of energy, such as hot water, need to put pressure on efficient operations. One of the factors that needs to be constantly under scrutiny is the maintenance and operation of the power plants. If the operation and maintenance are not carried out in an efficient manner, the resource will not be utilized to its fullest potential and the power plant can be prone to serious problems. Maintenance is generally divided into three categories, reactive, time-based and condition-based, where the second and third are often combined as simply preventive maintenance. To schedule or monitor such tasks, a wide range of methods is available, where parts are inspected or monitored to determine when the part is to be serviced, repaired or replaced. The maintenance procedures of fossil fuel power plants have been under development for much longer than those of geothermal power plants, which is a sector still in its infancy compared to the fossil fuel industry. The conditions geothermal power plants operate under are however very different in nature from those observed in fossil fuel power plants. One of the biggest difference is the steam purity, where the steam used in fossil fuel power plants is much cleaner than the steam used in geothermal power plants. As a result, solutions already provided to the fossil fuel energy sector can only be used in the geothermal sector in a very limited way.

This article explores solutions that are wanted by chief engineers at the Icelandic geothermal power plants, and that are believed to further improve the maintenance operations. This was done by conducting thorough interviews on site with chief maintenance engineers. A need analysis was then conducted amongst the total population of maintenance engineers responsible for 71.5% of all geothermal energy utilization in Iceland in an attempt to prioritize the requirements put forward by power plant engineers for further development.

2. Maintenance

Most people face reactive maintenance on a daily basis. However, when a breakdown leads to downtime of a factory, or risks lives, as is the case on airplanes, more advanced methods are equipped. Various maintenance methods have been recorded from ancient history, where large scale constructions were kept up and running. Proper maintenance can lead to financial savings even though it can require financial expenditures to begin with. Often, downtime can be very expensive for companies since no output is derived from their operations. A company might therefore minimize and plan downtime by replacing certain parts frequently, or even learn how to predict the failure of given parts. Different management methods for maintenance have been developed throughout history. Failure driven, often regarded to as reactive maintenance, time based maintenance and condition based maintenance are claimed to be the most employed methods [2–5]. It has been estimated that 55% of maintenance methods in the average maintenance program is reactive, 31% is time-based, 12% is predictive and the last 2% accounts for other methods [6]. The main methods will be discussed in further detail in the following sections.

2.1. Reactive Maintenance

Studies, as recently as in 2000, show that this method is still the most dominant one. It has been stated that majority of maintenance activities on average facilities are considered reactive [6]. The advantages of such methods are that a minimum amount of staff is required for the program [6]. However, unplanned costs related to increased downtime, increased labor cost because of downtime, cost for repair of equipment and perhaps secondary equipment which got damaged along with the primary failure are the examples of disadvantages of reactive maintenance [6]. Reactive maintenance essentially aims to restore a given system to a functional state after it has failed [7].

2.2. Preventive Maintenance

As described, reactive maintenance simply waits for parts to become inoperable. This has proven costly for companies for reasons described above. The quality of equipment is bound to deteriorate with time, and eventually break down. However, by looking retrospectively at breakdown statistics, failure can be minimized. Time based and condition based maintenance are considered to be preventive. The following sections will discuss these methods in greater detail.

2.2.1. Time Based Maintenance

Time-based (also known as calendar based) maintenance tries to schedule maintenance at predetermined intervals, such as amount of produced goods, hours of running machine, mileage or condition [8]. It is, however, considered to be the second least cost efficient maintenance method after reactive maintenance [9]. The aircraft industry however had some problems with time based maintenance. It was shown that time based maintenance was a difficult approach in an industry as rapidly developing as the aircraft industry, also, it was shown that failure likelihood was not expected to rise with ageing parts [10]. Time-based maintenance is generally based on the assumption that one rule fits all. However, each system is operated in a unique environment and is subjected to different conditions. After all, more efficient maintenance is less costly for airlines. The airline industry is an example of an industry that has to have maintenance management issues very clear, since during their business hours, a minor failure can have catastrophic effects, for human lives and subsequently their business. In 1965, the first computerized time-based maintenance system was created by Mobil Oil to manage lubrication on mobile equipment, it was the Midec program [8].

2.2.2. Condition Based Maintenance

Like time-based maintenance, condition based maintenance (CBM) seeks to replace parts before they fail. However, the difference is that instead of the rule of scheduled intervals or amount of use, CBM in modern time is a lot based on observation data [8]. Maintenance is therefore scheduled when a certain condition has been met within the system under study, be it pressure, vibration or anything else that could indicate failure in near future [11]. CBM systems are often not used within industries even though the possibility may seem obvious, this has been speculated to because of the low maturity level of the CBM systems [12].

3. Geothermal Power Plants Studied

In Iceland, there are three energy companies that operate geothermal power plants. They are, HS (Hitaveita Sudurnesja) Energy, Reykjavik Energy and Landsvirkjun. Reykjavik Energy, and Landsvirkjun contributed to this research. In total, these companies operate four geothermal power plants. However, each company operates their power plants under different geological condition. This section will provide a description of the power plants under study. It will further outline the different geological conditions at the different sites where the power plants are located. Data was gathered from the literature as well as from the energy companies. The data from Hellisheidi is from brine water leaving the separators, this is also the case for the data at Nesjavellir [13,14]. The chemical composition data for Krafla was also gathered from an ISOR (Islenskar Orkurannsoknir) report, from the KG-26 hole after it was deepened to 2000 m [15].

3.1. Hellisheidi

The Hellisheidi geothermal power plant is owned by Reykjavik Energy and began its electric production in 2006 [16]. The plant is located on the southern part of the Hengill geothermal field, a detailed location can be seen on Figure 1. It produces approximately 303 MW of electric power and 133 MW of hot water through a double flash process. Around 50 wells have been drilled to harness hot water for the power production [17]. Reykjavik Energy provided data about the fluid chemical composition when it leaves the separators. One can see that the fluid consists mostly of SiO_2 (822 mg/kg), Na (213 mg/kg), Cl (170 mg/kg), K (38.4 mg/kg), and SO_4 (19 mg/kg) [13].

Figure 1. Location of the power plants under study. Locations of the plants are as follows: (**1**) Krafla; (**2**) Hellisheidi; (**3**) Nesjavellir and (**4**) Bjarnarflag. Images retrieved from Google maps software.

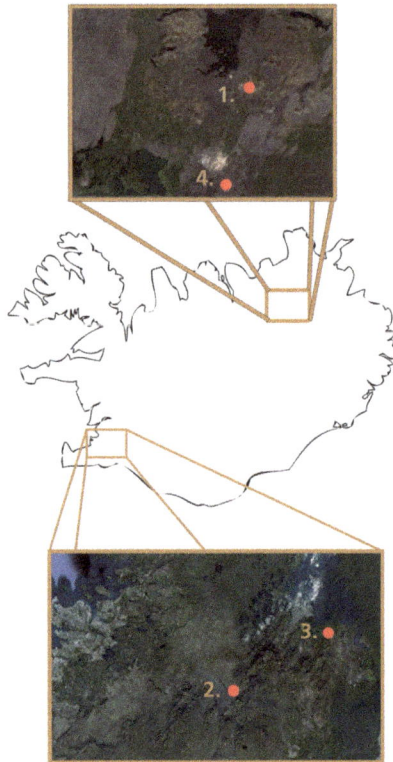

3.2. Nesjavellir

Also located on the Hengill geothermal field, Nesjavellir geothermal power plant produces 120 MWe and 300 MWt. Experimental wells were drilled, where each well was providing up to 60 MWt, with a usable 30 MWt. Construction of the plant began in 1987, and the first phase was completed in 1990. In the same year, four holes, generating 100 MWt were connected to the production. in

1995 an additional hole was drilled and connected and the production capacity increased to 840 liters per second. Today, 26 holes have been drilled. Temperatures at Nesjavellir have been recorded as high as 380 °C. It is estimated that the current production can continue for the next 30 years [18]. The brine at Nesjavellir consists mostly of Silicon dioxide (SiO_2), Sodium (Na), Chlorine (Cl) and Sulfate (SO_4).

3.3. Krafla

No geothermal power plant is located further north in Iceland than Krafla. The plant is currently producing 60 MWe of power from two 30 MW Mitsubishi turbines. Construction began in 1974 when test wells were drilled. In 1975 construction of the plant began and in 1977 production began. Initially the plant only operated using one turbine, the second turbine was installed in 1996 and began producing in 1997 [19]. The brine at Krafla consists mostly of SiO_2 (790 mg/kg), Cl (608 mg/kg), Na (356). Of gases, the plant can be expected to release around 40 mg/kg of H_2S and 235 of CO_2.

3.4. Bjarnarflag

In 1969, Landsvirkjun began operations in Bjarnarflag, Icelands oldest geothermal power plant. The plant is also the smallest operated within the country, producing 3 MW. The power plant uses one back pressure turbine with a single Curtis wheel for its production. Plans are currently underway to increase the capacity of the power plant up to 90 MW [20].

4. Methodology

To identify which maintenance systems are used at the Icelandic geothermal power plants, interviews were carried out. The head of power plant operations and the technical supervisor at Reykjavik Energy were interviewed, as well as the head of maintenance management at Landsvirkjun. These two companies are responsible for 71.5% of all energy production from geothermal resources in Iceland. In addition, a quantitative Kano need analysis was carried out among specialists and heads of operations within the energy companies, in order to identify which features are considered mandatory and which would improve the maintenance efficiency further. There were therefore more substantially more participants in the Kano survey than were interviewed. The interviews were conducted in order to get a sense of what to include in the Kano survey. The interviews were focused on the following issues:

- Identify what measures are currently taken in the geothermal power plants when it comes to maintenance;
- How data is used to improve maintenance;
- Identify which solutions are desirable for the power plants.

A quantitative Kano model [21] was utilized to perform a need analysis with regards to the maintenance management systems at the Icelandic geothermal power plants. Interviews were also conducted at the power plants with chief engineers who are in charge of operations and maintenance. A specialist in questionnaires from the University of Iceland was consulted before the study was conducted to provide support with the construction of the questionnaire.

4.1. Interviewees and Population

To gain further understanding of the requirements by the energy companies, detailed interviews at the power plants were conducted. Those include the chief operations engineers at the power plants who are in direct contact with operations and maintenance on site. The purpose of the interviews was to identify what solutions are already being used, and identify which solutions are portrayed as attractive by the chief engineers. The input from chief engineers and maintenance staff proved essential when the Kano questionnaire was constructed. The proportion of answers retrieved from the energy companies amounted to 37.5% of employees directly engaged in operation and maintenance issues. The answers include the total population of heads of operations, engineers with high level of expertise who are in charge of large maintenance activities. The total amount of answered questionnaires amounts to 12, from the employees mentioned above. We assume that the translations from the questionnaire do not have effects on the results.

4.2. The Kano Model

The Kano model of customer satisfaction was initially introduced in 1984 [22]. The survey based method is used to analyze given qualities of a product and how customers may perceive them. The model classifies three different product requirements which customers react to in a different manner when met [23]. First are so called "Must-be requirements". When not fulfilled, dissatisfaction is experienced by the user. However, those requirements are taken for granted by the user, therefore, their fulfillment does not increase the customer satisfaction. As an example of a must-be requirement as defined by Kano is the Internet connection ability in smartphones. When a customer buys a smartphone, he expects it to have the ability to connect to the Internet wirelessly. Failure to meet this requirement results in user dissatisfaction. The second criteria requirements are so called "One dimensional requirements". Those requirements have a linear connection to the satisfaction of the customer. As the requirement is fulfilled in an efficient manner, the customer becomes more satisfied. As an example of this may be the fuel usage of a car. The less fuel used, the more satisfied will the customer become. The last requirements can be regarded as the most important [23]. Those requirements are regarded as "attractive requirements". Such a requirement was not expected by the user and its absence would therefore not result in less satisfaction. Its appearance however increases the customer satisfaction greatly. There seems to be a tendency for "attractive requirements" to become "must-have" requirements over time. For example, the ability to interact with the cellular telephones through a touchscreen was considered an attractive requirement when it originally became available on the public market. However, over time, this functionality has become a "must-have" requirement in many countries. Figure 2 shows the relationship between those requirements and customer satisfaction.

Figure 2. Relationship between customer satisfaction and requirement fulfillment according to Kano models [22].

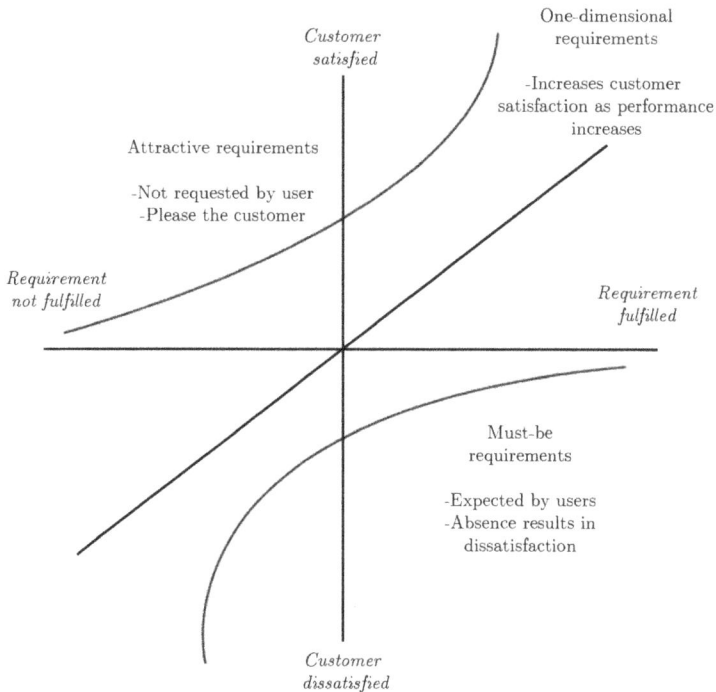

Kano modeling can prove helpful to product development. A product that already is fulfilling a must-be requirement should perhaps not be developed to fulfill that requirement further as it would not increase the user satisfaction. A Kano model can provide a better perspective of the product that is under development by showing which requirements or criteria improve the customer satisfaction the most. During the development stage, some trade offs may be inevitable. A Kano model can assist with such trade offs by showing which feature results in the greatest user satisfaction.

The method of using the Kano model is gathered straight from the field of product development. Despite of being qualitative in nature, Kano models have shown to be an effective tool in the product development process. However, quantitative versions of the Kano model have been presented, such as the Analytical Kano Model, or A-Kano [24]. In this study we use a quantitative analysis of the Kano model presented by Wang and Ji [21]. Upcoming sections will describe the methodology.

4.2.1. Quantitative Kano Model—The Questionnaire

After visualizing the market segment that is to be studied, a questionnaire is constructed. The questionnaire consists of questions about the functional requirements of a product. Each functional requirement consists of two questions, one functional and one dysfunctional. For example, if being asked about the weight of a cellular telephone, the customer might be asked "If the phone is as light as a matchbox, how do you feel?" and then subsequently "If the phone is heavier than a matchbox, how do you feel?". Each question has five possible outcomes, (1) I like it that way; (2) It must be that way; (3) I am neutral; (4) I can live with it that way; (5) I dislike it that way. A form as is shown in Table 1 is then used to evaluate each functional requirement, be it attractive, one-dimensional, must-be, indifferent, reverse or questionable. An experienced researcher in the field of psychology provided guidance when the questionnaire was constructed. This was done to improve the structure and clarity of the questionnaire as well as to make the questions non-biased. It was decided to randomize the order of the questions, that is, the functional and dysfunctional form of each question does not come in perfect sequence. The questions are instead in a random order. This was done to avoid the respondents to answer one question with relation to the other. The questionnaire was then distributed amongst specialists and heads of operations within all geothermal power plants operated by Reykjavik Energy and Landsvirkjun, which were all heads of operations and 37.5% of operation and maintenance employees within the companies, and at the same time the majority of such employers in Iceland.

Table 1. The table used to evaluate the classification of the functional requirements (FR) by the customer [25].

CR's		Dysfunctional				
		1. Like	2. Must be	3. Neutral	4. Live with	5. Dislike
Functional	1. Like	Q	A	A	A	O
	2. Must-be	R	I	I	I	M
	3. Neutral	R	I	I	I	M
	4. Live with	R	I	I	I	M
	5. Dislike	R	R	R	R	Q

A = Attractive; M = Must-be; R = Reverse; O = One-dimensional; I = Indifferent; Q = Questionable.

4.2.2. Quantitative Kano Model—Computation

Based on the findings of the questionnaire and subsequently identifying the nature of the combined answers (Attractive, one-dimensional *etc.*), it is possible to calculate two values, namely customer satisfaction (CS) and customer dissatisfaction (DS). the CS value can be expressed as [21]:

$$CS_i = \frac{f_A + f_O}{f_A + f_O + f_M + f_I}$$

Let f_A denote the number of attractive, f_O the number of one-dimensional, f_M the number of must-be and f_I indifferent responses. Similarly, to calculate the DS_i the following equation can be used:

$$DS_i = \frac{f_O + f_M}{f_A + f_O + f_M + f_I}$$

Subsequently, two points are located for each functional requirements (FR). These points define the customer satisfaction if the FR is fully implemented or fully excluded from the product. These points can be plotted as $(1, CS_i)$ and $(0, -DS_i)$ [21]. To find the relationship functions, one must first identify if the FR is a must-be, one-dimensional or attractive. This is done by finding the mode of the answers for that particular FR. The relationship function can be shown as $S = f(x, a, b)$, where S is the customer satisfaction, x the level of fulfillment, a and b are the adjustment parameters for the Kano categories of customer requirements. For one dimensional attributes the function is $S = a_1x + b_1$ where a_1 denotes the slope and b_1 is the DS value when customer requirement (CR) (x) is at 0. Entering CS and DS points into the equation we get $a_1 = CS_i + DS_i$ and $b_1 = DS_i$. The function for one-dimensional product attributes can be seen as: [21]:

$$S_i = (CS_i - DS_i)x_i + DS_i$$

If the CR is an attractive attribute, the function can be seen to be exponential. the function is therefore modified to be $S = a_2e^x + b_2$. However, we now get $a_2 = (CS_i - DS_i)/e - 1$ and $b_2 = -(CS_i - eDS_i)/e - 1$. We can therefore see that the function for such attributes is [21]:

$$S_i = \frac{CS_i - DS_i}{e - 1}e^{x_i} - \frac{CS_i - eDS_i}{e - 1}$$

For must-be attributes, the function can also be estimated using an exponential function. In the case of must-be attributes the function is $S = a_3(-e^{-x} + b_3)$. We acquire a_3 and b_3 by using [21]:

$$a_3 = \frac{e(CS_i - DS_i)}{e - 1}$$

and

$$b_3 = \frac{eCS_i - DS_i}{e - 1}$$

The function for must-be attributes can therefore be plotted as [21]:

$$S_i = -\frac{e(CS_i - DS_i)}{e - 1}e^{-x} + \frac{eCS_i - DS_i}{e - 1}$$

4.3. Functional Requirements

After interviewing chief engineers at the power plants, functional requirements were identified. These requirements were identified because of either (1) current lack of fulfilling the requirements using current solutions; (2) the requirements had not been attempted to be fulfilled in any solution currently used by the power plants. The requirements were based on the possibilities of using data currently gathered by the power plants. The data is often in the form of time between failures, type of maintenance procedures and "real time" condition monitoring. The functional requirements to be investigated are shown below (dysfunctional and functional version of each requirement is presented in the Appendix):

1. Know the effect of one maintenance procedure on other parts in the power plant. This can apply later in time. For example, a maintenance procedure is carried out at one time, which causes an unusual failure later in time in a different part of the plant. Would it prove valuable to know the connection between these factors;

2. Knowing the effects of postponing the maintenance procedure on mechanical components;

3. Have detailed, pre made, protocols for large scale maintenance procedures;

4. Shorter time to document maintenance procedures;

5. Provide suppliers with information about the predicted failure of certain parts in order to shorten waiting times;

6. Predict individual workers (or workers unit) workload based on predicted mechanical failures;

7. Plan maintenance according to the predicted state of the part instead of planning for all possible outcomes;

8. Base power plant inventory on failure predictions.

These requirements provided the basis for the questionnaire.

5. Results

From the interviews, it was evident that the main focus of the chief engineers were the intervals between major maintenance on the steam turbines. Currently, the turbines are scoped annually and full, planned stops are conducted quadrennially. It was seen that two requirements were frequently mentioned that are considered to be of major importance to improve the maintenance procedures of the power plants. Firstly, risk assessment method to determine the operational effects if a turbine is not maintained on a scheduled time, instead postponing the maintenance for some period. Secondly, detailed protocols for maintenance based on the predicted condition of the turbines are needed.

For example, if a certain condition is observed, clear protocols are currently not available to address the known condition but are instead tailor made by chief engineers for each case. Detailed protocols, more similar to protocols known in the medical or aerospace sector, are needed. Where protocols are pre-defined and are deployed based on some observed condition. We also identified that inventory was mentioned and the possibility to further minimize it. Twelve responses were gathered from staff members highly involved in maintenance procedures at the power plants. Even though the sample is relatively small, it represents the majority of employees highly involved in daily maintenance operations at the power plants studied. The Kano analysis of the data shown in Table 2 reveals that the greatest customer satisfaction will be reached if the condition of a part which requires major maintenance activities can be predicted to some extent. The attractive requirements are shown in Figure 3, the one dimensional requirements are shown in Figure 4.

Shortening the time which it takes to document maintenance was also found to offer high customer satisfaction if shortened. The ability to know the operational and operational effects of postponing maintenance procedures was found to offer linear relationship to customer satisfaction. In fact, if that particular functional requirement is addressed fully, it will provide almost the same customer satisfaction as planning major maintenance procedures according to their known, or predicted condition. However, addressing the attractive requirement of planning major maintenance according to known or predicted condition will bring more satisfaction earlier than the one dimensional requirement of knowing the effects of postponing. The attractive requirement of having predefined protocols for major maintenance activities did not bring as high customer satisfaction as other functional requirements, even though its importance was discussed in some detail by the chief engineers. One of the functional requirements (Base power plant inventory on failure predictions) which was to be studied returned mostly questionable results from the Kano table. Therefore, no tangible results were to be calculated for that particular functional requirement.

Table 2. Function calculations for customer requirements.

Customer Requirements	CS Point	DS Point	a	b	f(x)	S = af(x) + b
One-dimensional						
(1) Relationship of effects	(1, 0.9)	(0, −0.8)	1.7	−0.8	x	$f(x) = 1.7x − 0.8$
(2) Effects of postponing	(1, 1)	(0, −0.57)	1.57	−0.57	x	$f(x) = 1.57x − 0.57$
Attractive						
(3) Predefined detailed protocols	(1, 0.43)	(0, −0.38)	0.47	−0.85	e^x	$f(x) = 0.47e^x − 0.85$
(4) Short documentation time	(1, 1)	(0, −0.27)	0.73	−1.0	e^x	$f(x) = 0.74e^x − 1$
(5) Supplier need avareness	(1, 0.88)	(0, −0.13)	0.58	−0.71	e^x	$f(x) = 0.58e^x − 0.71$
(6) Knowledge of future workload	(1, 0.89)	(0, −0.12)	0.58	−0.70	e^x	$f(x) = 0.58e^x − 0.7$
(7) Plan maintenance according to condition	(1, 0.75)	(0, −0.42)	0.60	−1.1	e^x	$f(x) = 0.86e^x − 1.1$

Figure 3. Attractive functional requirements retrieved using a Kano analysis. Functions are shown in Table 2.

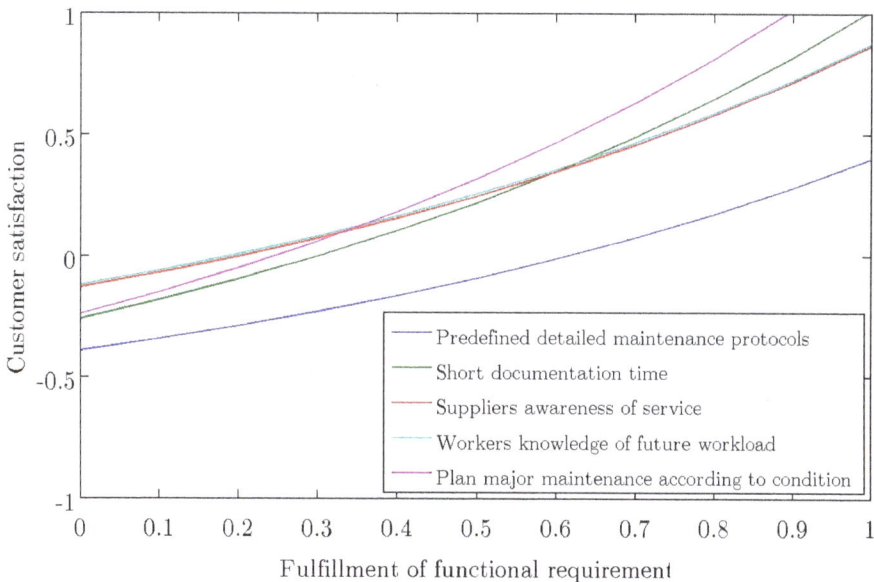

Legend:
- Predefined detailed maintenance protocols
- Short documentation time
- Suppliers awareness of service
- Workers knowledge of future workload
- Plan major maintenance according to condition

Axes: Customer satisfaction (vertical), Fulfillment of functional requirement (horizontal).

Figure 4. One dimensional functional requirements obtained using linear functions in Table 2.

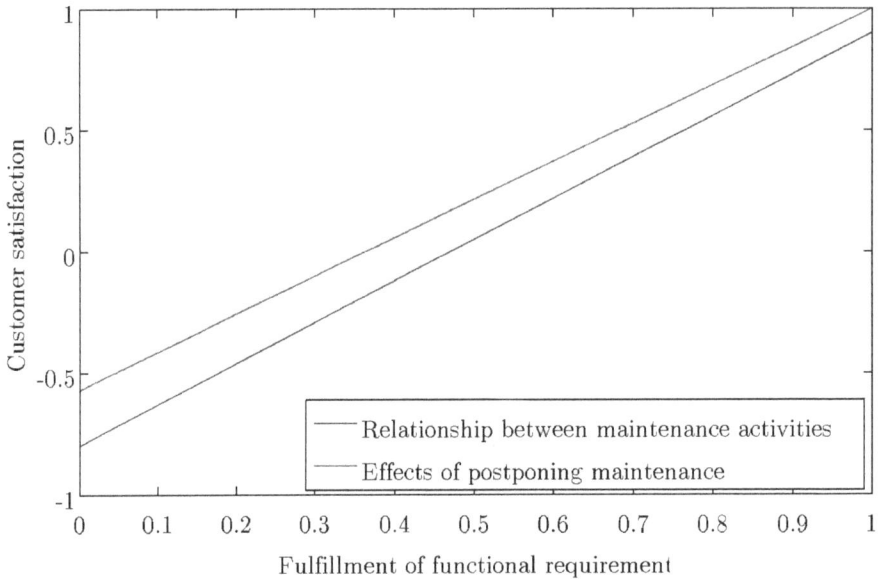

In this study, both the interviews and the Kano analysis were used to prioritize which solutions or methodologies should be developed by, or for, the power plants. Based on this study, the following prioritizations can be made:

- Develop a method where the condition of a part (such as a turbine) can be predicted before it requires maintenance. The maintenance is then conducted and planned based on such predictions;
- Shorten the time needed for standard documentation by the employees. This solution should also allow for easier collection of maintenance data;
- Risk analysis for important parts. It frequently mentioned by chief engineers how valuable the knowledge of postponing maintenance could be. Developing the methodology would be result in high customer satisfaction according to the Kano analysis, which further confirmed what had previously been seen in the interviews.

As can be seen in Table 2, the results for one-dimensional and attractive requirements are shown but must-be requirements are absent. This can be explained by the fact that solutions that are currently being addressed or fulfilled to some extent were intentionally not included in the Kano analysis.

6. Conclusions

In this study, we investigated which functional requirements desired by the engineers working at the two largest power companies in Iceland to improve their maintenance procedures if available. The desired solutions are currently not available, or not being deployed by the geothermal sector in

Iceland. The Kano model was found useful to further understand and formalize what had previously been observed in meetings on site. The previously mentioned results should serve as a roadmap for upcoming steps in product development for the geothermal sector and which solutions are needed by the power plants to further improve their maintenance procedures. Using the approach of investigating the power plants, their needs and requests for solutions should in essence provide greater likelihood of producing solutions that will be accepted and used by the plants than addressing needs that are perhaps currently being addressed (must-have requirements). The reason for functional requirement number 8, (planning of inventories based on failure predictions) not showing tangible results is most likely to be found in the formulation of the functional and dysfunctional forms of the question in the questionnaire. The results from the Kano questionnaire as well as the interviews underline the possibilities in predictive maintenance in the geothermal sector. A great deal of data is currently being gathered, which has the potential to serve as the backbone in predictive modeling. It was seen that the greatest interest from the power companies is currently in the field of predictive maintenance. It is therefore likely that solutions addressing the desired functional requirements by the power companies will be deployed by the Icelandic geothermal sector. Chief engineers at the power plants emphasized the importance of having pre-defined protocols for most maintenance activities. The Kano model however showed that fulfilling that particular requirement would return the lowest customer satisfaction of all the functional requirements analyzed. This may be because of different views of chief engineers and the maintenance staff. This study can serve as a guide for industries looking to serve the geothermal sector in the field of maintenance. The study indicates which requirements are sought after and expected to further improve the maintenance procedures in Icelandic geothermal power plants.

Acknowledgments

Our gratitude goes to GEORG (Geothermal Research Group) for financial support. Saemundur Gudlaugsson, Gudmundur Hjaltalin at Reykjavik Energy and Thrandur Rognvaldsson and Steinn A. Stensson at Landsvirkjun for assisting with the research. Also, the support from Ragna Benedikta Gardarsdottir, Associate Professor in psychology from the University of Iceland is acknowledged.

Author Contributions

All authors contributed to the conceptualisation of the article. Reynir S. Atlason carried out the main body of research. Gudmundur V. Valsson and Runar Unnthorsson reviewed the work continually.

Appendix

Below are the questions used in the Kano questionnaire. First is the functional requirement shown, then the functional version of the question, followed by the dysfunctional version.

1. Know the effect of one maintenance procedure on other parts in the power plant. This can apply later in time. For example, a maintenance procedure is carried out at one time, which causes an unusual failure later in time in a different part of the plant. Would it prove valuable to know the connection between these factors.

   Functional: "How would you feel if you knew the effects of your work on other parts of the power plant?" (2) (The number shown in brackets behind each question indicates when it appeared in the questionnaire.)
   Dysfunctional: "How would you feel if the causes of failures would not be known, as long as the failure is solved?" (16)

2. Knowing the effects of postponing the maintenance procedure on mechanical components.

   Functional: "How would you feel if you knew the effects on the equipment if maintenance is postponed?" (6)
   Dysfunctional: "how would you feel if maintenance is carried out on predefined times, without exceptions?" (9)

3. Have detailed, pre made, protocols for large scale maintenance procedures.

   Functional: "How would you feel if precise, exact protocols, describing what should be done and how, would be existent for large maintenance procedures?" (13)
   Dysfunctional: "How would you feel if each large maintenance job would be planned individually as a single occurrence?" (16)

4. Shorter time to document maintenance procedures.

   Functional: "How would you feel if it would only take you approximately one minute to document a standard maintenance procedure?" (5)
   Dysfunctional: "How would you feel if it would take you more than five minutes to document a standard maintenance procedure?" (12)

5. Provide suppliers with information about the predicted failure of certain parts in order to shorten waiting times.

   Functional: "How would you feel if suppliers could foresee your orders and plan accordingly?" (7)
   Dysunctional: "How would you feel if suppliers react to orders as they are made (but not before)?" (14)

6. Predict individual workers (or workers unit) workload based on predicted mechanical failures.

   Functional: "How would you feel if you knew your workload ahead of time?" (4)
   Dysfunctional: "How would you feel if your workload is known when (not before) jobs are assigned to you?" (8)

7. Plan maintenance according to the predicted state of the part instead of planning for all possible outcomes.

   Functional: "How would you feel if you could plan maintenance according to the condition of the equipment in question?" (3)

Dysfunctional: "How would you feel if you needed to prepare for every possible outcome when maintenance is conducted?" (10)

8. Base power plant inventory on failure predictions.

Functional: "How would you feel if you could plan the power plants inventory based on predicted failures?" (1)
Dysfunctional: "How would you feel if the inventory is constantly well loaded and guaranteed that spare parts are always available?" (11)

Conflicts of Interest

The authors declare no conflict of interest.

References

1. Lambert, J.G.; Hall, C.A.; Balogh, S.; Gupta, A.; Arnold, M. Energy, eroi and quality of life. *Energy Policy* **2014**, *64*, 153–167.
2. Davies, A. *Management Guide to Condition Monitoring in Manufacture*; Manufacturing Series; Institution of Production Engineers: London, UK, 1990.
3. Gits, C. Structuring maintenance control systems. *Int. J. Oper. Prod. Manag.* **1994**, *14*, 5–17.
4. Niebel, B.W. *Engineering Maintenance Management*, 2nd ed.; Industrial Engineering Series; CRC Press: Boca Raton, FL, USA, 1994.
5. Pintelon, L.; Gelders, L. Maintenance management decision making. *Eur. J. Oper. Res.* **1992**, *58*, 301–317.
6. Sullivan, G.; Pugh, R.; Melendez, A.; Hunt, W. *Operations & Maintenance Best Practices*; U.S. Department of Energy: Washington, DC, USA, 2002.
7. Gits, C. Design of maintenance concepts. *Int. J. Prod. Econ.* **1992**, *24*, 217–226.
8. Levitt, J. *Complete Guide to Preventive and Predictive Maintenance*; Industrial Press: New York, NY, USA, 2003.
9. Cosar, E. A Wireless Toolkit for Monitoring Applications. Master's Thesis, Helsinki University of Technology, Helsinki, Finland, 2009.
10. Nowlan, F.S.; Howard, F.H. Reliability-Centered Maintenance. Department of Defence, Australian Government: Canberra, Australia, 1978.
11. Reichard, K.; van Dyke, M.; Maynard, K. Application of sensor fusion and signal classification techniques in a distributed machinery condition monitoring system. In Proceedings of the SPIE International Society for Optical Engineering, Orlando, FL, USA, 25–28 April 2000; pp. 329–336.
12. Bengtsson, M.; Olsson, E.; Funk, P.; Jackson, M. Design of condition based maintenance system—A case study using sound analysis and case-based reasoning. In *Condition Based Maintenance Systems—An Investigation of Technical Constituents and Organizational Aspects*; Malardalen University: Eskilstuna, Sweden, 2004; p. 57.

13. Gunnlaugsson, E. *Chemical Composition of Separator Fluids at Hellisheidi Geothermal Power Plant*; Technical Report; Reykjavik Energy: Reykjavik, Iceland, 2013.

14. Gunnlaugsson, E. *Chemical Composition of Separator Fluids at Nesjavellir Geothermal Power Plant*; Technical Report; Reykjavik Energy: Reykjavik, Iceland, 2013.

15. Fridleifsson, G.; Ármannsson, H.; Mortensen, A.K. *Geothermal Conditions in the Krafla Caldera with Focus on Well KG-26*; Iceland Geosurvey: Reykjavik, Iceland, 2006.

16. Hellisheidi Geothermal Plant. Available online: http://www.or.is/en/projects/hellisheidi-geothermal-plant (accessed on 30 June 2014).

17. Hellisheidavirkjun. Available online: http://www.or.is (accessed on 27 June 2014).

18. Nesjavellir Power Plant. Available online: http://www.on.is/en/power-plants (accessed on 30 June 2014).

19. Landsvirkjun. Krafla Geothermal Power Plant. Available online: http://www.landsvirkjun.is/Fyrirtaekid/Aflstodvar/kroflustod (accessed on 30 June 2014).

20. Mannvit. Bjarnarflag Geothermal Power Plant. Available online: http://www.mannvit.com/GeothermalEnergy/ProjectExampleinfo/bjarnarflag-geothermal-power-plant (accessed on 30 June 2014).

21. Wang, T.; Ji, P. Understanding customer needs through quantitative analysis of kano's model. *Int. J. Qual. Reliab. Manag.* **2010**, *27*, 173–184.

22. Kano, N.; Seraku, N.; Takahashi, F.; Tsuji, S. Attractive quality and must-be quality. *J. Jpn. Soc. Qual. Control* **1984**, *14*, 39–48.

23. Sauerwein, E.; Bailom, F.; Matzler, K.; Hinterhuber, H.H. The kano model: How to delight your customers. *Int. Work. Semin. Prod. Econ.* **1996**, *1*, 313–327.

24. Xu, Q.; Jiao, R.J.; Yang, X.; Helander, M.; Khalid, H.M.; Opperud, A. An analytical kano model for customer need analysis. *Des. Stud.* **2009**, *30*, 87–110.

25. Berger, C.; Blauth, R.; Boger, D.; Bolster, C.; Burchill, G.; DuMouchel, W.; Pouliot, F.; Richter, R.; Rubinoff, A.; Shen, D.; *et al.* Kano's methods for understanding customer-defined quality. *Cent. Qual. Manag. J.* **1993**, *2*, 3–35.

MDPI AG
Klybeckstrasse 64
4057 Basel, Switzerland
Tel. +41 61 683 77 34
Fax +41 61 302 89 18
http://www.mdpi.com/

Energies Editorial Office
E-mail: energies@mdpi.com
http://www.mdpi.com/journal/energies